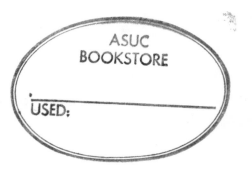

THIRD EDITION

TECHNICAL WRITING
Situations
and Strategies

THIRD EDITION

TECHNICAL WRITING

Situations and Strategies

Michael H. Markel

Boise State University

St. Martin's Press
New York

Senior editor: Mark Gallaher
Development editor: Bob Weber
Project management: Gene Crofts
Production supervisor: Alan Fischer
Text design: Paula Goldstein
Cover design: Leon Bolognese
Cover photo: J. A. Kraulis / Masterfile
Software editor: Kim Richardson

Library of Congress Catalog Card Number: 90-71634

Manufactured in the United States of America.
54321
fedcba

For information, write:
St. Martin's Press, Inc.
175 Fifth Avenue
New York, NY 10010

ISBN: 0-312-04833-5
      0-312-06756-9 (Instructor's Edition)

ACKNOWLEDGMENTS

Figure 3.1: Reproduced by permission from Honeywell Inc.

Figure 4.2: Pages from *Cumulative Index to Nursing & Allied Health Literature* (1990). "Searching CINAHL Effectively," parts A & B, pp. VI–VII. CINAHL Information Systems, Glendale, CA 91209. Copyright © 1990 by CINAHL Information Systems. Reprinted with permission.

Chapter 4: Excerpt "A Brief History of Television" based on G. McComb, *Troubleshooting and Repairing VCRs*, 2nd ed., pp. 21–23. Copyright © 1991 by TAB/McGraw-Hill, Inc. Reproduced with permission of McGraw-Hill, Inc.

Figure 10.8: Passage from Robert Jastrow, *Until the Sun Dies*, pp. 104–105. Copyright © 1977 by Robert Jastrow. Reprinted by permission.

Figures 11.1 and 11.2: Graphs and table from C. F. Schmid and S. E. Schmid, *Handbook of Graphic Presentation*, 2nd ed., p. 147, pp. 8–9, and fig. 2-11, p. 25. Copyright © 1979 by John Wiley & Sons, Inc. Reprinted by permission of John Wiley & Sons, Inc.

Figure 11.7: Maintenance schedule from G. McComb, *Troubleshooting and Repairing VCRs*, 2nd ed., fig. 5-1, p. 133. Copyright © 1991 by TAB/McGraw-Hill, Inc. Reproduced with permission of McGraw-Hill, Inc.

Figure 11.21: Graphs from M. E. Spear, *Practical Charting Techniques*, fig. 3-1, p. 56. Copyright © 1969 by McGraw-Hill, Inc. Reproduced with permission of McGraw-Hill, Inc.

Acknowledgments and copyrights are continued at the back of the book on page 680, which constitutes an extension of the copyright page.

BRIEF CONTENTS

To My Parents

PREFACE

The Third Edition of *Technical Writing: Situations and Strategies* has two goals: to be a clear and comprehensive teaching tool for technical writing instructors and at the same time to prepare students to approach the working world's various writing situations. Behind these goals is the belief that the best way to learn to write is to write and rewrite. Accordingly, the text includes numerous samples of technical writing along with dozens of writing and revising exercises that enable students to apply what they have learned in realistic technical writing situations.

The difficult part of technical writing is to solve the simple, overriding problem involved in every kind of fact-based communication. This problem might be called "How to Write about Your Subject, Not about Yourself," or "How to Write to Your Readers, Not to Yourself." When technical writing fails, the most common cause is that the writers did not consider *why* and *for whom* they were writing. Too many people write as though they are still in school and their readers are their instructors. The results, of course, are all too predictable: long, complicated documents that few people have the time, interest, or knowledge to read.

Technical Writing: Situations and Strategies is designed to help students analyze the writing situation—their audience and purpose in communicating—and then to decide how this situation should determine a document's development, structure, and style.

New to the Third Edition

Before providing an overview of the book's contents and organization, I would like to point out the major changes in the Third Edition. Most notably, there are two new chapters: on ethics and on page design.

In previous editions, the treatment of ethics consisted of brief discussions that focused on rhetorical strategies and kinds of documents. In discussing proposals, for example, I pointed out the ethical implications of promising more than you can deliver. The Third Edition expands such discussions and adds Chapter 2, The Ethics of Technical Writing. In approaching this subject I have chosen to base the discussion on the literature of business ethics, because most readers will confront ethical dilemmas as they write for business, industry, or government. The chapter begins by describing various ethical standards and providing a theoretical framework of ethics based on a writer's obligations to different stakeholders, such as the employer and the general public. Then I examine basic ethical issues that a technical writer is likely to face: copyright and plagiarism, the fair use of language and graphics, trade secrets, whistleblowing, and codes of conduct. One student asked me, "Is the chapter intended to make people ethical?" No. But students of mine who have used this chapter found that it helped them understand and think clearly about their ethical obligations as writers.

Chapter 12, Page Design, is my response to current research suggesting that design elements exert a critical influence on readability and rhetorical effect. As in the case of the ethics material, my previous two editions did treat some design elements, such as indentation, spacing, and the use of headings and lists. The new chapter, however, is more comprehensive and more orderly; it covers all the basic design decisions that a technical writer must make in creating effective documents.

In addition to these two new chapters, I have expanded the discussion of numerous important subjects:

☐ analyzing the audience, based on an audience-profile sheet
☐ carrying out experiments and making observations as a means of gathering information
☐ understanding the writing process
☐ writing collaboratively
☐ using the word processor effectively
☐ using new technology, such as the fax machine
☐ improving the coherence and cohesion of writing
☐ using Toulmin logic in causal discussions

I have retained a number of writing samples and exercises that reviewers and users of previous editions found valuable, but I have replaced many others and have expanded the total number throughout the book.

Instructors and students who use IBM or Macintosh word-processing programs should be pleased by the addition of two new software items:

■ *Revision Exercise Disk to Accompany Technical Writing* contains eleven documents for revision (making it unnecessary to keyboard the imperfect examples). Ranging from memos to a completion report, the documents let students edit, revise, and manipulate both text and formats. The

Instructor's Manual includes these documents in hard copy, along with suggestions for using the disk.

- *The St. Martin's Technical Writing Hotline* is an online pop-up reference system that gives immediate access to information about style, usage, punctuation, and mechanics.

Organization of the Text

The Third Edition reorganizes the text in several important ways. Now the introductory chapter is followed immediately by the chapter on ethics in technical writing. The ethics chapter appears early in the book because I feel that it informs everything else about technical writing. Ethical behavior is not an afterthought or a problem to be considered along with rhetorical concerns; rather, it is the foundation, the starting point of all technical writing.

Following these two chapters, the text is divided into four main parts:

- Part One, The Technical Writing Process, consists of five chapters. Chapter 3 is the expanded discussion of audience and purpose. Chapter 4 discusses ways of finding and using information. Chapter 5 covers the writing process, including computer technology and collaboration. The previous edition's discussion of style is now expanded and divided into two chapters aimed at developing cohesive, coherent prose: Chapter 6 focuses on sentence-level matters, and Chapter 7 considers larger elements of style.
- Part Two, Techniques of Technical Writing, also consists of five chapters. Chapter 8 covers definitions; Chapter 9 looks at descriptions; and Chapter 10 greatly expands the discussion of analysis as an organizational technique. Chapters 11 and 12 then focus on the visual aspects of communication — graphics and page design.
- Part Three, Common Applications, covers memos (Chapter 13), letters (Chapter 14), job-application materials (Chapter 15), instructions and manuals (Chapter 16), technical articles (Chapter 17), and oral presentations (Chapter 18). In previous editions, these applications were treated later in the text; the new arrangement reflects the current increased emphasis on shorter forms of communication and on instructional writing.
- Part Four, Technical Reports, retains the structure of previous editions: a discussion of formal elements of a report (Chapter 19) is followed by three chapters covering proposals, progress reports, and completion reports (Chapters 20 through 22).

Finally, the book concludes with four appendixes: (A) a handbook of style, punctuation, and mechanics; (B) a guide to documenting sources; (C) a list of Postal Service abbreviations for the states; and (D) a selected, up-to-date bibliography.

Acknowledgments

All the examples in the book—from single sentences to complete reports—are real. Some were written by my students at Clarkson College in Potsdam, New York. Others were written by my students at Drexel University in Philadelphia, either while they were in school or while they were off campus working as co-op students. Still others were written by my students at Boise State University. In addition, many examples were written by engineers, scientists, and businesspersons with whom I have worked as a consultant for some twenty years. Because much of the information in their writing is proprietary, I have silently changed brand names, company names, and other identifying nouns. I sincerely thank these dozens of people—students and professionals alike—who have graciously allowed me to reprint their writing. They have been my best teachers.

The Third Edition of *Technical Writing: Situations and Strategies* has benefited greatly from the perceptive observations and helpful suggestions of my fellow instructors throughout the country, many of whom generously completed extensive questionnaires about the previous edition. Among my reviewers I thank Susan K. Ahern, University of Houston, Downtown; James Franklin, Mercer County Community College; John Frederick Reynolds, Old Dominion University; and Jane M. Tanner, University of Wisconsin, Milwaukee.

I would like to pay special tribute to one reader, Stephen C. Brennan of Louisiana State University, whose expertise in technical writing and editing has helped me clarify the text and weed out numerous errors. Steve's help has made this a much better book.

At St. Martin's Press, I thank Senior Editor Mark Gallaher and Development Editor Bob Weber for their perceptive, careful, encouraging, and eminently reasonable help.

I want to thank, too, my department chair, Carol Martin, and my dean, Daryl Jones, here at Boise State University. Both have assisted me by providing material resources and by creating a congenial, productive atmosphere for the teaching and study of technical writing. They exemplify the highest standards of collegiality.

My greatest debt, however, is to my wife, Rita, who over the course of many months and, now, three editions has helped me say what I mean. To her should go praise for the good things in this book. For not always following her advice, I take responsibility.

Mike Markel

CONTENTS

4 Finding and Using Information   56

5 The Writing Process   98

10   Organizing Discussions   239

11   Graphics   273

12   Page Design   314

PART THREE ■ Common Applications   341

PART FOUR ■ Technical Reports   511

19   Formal Elements of a Report   512

20   Proposals   547

21   Progress Reports   576

Appendix D   **Selected Bibliography   676**

Index   681

Introduction to Technical Writing

1

Technical writing conveys specific information about a technical subject to a specific audience for a specific purpose. Much of what you read every day is technical writing—textbooks (including this one), the owner's manual for your car, cookbooks. The words and graphics of technical writing are meant to be practical: that is, to communicate a body of factual information that will help an audience understand a subject or carry out a task. For example, an introductory biology text helps students understand the fundamentals of plant and animal biology and carry out basic experiments. An automobile owner's manual describes how to operate and maintain that particular car.

Many kinds of technical writing are produced by professional technical writers; other kinds are produced by technical people who write as part of their daily work.

A technical writer is hired to write documents, including manuals, proposals, reports, sales literature, letters, journal articles, and speeches. Technical writing as a career began around the time of World War II, when people were needed to write user's manuals and maintenance manuals for military hardware and systems. In the last fifteen years, however, the number of technical writers has grown tremendously, largely because of the growth of the computer industry and related high-technology industries.

In the early days of high technology, before around 1970, technical writing was often an afterthought. A manufacturer would market large computer systems without any instruction manuals at all. The company would send out a technician to install the system, and while he was there he would try to explain how to use it. Naturally, this wasn't a very effective or efficient way to educate the purchaser, who had to resort to calling the manufacturer to ask questions.

As more companies entered the richly profitable field, manufacturers learned that customers were frustrated when the product came without a user's manual, or when the manual arrived six months after the system, or when the manual was useless because of its poor quality. To keep their customers, high-tech companies started to pay more attention to their documentation, as these manuals are called.

In most companies today, technical writers work closely with the design engineers and the legal and marketing staffs in creating a new product. The emphasis today is on "user friendly" products, and no product is friendly if the user can't figure out how to use it. Because technical writers are valued members of a professional organization, their salaries and prestige have grown substantially. At many companies, such as IBM, technical writers receive the same salary as hardware and software engineers.

Although there are many thousands of technical writers, most writing on the job is done by technical people who write. Engineers, scientists, businesspersons, and other technically trained people are sometimes surprised to learn how much writing they do on the job. According to a number of recent surveys, technical professionals can expect to devote at least a fifth of their time to writing. And as they advance, the percentage of time spent writing increases.

Companies now expect their technical people to write. As personal computers have become less expensive, they have become much more common at even the smallest job sites. Some companies give their technical people personal computers to take home with them so they have a compatible system to use in the evenings and on weekends. Businesspersons and engineers often carry laptop computers when traveling and work on the trip report on the airline flight home.

As technological aids to communication become cheaper, smaller, and more powerful, and as writing and graphics software improve, technical people can write more and better. With easy-to-use word-processing software, spell checkers, style programs, and graphics packages, technical people can control the entire design and production of their documents without relying on graphic artists and on secretaries. Therefore, the technical person's value to the company will depend more and more on the quality of the writing.

The Role of Technical Writing in Business and Industry

The working world depends on written communication. Virtually every action taken in a modern organization is documented in writing (whether on paper or in a computer's memory). The average company has dozens of different forms to be filled out for routine activities, ranging from purchasing office supplies to taking inventory. If you need some filing cabinets at a total cost of $300, for instance, you can purchase them after an office manager signs a simple form. But expenditures over a specified amount—such as $2,000 or $10,000—require a brief report showing that the purchase is necessary and that the supplier you want to buy from is offering a better deal (in terms of price, quality, service, etc.) than the competition. Similarly, most technicians record their day-to-day activities on relatively simple forms. But when these same technicians travel to a client's plant to teach a class on a new procedure, they have to write up a detailed memo or report upon their return. That report describes the purpose of the trip, analyzes any problems that arose, and suggests ways to make future trips more successful. When new processes and procedures are introduced at an organization, someone has to write up detailed step-by-step instructions and policy and procedure statements.

When a major project is being contemplated within an organization, a proposal must be written to persuade management that the project would be both feasible and in the best interest of the organization. Once the project is approved and under way, progress reports must be submitted periodically to inform the supervisors and managers about the current status of the project and about any unexpected developments. If, for instance, the project is not going to be completed on schedule or is going to cost more than anticipated, the upper-level managers must be able to reassess the project in the broader context of the organization's goals. Often, projects are radically altered on the basis of the information contained in progress reports. When the project has ended, a completion report must be written to document the work and enable the organization to implement any recommendations.

In addition to all this in-house writing, every organization must communicate with other organizations and the public. The letter is the basic format for this purpose. Inquiry letters, sales letters, goodwill letters, and claim and adjustment letters are just some of the types of daily correspondence. And if a company performs contract work for other companies, then proposals, progress reports, and completion reports are called for again.

Professionals also communicate with other professionals through technical writing. Every year, scientists, engineers, businesspersons, academics, and numerous others write hundreds of thousands of articles for trade and professional journals.

The world of business and industry is a world of communication.

If you are taking a technical writing course now, it was probably created to meet the needs of the working world. You have undoubtedly seen articles in newspapers and magazines about the importance of writing and other communication skills in a professional's career. The first step in *getting* a professional-level position, in fact, is to write an application letter and a résumé. If there are other candidates as well qualified as you—and in the job market today there almost certainly are—your writing skills will help determine whether an organization decides to interview you. At the interview, your oral communication skills will be examined along with your other qualifications. Once you start work, you will write memos, letters, and short reports, and you might be asked to contribute to larger projects, such as proposals. During your first few months on the job, your supervisors will be looking at both your technical abilities and your communication skills. The facts of corporate life today are simple: if you can communicate well, you are valuable; if you cannot, you are much less valuable. The tremendous growth in continuing-education courses in technical and business writing reflects this situation. Organizations are paying to send their professionals back to the classroom to improve their writing skills. New employees who can write and speak effectively bring invaluable skills to their positions.

Characteristics of Effective Technical Writing

Technical writing is meant to get a job done. Everything else is secondary. If the writing style is interesting, so much the better. But there are six basic characteristics of technical writing:

1. clarity
2. accuracy

3. comprehensiveness
4. accessibility
5. conciseness
6. correctness

Clarity

The primary characteristic of effective technical writing is clarity. That is, the written document must convey a single meaning that the reader can understand easily.

The following directive, written by the British navy (*Technical Communication* 1990), is an example of what you *don't* want to do:

> It is necessary for technical reasons that these warheads should be stored upside down, that is, with the top at the bottom and the bottom at the top. In order that there may be no doubt as to which is the top and which is the bottom, for storage purposes, it will be seen that the bottom of each warhead has been labeled with the word TOP.

Unclear technical writing is expensive. The cost of a typical business letter today, for instance, counting the time of the writer and the typist and the cost of equipment, stationery, and postage, is more than eight dollars. Every time an unclear letter is sent out, the reader has to write or phone to ask for clarification. The letter then has to be rewritten and sent again to confirm the correct information. On a larger scale, clarity in technical writing is essential because of the cooperative nature of most projects in business and industry. One employee investigates one aspect of the problem, while other employees work on other aspects of it. The vital communication link among the employees is usually the report. One weak link in the communication chain can jeopardize the entire project. For example, consider the case of an electronics manufacturing company trying to decide whether to develop a new product. The marketing department is in charge of determining the size of the market. If that report recommends against producing the product but is misinterpreted, the company might begin research and development. Such a breakdown could represent a huge financial loss for the organization.

More important, unclear technical writing can be dangerous. Poorly written warnings on bottles of medication can kill people, as can unclear instructions on how to operate machinery safely. A carelessly drafted building code tempts contractors to save money by using inferior materials or techniques. The 1986 space-shuttle tragedy might have been prevented had the officials responsible for deciding whether to launch received clear reports of the safety risks involved that day.

Accuracy

Inaccurate writing causes as many problems as unclear writing. In one sense, accuracy is a simple concept: you must record your facts carefully.

Technical writing is objective, free of bias

If you mean to write *4,000*, don't write *40,000*; if you mean to refer to Figure 3.1, don't refer to Figure 3.2. A slight inaccuracy will at least confuse and annoy your readers. A major inaccuracy can be dangerous.

In another sense, however, accuracy is a question of ethics. Technical writing must be as objective and free of bias as a well-conducted scientific experiment. If the readers suspect that you are slanting information—by overstating the significance of a particular fact or by omitting an important point—they have every right to doubt the validity of the entire document. Technical writing must be effective by virtue of its clarity and organization, but it must also be reasonable, fair, and honest.

Comprehensiveness

A comprehensive technical document provides all the information the readers will need. It describes the background so that readers who are unfamiliar with the project can understand the problem or opportunity that led to the project. It clearly describes the methods used to carry out the project, and it states the principal findings—the results and any conclusions and recommendations.

Comprehensiveness is crucial for two reasons. First, the people who will act on the document need a complete, self-contained discussion so that they can apply the information effectively, efficiently, and safely. Second, they need an official company record of the project, from its inception to its completion.

For example, a scientific article reporting on an experiment comparing the reaction of a new strain of bacterium to two different compounds will not be considered for publication unless the writer has fully described the methods used in the experiment. Because other scientists should be able to replicate the researcher's methods, every detail must be included, even the names of the companies from which the researcher obtained all the materials.

Or consider a report recommending that a company network its computers. The company will probably want to analyze the recommendations in detail before committing itself to such an expensive and important project. The team charged with studying the report will need all the details. If the recommendations are implemented, the company will need a single, complete record in case changes have to be made several months or years later.

Accessibility

Accessibility refers to the ease with which readers can locate the information they seek. One of the major differences between technical writing and other kinds of nonfiction is that most technical documents are made up of small, independent sections. Some readers are interested in only one or two sections; other readers might want to read several or most of the sections. Because relatively few people will pick up the document and start reading from the first page all the way through, the writer's job is to make the various parts of the document accessible. That is, the readers should not be forced to flip through the pages to find the appropriate section.

Conciseness

To be useful, technical writing must be concise. The longer a document is, the more difficult it is to use, for the obvious reason that it takes more of the reader's time. In a sense, conciseness works against clarity and comprehensiveness. If a technical explanation is to be absolutely clear, it must describe every aspect of the subject in great detail. To balance the claims of clarity, conciseness, and comprehensiveness, the document must be just long enough to be clear—given the audience, purpose, and subject—but not a word longer. You can shorten most writing by 10–20 percent simply by eliminating unnecessary phrases, choosing short words rather than long ones, and using economical grammatical forms.

The battle for concise writing, however, is often more a matter of psychology than of grammar. A utility company, for instance, recently experienced a fairly serious management problem caused by its employees' inability to understand their readers' psychology. Each branch manager of the company was required to submit a semiannual status report, informing the corporate managers about any technical problems during the previous six months. The length of the report was clearly specified as three pages, maximum.

As it turned out, the corporate headquarters was paralyzed for almost a month after the status reports came in. They averaged 17 pages. The branch managers had read the "three pages, maximum" directive; they just didn't believe it. They wanted to impress their supervisors. So each put a good many weeks into creating reports that covered everything. No detail was too small to be included. In their attempt to produce the ultimate status report, they forgot the simple fact that somebody had to read it. From the reader's point of view, few things are more frustrating than having to read 17 pages that say, in effect, that no problems occurred. The corporate managers had to spend hours chopping the 17-page reports down to 3 pages so that they could be filed for future reference. Justifiably enough, these managers felt they were doing the branch managers' jobs.

The utility company hired a writing consultant to make one point to the branch managers: not only is it acceptable to write three-page status reports, but more than three pages is less than good. Once this point was communicated, the rest of the time was spent discussing ways to get all the necessary information into three pages.

Correctness

Good technical writing is correct: it observes the conventions of grammar, punctuation, and usage, as well as any appropriate format standards.

Many of the "rules" of correctness are clearly important. If you mean to write, "The three inspectors—Bill, Harry, and I—attended the session," but you use commas instead of dashes, your readers might think six people (not three) attended. If you write, "While feeding on the algae, the researchers captured the fish," some of your readers might have a little trouble following you—at least for a few moments.

Most of the rules, however, make a difference primarily because readers will judge your writing on how it looks and sounds. Sloppiness and grammar errors make your readers doubt the accuracy of your information, or at least lose their concentration. You will still be communicating, but the message won't be the one you had intended. As a result, the document will not achieve its purpose.

Technical writing is meant to fulfill a mission: to convey information to a particular audience or to persuade that audience to a particular point of view. To accomplish these goals, it must be clear, yet accurate, and complete, yet easy to access. It must be economical and correct.

The rest of this book describes ways to help you say what you want to say.

REFERENCE

Technical communication. 1990. 37, no. 4: 385.

2

The Ethics of
Technical Writing

A power company in a major American city faced a problem recently: its obsolete generating equipment produced illegal levels of air pollution. Unfortunately, the company didn't have the many millions of dollars needed to replace the old equipment or even to add the scrubbers that would lower the pollution to legal levels.

The Environmental Protection Agency, fully aware of the pollution levels, routinely fined the utility: $250, four times a year. Why were the penalties so low? Because the law recognized that older power companies would not be able to meet modern pollution standards and continue to supply power to the city, as well as to continue to employ thousands of people. The utility industry and the lawmakers in effect had worked out a plan to replace the older technology with newer equipment over a period of decades, while enabling the companies to stay in business.

What does this have to do with you? As a technical writer or a technical person who writes, you are likely to be involved with problems like this — problems that have an ethical dimension. Everyone knows why pollution is undesirable, but getting rid of it — or lowering it to acceptable levels — is often difficult and expensive. In business, government, and industry, people work every day to solve such ethical problems in ways that address the needs of different constituencies: in this case, the homeowners and businesspersons who need the power, the company employees who need paychecks, and everyone who breathes the city's air.

As a working professional, you will have to make decisions about how to treat clients, customers, and other organizations. You will have to make decisions about how your organization deals with the government regulatory bodies that oversee your industry. You will have to make decisions about how the actions of your organization affect the environment. All these decisions will be communicated in your writing and speaking.

To make wise decisions about how to deal with ethical problems, a technical writer can benefit from a basic discussion of professional ethics. That is what this chapter is intended to provide. This chapter briefly introduces ethics and explains the standards commonly used to decide ethical questions. The chapter then addresses specific issues that a technical writer routinely faces — plagiarism, trade secrets, the fair use of language and visuals in product information and advertising, and whistle-blowing. Finally, the chapter discusses codes of conduct.

Throughout the rest of this book, you will also see other discussions of ethics. In the chapter on proposals, for example, there is a discussion of some of the ethical issues involved in writing proposals.

A Brief Introduction to Ethics

Televangelists Jim and Tammy Bakker and Jimmy Swaggart, hotel tycoon Leona Helmsley, Secretary Samuel Pierce of the Department of Housing and Urban Development, Congressmen Jim Wright and Barney Frank, Canadian sprinter Ben Johnson, financiers Ivan Boesky and Michael Milken. These names suggest the range of persons whose business and personal affairs have been the subject of ethical investigation in recent years.

Some bigger and more complex ethical questions in recent headlines include Iran-contra and the question of lying to Congress; Exxon's response to the Alaskan oil spill; sanctions against organizations operating in South Africa; Union Carbide's handling of the 1984 tragedy in Bhopal, India; the 1986 *Challenger* explosion and the defective O-rings; the rights of the unborn versus the rights of the pregnant woman; the rights of the minority applicant versus the rights of the nonminority applicant. The list goes on and on.

But what exactly is ethics? Some people equate ethical conduct and legal conduct; if an act is legal, it is ethical. Most people, however, believe that ethical standards are more demanding than legal standards. It is perfectly legal, for example, to try to sell an expensive life insurance policy to an impoverished elderly person who has no dependents and therefore no need for such a policy, yet many people would consider the attempt unethical.

When businesspersons were asked to give their definition of ethics, half of them answered, "What my feelings tell me is right." One quarter of the respondents answered, "What is in accord with my religious beliefs." And most of the rest of the businesspersons said, "What conforms to 'the Golden Rule'" (Baumhart 1968). While philosophers cannot agree on what constitutes ethical conduct, most would agree to the following definition: ethics is the study of the principles of conduct that apply to an individual or a group.

What are the basic principles used to think through an ethical problem? Ethicist Manuel G. Velasquez (1988) outlines three kinds of moral standards useful in confronting ethical dilemmas:

- *Rights.* The standard of rights concerns the basic needs and welfare of particular individuals. If a company has agreed to provide continuing employment to its workers, the standard of rights requires that the company either keep the plant open or provide adequate job training and placement services.

- *Justice.* The standard of justice concerns this question: how can the positive and negative effects of an action or policy be distributed fairly among a group? For example, the standard of justice would suggest that the expense of maintaining a highway be borne primarily by persons who use that highway. However, since everyone benefits from the highway, it is just that general funds also be used for maintenance.

- *Utility.* The standard of utility concerns this question: what will be the effects, both positive and negative, of a particular action or policy on the general public? If, for example, a company is considering closing a plant, the standard of utility requires that the company's leaders consider not only any savings they will reap from shutting down the plant, but also the financial hardships on the unemployed workers and the economic impact on the rest of the community.

Although it is best to think about the implications of any serious act in terms of all three standards, often they conflict. For instance, from the point of view of utility, no-fault car insurance laws—under which people may not sue for damages under a certain dollar amount—are good because they reduce the number of nuisance suits that clog up the court system and increase everyone's insurance costs. However, from the point of view of justice, these laws are unfair because they force innocent drivers to pay for the repairs.

In conflicts among the three standards, the standard of rights is generally considered the most important, and the standard of justice the second most important. However, simply ranking the three standards cannot solve all ethical problems. If an action or policy were to greatly affect the general public, the standard of utility—the "least important" standard— might outweigh the standard of rights. For instance, if the power company has to cross your property to repair a transformer, the standard of utility (the need to get power to all the people affected by the problem) takes precedence over the standard of rights (your right to private property). Of course, the power company is obligated to respect your rights, insofar as possible, by explaining what it wants to do, trying to accommodate your schedule, and repairing any damage to your property.

Ethical problems are difficult to resolve precisely because there are no rules to determine when one standard outweighs another. In the example of the transformer, how many customers have to be deprived of their power before the power company is morally entitled to violate the property owner's right of privacy? Is it a thousand? Five-hundred? Ten? Only one?

Most people do not debate the conflict among rights, justice, and utility when they confront a serious ethical dilemma; instead, they do what they think is right. Perhaps this is good news. However, the quality of ethical thinking varies dramatically from one person to another, and the consequences of superficial ethical thinking can be profound. For these reasons, ethicists have described a general set of principles that can help people organize their thinking about the role of ethics within an organizational context. This set of principles is a web of rights and obligations that connect an employee, an organization, and the world in which it is situated.

For example, in the modern era it is generally acknowledged that employees enjoy three basic rights in exchange for their labor—fair wages, safe and healthy working conditions, and due process in the handling of such matters as promotions, salary increases, and firing. Although there is still serious debate about employee rights, such as the freedom from surreptitious surveillance and unreasonable search in the case of drug investigations, the question almost always concerns the extent of the employees' rights, not the existence of the basic rights. For instance, there is disagreement about whether hiring undercover investigators to discover drug users at the job site is an unwarranted intrusion on the employees' rights, but there is no debate about the principle of exemption from unwarranted intrusion.

Your Basic Obligations as an Employee

In addition to enjoying rights, an employee must accept certain obligations. These obligations can form a clear and reasonable framework for examining the ethics of technical communication. The following discussion outlines three sets of obligations:

- ☐ your obligations to your employer
- ☐ your obligations to the public
- ☐ your obligations to the environment

Your Obligations to Your Employer

You will be hired to further your employer's aims and to refrain from any activities that run counter to those aims. Specifically, you will have four obligations:

- ■ "*Competence and Diligence.*" Competence refers to your skills; you should have the training and experience to do the job adequately. Diligence simply means hard work; you are obligated to devote your energies to the task.

- ■ "*Honesty and Candor.*" You should not steal from your employer. Stealing involves such obvious practices as embezzlement but also includes such common occurrences as "borrowing" office supplies and padding expense accounts. Candor means truthfulness; you should report problems to the employer that might threaten the quality or safety of the organization's product or service. If, for instance, you have learned that a chemical that your company wants to make could pollute the drinking water, you are obligated to inform your supervisor, even though the news may be displeasing.

    Another kind of problem involving honesty and candor concerns what Sigma Xi, the Scientific Research Society, calls trimming, cooking, and forging (*Honor in Science*, 1986, 11). "Trimming" is the smoothing of irregularities to make the data look extremely accurate and precise. "Cooking" is retaining only those results that fit the theory and discarding the others. And "forging" is inventing some or all of the research data, and even reporting experiments that were never performed. In carrying out research, an employee might feel some pressure to report positive and statistically significant findings, but must resist the temptation.

- *Confidentiality.* You should not divulge company business outside of the company. If a competitor knew that your company was planning to introduce a new product, it might introduce its own version of that product, robbing you of your competitive edge. Many other kinds of privileged information, such as internal problems of quality control, personnel matters, relocation or expansion plans, and financial restructuring, also could be used against the company. A well-known problem of confidentiality involves insider information; an employee who knows about a development that will increase the value of the company's stock secretly buys the stock before the information is made public, thus reaping an unfair (and illegal) profit.

- *Loyalty.* You should act in the employer's interest, not in your own. Therefore, it is unethical to invest heavily in a competitor's stock, because that could jeopardize your objectivity and judgment. For the same reason, it is unethical to accept bribes or kickbacks. It is unethical to devote considerable time to moonlighting—that is, performing an outside job, such as private consulting—because the outside job can lead to a conflict of interest and because the heavy workload can make you less productive in your primary position.

Your Obligations to the Public

Every organization that offers products or services is obligated to treat its customers fairly. As a representative of an organization—and especially as an employee communicating technical information—you will frequently confront ethical questions.

In general, an organization is acting ethically if its product or service is both safe and effective. It must not injure or harm the consumer, and it must fulfill its promised function. However, these common-sense principles provide little guidance for dealing with the complicated ethical problems that arise routinely.

Product-related accidents are commonplace, and many of them are caused by poor design. The major cause of death for people under the age of thirty, for example, is automobile accidents. A quarter-million Americans are injured each year by power tools, and several hundred die from their injuries. Almost two million a year are injured in home-construction projects; more than a thousand die. The financial losses from injuries total many billions of dollars each year.

Even more commonplace, of course, are product and service failures: the items are difficult to assemble or operate; they don't do what they are supposed to do; they break down; or they require more expensive maintenance than indicated in the product information.

Who is responsible: the company that supplies the product or service, or the consumer who purchases it? In individual cases of injury or product failure, blame is sometimes easy enough to fix. A person who operates a chain saw without reading the safety warnings and without seeking any instruction in how to use it is surely to blame for any injuries caused by the normal operation of the saw. On the other hand, a manufacturer who knew that the chain on the saw is likely to break when used under certain circumstances, but failed to remedy this problem or warn the consumer, would be responsible for any resulting accidents.

However, cases such as these do not outline a rational theory that can help people understand how to act ethically in fulfilling their obligations to the public. There is no clear consensus on this issue, but today there are three main theories that describe obligations to the public:

- *The Contract Theory.* The contract theory holds that when a person buys a product or service, he or she is entering into a contract with the manufacturer. The manufacturer (and by implication the employee representing the manufacturer) has four main obligations:

   1. to make sure the product or service complies with the contract in several respects: it should do what its advertisements say it can, it should operate a certain period of time before needing service or maintenance, and it should be at least as safe as the product information states and the advertising suggests
   2. to disclose all pertinent information about the product or service, so that the potential consumer can make an informed decision on whether to purchase it
   3. to avoid misrepresenting the product or service
   4. to avoid coercion, like that illustrated by the unethical funeral director who takes advantage of a consumer's grief to sell an expensive funeral

   Critics of the contract theory argue that the typical consumer cannot understand the product as well as the manufacturer does, and that consumer ignorance invalidates the "contract."

- *The "Due-Care" Theory.* The "due-care" theory places somewhat more responsibility on the manufacturer than the contract theory does. This theory holds that the manufacturer knows more about the product or service than the consumer does, and thus has a greater responsibility to make sure the product or service complies with all its claims and is safe. Therefore, in designing, manufacturing, testing, and communicating about a product, the manufacturer has to make sure the product will be safe and effective when used according to the instructions. However, the manufacturer is not liable when something goes wrong that it could not foresee or prevent.

   Critics of the "due-care" theory argue that because it is almost impossible to determine whether the manufacturer has in fact exercised due care, the theory offers little of practical value.

■ *The Strict-Liability Theory.* The strict-liability theory, which goes one step further than the due-care theory, holds that the manufacturer is at fault when any injury or harm occurs from any product defect, even if it has exercised due care and could not possibly have predicted or prevented the failure or injury. This theory assumes that the only way to assess blame is to hold the manufacturer responsible and thereby force it to assume all liability costs. This way, society subsidizes the product when the manufacturer builds these costs into the price of the product.

Critics of the strict-liability theory hold that, although it might be practical, it is unfair because no person or organization should be held liable for something that is not its fault.

Your Obligations to the Environment

Perhaps the most important lesson we have learned in the last decade is that we are polluting and depleting our limited natural resources at an unacceptably high rate. The overreliance on fossil fuels not only deprives future generations of their use but also causes terrible pollution problems that many scientists believe are irreversible, such as global warming. Everyone—government, business, and individuals—must work to preserve the fragile ecosystem, to ensure the survival not only of our own species but also of the other species with which we share the planet.

But what does this have to do with you as a technical writer? Technical writers, in their daily work, do not cause pollution or deplete the environment in any unique or extraordinary way. Yet because of the nature of your work, often you will know how your organization's actions affect the environment. For example, if you are working on a proposal that is to be submitted to the federal government, you might also contribute to the creation of the environmental-impact statement.

Writers should treat every actual or potential occurrence of environmental damage seriously. You should alert your supervisor to the situation and try to work to reduce the damage. The difficulty, of course, is that protecting the environment is expensive. Clean fuels cost more than dirty ones. Disposing of hazardous waste properly costs more (in the short run) than merely dumping it. Organizations that want to cut costs may be tempted to cut corners on environmental protection.

Common Ethical Issues Faced by Technical Writers

As a working professional you will likely confront both simple and complicated ethical problems. Should you use company pens and stationery for personal business? This is obviously a simple ethical question. But if you accept an attractive job with one of your previous employer's competitors, only to discover that the new company wants you to divulge information about your previous employer's products, then you are facing a much more complicated ethical dilemma.

Most of the difficult ethical dilemmas involve a conflict between two competing principles. For example, in the question of the new company's wanting you to provide secret information about a competitor, the dilemma is between loyalty to your current employer and the desire to be honest (not to steal information, in this case). The following discussion suggests some ways to resolve several common ethical issues faced by people in their roles as writers: plagiarism, trade secrets, the fair use of language and visuals in product information and advertising, and whistleblowing.

Plagiarism and Copyright Violation

Plagiarism and copyright violation are complicated issues, especially in modern technical writing.

Plagiarism, of course, is the practice of using someone else's words or ideas without crediting the source. Many organizations treat authorship of internal documents, such as memos and most reports, casually; that is, if the organization asks you to update an internal procedures manual, it expects you to use any material from the existing manual, even if you cannot determine the original author.

Organizations tend to treat the authorship of published documents, such as external manuals or journal articles, more seriously. Although the authors of some kinds of published technical documents are not listed, many documents such as user's guides do acknowledge their authors. However, what constitutes authorship can be a complicated question, because most large technical documents are produced collaboratively, with several persons contributing text, another doing the graphics, still another reviewing for technical accuracy, and finally someone reviewing for legal concerns. Problems are compounded when a document goes into revision, and parts of original text or graphics are combined with new material.

The best way to determine authorship is to discuss it openly with everyone who contributed to the document. Some persons might deserve to be listed as authors; others, only credited in an acknowledgment section. To prevent charges of plagiarism, the wisest course is to be very conservative: if there is any question about whether to cite a source, cite it.

For a more detailed discussion of plagiarism, see Appendix B, Documentation of Sources.

A related problem involves copyright violation. Copyright law provides legal protection to the author of any document, whether it be published or unpublished, and whether the author be an individual or a corporation. Unfortunately, some companies will take whole sections of another company's product information or manual, make cosmetic changes, and publish it themselves. This, of course, is stealing.

But the difference between stealing and learning from your competitors can be subtle. Words are protected by copyright, but ideas aren't. Hundreds of companies produce personal computers, for instance, all of which come with user's guides. Rare is the manufacturer who doesn't study the competitors' user's guides to see how a feature or task is described. Inevitably, a good idea spreads from one document to another, and then to another. If one manual contains a particularly useful kind of troubleshooting guide, pretty soon a lot of others will contain similar ones. Even though this process of imitation tends to produce a dull uniformity, it can improve the overall quality of the document. Under no circumstances, however, should you violate copyright by using another organization's words.

Trade Secrets

What is a trade secret? According to the law, a trade secret is "any confidential formula, pattern, device, or compilation of information that is employed in a business and that gives the business the opportunity to gain a competitive advantage over those who do not know or use it" (Beauchamp and Bowie 1988, 264). For example, the formula for Coca-Cola® syrup is a trade secret, even though the individual ingredients of the syrup are commonly available. According to the law, a scientist who quit working for Coca-Cola® and joined a competing soft-drink manufacturer could not legally divulge the formula to the new company. Coca-Cola® owns that information.

Often, however, trade secrets are very difficult to define. If, for instance, during a nine-year career as a systems analyst with a company, you devise a unique approach to structured programming that your company uses in many of its products, who owns that approach: you or the company? That approach gives the company a competitive advantage, and so it would seem to be a trade secret. Yet you devised that approach, and one of your fundamental rights is the ability to work for whomever you choose.

An additional complication is that you may disclose information unintentionally. If you work at a company for several years, the line blurs between what information you bring to the company and what information you gain from working there. Through your work experience you develop a way of thinking, a way of approaching problems. You might find it impossible to state what is your company's legitimate trade secret and what is your own thinking.

Many companies are trying to cut down the number of problems involving trade secrets by developing new management practices, such as entering into contracts with certain key individuals explicitly forbidding the use of some trade secrets, restricting the number of persons exposed to the crucial information, and preventing employees from engaging in outside consulting. Other companies use positive incentives, such as generous salaries and postretirement consulting contracts, to keep their employees from wanting to work for their competitors.

Because employees today tend to change jobs much more frequently than they did in past decades, companies will find it increasingly difficult to keep information secret. The law does not provide clear, precise guidelines on the question of what constitutes a trade secret, and therefore the question will likely remain an ethical dilemma that you must try to resolve.

The Fair Use of Language and Graphics

As an employee who writes or who creates graphics, you will often face ethical dilemmas as you try to communicate fairly and effectively. In writing a proposal, for example, you might feel pressure to exaggerate claims about your company's expertise or experience, or to minimize or even ignore disadvantages of the proposed plan. As Louis Perica (1972) has pointed out, it is relatively simple to manipulate drawings and photographs technically to influence the reader unfairly. Now that photographs are routinely digitized, the potential for this kind of abuse is increased.

Ethical problems are particularly common in product information, from descriptions in sales catalogs to specification sheets, operating instructions, and manuals. As Powledge (1980) notes, in most cases, the cause of the ethical dilemma is that in creating product information you are doing two things at once: describing and advertising the product. These two functions are not only different, they are often in conflict. (Ironically, most advertisers maintain they are merely providing product information, even though most ads are light on facts.)

Employees report that sometimes they are asked to lie or mislead the reader. Lying—providing false information—is obviously unethical in these situations. If your company's own tests of the disk drive it sells show a mean time between failures (MTBF) of 1,000 hours, but the competitor's figure is 1,500 hours, you might be pressured by a supervisor to simply lie, to say that the MTBF is 1,750 hours. Your responsibility is to resist this pressure, even by going over the supervisor's head if necessary.

Providing misleading information is a little more complicated, but it amounts to the same thing ethically. A misleading statement or graphic, while not actually a lie, enables or even encourages the reader to believe false information. For instance, a product-information sheet for a computer system is misleading if the accompanying photograph of the unit includes a modem when the modem is sold separately and is not part of the purchase price of the system.

Information can mislead the reader in a number of other ways. The writer can try to scare the reader by falsely suggesting that a product—or a particular brand of the product—is necessary. For example, a flashlight manufacturer misleads in suggesting that only its own brand of batteries will power the flashlight. Another misleading technique is to ignore the negative features of a product. For instance, if an information sheet for a portable compact disc player suggests that it can be used by joggers, without mentioning that the bouncing will probably make it skip, that is misleading. A third misleading technique is the use of legalistic constructions that don't mean what they seem to mean. For instance, it is unethical to write, "The 3000X was designed to operate in extreme temperatures, from -40 degrees to 120 degrees Fahrenheit," if in fact the unit cannot reliably operate in those temperatures. The fact that the statement might actually be accurate—the unit was *designed* to operate in those temperatures—doesn't make it any less misleading.

William James Buchholz (1989) describes three other characteristics of writing that can mislead readers, regardless of the writer's intentions: (1) abstractions and generalities (a company's assertion that "We care about you"), (2) jargon ("user friendly," "ergonomically designed," "state of the art"), and (3) euphemism ("We carried out extensive market research," when in fact they made a few phone calls).

Whistleblowing

Whistleblowing is the practice of going public with information about serious unethical conduct within your company. For example, an engineer is blowing the whistle in telling a government regulatory agency or a newspaper that quality-control tests on one of the products the company sells have been faked.

In deciding whether to blow the whistle, you must choose between loyalty to your employer and to your own standards of ethical behavior. Some people believe that an employee owes complete loyalty to the employer. The former president of General Motors, James M. Roche, for example, has written, "Some of the enemies of business now encourage an employee to be *disloyal* to the enterprise. They want to create disharmony, and pry into the proprietary interests of the business. However this is labeled—industrial espionage, whistle blowing, or professional responsibility—it is another tactic for spreading disunity and creating conflict" (Beauchamp and Bowie 1988, 262).

Yet most reasonable people believe that workers should not be asked to steal or lie or take actions that could physically harm others. Where does loyalty to the employer end, and the employee's right to take action begin? And what should an employee do before blowing the whistle?

Ethicist Manuel Velasquez (1988) outlines four questions that you should consider carefully before taking any action:

☐ Do you understand the situation fully, and are your facts accurate? Sometimes employees' grievances are the result of incomplete or inaccurate information.

☐ What exactly is the ethical problem involved? You should be able to explain what is unethical about the practice and state just who is being harmed by it.

☐ Is the ethical problem serious or trivial? The more serious the ethical problem, the more you are justified or even obligated to act.

☐ Would you be more effective in stopping the unethical practice if you worked privately within the organization or blew the whistle? Think about the implications—for society, for the company, and for your-self—of each of the paths.

These questions suggest that you should not blow the whistle casually. You should make every effort to solve the problem internally before going public. A number of organizations have instituted procedures to try to encourage employees to bring ethical questions to management. Among the more common means are to use anonymous questionnaires and to establish an ombudsman position: a person within the company whose job is to bring ethical grievances to management's attention. An ombuds-man who feels that management has not responded satisfactorily to the situation is empowered to report the information freely.

However, most companies still have no formal procedures for han-dling serious ethical questions. For this reason, whistleblowing remains risky. Although the federal government and about half the states have laws intended to protect whistleblowers, these laws are not highly effective. It is simply too easy for the organization to penalize the whistleblower—subtly or unsubtly—through negative performance appraisals, transfers to undesirable locations, or isolation within the company. For this reason, many people feel that an employee who has unsuccessfully tried every method of alerting management to a serious ethical problem would be wise simply to resign rather than face the professional risks of whistle-blowing. Of course, resigning quietly is much less likely to force the organi-zation to remedy the situation. As many ethicists say, doing the ethical thing does not always advance a person's career.

Codes of Conduct

The kinds and numbers of problems that lead to whistleblowing will prob-ably increase along with heightened public interest in and understanding of ethical behavior. To try to reduce the incidence of unethical behavior, many businesses and professional organizations have created professional codes of conduct. As of 1986, three out of four of the country's 1,500 largest corporations had them, as had virtually all professional organiza-tions (Brockmann and Rook 1989, 91).

Codes of conduct vary greatly from organization to organization, but all include statements of ethical principles that the employees of the com-pany or members of the organization should follow. Some are brief and general, offering only guidelines for proper behavior. The Code for Communicators written by the Society for Technical Communication, for example, is less than 200 words long, and consists of statements such as the following: "Because I recognize that the quality of my services directly affects how well ideas are understood, I am committed to excellence in performance and the highest standards of ethical behavior."

Other organizations' codes go into great detail in describing proper and improper behavior, and actually stipulate penalties for violating the principles. These codes could be thought of more as sets of rules than as general guidelines. The American Society of Mechanical Engineers' code, for instance, specifies procedures the Society is to follow in cases of complaints about ethical violations. These procedures are quite specific about the proper functioning of its Professional Affairs and Ethics Committee, even indicating, for instance, that the complaint must be acknowledged "by Certified Mail" (Chalk 1980, 193).

Do codes of conduct really encourage ethical behavior? This is not an easy question to answer, of course, because there are no statistics on how many people were inspired or frightened into acting ethically by a code of conduct. However, many ethicists and officers in professional societies are skeptical. Because codes have to be flexible enough to cover a wide variety of situations, they tend to be so vague that it is virtually impossible to say that a person in fact violated one of their principles.

A study conducted by the American Association for the Advancement of Science in 1980 (Chalk 1980) found that whereas most professional organizations in engineering and the sciences have codes that provide for hearing cases of unethical conduct, relatively few such cases have ever been brought before the organizations. Many societies reported, for example, that only three or four allegations were ever lodged against individuals, and most of the societies have never taken disciplinary action—censure or expulsion—against any of their members for ethical violations.

If codes of conduct are not often systematically enforced, do they have any real value? Critics point out that almost no organization is willing to support someone who brings a charge. Like a whistleblower, an accuser within an organization is likely to face overt or covert punishment; thus self-interest compels many people to remain silent, regardless of the high-sounding statements in the company's code of conduct. Few professional organizations have ever come to the financial aid of an accuser who has lost a job because of a justified allegation.

For this reason, codes of conduct sometimes seem to be public-relations tools intended to persuade not only employees and organization members, but also the general public and the government, that the organization polices its own members, so that the federal government will not impose its own regulations.

Ethicist Jack N. Behrman finds the greatest value in merely writing a code, because an organization thereby clarifies its own position on ethical behavior (1988, 156). Of course, distributing the code might also have the effect of fostering an increased awareness of ethical issues, in itself a positive result.

EXERCISES

1. Research an important public event that involved ethical matters, such as the Exxon oil spill (1989) or the *Challenger* tragedy (1986), and write a memo to your instructor analyzing the following two factors:
   a. The ways in which elements of the incident represented a violation of ethical standards.
   b. The ways in which the organization (or organizations) applied ethical standards in its public statements and actions in the aftermath of the incident.

2. Research your college or university's code of conduct, and write a memo to your instructor describing and evaluating it. From your experience, does it appear to be widely publicized, enforced, and adhered to?

3. Research the code of conduct that applies to the field you are studying. Write a memo to your instructor describing and evaluating the code. How effective do you think it would be in preventing ethical lapses and in punishing people who have knowingly violated its precepts?

4. Following is a sample case that draws upon a number of the theoretical issues developed in this chapter. The protagonist in the case is a technical writer who must confront a set of facts that are a lot less conclusive than she would like. Any decision she will make will have both positive and negative outcomes.

> The town of Acton, Ohio (population 6,500), like many other small communities in the Rust Belt, has suffered economically during the last decade. Much of its infrastructure is old and in need of repair, and the town has a shrinking tax base. Young people routinely leave the area after high school in search of better jobs.
>
> The main employer in Acton is Diversified Construction Materials, which employs over 1,000 people from Acton and surrounding communities. Like Acton, Diversified has known better times. Its products are known for their high quality, but foreign manufacturers and domestic manufacturers who have moved their production facilities to third-world countries are undercutting Diversified's prices and gaining market share.
>
> However, the Research and Development Department at Diversified has just formulated a new type of blown insulation that the company thinks will perform as well as fiber glass but beat its price. This new substance promises to be a major part of Diversified's highly regarded insulation products. A number of retailers have placed large orders of the new insulation based on Diversified's exhibits at trade shows and some preliminary advertisements in industrial catalogs.
>
> As the head technical writer at Diversified, Susan Taggert oversees the creation of all the product information for the insulation. As she normally does in such cases, she gathers all the documentation from R&D and other pertinent materials in the company to study before mapping out a strategy.
>
> About one week into the project, Taggert discovers from laboratory notebooks that three of the seven technicians participating in the project experienced abnormally high rates of absence from work during their four months developing the insulation. One of the three technicians

requested to be transferred from the project at the end of the first month. His request was granted.

Calling Diversified's Personnel Department, Taggert learns that all three of the technicians complained of the same condition, bronchial irritation of varying degrees of severity, but that the irritation ceased two to three days after the last exposure to the insulation. Apparently some compound in the insulation, which the company physician could not immediately identify, affected some of the technicians who worked closely on it.

Taggert goes to the Vice President of Operations, Bill Mondale, who is in charge of the introduction of all new products. Taggert presents her information to Mondale and suggests that the company find out what is causing the bronchial irritation before it ships any of the product. Although the irritation does not appear to be serious, there are no data on the potential effects of long-term exposure to the insulation when used in houses or offices.

Mondale points out Diversified's tight deadline; delivery is scheduled in less than two weeks. Determining the cause of the irritation could take weeks or months and cost many thousands of dollars. Taggert points out the financial risks involved in selling a product that poses a health risk. Mondale responds that that is a risk the company will have to take, and adds that the product is in compliance with all applicable federal guidelines. The company has staked its reputation—and its third-quarter profits—on the insulation. He directs Taggert to proceed with the product literature as quickly as possible and not to spend any more time worrying about the health hazard.

What should Susan Taggert do?

REFERENCES

Baumhart, Raymond. 1968. *An honest profit: What businessmen say about ethics in business.* New York: Holt, Rinehart and Winston, 11–12.

Beauchamp, Tom L., and Norman E. Bowie. 1988. *Ethical theory and business,* 3rd ed. Englewood Cliffs, N.J.: Prentice-Hall.

Behrman, Jack N. 1988. *Essays on ethics in business and the professions.* Englewood Cliffs, N.J.: Prentice-Hall.

Brockmann, R. John, and Fern Rook, eds. 1989. *Technical communication and ethics.* Washington, D.C.: Society for Technical Communication.

Buchholz, William James. 1989. "Deciphering professional codes of ethics." *IEEE Transactions of Professional Communication* 32.2: 62–68.

Chalk, Rosemary. 1980. *AAAS professional ethics project: Professional ethical activities in the scientific and engineering societies.* Washington, D.C.: American Association for the Advancement of Science.

Honor in science. 1986. New Haven, Conn.: Sigma Xi. The Scientific Research Society.

Perica, Louis. 1972. "Honesty in technical communication." *Technical Communication* 15: 2–6.

Powledge, Tabitha M. 1980. "Morals and medical writing." *Medical Communications* 8.1: 1–10.

Velasquez, Manuel G. 1988. *Business ethics: Concepts and cases.* 2d ed. Englewood Cliffs, N.J.: Prentice-Hall.

3

Getting Started: Audience, Purpose, and Strategy

Chapter 1 describes technical writing as writing about a technical subject that conveys specific information to a specific audience for a specific purpose. In other words, the content and form of any technical document are determined by the situation that calls for that document: your audience and your purpose. Understanding the writing situation helps you devise a strategy to meet your readers' needs—and your own.

Understanding the Writing Situation

The concepts of audience and purpose are hardly unique to technical writing: apart from idle chit-chat, most everyday communication is the product of the same two-part environment. When you write, of course, you must be more precise than you are in conversation: your audience cannot stop you to ask for clarification of a complex concept, nor can you rely on body language and tone of voice to convey your message.

Everyday examples of the two-part writing situation abound. For instance, if you have recently moved to a new apartment and wish to invite a group of friends over, you might duplicate a set of directions explaining how to get there from some well-known landmark. In this case, your writing situation would be clearly identifiable:

Audience   the group of friends
Purpose   to direct them to your apartment

Given your comprehension of this writing situation, you would include in your directions the information that each member of the group might need in order to get to your apartment and would arrange this information in a useful sequence.

When a classified advertisement describes a job opening for prospective applicants, the writing situation of the advertiser can be broken down similarly:

Audience   prospective applicants
Purpose   to describe the job opening

When a notice is posted on a departmental bulletin board advising students interested in enrolling in a popular course to sign up on a waiting list, the department's writing situation also conforms to the two-part model:

Audience   students interested in enrolling in the course
Purpose   to advise signing up on the waiting list

Once you have established the two basic elements of your writing situation, you must analyze each before deciding upon the document's content and form. Effective communication satisfies the demands of both elements of the situation. Although you might assume that purpose would be your primary consideration, an examination of audience is in fact more helpful as a first step: an exploration of its characteristics often influences the course you take in refining your initial, general conception of purpose. The separation of audience and purpose is to some extent artificial—you cannot realistically banish one from your mind as you contemplate the other—but it is nonetheless useful. The more thoroughly you can isolate first audience and then purpose, the more exactly you will be able to tailor your document to the situation.

Analyzing Your Audience

Identifying and analyzing the audience can be difficult. You have to put yourself in the position of people you might not know reading something you haven't yet started to write. Most writers want to concentrate first on content—the nuts and bolts of what they have to say. Their knowledge and opinions are understandably what they feel most comfortable with. They want to write something first and shape it later.

Resisting this temptation is especially important. Having written mostly for teachers, students do not automatically think much about the audience. Students in most cases have some idea of what their teachers want to read. Further, teachers establish guidelines and expectations for their assignments and usually know what students are trying to say. The typical teacher is a known quantity. In business and industry, however, you will often have to write to different audiences, some of whom you will know little about and some of whom will know little about your technical field.

In addition, these different audiences might well have very different purposes in reading what you have written. A *primary audience*, for example, would consist of people who have to act on your recommendations. An executive who decides whether to authorize building a new production facility is a primary reader. So is the treasurer who has to plan for paying for it. A *secondary audience* would consist of people who need to know what is being planned, such as salespeople who want to know where the facility will be located, what products it will produce, and when it will be operational.

The first step in analyzing your audience is to make general assumptions about them by classifying them into one of several basic categories. The second step is to ask yourself specific questions about each reader you can identify. These steps are discussed in the following sections.

The Basic Categories of Readers

Although everyone is unique, a useful first step is to try to classify each reader on the basis of his or her knowledge of your subject. In general, every reader could be classified into one of four categories:

- ☐ expert
- ☐ technician
- ☐ manager
- ☐ general reader

Of course, these categories are slippery. For example, everybody is a general reader in almost every field outside of his or her own specialty. On the job, however, most people fit into one or perhaps two categories. Ellen DeSalvo, for example, a Ph.D. in materials engineering, would be an expert in that particular field. Perhaps she is also the manager of the Materials Groups at her company. Off the job, she might be an expert on native American folk art, having studied it seriously for many years as a hobby.

For this reason, classifying readers into categories is simply a first step in helping you understand how you will have to tailor a particular document. It is not a useful way to understand your readers as people; that part of audience analysis is discussed later in this section.

The Expert

The expert is a highly trained individual with an extensive theoretical understanding of the field. Often the expert carries out basic or applied research and communicates those research findings. A physician who is trying to understand how the AIDS virus works and who delivers papers at professional conferences and writes research articles for professional journals is obviously an expert in that field. An engineer who is trying to devise a simpler and less expensive test for structural flaws in composite materials is an expert. A forester who is trying to plan a strategy for dealing with the threat of fires in the forest is an expert. In short, almost everyone with a postgraduate degree—and many persons with an undergraduate degree in a technical field—is an expert in one area. However, it is useful to remember that not everyone with a lot of degrees is an expert, and many experts have no formal advanced training.

Because experts share a curiosity about their subject and a more-or-less detailed understanding of the theory in their field, they usually have no trouble understanding technical vocabulary and formulas. Therefore, when you write to them, you can get to the details of the technical subject right away, without spending any time sketching in the fundamentals. In addition, most experts are comfortable with long sentences, if the sentences are well constructed and no longer than necessary. Like all readers, experts appreciate graphics, but they can understand more sophisticated diagrams and graphs than most readers.

Following is a passage from an article on a geologic phenomenon called angle of repose (Mann and Kanagy 1990, 358). This passage illustrates the needs and interests of the expert reader.

Theory

Although a complete theoretical analysis of the angle of repose is not yet possible, certain components are well recognized that ultimately will be included

(Rowe, 1962; Statham, 1974; Parsons, 1988). One additional factor, the acceleration of gravity, has been ignored because it is virtually a constant for modern conditions and because its effect on repose angle vanishes for cohesionless material. Earth's surface acceleration of gravity is known to depend on latitude and to have changed over geologic time for various reasons (Kanagy and Mann, unpublished).

The effective acceleration of gravity, g, can be expected to decrease back into Earth's past for several reasons. Tidal evolution in the Earth-Moon system implies a significantly smaller Earth-Moon separation in the past (Walker and Zahnle, 1986). Conservation of angular momentum thus leads to a faster rotation ratio, ω, for Earth; a larger ω effectively decreases g through centrifugal effects. Tidal effects likewise generally reduce the effective value of g. For a spherical Earth of angular velocity ω, mass M_E, and radius R, the effective acceleration of gravity at latitude ϕ is given by

$$g \approx \frac{GM_E}{R^2} - \omega^2 \, (R \cos^2\phi), \tag{1}$$

where G is the constant of gravitation. Empirical evidence noted by Wells (1963) first and by many others since (e.g., Walker and Zahnle, 1986) confirms changes in Earth's angular velocity.

This passage exhibits several characteristics of technical writing addressed to the expert reader.

First, this passage assumes a very high level of technical expertise. The general reader would be unable to follow this discussion because of its vocabulary ("angle of repose" in the first sentence), its reference to theory ("conservation of angular momentum" in paragraph 2), and its citations of technical literature. The equations, too, are well beyond the general reader's abilities.

Second, this passage uses fairly long sentences, not because the writers want to show off, but because the material is complicated. The first sentence, for example, is a relatively sophisticated complex sentence; the second sentence, while technically a simple sentence, contains a parenthetical element in the first clause and twin subordinate clauses.

The Technician

The technician has practical, "hands-on" skills. The technician takes the expert's ideas and turns them into real products and procedures. The technician fabricates, operates, maintains, and repairs mechanisms of all sorts, and sometimes teaches other people how to operate them. An engineer who is having a problem with an industrial laser will talk over the situation with a technician. After they agree on a possible cause of the problem and a way to try to fix it, the technician will go to work.

Like experts, technicians are very interested in their subject, but they know less about the theory. They work with their heads, but also with their hands. Technicians have a wide variety of educational backgrounds; some have a high-school education, others have attended trade schools or earned an associate's degree or even a bachelor's degree.

When you write to technicians, keep in mind that they do not need complex theoretical discussions. They want to finish the task safely, effectively, and quickly. Therefore, they need schematic diagrams, parts lists, and step-by-step instructions. Most technicians prefer short or medium-length sentences and common vocabulary, especially in documents such as step-by-step instructions.

The installation instructions ("Tradeline" 1981) for a heating thermostat are shown in Figure 3.1. Notice that point 3 in the installation instructions explicitly states, "Installer must be a trained, experienced service technician." The instructions for mounting and adjusting the thermostat call for knowledge, experience, and even some tools that the general reader does not possess.

These instructions exhibit several characteristics of writing addressed to a technician.

First, the instructions assume a highly knowledgeable and skilled audience. For instance, the reader must know how to read schematic and wiring diagrams and how to use a spirit level and an ammeter.

Second, the instructions use a number of different kinds of graphics: the schematic and wiring diagrams, a labeled drawing, and a table. Also, the writers have used layout features—caution boxes, numbered steps—to help the reader install the thermostat safely and effectively.

Third, the instructions use relatively brief sentences and common vocabulary ("about" rather than "approximately").

The Manager

The manager is harder to define than the technical person, for the word *manager* describes what a person does more than what a person knows. A manager makes sure the organization operates smoothly and efficiently. The manager of the procurement department at a manufacturing plant is responsible for seeing that raw materials are purchased and delivered on time so that production will not be interrupted. The manager of the sales department of that same organization is responsible for seeing that salespeople are out in the field, creating interest in the products and following up leads. In other words, managers coordinate and supervise the day-to-day activities of the organization.

Upper-level managers, known as executives, are responsible for longer-range concerns. They have to foresee problems years ahead. Is current technology at the company becoming obsolete? What are the newest technologies? How expensive are they? How much would they disrupt operations if they were adopted? What other plans would have to be postponed or dropped altogether? When would the conversion start to pay for itself? What has been the experience of other companies that have adopted the new technologies? Executives are concerned with these and dozens of other questions that go beyond the day-to-day managerial concerns.

Because management is a popular college major, many managers today have studied general business, psychology, and sociology. Often, however, managers start out in a technical area. An experienced chemical engineer, for instance, might manage the engineering division of a consulting com-

TRADELINE

T822D HEATING THERMOSTAT

APPLICATION

The T822D 2-wire, mercury switch thermostat provides low voltage control of heating systems. The T822D features an adjustable heat anticipator.

INSTALLATION

WHEN INSTALLING THIS PRODUCT...

1. Read these instructions carefully. Failure to follow them could damage the product or cause a hazardous condition.

2. Check the ratings given in the instructions and on the product to make sure the product is suitable for your application.

3. Installer must be a trained, experienced service technician.

4. After installation is complete, check out product operation as provided in these instructions.

> ### CAUTION
> Disconnect power supply before beginning installation to prevent electrical shock or equipment damage.

LOCATION

Locate the thermostat on an inside wall about 5 ft [1.5 m] above the floor in an area with good air circulation at average temperature.

Do not mount the thermostat where it may be affected by—
- —drafts, or dead spots behind doors and in corners.
- —hot or cold air from ducts.
- —radiant heat from the sun or appliances.
- —concealed pipes and chimneys.
- —unheated areas behind the thermostat.

WIRING

Disconnect power supply before beginning installation to prevent electrical shock or equipment damage. All wiring must comply with local codes and ordinances. See Figs. 2-5 for internal schematic and typical connection diagrams.

For new installations, run low voltage thermostat wire to the chosen location. For replacement applications, check the old thermostat wires for frayed or broken insulation. Replace any wires in poor condition.

MOUNTING

The T822D Heating Thermostat is designed to be mounted on a wall or vertical outlet box.

1. Grasp the thermostat cover at the top and bottom with one hand. Pull outward on the top of the thermostat cover until it snaps free of the base. Remove the red plastic shipping pin from the thermostat.

2. Pull about 4 in. [101.6 mm] of wire through the wall or into the outlet box. Attach the thermostat wires to the appropriate terminals on the back of the thermostat base (Fig. 1). Push any excess wires back through hole and plug any opening to prevent drafts that may affect thermostat performance.

3. Set the adjustable heat anticipator. See SETTING, Heat Anticipator.

4. Fasten the thermostat to the wall or outlet box with a screw through the top mounting hole (Fig. 1).

NOTE: It may be necessary to move the set point lever to uncover the mounting hole.

5. Place a spirit level across the top of the thermostat. Adjust the thermostat until it is leveled. Start a screw in the center of the bottom mounting hole.

> ### IMPORTANT
> These thermostats are calibrated at the factory mounted at true level. Any inaccuracy in leveling will cause control deviation.

6. Recheck thermostat leveling and tighten the mounting screws.

7. Replace the thermostat cover.

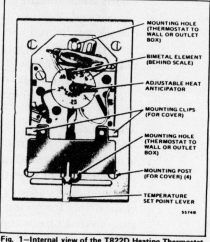

Fig. 1—Internal view of the T822D Heating Thermostat.

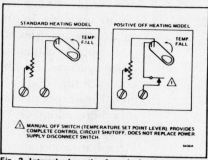

Fig. 2—Internal schematic of standard and positive OFF T822D.

L.M.
Rev. 8-81

Form Number
60-0313—5

FIGURE 3.1 **Heating Thermostat Installation Instructions** (Reproduced by permission from Honeywell Inc.)

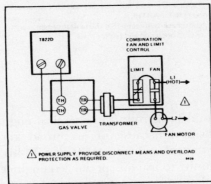

Fig. 3—T822D in typical oil heating system.

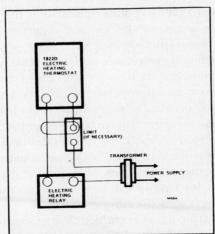

Fig. 4—T822D in typical gas heating system.

Fig. 5—Typical wiring hookup for T822D in an electric heat application.

SETTING
HEAT ANTICIPATOR

— IMPORTANT —

1. This thermostat is equipped with an adjustable heat anticipator and will operate properly ONLY IF THIS HEATER IS ADJUSTED TO MATCH THE CURRENT DRAW OF THE PRIMARY VALVE OR RELAY.

2. Use this thermostat only with controls that have current ratings within the rating of the heat anticipator.

3. Do not use the T822 thermostat on Powerpile (millivolt) systems. The TS822 thermostat was designed for use on Powerpile (millivolt) systems.

See table below for anticipator settings for some primary controls. The current rating of the primary control is usually stamped on the control nameplate. Move the indicator to correspond with this rating on the scale, and the anticipator will be properly adjusted.

RECOMMENDED HEAT ANTICIPATOR SETTINGS

CONTROL			SETTING (amperes)
L8124A-C;	R8182D,F,H,J;	V800A-C;	
L8148E;	R8184G,K,M;	V801;	0.2
	R8404A;	V810A-C	
	S86C		0.4
	S86A;	VR800;	0.6
		VR810	
		VR850	1st stage 0.6 2nd stage 0.2

If setting and current rating are not available, the thermostat should be wired into the system but not mounted (refer to instructions for wiring). If thermostat has already been mounted, remove it from wall, leaving wires connected.

1. Restore power to the system.

2. Connect an ac ammeter (about 0 to 2.0 A range) between the terminals on the back of the thermostat.

3. Move the temperature setting lever to a low setting so the switch contacts are broken. In cold weather installations it may be necessary to hold the switch so the contacts break.

4. Allow the system to operate for one minute and read the ammeter.

5. Set the anticipator to the meter reading. The anticipator may require further adjustment for best performance. To lengthen burner-on time, move the indicator in the direction of the "LONGER" arrows—not more than half a scale marking at a time. To shorten burner-on time, move indicator in opposite direction.

TEMPERATURE SETTING

The temperature setting lever (Fig. 1), when set at the desired set point on the temperature scale, controls the heating system at this point. On positive OFF models, the control circuit is broken when the lever is moved to the extreme left end of the scale.

2

FIGURE 3.1 *Continued*

CHECKOUT AND CALIBRATION ————————

CAUTION
Do not check operation by shorting across terminals of system primary controls—this will damage the heat anticipator.

Observe system operation for at least one automatic cycle. Make certain that the system comes on and turns off in response to the temperature setting lever. Check for proper operation of positive OFF switch and system switch, if used.

CALIBRATION

These thermostats are calibrated at the factory and should not need recalibration. If the thermostat seems out of adjustment, first check for accurate leveling.

To check calibration, proceed as follows:

1. Move the temperature setting lever to the left end of the temperature scale. (This is the low end.) Do not move the lever all the way to OFF on models with a positive OFF switch. Wait at least 5 minutes.

2. Remove the thermostat cover. Move the setting lever until the switch just makes contact; the mercury in the switch will drop to the contact end of the tube.

3. Replace cover and wait 5 minutes for the cover and the thermostat to lose the heat it has gained from your hands. If the thermometer pointer and the setting lever indicator read approximately the same, no recalibration is needed.

If recalibration appears necessary, proceed as follows:

1. Place the temperature setting lever at the same setting as the thermometer. Remove cover.

2. Slip open-end wrench, Part No. 104994A, onto the hex nut under the coil. Holding the setting lever so it does not move, turn the wrench clockwise ⌢ until the switch just breaks contact. Remove wrench and replace cover.

3. Move the setting lever to a low setting. Wait at least 5 minutes for temperature to stabilize.

4. Slowly move the setting lever until it reads the same as the thermometer.

5. Remove cover. Holding the setting lever so it does not move, reinsert wrench and carefully turn counterclockwise ⌢ until the mercury switch just makes contact, NO FARTHER.

6. Recheck calibration. Set thermostat system switch for desired operation and replace cover.

FIGURE 3.1 _Continued_

pany. Although he has a solid background in chemical engineering, he earned his managerial position because of his broad knowledge of engineering and his ability to deal well with colleagues. In his daily work he might have little opportunity to use his specialized engineering skills.

Although generalizing about the average manager's background is difficult, identifying the manager's needs is easier. Managers want to know the "bottom line." They have to get a job done on schedule; they don't have time to study and admire a theory the way an expert does. Rather, managers have to juggle constraints—money, data, and organizational priorities—and make logical and reasonable decisions quickly.

When you write to a manager, try to determine his or her technical background and then choose an appropriate vocabulary and sentence length. Regardless of the individual's background, however, focus on the practical information the manager will need. For example, a research-and-development engineer who is describing to the sales manager a new product line should provide some theoretical background so that the sales representatives can communicate effectively with potential clients. For the most part, however, the description will concentrate on the product's capabilities and its advantages over the competition.

The executive summary from a report entitled "Manufacturing Bidirectional Valves: A Feasibility Study" (D'Ottavi 1990) is presented in Figure 3.2. This executive summary is characteristic of technical writing addressed to the manager. It ignores the technology itself—how the valves work—and concentrates on what the manager cares about: the demand for the product, the costs of researching and developing it, and its market value. The

1. EXECUTIVE SUMMARY

Advances in biomechanical engineering have created a demand for small, quick-acting, bidirectional hydraulic valves. Presently, no valves exist commercially that can satisfy bidirectional flow requirements. Experts speculate that federal and private funding for biomechanical projects will exceed $2 million by 19XX. Other specialty valves similar to bidirectional valves have proven to be highly patentable, resulting in fifteen years or more of exclusive rights to manufacture and sell. Comparable valves can cost between $100 and $300.

This report describes an investigation that examined the demand, the production costs, and the market value for bidirectional valves.

Presently, the demand for these valves is low, due to the fairly limited number of applications. In addition, their future use, although expected to be high, is uncertain at this time.

The initial investment to research and develop bidirectional valves is estimated to be $42,500. The cost to manufacture the valves is estimated at $81 each.

The estimated market value is $200 each. According to a break-even analysis, revenue from 360 valves would equal the initial investment costs plus the costs per valve. These estimates are very conservative, since the actual production costs would be significantly lower if a high volume of valves were manufactured.

Because bidirectional valves are highly versatile (i.e., they can also be used and sold as unidirectional valves), investment in the research and development of bidirectional valves would be a relatively low risk. In addition, the initial investment costs are no higher than those for a unidirectional valve. Considering that technological development of devices that use bidirectional valves could increase dramatically, a valve prototype available for customer testing and evaluation could prove highly profitable.

I therefore recommend that bidirectional valves be researched and developed. However, because of the uncertain future demand, I recommend delaying their manufacture.

FIGURE 3.2   An Executive Summary

writer uses a simple vocabulary; the only term that the general reader might not understand is "break-even analysis," which the intended reader would surely understand.

The General Reader

Occasionally you will have to address the general reader, sometimes called the layperson. In such cases, you must avoid technical language and concepts and translate jargon—however acceptable it might be in your specialized field—into standard English idiom. A nuclear scientist reading about economics is a general reader, as is a homemaker reading about new drugs used to treat arthritis.

The layperson reads out of curiosity or self-interest. The average article in the magazine supplement of the Sunday paper—on attempts to increase the populations of endangered species in zoos, for example—will attract the general reader's attention if it seems interesting and well written. The general reader may also seek specific information that will bring direct benefits: someone interested in buying a house might read articles on new methods of alternative financing.

In writing for a general audience, use a simple vocabulary and relatively short sentences when you are discussing areas that might be confusing. Use analogies and examples to clarify your discussion. Sketch in any special background—historical or ethical, for example—so that your reader can follow your discussion easily. Concentrate on the implications for the general reader. For example, in discussing a new substance that removes graffiti from buildings, focus on its effectiveness and cost, not on its chemical composition.

Following is a brief definition (Leone 1986, 72) of the term "hard disk" addressed to computer novices. The term "floppy disk" has already been described.

> A *hard disk* is made of a firm plastic-like or metal material, much like a phonograph record. By contrast, a floppy disk would be as limp as a piece of paper if you took the disk out of its jacket.
>
> The other difference between a hard disk and a floppy disk is that a hard disk is permanently installed in the computer. If you have a computer with a hard disk, you do not have to put a disk in or take it out every time you use the computer, as you have to when you use a computer that uses floppy disks.
>
> Since a hard disk is rigid, it is possible to record more information on it. Because of the increased amount of information a hard disk can hold, it is possible to have many programs on one hard disk, rather than having to use many floppy disks. For example, one 10 MB hard disk can hold as much information as 26 floppy disks.

This passage shows several of the characteristics of technical writing addressed to the general reader.

First, this passage assumes that the reader knows little or nothing about the subject. The reader need know only that a hard disk is a kind of disk, a storage medium used in most personal computers.

Second, this passage builds on what the reader already knows. In this case, the reader has just read a brief definition of the term "floppy disk." Therefore, the passage is an extended comparison/contrast description of the hard disk and the floppy. In addition, the passage uses analogies like those used in informal conversation (and poetry); the first paragraph uses a metaphor—"much like a phonograph record"—and a simile—"as limp as a piece of paper"—to help the reader visualize the concepts.

Third, this passage concentrates on what is likely to interest the reader rather than on the technical details. The reader learns that a hard disk is made of "a firm plastic-like or metal material," but that is all the writer says about the composition of the disk. And the bulk of the description focuses on the advantages of the hard disk over the floppy: ease of use and greater storage capacity.

Fourth, this passage has an informal tone. Throughout, the writer addresses the reader as "you" and confines herself to common vocabulary: "uses" instead of "utilizes" or "employs," and "put a disk in" rather than "load" or "boot" it.

The Individual Characteristics of Readers

Classifying your readers into categories is a good first step, but as mentioned earlier, it is a very imprecise process. Sometimes it's difficult to be more precise because you don't know who the readers will be. You might be writing a report for your supervisor, who is likely to distribute it—but you don't know to whom. Or you might be addressing an audience of several hundred or even several thousand. Although these many readers share some characteristics, they are not a unified group.

Still, it is a good idea to try to find out as much information as you can about the characteristics of your individual readers.

Eight Questions to Ask about Your Readers

To refine your analysis of your audience, ask yourself the following eight questions about each person whom you know will be getting a copy of your document:

1. What is the reader's name and job title?

     When writing to a number of people in your organization, take out a copy of the organization chart and circle the boxes that represent your readers. This process will give you a clear idea of your readers' different technical and managerial areas, useful information as you plan the content, organization, and style of the document.

2. What are the reader's chief responsibilities on the job?

     In technical writing, you should be less interested in expressing yourself than in helping your readers do their jobs. For each of your primary readers, write down in a sentence his or her major job responsibility. Then think about how your document will help that person accomplish it. For example, if you are writing a feasibility study of several means of cooling the air for a new office building, you know that your reader—an upper-level manager—will have to worry about utility costs over the lifetime of the cooling system. Therefore, in your report you will have to make clear how you are estimating future utility costs.

3. What is the reader's educational background?

The earlier discussion of categories of readers made the point that a person can be classified as an expert, a technician, a manager, or a general reader. At this stage of audience analysis, however, you should try to go a little further. Think not only about the degree the person has earned, such as a Bachelor of Science in Nutrition; think about what kinds of courses, and concentrations of courses, the person has taken. For example, a Metallurgical Engineer who earned a B.S. in 1971 has a much different technical background from the person who earned the same degree in 1989, because the field has changed so much in recent years.

Don't forget, too, formal and informal course work on the job; business and industry spend much more on education than colleges and universities do. A reader who has not studied business in college might have studied it in on-site workshops or training programs.

Knowing your reader's educational backgrounds will help you determine how much background to provide, what level of vocabulary to use, and whether to provide such formal elements as a glossary or an executive summary.

How do you find out your readers' educational backgrounds? Unfortunately, it isn't easy. You cannot ask people to send you their current résumés. But you can try to learn what you can in conversation with your colleagues. The longer you stay at your organization—and the more you talk with people—the more you will find out.

4. What is your reader's professional background (previous positions or work experience)?

Although formal education in school and coursework on the job are important, a person's professional background is at least as significant. The longer a person has been out of school, of course, the more critical the work experience becomes. Working in a field is the best form of education.

A reader whose career spans a decade or more likely has had a variety of professional experiences. A manager in personnel might have worked in accounting or finance for several years. An experienced nurse might have represented her hospital on a community committee to encourage citizens to give blood, might have worked with the hospital administration to choose vehicles for the emergency medical staff, or might have contributed to the planning for the hospital's new delivery room or hospice. In short, her range of experience might have provided one or several areas of competence or even expertise.

If one of your readers has a background in an area related to your subject, you might want to adjust the content and style of your document accordingly. In addition, you should ask that person about tactics to try or to avoid, both in carrying out the project and in writing it up.

5. What is the reader's attitude toward the subject of the document?

Try to determine whether the reader's attitude toward your subject and your approach is positive, negative, or neutral. If possible, discuss the subject thoroughly with your primary readers.

If your reader is neutral or positively inclined toward your subject, simply put together the document so that it responds to the other needs of your readers; make sure the vocabulary, level of detail, organization, and style are appropriate for them.

If your readers are hostile to the subject or your approach to it, you have to be more careful. If one of your primary readers, for instance, opposes your recommendation to network the personal computers in your office, you should try to find out what the objections are and then answer them effectively. Explain clearly why the objections are either not valid or not as important as the benefits of networking. In addition, you might organize the document so that the actual recommendation follows the explanation of the benefits. This strategy encourages the hostile reader to understand your argument rather than reject it out of hand.

An even more delicate situation occurs when you have to argue that a present policy or procedure is ineffective—and one of your primary readers established it. Naturally, you always want to be diplomatic when you discuss the shortcomings of the present system, but you have to be especially careful if there is a chance of offending one of your readers. People get very defensive about their ideas, and sometimes they get most defensive about their worst ideas. You don't want someone to dig in and do everything to oppose you just because you were tactless.

When you might offend such a reader, don't write, "The present system for logging the customer orders is completely ineffective." Instead state, "While the present system has worked well for many years, new developments in electronic processing of orders might enable us to improve the speed and reduce the errors substantially."

6. What will the reader do with the document?

Will he or she file it, skim it, read only a portion of it, study it carefully, modify it and submit it to another reader, attempt to implement recommendations?

Just as you have a purpose in writing the document, each of your readers will have a purpose in reading it. If only one of your

15 readers needs detailed information, you have to provide it, but you can't make the other 14 people wade through it. You have to determine some way to make it available but clearly set off from the rest of the document; an appendix might work well. If you know that your reader wants to use your status report as raw material for a status report to a higher-level reader, try to write your report so that it requires little rewriting. You might even use your reader's own writing style—and even submit the computer disk so that your work can be merged with the new document without the need for retyping.

7. **What are the reader's likes and dislikes that might affect how he or she reacts to the document?**

   As much as we like to think that our actions are guided purely by reason, experience tells us this isn't true. All readers have preferences and biases about writing, some rational and reasonable, some not. One person really hates the first-person pronoun *I*. Another finds the word *interface* distracting when the writer isn't discussing computers. A third reader appreciates a descriptive abstract on the title page of documents. A fourth reader relies on executive summaries. The more you can learn about your readers' likes and dislikes, the greater the chances you can please them. One good way to find out a person's likes and dislikes is to read one of his or her own documents. With some careful study, you will find that everyone's writing is as distinct as a fingerprint.

   Common sense suggests that you should accommodate as many of the readers' preferences as you can. Sometimes, of course, you can't, because the preferences contradict one another or special demands of the subject won't permit it. In any case, never forget that technical writing is intended to get a job done; it is not a personal statement. Whenever possible, try to avoid alienating or distracting your readers.

8. **How will the readers' physical environment affect how you package the document?**

   Most novels look essentially the same, except that some are published in hard cover, some in soft. One exception is the small paperback, sometimes called a pocket book, designed for portability. Technical documents often require special formats. Some will be used in demanding environments, such as on-board ships or aircraft or in garages, where they might be exposed to wind, salt water, and grease. Special waterproof bindings, oil-resistant or laminated paper, coded colors, unusual size paper—these are just some of the special factors that might have to be considered. A user's manual for a computer system, for example, has to lie flat on a table for a long period; it should be produced in a ring binder or with wire or plastic teeth, not glued like a paperback book.

Figures 3.3 to 3.5 show an audience profile sheet that can help you record your readers' characteristics. See the text discussion on page 44.

Audience Profile Sheet

1. Reader's Name _____ Reader's Job Title _____

2. Kind of reader. Primary _____ Secondary _____

3. Reader's chief responsibilities on the job _____
   _____
   _____
   _____
   _____

4. Reader's educational background.

   Formal Education _____
   _____

   Training Courses and Workshops _____
   _____

5. Reader's professional background (previous positions or work experience) _____
   _____
   _____

6. Reader's attitude toward the subject of the document. Positive _____ Neutral _____ Negative _____

   Why? In what ways? _____
   _____
   _____

7. Reader's use for the document. Skim it _____ Study it _____

   Read a portion of it _____ Which portion? _____

   Modify it and submit it to another reader _____ Attempt to implement recommendations _____

   Other? ____ Explain _____
   _____

8. Reader's likes _____
   _____

   Reader's dislikes _____
   _____

9. Reader's physical environment _____
   _____

FIGURE 3.3   Audience Profile Sheet
(See page 44.)

Audience Profile Sheet

1. Reader's Name _____Harry Becker_____   Reader's Job Title _Manager, Drafting/Design Dept._

2. Kind of reader.   Primary __X__   Secondary _____

3. Reader's chief responsibilities on the job _Supervises a staff of twelve draftspersons. Approves or disapproves all requests for capital expenditures over $1,000 coming from his dept. Works with the employees to help them make the best case for the purchase. After approving or disapproving the request, forwards it to Tina Buterbaugh, Manager, Finance Dept., who maintains all capital expenditure records._

4. Reader's educational background.

   Formal Education _B.S. Architectural Engineering, Northwestern, 1981_

   Training Courses and Workshops _CAD/CAM Short Course 1988; Motivating Your Employees Seminar, 1989; Writing on the Job Short Course, 1990_

5. Reader's professional background (previous positions or work experience) _Worked for two years in a small architecture firm. Started here ten years ago as a draftsperson. Worked his way up to Assistant Manager, then Manager._

6. Reader's attitude toward the subject of the document.   Positive __X__   Neutral ____   Negative ____

   Why? In what ways? _Strongly in favor of the request, but he knows that his department has authorized a lot of computer expenditures, and is skeptical because of an unwise purchase last year. His attitude is, if someone wants to buy it, let them make a good case for it._

7. Reader's use for the document.   Skim it _____   Study it __X__

   Read a portion of it _____   Which portion? _____

   Modify it and submit it to another reader ____X____   Attempt to implement recommendations _____

   Other? ___   Explain _____

8. Reader's likes _Straightforward, simple, lots of evidence, clear structure_

   Dislikes _Long, complicated documents full of jargon_

9. Reader's physical environment _normal_

FIGURE 3.4   Sample Audience Profile Sheet   *(See page 44.)*

Audience Profile Sheet

1. Reader's Name _____ Tina Buterbaugh _____   Reader's Job Title _____ Manager, Finance Dept. _____

2. Kind of reader.   Primary _____   Secondary __X__

3. Reader's chief responsibilities on the job In general, in charge of long-range financial planning. In this case, she records all capital expenditures and, at the end of the year, will decide the department's capital budget for next year.

4. Reader's educational background.
   Formal Education B.S. Finance, USC, 1979; MBA Wharton, 1983

   Training Courses and Workshops Don't know, but she is on the training and development committee.

5. Reader's professional background (previous positions or work experience) Not sure, but I know she worked for ten years as Director of Finance for a large retailer.

6. Reader's attitude toward the subject of the document.   Positive _____   Neutral __X__   Negative _____
   Why? In what ways? She has no feelings about proposals this small ($4,000), and she knows Harry Becker is good at forcing the writer to make a good case, but the company has spent a lot lately on computers. She expects Harry to ask for a bigger capital budget next year.

7. Reader's use for the document.   Skim it __X__   Study it _____
   Read a portion of it _____   Which portion? _____
   Modify it and submit it to another reader _____   Attempt to implement recommendations _____
   Other? ___   Explain She will probably spend no more than 5 minutes on it.

8. Reader's likes Same as Harry Becker

   Dislikes Long, complicated documents full of jargon

9. Reader's physical environment normal

FIGURE 3.5   Sample Audience Profile Sheet   *(See page 44.)*

The Audience Profile Sheet

To help you remember all the information you need about your different readers, you might create a form with questions already written out. You could then fill out the form for each of your primary readers and for each of your secondary readers (if there are only a few). Figure 3.3 is an example of an audience profile sheet.

To understand how to use the audience profile sheets, assume you work in the drafting department of an architectural engineering firm. You know that the company's computer-assisted design (CAD) equipment is out of date, and that recent CAD technology would make it easier and faster for the draftspersons to do their work. You want to persuade your company to authorize the purchase of a CAD workstation costing roughly $4,000.

Your primary reader is Harry Becker, the Assistant Manager in the Finance Department. Your secondary reader is Tina Buterbaugh, Manager of the Finance Department. Figures 3.4 and 3.5 are completed audience profile sheets: one for Harry Becker and one for Tina Buterbaugh.

Determining Your Purpose

Once you have identified and analyzed your audience, reexamine your general purpose in writing. Ask yourself this simple question: "What do I want this document to accomplish?" When your readers have finished reading what you have written, what do you want them to *know*, or *believe*? Think of your writing not as an attempt to say something about the subject but as a way to help others understand it or act on it.

To define your purpose more clearly, think in terms of verbs. Try to isolate a single verb that represents your purpose and keep it in mind throughout the writing process. (Of course, in some cases a technical document has several purposes, and therefore you might want to choose several verbs.) Here are a few examples of verbs that indicate typical purposes you might be trying to accomplish in technical documents. The list has been divided into two categories: verbs used when you primarily want to communicate information to your readers, and verbs used when you want to convince them to accept a particular point of view.

"Communicating" Verbs	*"Convincing" Verbs*
to explain	to assess
to inform	to request
to illustrate	to propose
to review	to recommend
to outline	to forecast
to authorize	
to describe	

This classification is not absolute. For example, "to review" could in some cases be a "convincing" verb rather than a "communicating" verb: one writer's review of a complicated situation might be very different from another's review of the same situation. Following are a few examples of how these verbs can be used in clarifying the purpose of the document. (The verbs are italicized.)

1. This report *describes* the research project to determine the effectiveness of the new waste-treatment filter.
2. This report *reviews* the progress in the first six months of the heat-dissipation study.
3. This letter *authorizes* the purchase of six new word processors for the Jenkintown facility.
4. This memo *recommends* that we study new ways to distribute specification revisions to our sales staff.

As you devise your purpose statement, remember that your real purpose might differ from your expressed purpose. For instance, if your real purpose is to persuade your reader to lease a new computer system rather than purchase it, you might phrase the purpose this way: "to explain the advantages of leasing over purchasing." Many readers don't want to be "persuaded"; they want to "learn the facts."

Planning a Strategy

Once you have analyzed your audience and determined your purpose, the next step is to put the two together and create a strategy for writing the document. *Strategy* refers to your game plan: how you are going to create a document that accomplishes your goals.

Specifically, devising a strategy involves the following four steps:

- ☐ accommodating the multiple audience
- ☐ determining the constraints
- ☐ devising a strategy
- ☐ checking the strategy with your primary reader

Accommodating the Multiple Audience

Twenty years ago, there was little difference between writing to a single reader and writing to multiple readers. Most readers, whether technical people or managers, shared a basic understanding of the organization's product or service. In fact, most managers had risen from the technical staff. Today, however, the knowledge explosion and the increasing complexity of business operations have opened a fairly wide gap between what the technical staff knows and what the managers know—and possibly a gap within your audience.

Another complicating factor is technology. The photocopy machine has made it simple and inexpensive to send copies of most documents to dozens of people; electronic mail makes it even simpler and less expensive. These technologies may expand your audience greatly over what it once may have been.

If you think your document will be read by a number of readers, consider making it modular, that is, a document with different components addressed to different kinds of readers. A report, for example, will contain an executive summary for the managers who don't have the time, the knowledge, or the desire to read the whole report. It might also contain a full technical discussion for expert readers, an implementation schedule for technicians, and a financial plan in an appendix for budget officers. Strategies for accommodating the multiple audience are discussed in Parts Three and Four, which treat the different kinds of technical writing documents.

Determining the Constraints

As a student writer, you work within a number of limits, or constraints: the amount of information you can gather for a paper, the length and format, the due date, and so forth. In business, industry, and government, you will be working within similar constraints. Following is a discussion of three sets of constraints: organizational, informational, and time.

Organizational Constraints

Organizational constraints are those limitations placed on you, either implicitly or explicitly, by the organization for which you work or the organization to which you are writing.

For example, many organizations have guidelines about how to present different kinds of information. All tables and figures are presented at the end of the report. Or the names of the people getting copies of your memo are presented in alphabetical order, or by hierarchy. The best way to determine the organizational constraints is to find out if there is a company style guide. If not, the company might use some externally prepared style guide, such as the *Chicago Manual of Style* or the *U.S. Government Printing Office Style Manual*. If your organization has no official rule book, check similar documents on file to see what other writers have done, and of course talk with more-experienced coworkers.

If you are writing to another organization, learn what you can about their preferences. If you are responding to an invitation for a proposal, investigate such matters as length, organization, and style. Many government agencies, for example, require that potential suppliers submit their proposals in two separate parts: the technical proposal and the budget.

Informational Constraints

In addition to any organizational constraints, you might face constraints on the kind or amount of information you can use in the document.

Most often, the situation is that the information you need is not available. For example, you might want to recommend a piece of equipment for your organization, but you can't find any objective evidence that it will do the job. You have advertising brochures and testimonials from satisfied users, but apparently nobody has performed the kinds of controlled experiments that would convince the most skeptical reader.

What do you do in situations like this? You tell the truth. You state exactly what the situation is, weighing the available evidence and carefully noting what is missing. Your most important credential on the job is credibility; you will lose it if you unintentionally suggest that your evidence is better than it really is. In the same way, you don't want your readers to think you don't realize your information is incomplete; they will doubt your technical knowledge.

Time Constraints

The first piece of information you need, of course, is the deadline for the document. Sometimes, a document will have a number of deadlines—for intermediate reviews, progress and status reports, and so forth. Once you know when a particular document (or part of the document) is due, work backwards to devise a schedule. For example, if a major report is due on a Friday, you know that you will have to have it essentially done by the day before, because you want to let it sit overnight before you proofread it for the last time. If it has to be done by Thursday, you might want to give yourself at least four days for drafting and revising. That means you want to finish your outlining by the previous Friday.

People never think they have enough time to do their jobs, writers especially. In fact, many managers don't allot enough time for their employees to do a good job putting together all the necessary memos, letters, reports, and manuals. Most working professionals resign themselves to taking work home in the evenings or on the weekends. Some people come in early to do their writing before things get hectic.

Keep in mind too that few people can accurately estimate the time a particular task will take. As you probably know intuitively, things almost always take longer than people estimate. People get sick, information gets lost or delayed in the mail—whatever can go wrong will go wrong, as Murphy's Law states it. When you collaborate, the number of potential problems increases dramatically, because when one person is delayed, other people may lack necessary information to proceed, causing a snowball effect. In addition, friction between people who are working hard can slow down a project substantially.

Some people build a fudge factor into the writing schedule. When I have to do a big project, I carefully figure out just how long every part of the project will take—then I add 50 percent. I still end up putting in some late nights, but I usually make the deadline.

Devising a Strategy

You have analyzed your audience, determined your purpose, and figured out what constraints you will be working under. Now it is time to put all these factors together and start to think about what the document will eventually look like, and how you are going to get from here to there.

For illustration, let's go back to the writing situation described earlier in this chapter: you want to persuade your two readers to authorize the purchase of some CAD equipment. Your primary reader, Harry Becker, is the head of your department. Your secondary reader is Tina Buterbaugh, the head of Finance. Your audience analysis suggests that only Harry Becker knows much about CAD equipment. However, because Tina Buterbaugh has been at the company for a number of years, she probably understands the basics of the technology: what the equipment is used for, why it makes draftspersons much more efficient, and so forth.

The readers' personal preferences about writing style indicate that you should be straightforward, direct, and objective. You should avoid technical vocabulary and unnecessary technical details, but you have to make clear the practical advantages of the new equipment. It would be a good idea to include an executive summary for Tina Buterbaugh's convenience.

Together, the two audience profiles suggest that you also consider recent history at the company. Ordinarily, a recommendation to spend $4,000 would be routine. However, in light of the networking that wasted $50,000, you know that Harry has become very careful in authorizing any capital expenditures, even small ones, because he knows that next year's budget will depend, to a great extent, on how wise this year's purchases turn out to be.

Next, you define your purpose in writing. In this case, the purpose is clear: to convince Harry Becker that your proposal is reasonable, so that he will authorize the purchase of the equipment.

What do your audience and purpose tell you about what to include in the document? Because your readers will be particularly careful about large capital expenditures, you must clearly show that the type of equipment you want is necessary and that you have recommended the most effective and efficient model. Your proposal will have to answer a number of questions that will be going through your readers' minds:

☐ What system is being used now?
☐ What is wrong with that system, or how would the new equipment improve our operations?

☐ On what basis should we evaluate the different kinds of available equipment?

☐ What is the best piece of equipment for the job?

☐ How much will it cost to purchase (or lease), maintain, and operate the equipment?

☐ Is the cost worth it? At what point will the equipment pay for itself?

☐ What benefits and problems have other purchasers of the equipment experienced?

☐ How long would it take to have the equipment in place and working?

☐ How would we go about getting it?

You start to think of any constraints that might affect how you write the proposal. Although your company doesn't have a style manual that describes internal proposals, the format has remained fixed over the years, and there are plenty of models to consult.

As far as you know, there should be no problem getting the necessary information because the CAD system you are interested in has been on the market for some months and has been reviewed in several professional journals. In addition, you know draftspersons at other companies who have purchased it and will talk with you about their experience and let you try out the equipment.

The only time constraint is the familiar problem: finding the time to write it. There is no external deadline, because this is a self-generated project. You will probably have to work on it for a few evenings at home.

Checking the Strategy with Your Primary Reader

You have a good understanding of your audience, purpose, and strategy, and a general outline is starting to take shape in your mind. But before you actually start to write an outline or gather all the information you will need, it's a good idea to spend another 10 or 15 minutes making sure your primary reader agrees with what you're thinking. You don't want to waste days or even weeks working on a document that won't fulfill its purpose.

Submitting to your primary reader a statement of your understanding of the audience, purpose, and strategy is a means of establishing an informal contract. While it is true that people can change their minds later—in response to new developments or after more thought—provisional approval is better than no approval at all. Then, if your assignment is subsequently revised, at least you can prove that you have been using your time well.

Some writers are reluctant to submit such a statement to the primary reader, either because they don't want to be a nuisance or because they're worried that it might reveal a serious misunderstanding of the assignment. There must be some people out there who don't want to be "bothered," but I haven't met any. Why should they object? It doesn't take more than 30 seconds to read your brief statement, and if there *has* been a misunderstanding, it's far easier to remedy at this early stage. It's in everyone's interest to have the document come out right the first time around.

What should this statement look like? It doesn't matter, as long as you clearly and briefly state what you are trying to do. Here is an example of the statement you might submit to your boss about the CAD equipment.

Harry:

Tell me if you think this is a good approach for the proposal on CAD equipment.

Outright purchase of the complete system will cost more than $1,000, so you would have to approve it. (I'll provide leasing costs as well.) I want to show that drawing by hand is costly and time consuming, and that the CAD system could save us some money and increase our output. I'll be thorough on how good the equipment is, with independent evaluations in the literature, as well as product demonstrations. The proposal should specify what the procedure we are currently using is costing us and show how much we can save by buying the recommended system.

I'll call you later today to get your reaction before I begin researching what's available.

Here is another example of a statement to the boss.

To: John Binder, Director of Research
From: Bill Weisman, R&D Staff
Subject: Yearly Status Report
Date: May 21, 19XX

Before I begin drafting the yearly status report, I want to make sure I understand yesterday's discussion of the report's goals.

The readers will be the top management, some of whom are not engineers. The report should cover all of the department's activities for the year, but should concentrate on the photovoltaic cells project. The purpose is to describe our activities and recommend that we hire another chemist.

The report is due June 1, but you would like to look at a draft the day before.

Please let me know if you'd like me to follow a different direction.

Once you have received your primary reader's approval, you can feel confident in starting to gather and interpret your information and then in writing your document. These tasks will be discussed in the next two chapters.

WRITER'S CHECKLIST

Following is a checklist of questions you should ask yourself when you analyze your audience and purpose. Keep in mind as you think about your audience that your document might be read by one person, several people, a large group, or several groups with various needs.

1. What is the reader's name and job title?

2. What are the reader's chief responsibilities on the job?

3. What is the reader's educational background?

4. What is your reader's professional background (previous positions or work experience)?

5. What is the reader's attitude toward the subject of the document?

6. What will the reader do with the document: file it, skim it, read only a portion of it, study it carefully, modify it and submit it to another reader, attempt to implement recommendations?

7. What are the reader's likes and dislikes that might affect his or her reaction to the document?

8. How will your reader's physical environment affect how you write and package the document?

9. What is your purpose in writing? What is the document intended to accomplish?

10. Is your purpose consistent with your audience's needs?

11. How does your understanding of your audience and of your purpose determine your strategy: the scope, structure, organization, tone, and vocabulary of the document?

12. Are there any organizational constraints that you have to accommodate?

13. Are there any informational constraints that you have to accommodate?

14. Are there any time constraints?

15. Have you checked with your primary reader to see if he or she approves of your strategy for the document?

EXERCISES

1. Choose two articles on the same subject, one from a general-audience periodical, such as *Reader's Digest* or *Newsweek*, and one from a more technical journal, such as *Scientific American* or *Forbes*. Write a memo to your instructor comparing and contrasting the two articles from the point of view of the authors' assessment of the writing situation: the audience and the purpose. As you plan the memo, keep in mind the following questions:
   a. What is the background of each article's audience likely to be? Does either article require that the reader have specialized knowledge?
   b. What is the author's purpose in each article? In other words, what is each article intended to accomplish?
   c. How do the differences in audience and purpose affect the following elements in the two articles?
      (1) scope and organization
      (2) sentence length and structure
      (3) vocabulary
      (4) number and type of graphics
      (5) references within the articles and at the end

2. Choose a 200-word passage from a technical article addressed to an expert audience. Rewrite the passage so that it is clear and interesting to the general reader.

3. Audience is the primary consideration in many types of nontechnical writing. Choose a magazine advertisement for an economy car, such as a Hyundai, and one for a luxury car, such as a Mercedes. In a memo to your instructor, contrast the audiences for the two ads by age, sex, economic means, lifestyle, etc. In contrasting the two audiences, consider the explicit information in the ads — the writing — as well as the implicit information — hidden persuaders such as background scenery, color, lighting, angles, and the situation portrayed by any people photographed.

4. The Greenlawn letter reproduced on the following pages was written by a branch manager of a lawn service company to a homeowner who inquired about the safety of the lawn-care service. The homeowner specifically mentioned in her letter that she has an infant who likes to play outside on the lawn. Assume that you have recently been hired by Greenlawn's Customer Service Department. Your supervisor has asked you to review some of the company's recent letters to customers. Write a memo to your supervisor evaluating the way the letter responds to the concerns of its reader.

GREENLAWN   3220 Orange Blossom Trail, Orlando, FL 32616

April 5, 19XX

Mrs. John Smith
200 Delaware Street
Orlando, FL 32919

Dear Mrs. Smith:

    Thank you for inquiring about the safety of the Greenlawn program. The materials purchased and used by professional landscape companies are effective, nonpersistent products that have been extensively researched by the Environmental Protection Agency. Scientific tests have shown that dilute tank-mix solutions sprayed on customers' lawns are rated "practically nontoxic," which means that they have a toxicity rating equal to or lower than such common household products as cooking oils, modeling clays, and some baby creams. Greenlawn applications present little health risk to children and pets. A child would have to ingest almost 10 cupfuls of treated lawn clippings to equal the toxicity of one baby aspirin. Research published in the American Journal of Veterinary Research in February 1984 demonstrated that a dog could not consume enough grass treated at the normal rate of application to ingest the amount of spray material required to produce toxic symptoms. The dog's stomach simply is not large enough.

    A check at your local hardware or garden store will show that numerous lawn, ornamental, and tree-care pesticides are available for purchase by homeowners either as a concentrate or combined with fertilizers as part of a weed and feed mix. Label information shows that these products contain generally

Mrs. John Smith
Page 2
April 5, 19XX

the same pesticides as those programmed for use by professional lawn-care companies, but contain higher concentrations of these pesticides than found in the dilute tank-mix solutions applied to lawns and shrubs. By using a professional service, homeowners can eliminate the need to store pesticide concentrates and avoid the problems of improper overapplication and illegal disposal of leftover products in sewers or household trash containers.

Greenlawn Services Corporation has a commitment to safety and to the protection of the environment. We have developed a modern delivery system using large droplets for lawn-care applications. Our applicators are trained professionals and are licensed by the state in the proper handling and use of pesticides. Our selection of materials is based on effectiveness and safety. We only apply materials that can be used safely, with all applications made according to label instructions.

On the basis of these facts, I am sure that you will be pleased to know that the Greenlawn program is a safe and effective way to protect your valued home landscape. I have also enclosed some additional safety information. I encourage you to contact me directly should you have any questions.

Sincerely,

Helen Lewis

Helen Lewis
Branch Manager

5. Each of the situations described below explains a writing situation and the real purpose you wish to accomplish. For each of the situations, rewrite the purpose in a diplomatic, polite sentence that can be included early in the document.

    a. Your final paper in this course will be two days late because a large project in a course in your major took much longer than you had anticipated. Your real purpose in writing to your instructor is to persuade him or her not to penalize you for your lateness.

    b. You are introducing a section of a proposal to rebuild a section of highway. Your real purpose is to get the contract.

    c. You are writing an advertising brochure for your company, which specializes in custom landscaping for homeowners. Your real purpose is to persuade your readers to invite you out to their homes to give them a free estimate.

    d. You are writing to your dean requesting permission to substitute one course for another in your curriculum. Your real purpose is to persuade him or her to not force you to take the required course, which doesn't sound appealing to you.

    e. You are writing to your boss, who does not want to upgrade the company's personal computers. You want to persuade him that the inadequate memory in the personal computers is not only a nuisance but also a real hindrance for all of the employees in their daily work.

6. Fill out an audience profile sheet about the instructor of your course. What information can you not supply? What do the course syllabus and the information stated in class add to your understanding of the instructor's needs and expectations?

7. Fill out an audience profile sheet about yourself, as if you were the reader of your own writing. What are the major differences between your profile of your instructor and your self-profile?

REFERENCES

D'Ottavi, J. 1990. Manufacturing bidirectional valves: A feasibility study. Unpublished report.

Leone, N. L. 1986. *A mother's guide to computers*. Rochester, N.Y.: Lion.

Mann, J. C. and S. P. Kanagy II. 1990. Angles of repose that exceed modern angles. *Geology* 18.3: 358–361.

"Tradeline T822D Heating Thermostat." 1981. Minneapolis: Honeywell, Inc.

4

Finding and Using Information

Chapter 3 discussed planning a technical document: analyzing its audience and determining the purpose of the document, then planning a writing strategy that accommodates the audience's needs and your constraints.

Chapter 5 will discuss the writing process itself: prewriting, drafting, and revising. Prewriting includes finding information, generating ideas, and outlining the document. Drafting is the process of turning the outline into sentences and paragraphs. Revising involves turning that draft — a very rough draft — into a finished product by going over it numerous times, looking to fix different kinds of problems each time.

The present chapter, which covers the skills needed to find and use technical information, is a bridge between the planning and the actual writing. For most sophisticated technical-writing applications, such as reports and manuals, you will want to plan a strategy and then locate and study the technical information needed to carry out that strategy. Finally, you will begin the actual writing.

Research falls into two categories: primary and secondary. *Primary research* is the process of creating or observing information yourself. For example, if you are an agricultural researcher trying to improve the growth rate of a particular variety of potato, you might experiment on the effects of a new kind of soil nutrient, periodically measuring the size and weight of the potatoes. *Secondary research* is the process of collecting information that other people have created. For example, before you start to study that soil nutrient, you will probably read articles about its chemistry or about how it has worked for farmers with other crops or with potatoes in some other geographical area or under different soil conditions. You might also talk with other researchers about the soil nutrient. Only after you have learned from this secondary research that the soil nutrient is promising will you devote the necessary time and money to the experiment.

The relationship between primary and secondary research should be clear. A person who produces primary data might write it up in a report or a journal article, or merely talk with another interested researcher about it. Anyone who reads or listens to that information is doing secondary research.

This chapter is a brief overview of the standard techniques of finding and using information for technical documents.

Choosing a Topic

Whereas very few professionals get to choose their topics, you as a student may on occasion. This freedom is a mixed blessing. An instructor's request that you "come up with an appropriate topic" is for many people frustratingly vague; they would rather be told what to write about — how the Soviet

Union's *Sputnik* changed the American policy on space exploration, for instance—even though the topic might not interest them.

Provided you don't spend weeks agonizing over the decision, the freedom to choose your own topic and approach is a real advantage: if you are interested in your topic, you'll be more likely to want to read and write about it. Therefore, you will do a better job.

Start by forgetting topics such as the legal drinking age, the draft, and abortion. These topics don't relate to most practical writing situations. Ask yourself what you are *really* interested in: perhaps something you are studying at school, some aspect of your part-time job, something you do during your free time. Browse through three or four recent issues of *Time* or *Newsweek*, and you will find dozens of articles that suggest interesting, practical topics involving technical information. The June 18, 1990, *Newsweek*, for example contained the following articles:

Subject of the Article	*Possible Subject Areas for a Report*
Saying yes to taxes	The effects of the deficit on the economy
Potholes ahead on tobacco road	The ethical dilemma of tobacco subsidies The science of addiction
Keeping a deadly secret: radioactive uranium mines	The pathology of radiation-induced cancer The ethical issue of disclosing health hazards to workers The ethical issue of reparations to victims
A plan for German reunification	Military implications of shifting politics in Europe Economic implications of shifting politics in Europe Methods for combating East German industrial pollution
Trump: the fall	The future of junk bonds The banking industry's double standards for different kinds of borrowers
The doctor's suicide van	The ethics of euthanasia
Japan answers the 64-megabit question	The future of semiconductor technology
Fighting the greenhouse	Low-tech responses to the greenhouse effect The politics of the environment

Many of these topics are too broad for a brief research report of 10 to 15 pages, but they are good starting points. The article on federal subsidies to the tobacco industry, for example, could be narrowed to a topic on the pathology of addiction, or on the economics or politics of federal subsidies, or on the government's anti-smoking educational efforts, or on public criticism of tobacco companies' efforts to introduce new cigarette brands to narrow markets. The point is that good topics are all around. If you like music, you might write about the fate of vinyl records, or the legal struggles regarding digital audio tape, or the use of new materials in instruments, or the Japanese dominance in audio components. If you are interested in health care, you might write about hospices, or birthing rooms, or the nursing shortage, or the regulation of so-called quack medicine.

Once you have chosen a tentative topic for your research report, the next step is to see if you have information and time to do the job well.

Determining What You Already Know about the Topic

First, consider how much you already know about the topic. If you don't know very much, you will have to do a lot of background reading. Most students find that they have plenty to do just to make sure the document is well organized and well written; the extra research just takes too much time.

Before you commit yourself to a major writing project, try one of the following techniques to determine how much you already know about the topic:

- *Brainstorm.* Brainstorming reveals how much elementary research you would have to do just to understand the basics of your subject. If you cannot quickly list 20 or 30 points about the subject, you might need to spend many extra hours in the library. Brainstorming is discussed in more detail in Chapter 5.

- *Ask yourself the journalistic questions about the topic: who, what, when, where, why, and how.* If, for instance, you think you might be interested in writing about digital audio tape (DAT), test yourself with the following questions:

  □ Who invented DAT? Who uses it?
  □ What is DAT?
  □ When was DAT invented? When was it licensed? When is it expected to become popular?
  □ Where was it invented? Where is it used now? Where will it be used in the future?
  □ Why is DAT used? What advantages does it offer over analog audio tape or compact discs? Why is it not widely used now?
  □ How does DAT work? What is its principle of operation?

These are just a few questions that come to mind. If I didn't know the answers to most of them, I would think about whether I have the time to both do the research *and* write a good report.

■ *Talk with someone about it for ten minutes.* The mere process of trying to explain the subject to someone will help you understand how much you know about it. The person can help you even more by asking you questions.

Determining How to Deal with Time Limitations

The previous chapter discussed time constraints in writing on the job. Time limitations are, of course, an important factor in student writing too.

Getting library materials sometimes takes a long time. At any given moment, approximately ten percent of a library's holdings are missing: miscatalogued, misplaced, stolen, or checked out indefinitely by a faculty member. The fastest way to get hold of one of these items might be through interlibrary loan, which can take from a week to several months. Therefore, when you check to see whether your library has an item you need, you should actually put your hands on it. If it is a book, check it out of the library; if it is an article, photocopy it. Don't gamble; it might not be there next week when you want it.

Other methods of locating information also can require more time than you had anticipated. Getting the results of a mailed questionnaire might take a month. Scheduling interviews with busy people can often require weeks. Before you start any serious research, skim through your printed sources to determine if you will have enough time. Changing topics is much easier *before* you've put in a lot of hours.

Some writers like to have a precise topic in mind before they begin to search the literature. A few examples of precise topics follow:

1. the role of genetic engineering in the control of agricultural pests that attack tomatoes
2. the effect of salt-water encroachment on the aquifer in southern New Jersey
3. the market, during the coming decade, for videodisc instructional systems designed to be used in elementary schools

The advantage of a precise topic is that you can begin your research quickly, because you know what you're looking for.

Other writers prefer to begin their research with only a general topic. For example, you might want to write about computers but be unsure about what aspect to focus on. The basic research techniques described in this chapter will enable you to discover—quickly and easily—what aspects of computers will make the most promising topics. Perhaps you didn't know about the extensive research to help companies reduce the theft of computerized data. That topic might appeal to you. Or you might want to research new techniques in computer graphics. The advantage of being flexible about your topic is obvious: in 15 minutes you can discover dozens of possible topics from which to choose.

Finding Additional Information

For most technical-writing projects, you will need to do both primary and secondary research. For example, you might have read news reports claiming that the sound quality of compact discs is improved if highlighting-pen ink is applied to the outside edge of the disc. If you wanted to do a report on this subject, you would read secondary research in the library, studying the many articles in the audiophile magazines about the subject. Then, you would do primary research, testing whether people can actually tell the difference in sound quality.

The following discussion covers the basic tools for finding information to supplement what you already know about a subject. First it explains how to use the library. Then, it covers conducting interviews, creating questionnaires, and performing inspections and experiments.

Using the Library

You may find information in several different kinds of libraries. The local public library in a small city is a good source for basic reference works, such as general encyclopedias, but it is unlikely to have more than a few specialized reference works. Most college libraries have substantially larger reference collections and receive the major professional journals. Large universities, of course, have comprehensive collections. Many large universities have specialized libraries that complement selected graduate programs, such as those in zoology or architecture. Large cities often have special scientific or business libraries that you can use as well.

Reference Librarians

The most important information sources at any library are the reference librarians. Although a college student is expected to know how to use the library, reference librarians are there to help you solve special problems. They are invariably willing to suggest new ways to search for what you need—specialized directories, bibliographies, or collections that you didn't know existed. Perhaps most important, they will tell you if the library *doesn't* have the information you need and suggest other libraries to try. Reference librarians can save you a lot of time, effort, and frustration. Don't be afraid to ask them questions when you run into problems.

Card Catalogs

With few exceptions, every general library has a main card catalog, which lists almost all the library's books, microforms, audio and video materials, and so on. Some libraries list periodicals (newspapers and journals) in the main catalog; others keep separate "serials" catalogs. The library's nonfiction books and pamphlets, as well as many nonliterary resources, are usually entered on three types of cards (author, title, and subject) in the main catalog. Periodicals, whether in the main catalog or in the serials catalog, are listed by title; to find an article, you must know the name and date of the journal in which it appears (see the discussion of periodicals indexes below).

Accordingly, if you know the title or author of the nonserial item for which you are looking, you can determine easily whether the library has it. Your search is more difficult if you have no specific work in mind and are simply looking for a discussion of a topic by consulting the subject cards. (Some libraries separate their subject cards from their author and title cards.) In such cases you must determine the likeliest subject headings under which relevant publications might be classified. If you have trouble finding appropriate materials, a reference librarian should be able to suggest other subject headings to look under. The broader the subject, the greater the number of cards there will be that you will have to go through. However, broad subjects are often subdivided: *biology*, for example, might be subdivided into *cytology, histology, anatomy, physiology,* and *embryology,* following the range of general *biology* entries. Also, at the end of a range of subject cards, a "see also" card often suggests other subject headings under which to look.

Most libraries post a guide to their classification system, indicating the locations of the books in all the different subject areas. This guide is useful if you want to make sure you are using the correct terminology in searching for a subject area.

All three types of catalog cards provide the same basic information about an item, as Figure 4.1 shows. In addition to the bibliographic information—author, title, place of publication, publisher, number of pages, International Standard Book Number (ISBN), and cataloguing codes—each card lists the call number (which indicates where on the shelves the item is located) in the upper left-hand corner. Call numbers are based on one of two major classification systems. The older, the Dewey Decimal system, uses numbers (such as 519.402462) to designate subject areas. The newer system, the Library of Congress classification, uses combinations of letters and numbers (such as TA330.H68). Some libraries have converted completely from Dewey Decimal to Library of Congress; others are still in the process. In Dewey Decimal, for example, 519 is the category for engineering mathematics; in Library of Congress, engineering

Author ───────────
Call number ───────
Title ─────────
Publisher ───────

Number of chapters,
number of pages,
longest dimension,
and other features
(illustrations, maps,
bibliographies, etc.)

Card headings under
which the book is
catalogued

TA
330
H68

Hovanessian, Shahen A   1931-
    Computational mathematics in engineering / S. A. Hovanes-
sian.—Lexington, Mass. : Lexington Books, c1976.
    xv, 251 p. : ill. ; 24 cm.
    Includes bibliographies and index.
    ISBN 0-669-00733-1

    1. Engineering mathematics.   1. Title.
    TA330.H68   519.4'02'462   76-14667
                                                          MARC

    Library of Congress   76

TITLE CARD

TA   Computational mathematics in engineering
330   **Hovanessian, Shahen A**   1931-
H68   Computational mathematics in engineering / S. A. Hovanes-
         sian.—Lexington, Mass. : Lexington Books, c1976.
             xv, 251 p. : ill. ; 24 cm.
             Includes bibliographies and index.
             ISBN 0-669-00733-1

SUBJECT CARD

TA   ENGINEERING MATHEMATICS
330   **Hovanessian, Shahen A**   1931-
H68   Computational mathematics in engineering / S. A. Hovanes-
         sian.—Lexington, Mass. : Lexington Books, c1976.
             xv, 251 p. : ill. ; 24 cm.
             Includes bibliographies and index.
             ISBN 0-669-00733-1

FIGURE 4.1   Author, Title, and Subject Cards

mathematics is TA. Also posted in every library is a map that will direct you to the area where the books are shelved (if the stacks are open).

Card catalogs are changing in format. Some libraries are transferring their catalogs to microfiche or to computer-generated printed format. And some libraries have converted completely to online catalogs. Instead of walking around the card catalog area to find the appropriate tray of cards, you simply sit before a terminal and enter commands that call up the appropriate entries on the screen. There are a number of different online catalog systems, but they all involve the standard accessing categories of author, title, and subject.

Reference Books

Some books in the card catalog have call numbers preceded by the abbreviation _Ref._ These books are part of the reference collection, a separate grouping of books that normally may not be checked out of the library.

The reference collection includes the general dictionaries and encyclopedias, biographic dictionaries (_International Who's Who_), almanacs (_Facts on File_), atlases (_Rand McNally Commercial Atlas and Marketing Guide_), and dozens of other general research tools. In addition, the reference collection contains subject encyclopedias (_Encyclopedia of Banking and Finance_), dictionaries (_Psychiatric Dictionary_), and handbooks (_Biology Data Book_). These specialized reference books are especially useful when you begin a writing project, for they provide an overview of the subject and often list the major works in the field.

How do you know if there is a dictionary of the terms used in a given field, such as nutrition? As you might have guessed, you can find out in the reference collection. It would be impossible to list here even the major reference books in science, engineering, and business, but you should be familiar with the reference books that list the many others available. Among these guides-to-the-guides are the following:

Downs, R. B., and C. D. Keller. 1975. _How to do library research_, 2d ed. Urbana, Ill.: University of Illinois Press.
Guide to reference books. 1986. 10th ed., ed. E. P. Sheehy. Chicago: American Library Association.
Walford's guide to reference material. 1989. 5th ed. 3 vols., ed. A. J. Walford. London: Library Association.

Look under _Reference_ in the subject catalog to see which guides the library has. The most comprehensive is Sheehy's _Guide to Reference Books_ (also see its recent _Supplements_), an indispensable resource that lists bibliographies, indexes, abstracting journals, dictionaries, directories, handbooks, encyclopedias, and many other sources. The items are classified according to specialty (for example, _organic chemistry_) and annotated. One of the book's most useful features is that it directs you to other guides (such as Henry M. Woodburn's _Using the Chemical Literature: A Practical Guide_) geared to your own specialty. Read the prefatory materials in Sheehy; you can save yourself many frustrating hours.

Periodicals Indexes

Periodicals are the best source of information for most research projects because they offer recent discussions of limited subjects. The hardest aspect of using periodicals is identifying and locating the dozens of relevant articles that are published each month. Although there may be only half a dozen major journals that concentrate on your field, a useful article might appear in one of a hundred other publications. A periodical index, which is simply a listing of articles classified according to title, subject, and author, can help you determine which journals you want to locate. Figure 4.2 (pages 66 and 67) shows two explanatory pages from the *Cumulative Index to Nursing and Allied Health Literature.*

After using a periodical index you might want to find out more information about the journals containing the articles. *Ulrich's International Periodicals Directory,* which is updated every two years, listed 111,950 journals in 557 subject areas in its 1989 edition. *Ulrich's* is indexed by subject area and by title of journal and lists the circulation and publisher for each entry.

Some periodical indexes are more useful than others. A number of indexes — such as *Engineering Index* and *Business Periodicals Index* — are very comprehensive. However, if you are going to rely on a narrower, more specialized index when compiling a preliminary bibliography for your report, you should determine if it is accurate and comprehensive. The prefatory material — the publisher's statement and the list of journals indexed — will supply answers to the following questions:

1. Is the index compiled by a reputable organization? Many of the better indexes, such as *The Readers' Guide to Periodical Literature,* are published by the H. W. Wilson Company. Almost all the reputable indexes are sponsored by well-known professional societies or associations.

2. What is the scope of the index? An index is not very useful unless it includes all the pertinent journals. Scan the list of journals that the index covers. Also, determine if the index includes materials other than articles, such as annual reports or proceedings of annual conferences. Does the index include articles in foreign languages? In foreign journals? Ask the reference librarian if you have questions.

3. Are the listings clear? Does the index define the abbreviations, and do the listings provide enough bibliographic information to enable you to locate the article?

4. How current is the index? How recent are the articles listed, and how frequently is the index published? Don't restrict your search to the bound annual compilations; most good indexes are updated monthly or quarterly.

5. How does the index arrange its listings? Most of the better indexes are heavily cross-indexed: that is, their listings are arranged by subject, author, and (where appropriate) formula and patent.

SEARCHING CINAHL EFFECTIVELY

SUBJECT SEARCHING

The *CINAHL Subject Headings* (or "yellow pages") is the key to effective searching. It includes all valid subject headings and cross references arranged in alphabetic, tree structure, and permuted formats. The first step in every subject search should be to identify the most appropriate heading(s) from this list.

Alphabetic Section

Start with the *Alphabetic List* of the "yellow pages." Look up your topic to see if it is a valid heading. (Valid headings will appear in all capitals.) If it is not, a "see" reference may lead you to the appropriate heading for your topic. For example, when you look up Cancer, a cross reference "see NEO-PLASMS" tells you that articles on this topic appear under the valid heading NEOPLASMS, not under Cancer.

Once you've identified a valid heading, look it up in the Alphabetic Section to make sure it is the most appropriate heading for your search. Scope notes and indexing notes clarify the usage of many subject headings. "See also" references suggest additional related headings that might be useful. When you look up NEOPLASMS, you will find an indexing note to the effect that ANTINEOPLASTIC AGENTS is often used together with NEOPLASMS—drug therapy. A "see also" PRECANCEROUS CONDITIONS reminds you to consider that heading as well.

Tree Structures

Careful use of the *Tree Structures* can also help you to select the most appropriate heading(s). The *Tree Structures* list all CINAHL headings in a subject hierarchy which shows the relationship between broader and narrower terms.

Alphanumeric codes ("tree numbers") from the *Alphabetic List* guide you to the appropriate tree (C4 for NEOPLASMS). Browse the tree to locate the most specific heading(s) related to your topic. This is important since the indexers always assign the most specific subject headings available. If you are looking for articles on BREAST NEOPLASMS, you must search under that heading; relevant articles will *not* appear also under the general term NEOPLASMS.

Permuted

Consult the *Permuted List* if you cannot locate an appropriate subject heading in the *Alphabetic List*. Every significant word appearing in a CINAHL heading or cross reference is listed here alphabetically. It can be a very powerful tool when only one word of a heading is known.

Searching the White Pages

Once you have identified all of the appropriate subject headings for your search, it is time to turn to the white pages of the index. Journal articles, new books, and dissertations are listed in separate sections. In each case, subject headings appear in alphabetical order with appropriate citations underneath. Citations are further divided by subheadings (e.g., psychosocial factors, therapy, and trends) to group together related articles. Start searching in the most current bimonthlies and work backwards to earlier volumes. Be sure to check the list of subject headings for each year searched. The following diagrams explain bibliographic citations:

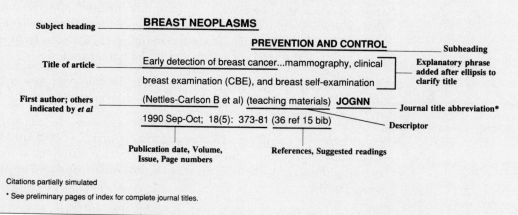

Journal Article

Subject heading —— **BREAST NEOPLASMS**

PREVENTION AND CONTROL —— Subheading

Title of article —— Early detection of breast cancer...mammography, clinical breast examination (CBE), and breast self-examination —— Explanatory phrase added after ellipsis to clarify title

First author; others indicated by *et al* —— (Nettles-Carlson B et al) (teaching materials) **JOGNN** —— Journal title abbreviation*

1990 Sep-Oct; 18(5): 373-81 (36 ref 15 bib) —— Descriptor

Publication date, Volume, Issue, Page numbers —— References, Suggested readings

Citations partially simulated

* See preliminary pages of index for complete journal titles.

FIGURE 4.2 Explanatory Pages from *Cumulative Index to Nursing & Allied Health Literature*

Book

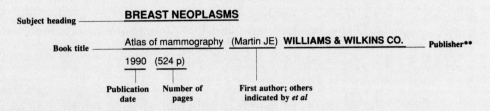

Subject heading — **BREAST NEOPLASMS**

Book title — Atlas of mammography (Martin JE) **WILLIAMS & WILKINS CO.** — Publisher**

1990 (524 p)

Publication date

Number of pages

First author; others indicated by *et al*

** See preliminary pages of index for complete list of book publishers (with addresses)

Interpreting Citations

Bibliographic citations provide the information needed to locate materials. To find journal articles in a library, make note of the article title, journal title, publication date, volume number, issue number (if any), and pages. To find books, you will need to know the book title, author, publisher, and publication date. And for dissertations, keep a record of the title, author, school attended, and date. The UMI number facilitates dissertation ordering.

Special features of CINAHL's bibliographic citations help you to judge the value of the materials indexed. Explanatory phrases are added to clarify vague titles. Descriptors indicate the form of an item (e.g., editorial, research) or the presence of some special data (e.g., statistics, care plans). The number of references and suggested readings are also included in the citation.

Dissertation

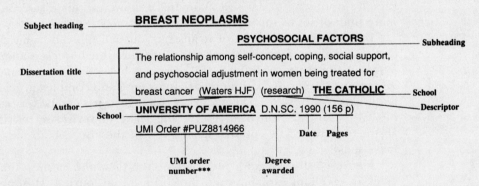

Subject heading — **BREAST NEOPLASMS**

PSYCHOSOCIAL FACTORS — Subheading

Dissertation title — The relationship among self-concept, coping, social support, and psychosocial adjustment in women being treated for breast cancer (Waters HJF) (research) **THE CATHOLIC** — School

Author — School — **UNIVERSITY OF AMERICA** D.N.SC. 1990 (156 p) — Descriptor

UMI Order #PUZ8814966

Date Pages

UMI order number***

Degree awarded

*** To order contact UMI at 300 North Zeeb Road, Ann Arbor, MI 48106; Phone (800) 521-3042

AUTHOR SEARCHING

Authors for journal articles, books, and dissertations are listed in separate sections. They are, however, searched identically. Search author names alphabetically in the white pages. The maximum number of authors listed for each citation is three. Additional authors are indicated by et al. The complete bibliographic citation appears only under the name of the first author, with "see" references for the second and third authors.

SEARCHING *for* BOOKS *and* DISSERTATIONS

The last two sections of the white pages list new books and nursing dissertations. These may be searched by subject heading and/or author. (See discussions above.)

FIGURE 4.2 *Continued*

The better subject indexes include *Applied Science and Technology Index* (New York: H. W. Wilson Company, 1913 to date), *Science Citation Index* (Philadelphia: Institute for Scientific Information, 1961 to date), and *Index Medicus* (Washington, D.C.: National Library of Medicine, 1960 to date). The *New York Times* and *Wall Street Journal* also publish indexes of their own articles.

Abstract Journals

Abstract journals not only contain bibliographic information for articles listed but also provide abstracts — brief summaries of the articles' important results and conclusions. (See Chapter 19 for a discussion of abstracts.) The advantage of having the abstract is that in most cases it will enable you to decide whether to search out the full article. The title of an article, alone, is often a misleading indicator of its contents.

Whereas some abstract journals, such as *Chemical Abstracts*, cover a broad field, many are specialized rather than general. *Adverse Reaction Titles*, for instance, covers research on the subject of adverse reactions to drugs. *Applied Mechanics Review*, too, is relatively narrow in scope. A major drawback of abstract journals is that they take longer to compile than indexes (about a year, as opposed to about two months); therefore, the listings are not as up-to-date. In many fields, however, this disadvantage is more than offset by the usefulness of the abstracts.

In evaluating the quality of an abstract journal, apply the same criteria that you apply to indexes, but ask one other question: how clear and informative is the abstract? To be useful, abstracts have to be brief (usually no more than 250 words), and summarizing a long and complex article in so few words requires great skill. Poorly written abstracts can be very confusing. Most abstract journals reserve the right to revise or rewrite the abstracts provided by authors. Read a few of the abstracts to see if they make sense.

Figure 4.3 shows a guide included in *Energy Research Abstracts* that provides an excellent explanation of how to use that journal. Most abstract journals are organized in this way.

Government Publications

The United States government publishes about twenty-thousand documents annually. In researching any field of science, engineering, or business, you are likely to find that a government agency has produced a relevant brochure, report, or book.

Government publications usually are not listed in the indexes and abstract journals. The *Monthly Catalog of United States Government Publications*, put out by the U.S. superintendent of documents, provides extensive access to these materials. The *Monthly Catalog* is indexed by author, title, and subject. If you are doing any research that requires information published by the government before 1970, you should know about the *Cumula-*

How To Read A Citation

The coverage of literature in this publication includes several document types, ranging from technical reports to journal articles to books. The principal data elements included in these citations are:

1. **Abstract number** within volume.
2. **Report number** identification for report-type literature.
3. **Title and subtitle** (non-English title may appear in parentheses, if applicable).
4. **Author(s).** First 10 names in the data record are printed, then "et al." is listed.
5. **Author affiliation.** Only first one is listed, in parentheses after author(s) to which it applies.
6. **Collaboration,** if present.
7. **Corporate author(s)** identifying corporation responsible for document.
8. **Patent assignee and number** for citations of patent documents or applications.
9. **Journal title, volume, and issue** for citations of journal articles.
10. **Date of publication.** If not known, a processing date is in brackets.
11. **Number of pages** or page range.
12. **Language** of document if non-English.
13. **Monograph title** if citation is an analytic (part, chapter, or paper) of a larger monograph.
14. **Publisher's name and location** for documents published by a corporate or commercial source.
15. **Sponsoring organization.**
16. **Contract or grant number.**
17. **Secondary identifying number;** may be a conference number.
18. **Conference title, location, and date,** if applicable.
19. **Order number.** The "DE" order number may be used for ordering from NTIS or OSTI, as appropriate. The "TI" prefix is valid only at OSTI.
20. **Sources of availability** from which a copy of the document may be obtained; usually appear as abbreviations. (See information on following page.)
21. **Drop note** or explanatory statement.
22. **Abstract.**
23. **Subject descriptors.** Listed only if no abstract or only a brief statement is included.

Sample Citations

Report

1849   (DOE/ER/40438–T1) [Development of a hydrogen and deuterium polarized gas target for application in storage rings]: Progress report. Haeberli, W. Phys. VI collaboration. Wisconsin Univ., Madison (USA). Dept. of Physics. [1989]. 12p. Sponsored by DOE Energy Research. DOE Contract FG02-88ER40438. Order Number DE89007246. Available from NTIS, PC A03/MF A01 - OSTI; GPO Dep.

This paper briefly discusses the Wisconsin test facility for storage cells; results of target tests; the new UHV...

Report Analytic

18500   (INIS-SU—69, pp. 30-32) Transition energies in Ne-like Ions. Correlation effects. Vainshtejn, L.A. AN SSSR, Moscow. Fizicheskij Inst. 1988. (In Russian). In *Experimental and theoretical physics. Collection.* Order Number DE89780060. Available from NTIS (US Sales Only), PC A03/MF A01; INIS.

Kratkie Soobshcheniya po Fizike., no. 6.

SILVER IONS/energy-level transitions; XENON IONS/energy-level transitions; CORRELATIONS; D STATES; E STATES;...

Journal Article

19055   A theoretical and experimental investigation of long-pulse, electron-beam-produced rare gas plasmas. Brake, M.L.; Repetti, T.E. (Energy Beam Interaction Lab. of the Dept. of Nuclear Engineering, Univ. of Michigan, Ann Arbor, MI (US)). *IEEE (Institute of Electrical and Electronics Engineers) Transactions on Plasma Science (USA),* 16(5): 581-589 (Oct 1988). (CONF-881020–: Symposium on radiation physics, Sao Paulo, Brazil, October 3, 1988).

Visible emission spectroscopy (380-650 nm) has been performed on intense, electron beam (1 kA, 300 ns, 300 kV) produced Ar, Kr,...

Patent

18045   Polarization of fast particle beams by collisional pumping. Stearns, J.W. To Dept. of Energy, Washington, DC. USA Patent 4,724,117. 9 Feb 1988. Filed date 19 Oct 1984. vp. Available from Patent and Trademark Office, Box 9, Washington, DC 20232.

A method for polarizing a fast beam of particles by collisional pumping, comprising the steps of generating a beam...

FIGURE 4.3   Guide to an Abstract Journal

tive Subject Index to the Monthly Catalog, 1900–1971 (Washington, D.C.: Carrollton Press, 1973–1976). This 15-volume index eliminates the need to search through the early *Monthly Catalogs*.

Government publications are usually cataloged and shelved separately from the other publications. They are classified according to the Superintendent of Documents system, not the Dewey Decimal or the Library of Congress system. See the reference librarian or the government documents specialist for information about finding government publications in your library.

State government publications are indexed in the *Monthly Checklist of State Publications* (1910 to date), published by the Library of Congress.

If you would like more information on government publications, consult the following four guides:

Lesko, M. 1986. *Lesko's New Tech Sourcebook.* New York: Harper & Row.
Morehead, J. 1983. *Introduction to United States public documents.* 3rd ed. Littleton, Col.: Libraries Unlimited.
Schmeckebier, L. F., and R. B. Eastin. 1986. *Government publications and their use.* Washington, D.C.: Brookings Institution.
Schwarzkopf, L. C. *Government reference books.* Littleton, Col.: Libraries Unlimited. A biennial publication.

Guides to Business and Industry

Many kinds of technical research require access to information about regional, national, and international business and industry. If, for example, you are researching a new product for a report, you may need to contact its manufacturer and distributors to discover the range of the product's applications. For information on where to write, consult one of the following two directories:

Thomas' register of American manufacturers. New York: Thomas. 1905 to date.
Poor's register of corporations, directors, and executives. New York: Standard and Poor's. 1928 to date.

These annual national directories provide information on products and services and on corporate officers and executives. In addition, many local and regional directories are published by state governments, local chambers of commerce, and business organizations. These resources are particularly useful for students and professionals who wish to communicate with potential vendors or clients in their own area.

Also valuable as resources for two major investment services — publications that provide balance sheets, earnings data, and market prices for corporations.

Moody's investor's service. New York: Moody's.
Standard and Poor's corporation records. New York: Standard and Poor's.

Two government periodicals contain valuable information about national business trends:

The Federal Reserve bulletin. Washington, D.C.: U.S. Board of Governors of the Federal Reserve System. A monthly publication that outlines banking and monetary statistics.
Survey of current business. Washington, D.C.: U.S. Office of Business Economics. Details general business indicators.

For more information on how to use business reference materials, consult the following guides:

Daniells, L. M. 1985. *Business information sources.* Berkeley: University of California Press.
Lavin, M. R. 1987. *Business information: How to find it, how to use it.* Phoenix, Ariz.: Oryx.
Strauss, D. W. 1988. *Handbook of business information: A guide for librarians, students, and researchers.* Englewood, Col.: Libraries Unlimited.

Online Searching

Most college and university libraries—and even some public libraries—have facilities for online information searching. Technological improvements in computer science, along with continuing growth in the amount of scholarly literature produced, ensure that in the foreseeable future online searching will be a major way to access information.

Even today, when computerized information retrieval is in its infancy, hundreds of millions of items are contained in the hundreds of different databases. A database is a machine-readable file of information from which a computer retrieves specific data. Libraries today lease access to different database services, with which they communicate by computer. The largest database service is DIALOG Information Services, which by 1990 offered over 400 different databases containing some 200 million bibliographic entries in a wide range of fields. For example, DIALOG carries Agricola, an agriculture index; Biosis, a biology index; Insurance Abstracts; Pharmaceutical News Index; Commerce Business Daily; Chemical Industry Notes; Aptic, an air pollution index; World Textiles; Mathfile; Historical Abstracts; and World Affairs Report.

Online searches are now standard in many fields, especially in science and technology. Although most online searches require the assistance of a trained librarian, the concept is relatively simple. You decide what key words (or phrases) to use in your search, and what limitations you wish to place on it. For example, you might wish to limit your search to articles, excluding other types of literature, or you might wish to retrieve only those articles published in the last two years. When you enter a key word, the screen will show you how many items in the database are filed according to that key word. You can ask the computer to combine several key words and show the number of items that include all the key words. For example, the key "solar energy" might elicit 735 entries. The key "home heating" might elicit 1,438. Combined, the two might elicit 459 entries—that is, 459 items dealing with solar energy *and* home heating.

By trying out various combinations of key words, you can come up with an effective search strategy. You can then ask the computer to print the bibliography—or in some cases the actual abstracts— at the terminal or at the computer site.

Online searching offers several advantages over manual searching:

- ☐ *It is faster.* The search can take as little as a few minutes.
- ☐ *It is more comprehensive.* The computer doesn't overlook any items: it follows your instructions to the letter.
- ☐ *It is more flexible.* You can devise your strategy at the terminal, modifying it effortlessly.
- ☐ *It is more up-to-date.* Databases are produced and updated more quickly than printed indexes or abstract journals.
- ☐ *It is more accurate.* The printout of items contains no mechanical errors or indecipherable handwriting.

However, computerized information retrieval has its disadvantages:

- ☐ *It is expensive.* An average search can cost $10 to $20 in computer time. Students usually have to assume at least part of this expense.
- ☐ *It can produce a lot of irrelevant information.* Unless you tell it not to, the computer will print out items you hadn't anticipated—such as speeches or films. In addition, even a well-planned search strategy can yield items that are not useful. For instance, an article filed under the key words "Japan" and "whale hunting" might relate to whale hunting by the Japanese—as you had hoped—but deal almost exclusively with Japanese consumption of whale products.
- ☐ *It is inconvenient.* Your library might not have access to the database you need, and in most cases you cannot perform the search without assistance.

For more information on database services, see the *Directory of Online Databases* (New York: Cuadra/Elsevier), a quarterly publication that describes thousands of different databases. The journal *Online* is also devoted to the subject. For more information on how to use databases, see Armstrong and Large's *Manual of Online Search Strategies* (Boston: G. K. Hall, 1988).

CD-ROM

CD-ROM, which stands for Compact Disc–Read Only Memory, is a new technology that is transforming the way large volumes of information are stored and accessed. Basically, the CD-ROM system is a combination of a personal computer and a compact disc player. Information is stored on a compact disc, and the user accesses the information through the keyboard on the personal computer.

Currently, the major use of CD-ROMs is to store research tools such as indexes, abstracting journals, and reference texts. However, full-text discs— discs holding the full text of journal articles and books—are coming on the market. One example is *Business Dateline,* a product from University Micro-films International, which contains 100,000 full-text articles from 180 United States and Canadian business publications since 1985.

The major features and advantages of CD-ROM are the following:

- *It can store tremendous amounts of information in a limited space.* A single compact disc, the same size as an audio compact disc, can hold 250,000 double-spaced typed pages, the equivalent of a 100-volume encyclopedia.

- *It can be searched in a variety of ways.* As with any digitized system, the user can use a number of search strategies. For instance, if you want to check for all references to the scientist James Watson in an encyclopedia stored on a disc, you simply type in his name. You can also access information by subject and by title. Different discs offer additional ways to search for information.

- *It can provide multimedia output.* The *Compton's MultiMedia Encyclopedia* is one example of the multimedia capabilities of CD-ROM. The disc holds the 1989 edition of the *Compton's Encyclopedia,* the *Merriam-Webster Intermediate Dictionary,* 15,000 illustrations, 45 animated sequences, and 60 minutes of sound. More and more standard reference resources are becoming available that exploit CD-ROM's ability to present text, digitized photographs and illustrations, animation, and even music.

- *It can output data in several forms.* Once you have located information that you want, you can see it on the screen, print it out with an attached printer, or transfer it to your own disk. Although you cannot manipulate the information that is on the CD-ROM disc—just as you cannot change the music on an audio compact disc—once you have transferred the information to your own computer disk, you can manipulate it as you would any computer file. You don't have to copy the information by hand or take the source to the photocopy machine, and you don't have to type your own notes.

- *It is relatively timely.* Once an institution purchases a CD-ROM disk, it gets periodic updates. The most common update interval is six months or a year, but for some timely materials, new discs are sent out every month. Although information on CD-ROM is more current than that in a printed format, it is less current than information on an online database, which is commonly updated weekly or even daily.

- *It is easy to use.* Unlike online searching, which requires the assistance of a trained librarian, using the CD-ROM is simple; most students pick it up in a few minutes.

- *It is relatively inexpensive.* In 1990, the average cost of the compact-disc driver was $800, while the software—the discs themselves—averaged less than $1,000 each. These prices are likely to fall further. The major cost difference between CD-ROM and online searching is that CD-ROM has a fixed cost, regardless of how many people use it, whereas online searching involves an additional charge every time it is used. CD-ROM is a self-contained package; there is no telephone usage charge because it is not connected electronically to an outside source.

In 1990, there were well over one million CD-ROM systems in use worldwide, and that number is estimated to be growing at a rate of 150 percent a year. CD-ROM is likely to become the dominant form of information storage in all research libraries; it has already become a standard technology in business and industry, where its relatively low cost and small size make it the best storage medium.

Because the same reference source often is available both online and on CD-ROM, the different online directories include the CD-ROM sources. Also see Ann Lathrop's *Online and CD-ROM databases in school libraries* (1989, Englewood, Col.: Libraries Unlimited).

This section on CD-ROM searching concludes the discussion of secondary searching. The following sections discuss primary research techniques.

Conducting Inspections and Experiments

Inspecting and experimenting are two common techniques for finding information. To inspect something is, of course, to look at it, either with or without the use of specialized tools. To experiment is to test a theory about why something happened, or whether something would happen. For instance, an experiment could be devised that tests whether cars with four-wheel steering handle better in the rain than do cars with conventional two-wheel steering.

Conducting Inspections

Regardless of what field you work in, you are likely to read many sentences that begin, "An inspection was conducted to determine. . . ." A competent auto mechanic can tell you whether you need a front-end alignment by inspecting the wear pattern on your tires. A civil engineer can often determine what caused the cracking in a foundation by inspecting the site. An accountant can learn a lot about the financial health of an organization by inspecting the company's financial records.

These people are simply looking at a physical site or an object or an artifact (the financial records) and applying their knowledge and professional judgment to what they see. Sometimes the inspection techniques are more complicated. For instance, an auto mechanic testing the emissions of your car needs sophisticated equipment. A civil engineer inspecting foundation cracking might want to test hunches by taking a soil sample back to the lab and testing it. An accountant checking the financial records might need to perform some computerized analyses of the information.

When you carry out an inspection, be sure to take good notes. Try to answer the appropriate journalistic questions—who, what, when, where, why, and how—as you inspect, or as soon after as possible. Where appropriate, photograph or sketch the site or print out the output from the computer-assisted inspections. You will probably need the data later for the document; it's a lot easier to collect it the first time around than to have to repeat the inspection.

Conducting Experiments

It can take months or even years to learn how to conduct the many kinds of experiments used in a particular field. Therefore, this discussion can only serve as a brief overview.

For most kinds of experiments, you start with a hypothesis, which is an informed guess about the relationship between two factors. For instance, if you want to determine the relationship between gasoline octane and miles-per-gallon, you can set up an experiment with the following hypothesis: "A car will get better mileage with 89 octane gas than with 87 octane gas."

Next, you set up the experiment itself. In most cases, you want to have an "experimental group" and a "control group"—two groups identical except for the condition you are studying: in this case, the gasoline used. The control group would be a car running on 87 octane. The experimental group would be an identical car running on 89 octane. The experiment would consist of identical driving—preferably in some sort of controlled environment such as a laboratory—over a given distance, such as 1,000 miles. At the end of the 1,000 miles you would measure how much gasoline each car used and calculate the miles-per-gallon. The results will either support or negate your original hypothesis.

Often, it is difficult to set up an experiment as neat as this one. For instance, you probably wouldn't be able to get two identical cars and run them for 1,000 miles in a lab. So you do the best you can: you run your own car with two or three tanks of 87 octane and calculate your miles-per-gallon. Then you run it on the same amount of 89 octane and do the same calculations. Of course, you try to make sure that you are subjecting the car to the same driving conditions to prevent an unwanted variable. The important point is that if you write up your experiment you must be absolutely clear about the experimental conditions.

Then, when you interpret the results, you must try to explain as carefully as possible any factors that might have affected the results. For example, when researchers study how people read and write, they can never achieve ideal experimental conditions because no two people are alike. Therefore, when the researchers explain their findings, they try to account for as many extraneous factors as they can, such as the age, sex, education, knowledge, and attitude of the people who participated in the experiment. Even though a good experiment is designed to minimize the effects of these extraneous factors, it cannot eliminate them. For this reason, experiments involving the behavior of animals — rats, monkeys, people — can never be as reliable as experiments involving inanimate objects.

Experiments can be classified into two categories:

- quantitative
- qualitative

A quantitative experiment yields statistical data that can be measured. The gas-mileage experiment, for example, might indicate that the 89 octane gas produces 7 percent greater miles-per-gallon than the 87 octane. On the other hand, a qualitative experiment, as its name suggests, provides information about what something is, or how good it is. For instance, if you wanted to study group dynamics in a classroom with a new seating arrangement, you could design a qualitative experiment in which you observed and recorded the classes and perhaps interviewed the students and the instructor about their reactions. Then you could do the same in a traditional classroom and study the results.

Some kinds of experiments can have both quantitative and qualitative elements. In the case of the classroom seating arrangements, you could create some quantitative measures, such as the number of times students talked with each other or the length of the interchanges between them. In addition, you could create questionnaires for eliciting the opinions of the students and the instructor. If you used these quantitative measures on a sufficiently large number of classrooms, you could gather valid quantitative information.

When you are doing qualitative experiments on the behavior of animals — from rats to people — be careful to avoid two common problems:

- the effect of the experiment on the behavior you are studying
- bias in the recording and analysis of the data

The first problem is similar to the well-known situation of a television camera crew covering a protest demonstration. When the cameras arrive on the scene, the behavior of the protesters changes. If you are studying the effects of the classroom seating arrangement, you have to try to minimize the effects of your being there. For instance, make sure that the camera is placed unobtrusively, and that it is set up before the students arrive, so they don't see the process. Still, any time you bring in a videotape camera, you can never be sure that what you witness is typical. Even an outsider who sits quietly taking notes can disrupt typical behavior.

The second problem, bias in recording and analyzing data, can occur because researchers want to confirm their hypotheses. If, for instance, you want to show that word processors help students write better, you are likely to see improvement where other people don't. For this reason, you should try to design the experiment so that it is *double blind*. That is, the students doing the writing that is to be studied shouldn't know what the experiment is about, so they won't change their behavior to confirm or deny the hypothesis. And when the data are being analyzed, they should be disguised so that you don't know whether you are studying the control group or the experimental. If the control group wrote in ink and the experimental group used word processors, the control papers should be typed on the word processor, so that all the papers look identical as you analyze the writing.

Running an experiment is relatively simple; the hard part is designing it so that it accurately measures what you want to measure.

Conducting Personal Interviews

The previous section mentioned that personal interviews are often used in qualitative experiments. Personal interviews are also an extremely useful tool when you need information on subjects too new to have been discussed in the professional literature or inappropriate for widespread publication (such as local political questions). Most students are inexperienced at interviewing and therefore reluctant to do it. Interviewing, like any other communication skill, requires practice. The following discussion explains how to make interviewing less intimidating and more productive.

Choosing a Respondent

Start by defining on paper what you want to find out. Only then can you begin to search for a person who can provide the necessary information.

The ideal respondent is an expert willing to talk. Many times your subject will dictate whom you should ask. If you are writing about research at your university, for instance, the logical choice would be a researcher involved in the project. Sometimes, however, you might be interested in a topic about which a number of people could speak knowledgeably, such as the reliability of a particular kind of office equipment or the reasons behind the growth in the number of adult students. Use directories, such as local industrial guides, to locate the names and addresses of potential respondents.

Once you have located an expert, find out if he or she is willing to be interviewed. On the phone or in a letter, state what you want to ask about; the person might not be able to help you but might be willing to refer you to someone who can. And be sure to explain why you have decided to ask the respondent: a well-chosen compliment works better than admitting that the person you really wanted to interview is out of town. Don't forget to mention what you plan to do with the information, such as write a report or give a talk. Then, if the person is willing to be interviewed, set up an appointment at his or her convenience.

Preparing for the Interview

Never give the impression that you are conducting the interview to avoid library research. If you ask questions that are already answered in the professional literature, the respondent might become annoyed and uncoop-erative. Make sure you are thoroughly prepared for the interview: research the subject carefully in the library.

Write out your questions in advance, even if you think you know them by heart. Frame your questions so that the respondent won't simply answer yes or no. Instead of "Are adult students embarrassed about being in class with students much younger than themselves?" ask, "How do the adult students react to being in class with students much younger than themselves?" Don't give the impression that all you want is a simple confirmation of what you already know.

Conducting the Interview

Arrive on time. Thank the respondent for taking the time to talk with you. Repeat the subject and purpose of the interview and what you plan to do with the information.

If you wish to tape-record the interview, ask permission ahead of time; taping makes some people uncomfortable. Have paper and pens ready. Even if you are taping the interview, you will want to take brief notes as you go along.

Start by asking your first prepared question. Listen carefully to the respondent's answer. Be ready to ask a follow-up question or request a clarification. Have your other prepared questions ready, but be willing to deviate from them. In a good interview, the respondent probably will lead you in directions you had not anticipated. Gently return to the point if the respondent begins straying unproductively, but don't interrupt rudely or show annoyance.

After all your questions have been answered (or you have run out of time), thank the respondent again. If a second meeting would be useful—and you think the person would be willing to talk with you further—ask now. If you might want to quote the respondent by name, ask permission now.

After the Interview

Take the time to write up the important information while the interview is still fresh in your mind. (This step is, of course, unnecessary if you have recorded the interview.)

It's a good idea to write a brief thank-you note and send it off within a day or two of the interview. Show the respondent that you appreciate the courtesy extended to you and that you value what you learned. Confirm any previous offers you made, such as to send a copy of the report.

Sending Letters of Inquiry

A letter of inquiry is often a useful alternative to a personal interview. If you are lucky, the respondent will provide detailed and helpful answers.

However, the person might not understand what information you are seeking or might not want to take the trouble to help you. Also, you can't ask follow-up questions in a letter, as you can in an interview. Although the strategy of the inquiry letter is essentially that of a personal interview—persuading the reader to cooperate and phrasing the questions carefully—inquiry letters in general are less successful, because the readers have not already agreed to provide information.

For a full discussion of inquiry letters, see Chapter 14.

Administering Questionnaires

Questionnaires enable you to solicit information from a large group of people. Although they provide a useful and practical alternative to interviewing dozens of people spread out over a large geographical area, questionnaires rarely yield completely satisfactory results. For one thing, some of the questions, no matter how carefully constructed, are not going to work: the respondents will misinterpret them or supply useless answers. In addition, you probably will not receive nearly as many responses as you had hoped. The response rate will almost never exceed 50 percent; in most cases, it will be closer to 10 or 20 percent. And you can never be sure how representative the response will be; in general, people who feel strongly about an issue are much more likely to respond than are those who do not. For this reason, be careful in drawing conclusions based on a small number of responses to a questionnaire.

When you send a questionnaire, you are asking the recipient to do a favor. If the questionnaire requires only two or three minutes to complete, of course you are more likely to receive a response than if it requires an hour. Your goal, then, should be to construct questions that will elicit the information you need as simply and efficiently as possible.

Creating Effective Questions

Effective questions are unbiased and clearly phrased. Avoid charged language and slanted questions. Don't ask, "Should we protect ourselves from unfair foreign competition?" Instead, ask, "Are you in favor of imposing tariffs on men's clothing?" Make sure the questions are worded as specifically as possible, so that the reader understands exactly what information you are seeking. If you ask, "Do you favor improving the safety of automobiles?" only an eccentric would answer, "No." However, if you ask, "Do you favor requiring automobile manufacturers to equip new cars with air bags, which would raise the price $300 per car?" you are more likely to get a useful response.

As you make up the questions, keep in mind that there are several formats from which to choose:

Multiple Choice

Would you consider joining a company-sponsored sports team?

　　　　Yes _____　　　　No_____

How do you get to work? (Check as many as apply.)

　　　　my own car _____

　　　　car/van pool _____

　　　　bus _____

　　　　train _____

　　　　walk _____

　　　　other _____　　(please specify)

The flextime program has been a success in its first year.

Agree strongly	Agree more than disagree	Disagree more than agree	Disagree strongly

Ranking

Please rank the following work schedules in order of preference. Put a "1" next to the schedule you would most like to have, a "2" next to your second choice, etc.

　　　　8:00–4:30 _____

　　　　8:30–5:00 _____

　　　　9:00–5:30 _____

　　　　flexible　　_____

Short Answer

What do you feel are the major <u>advantages</u> of the new parts-requisitioning policy?

　　　　1. _____

　　　　2. _____

　　　　3. _____

Short Essay

The new parts-requisitioning policy has been in effect for a year. How well do you think it is working?

_____

_____

Remember that you will receive fewer responses if you ask for essay answers; moreover, essays, unlike multiple-choice answers, cannot be quantified. A simple statement with a specific number in it—"Seventy-five percent of the respondents own two or more PCs"—helps you make a clear and convincing case. However thoughtful and persuasive an essay answer may be, it is subject to the interpretations of different readers.

After you have created the questions, write a letter or memo to accompany the questionnaire. A letter to someone outside your organization is basically an inquiry letter (sometimes with the questions themselves on a separate sheet); therefore, it must clearly indicate who you are, why you are writing, what you plan to do with the information, and when you will need it. (See Chapter 14 for a discussion of inquiry letters.) For people within your organization, a memo accompanying the questionnaire should provide the same information.

Figure 4.4 shows a sample questionnaire.

Sending the Questionnaire

After drafting the questions, you next administer the questionnaire. Determining whom to send it to can be simple or difficult. If you want to know what the residents of a particular street think about a proposed construction project, your job is easy. But if you want to know what mechanical engineering students in colleges across the country think about their curricula, you will need background in sampling techniques to isolate a representative sample.

Before you send *any* questionnaire, show it and the accompanying letter or memo to a few people whose backgrounds are similar to those of your real readers. In this way you can test the effectiveness of the questionnaire before "going public."

Be sure also to include a self-addressed, stamped envelope with questionnaires sent to people outside your organization.

Using the Information

Once you have gathered your books and articles, conducted your interviews, and received responses from your questionnaires, you should start planning a strategy for using the information. The first step in working with any information is to evaluate its source.

Evaluating the Sources

Authority is sometimes difficult to determine. Your respondents are likely to be authoritative (you would not deliberately have chosen suspect sources), but you must still judge the quality of the information they have

September 6, 19XX

To: All employees
From: William Bonoff, Vice-President of Operations
Subject: Evaluation of the Lunches Unlimited food service

As you may know, every two years we evaluate the quality and cost of the food service that caters our lunchroom. We would like you to help by sharing your opinions about the food service. Your anonymous responses will help us in our evaluation. Please drop the completed questionnaires in the marked boxes near the main entrance to the lunchroom.

1. Approximately how many days per week do you eat lunch in the lunchroom?

   0 ___   1 ___   2 ___   3 ___   4 ___   5 ___

2. At approximately what time do you eat in the lunchroom?

   11:30–12:30 ___   12:00–1:00 ___   12:30–1:30 ___   varies ___

3. Do you have trouble finding a clean table?   often ___   sometimes ___   rarely ___

4. Are the Lunches Unlimited personnel polite and helpful?

   always ___   usually ___   sometimes ___   rarely ___

5. Please comment on the quality of the different kinds of food you have had in the lunchroom.

   a. Hot meals (daily specials)   excellent ___   good ___   satisfactory ___   poor ___

   b. Hot dogs and hamburgers   excellent ___   good ___   satisfactory ___   poor ___

   c. Sandwiches   excellent ___   good ___   satisfactory ___   poor ___

   d. Salads   excellent ___   good ___   satisfactory ___   poor ___

   e. Desserts   excellent ___   good ___   satisfactory ___   poor ___

6. What foods would you like to see served that are not served now?

   _____

7. What beverages would you like to see served that are not served now?

   _____

8. Please comment on the prices of the foods and beverages served.

   a. Hot meals (daily specials)   too high ___   fair ___   a bargain ___

   b. Hot dogs and hamburgers   too high ___   fair ___   a bargain ___

   c. Sandwiches   too high ___   fair ___   a bargain ___

   d. Desserts   too high ___   fair ___   a bargain ___

   e. Beverages   too high ___   fair ___   a bargain ___

9. Would you be willing to spend more money for a better-quality lunch, if you thought the price was reasonable?   yes, often ___   sometimes ___   not likely ___

10. Please provide whatever comments you think will help us evaluate the catering service.

   _____

   _____

Thank you for your cooperation.

FIGURE 4.4   Questionnaire

provided. Your questionnaires, once you have discarded eccentric or clearly extreme responses, should contain useful material.

With books and articles, however, the question is not so clear-cut. Check to see if the author and publisher or journal are respected. Many books and journals include biographical sketches. Does the author appear to have solid credentials in the subject? Look for academic credentials, other books or articles written, membership in professional associations, awards, and so forth. If no biographical information is provided, consult a "Who's Who" of the field.

Evaluate the publisher, too. A book should be published by a reputable trade, academic, or scholarly house. A journal should be sponsored by a professional association or university. Read the list of editorial board members; they should be well-known names in the field. If you have any doubts about the authority of a book or journal, ask the reference librarian or a professor in the appropriate field to comment on the reputation.

Finally, check the date of publication. In high-technology fields, in particular, a five-year-old article or book is likely to be of little value (except, of course, for historical studies).

Skimming

Inexperienced writers often make the mistake of trying to read every potential source. All too commonly, they get only halfway through one of their several books when they realize they have to start writing immediately in order to submit their reports on time. Knowing how to skim is invaluable.

To skim a book, read the preface and introduction to get a basic idea of the writer's approach and methods. The acknowledgments section will tell you about any assistance the author received from other experts in the field, or about his or her use of important primary research or other resources. Read the table of contents carefully to get an idea of the scope and organization of the book. A glance at the notes at the ends of chapters or the end of the book will help you understand the nature of the author's research. Check the index for clues about the coverage that the information you need receives. Read a few paragraphs from different portions of the text to gauge the quality and relevance of the information.

To skim an article, focus on the abstract. Check the notes and references as you would for a book. Read all the headings and several of the paragraphs.

Skimming will not ensure that a book or article *is* going to be useful. A careful reading of a work that looks useful might prove disappointing. However, skimming can tell you if the work is *not* going to be useful: because it doesn't cover your subject, for example, or because its treatment is too superficial or too advanced. Eliminating the sources you don't need will give you more time to spend on the ones you do.

Taking Notes

Note taking is often the first step in actually writing the document. Your notes will provide the vital link between what your sources have said and what you are going to say. You will refer to them over and over. For this reason, it is smart to take notes logically and systematically.

If you do not have access to a word processor, buy two packs of note cards: one 4 inches by 6 inches or 5 inches by 8 inches, the other 3 inches by 5 inches. (There is nothing sacred about ready-made note cards; you can make up your own out of scrap paper.) The major advantage of using cards is that they are easy to rearrange later, when you want to start outlining the report.

On the smaller cards, record the bibliographic information for each source from which you take notes. For a book, record the following information:

> author
> title
> publisher
> place of publication
> year of publication
> call number

Figure 4.5 shows a sample bibliography card for a book.
For an article, record the following information:

> author
> title of the article
> periodical
> volume
> number
> date
> pages on which the article appears
> call number

Figure 4.6 shows a sample bibliography card for an article.

On the larger cards, write your notes. To simplify matters, write on one side of a card only and limit each card to a narrow subject or discrete concept so that you can easily reorder the information to suit the needs of your document.

Most note taking involves two different kinds of activities: paraphrasing and quoting.

Paraphrasing

A paraphrase is a restatement, in your own words, of someone else's words. "In your own words" is crucial: if you simply copy someone else's words—even a mere two or three in a row—you must use quotation marks.

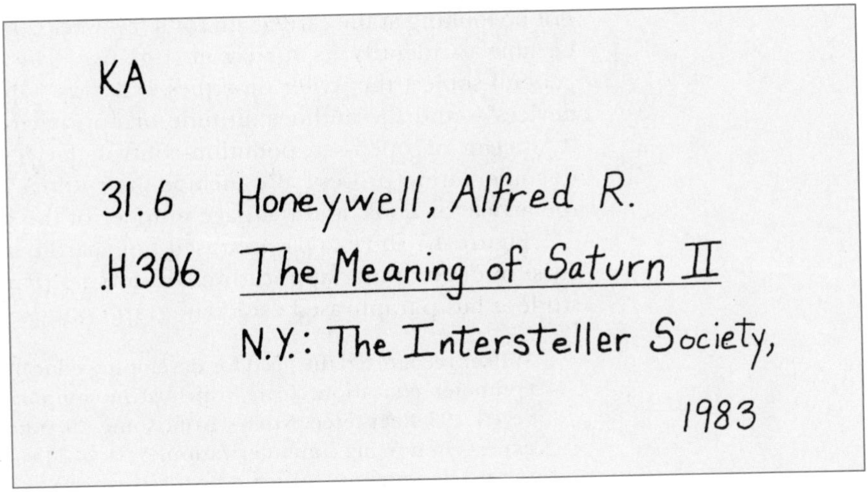

KA

31.6   Honeywell, Alfred R.
.H306   The Meaning of Saturn II

         N.Y.: The Intersteller Society,

                              1983

FIGURE 4.5   Bibliography Card for a Book

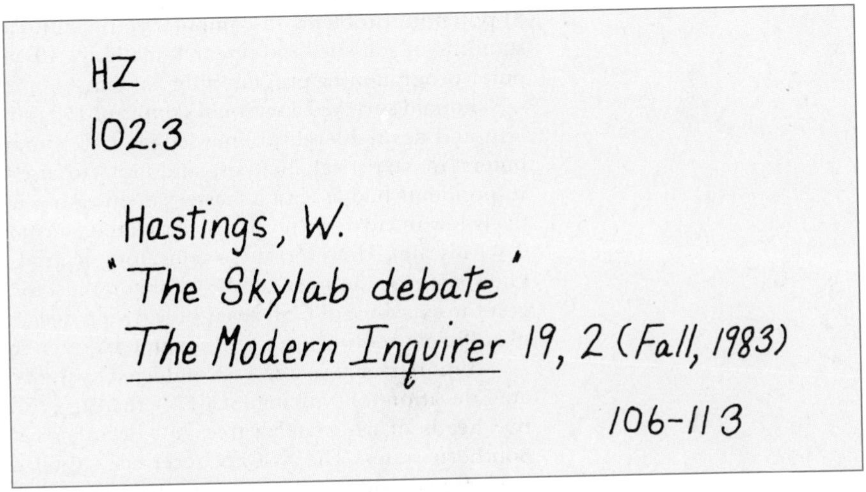

HZ
102.3

Hastings, W.
"The Skylab debate"
The Modern Inquirer 19, 2 (Fall, 1983)

                              106–113

FIGURE 4.6   Bibliography Card for an Article

What kind of material should be paraphrased? Any information that you think *might* be useful in writing the report: background data, descriptions of mechanisms or processes, test results, and so forth.

To paraphrase accurately, you have to study the original and understand it thoroughly. Once you are sure you follow the writer's train of thought, rewrite the relevant portions of the original. If you find it easiest to rewrite in complete sentences, fine. If you want to use fragments or merely list information, make sure you haven't compressed the material so much that you'll have trouble understanding it later. Remember, you might not be looking at the card again for a few weeks. Title the card so that you'll be able to identify its subject at a glance. The title should include the general subject the writer describes—such as "Open-sea pollution-control devices"—and the author's attitude or approach to that subject—such as "Criticism of open-sea pollution-control devices." To facilitate the later documentation process, also include the author's last name, a short title of the article or book, and the page number of the original.

Figure 4.7 shows a paraphrased note card based on the following discussion of computer applications in nursing (Parks et al. 1986: 105). The student has paraphrased each paragraph on a separate note card.

> Walker recognized the need for developing educational objectives for nursing computer education. To help prioritize computer learning needs, she surveyed 193 Registered Nurses from some 29 states who were recognized as experts in nursing computerization. A list of 11 categories of computer learning needs were rank ordered by these experts as follows: 1) fundamentals of data processing, 2) importance of nursing involvement, 3) overview of health care applications, 4) developing a systematized nursing data base, 5) systems analysis, 6) affective impact of computerization, 7) confidentiality/legal issues, 8) potential problems in computerized health care systems, 9) basic understanding of statistics and research methods, 10) change theory, and 11) computer programming/programming languages.
>
> Ronald surveyed a national sample of 159 nursing educators on their current and desired levels of knowledge in 16 computer areas (e.g., use of computers in statistical analysis, and how to use a terminal). Although all respondents had at least a master's degree, they evaluated themselves as relatively low in current knowledge across the 16 computer areas ($M = 1.04$ on a 0 to 4 scale). However, these educators aspired to relatively high levels of knowledge for the 16 areas ($M = 3.06$ on the 0 to 4 scale). These findings suggest the existence of high levels of learning needs for faculty, and presumably of sufficient motivation to engage in computer-related learning experiences.
>
> A more recent survey was conducted by the Southern Council on Collegiate Education for Nursing (SCCEN); the survey involved nursing administrative heads of associate degree and baccalaureate nursing programs in 14 Southern states. The SCCEN received a total of 257 replies (75 percent) regarding the administrators' personal interest in computer technology; these administrators also reported their opinions about their faculty's knowledge, experience, and interests concerning computer education. A large majority (83 percent) of the administrators reported little or limited personal knowledge about computers. However, the majority of the respondents reported using the computer, usually for administrative purposes.

Notice how a heading provides a focus for each card. The student has omitted the information he does not need, but he has recorded the necessary bibliographic information so that he can document his source easily or return to it if he wants to reread it. There is no one way to paraphrase: you have to decide what to paraphrase—and how to do it—on the basis of your analysis of the audience and the purpose of your report.

Parks

What nurses should be learning about computers

Walker's survey of experts in computers in nursing yielded a list of 11 priorities for nursing computer education. They range from data-processing basics and the role of computers in nursing to programming skills.

"Faculty and Student Perceptions . . ." p. 105.

Parks

Motivation of nursing educators to learn about computers

Ronald's survey of (at least) master's level nursing educators showed that although most were weak in computers, most wanted to learn much more.

"Faculty and Student Perceptions . . ." p. 105.

Parks

What do program administrators know about computers?

A Southern Council on Collegiate Education for Nursing survey of nursing-program administrators showed that, although they knew little or nothing about computers, they used them for administrative work.

"Faculty and Student Perceptions . . ." p. 105.

FIGURE 4.7   Paraphrased Notes

Quoting

On occasion you will want to quote a source, either to preserve the author's particularly well-expressed or emphatic phrasing or to lend authority to your discussion. In general, do not quote passages more than two or three sentences long, or your report will look like a mere compilation. Your job is to integrate an author's words into your own work, not merely to introduce a series of quotations.

The simplest form of quotation is an author's exact statement:

> As Jones states, "Solar energy won't make much of a difference in this century."

To add an explanatory word or phrase to a quotation, use brackets:

> As Nelson states, "It [the oil glut] will disappear before we understand it."

Use ellipses (three spaced dots) to show that you are omitting part of an author's statement:

Original Statement

> "The generator, which we purchased in May, has turned out to be one of our wisest investments."

Elliptical Quotation

> "The generator. . . has turned out to be one of our wisest investments."

For more details on the mechanics of quoting, see the entries under "Quotation Marks," "Brackets," and "Ellipses" in Appendix A.

For a discussion of the different styles of documentation, see Appendix B.

Summarizing

Summarizing, a more comprehensive form of taking notes, is the process of rewriting a passage to make it shorter while still retaining its essential message.

As you do your secondary research, some of the sources you skim will be useful, and you will want to summarize one or more of the passages to assist you in creating your own document. This is one of the two main reasons to summarize: to help you learn a body of information so that you can integrate it with other information.

The other major reason to summarize information is to create one or more short versions of your own document, so that readers with different backgrounds or needs can find the information they want. Most long technical documents contain several different kinds of summaries:

- ☐ a letter of transmittal that provides an overview of the document
- ☐ an abstract, or brief technical summary
- ☐ an executive summary, or brief nontechnical summary directed to the manager

In addition, many long documents contain a summary at the end that draws together a complicated discussion. These summaries are discussed in Chapter 19, Formal Elements of a Report.

The present discussion describes the process of summarizing printed information that you uncover in your research and are considering using in your own document. Although the technique described here will prove useful as you prepare the abstract and the different kinds of summaries of your own writing, the focus in this section is on extracting the essence of a passage by summarizing it.

The process of summarizing information consists of the following seven steps:

1. *Read the passage carefully.* The best way to save time is to make sure you understand the original passage before you try to summarize. Read it at least once, but preferably two or three times. Before you proceed, you should feel that you understand what the writer is saying.

2. *Reread the passage, underlining the key ideas.* Most writers put their main ideas in a few obvious places: titles, headings, topic sentences, transitional paragraphs, concluding paragraphs. The bodies of the paragraphs usually explain and exemplify the main points, and therefore are not the most likely places to find the general ideas of the passage.

3. *Combine the key ideas.* If possible, take a short break, then study what you have underlined. Try to rewrite the underlined ideas, in your own words. Don't worry about your grammar, punctuation, or style. If you will be submitting the summary to another reader, you can clean up the superficial problems later. At this point, you are just trying to see if you can reproduce the essence of the original.

4. *Check your draft against the original for accuracy and emphasis.* Reread the original to make sure your summary is accurate and reflects the writer's emphasis. Checking for accuracy covers a wide variety of matters, from getting individual statistics and names correct to ensuring that your version of a complicated concept represents the original faithfully. Checking for proper emphasis means getting the proportions right; if the original devotes 20 percent of its space to a particular point, your draft should not devote 5 percent or 50 percent of its space.

5. *If necessary, edit your summary for style.* If your purpose in summarizing is to learn the material yourself, you don't need to edit for style. However, if you will be submitting the summary to another reader, either as part of a longer document or as an independent document, now is the time to revise it for grammar, punctuation, usage, style, and spelling. The next chapter discusses the revision process.

6. *Record the bibliographic information carefully.* Even though a summary might contain all your own words, you still must cite it, because the main ideas are someone else's. Use bibliography cards, as described in the discussion on taking notes.

7. *If necessary, add key words.* If you are creating an abstract (see Chapter 19), or for some other reason readers will want to be able to retrieve your summary by using key words, you will want to list the key words at this point.

The following passage (based on McComb 1991, 21–23) is a narrative history of television technology addressed to the general reader. Following the passage are two summaries of it. The first summary is written in an informal shorthand; it is meant only for the person who actually wrote the summary. The second summary, which includes key terms, is fleshed out to be appropriate for another reader.

A BRIEF HISTORY OF TELEVISION

Though it seems as if television has been around for a long time, it's a relatively new science, younger than rocketry, internal medicine, and nuclear physics. In fact, some of the people that helped develop the first commercial TV sets and erect the first TV broadcast antennas are still living today.

The Early Years
The first electronic transmission of a picture was believed to be made by a Scotsman, John Logie Baird, in the cold month of February, 1924. His subject was a Maltese Cross, transmitted through the air by the magic of television (also called "Televisor" or "Radiovision" in those days) the entire distance of ten feet.

To say that Baird's contraption was crude is an understatement. His Televisor was made from a cardboard scanning disk, some darning needles, a few discarded electric motors, piano wire, glue, and other assorted odds and ends. The picture reproduced by the original Baird Televisor was extremely difficult to see—a shadow, at best.

Until about 1928, other amateur radiovision enthusiasts toyed around with Baird's basic design, whittling long hours in the basement transmitting Maltese Crosses, model airplanes, flags, and anything else that would stay still long enough under the intense light required to produce an image. (As an interesting aside, Baird's lighting for his 1924 Maltese Cross transmission required 2,000 volts of power, produced by a room-full of batteries. So much heat was generated by the lighting equipment that Baird eventually burned his laboratory down.)

Baird's electro-mechanical approach to television led the way to future developments of transmitting and receiving pictures. The nature of the Baird Televisor, however, limited the clarity and stability of images. Most of the sets made and sold in those days required the viewer to peer through a glass lens to watch the screen, which was seldom over seven by ten inches in size. What's more, the majority of screens had an annoying orange glow that often marred reception and irritated the eyes.

Modern Television Technology

In the early 1930's, Vladimir Zworykin developed a device known as the iconoscope camera. About the same time, Philo T. Farnsworth was putting the finishing touches on the image dissector tube, a gizmo that proved to be the forerunner to the modern cathode ray tube or CRT—the everyday picture tube. These two devices paved the way to the TV sets we know and cherish today.

The first commercially available modern-day cathode ray tube televisions were available in about 1936. Tens of thousands of these sets were sold throughout the United States and Great Britain, even though there were no regular television broadcasts until 1939, when RCA started what was to become the first American television network, NBC. Incidentally, the first true network transmission was in early 1940, between NBC's sister stations WNBT in New York City (now WNBC-TV) and WRGB in Schenectady.

Post-War Growth

World War II greatly hampered the development of television, and during 1941 and 1945, no television sets were commercially produced (engineers were too busy perfecting the radar, which, interestingly enough, contributed significantly to the development of conventional TV). But after the war, the television industry boomed. Television sets were selling like hotcakes, even though they cost an average of $650 (based on average wage earnings, that's equivalent to about $4,000 today).

Progress took a giant step in 1948 and 1949 when the four American networks, NBC, CBS, ABC, and Dumont, introduced quality, "class-act" programming, which at the time included Kraft Television Theatre, Howdy Doody, and The Texaco Star Theatre with Milton Berle. These famous stars of the stage and radio made people want to own a television set.

Color and Beyond

Since the late 1940's, television technology has continued to improve and mature. Color came on December 17, 1953 when the FCC approved RCA's all-electronic system, thus ending a bitter, four-year bout between CBS and RCA over color transmission standards. Television images beamed via space satellite caught the public's fancy in July of 1962 when Telstar 1 relayed images of AT&T chairman Frederick R. Kappell from the U.S. to Great Britain. Pay-TV came and went several times in the 1950's, 60's and 70's; modern-day professional commercial videotape machines were demonstrated in 1956 by Ampex; and home video recorders appeared on retail shelves by early 1976.

Following is the summary written in shorthand for the summarizer:

> The first electronic transmission of a picture: by Baird in 1924. The primitive equipment produced only a shadow. Baird's design modified by others in 1920's, but you had to look through a glass lens at a small screen that gave off an orange glow.
>
> Zworykin's iconoscopic camera and Farnsworth's image dissector tube—similar to CRT—led in 1936 to development of modern TV. Regular broadcasts, 1939, on the first network, NBC. Research stopped during WWII, but after that, sales grew, at about $650/set ($4,000 today).
>
> Color, 1953; satellite broadcasting, 1962; home VCRs, 1976.

This summary is approximately ten percent the length of the original. The average length of most summaries is 5 to 10 percent.

Here is the version of the summary intended for another reader:

> The first electronic transmission of a picture was produced by Baird in 1924. The primitive equipment produced only a shadow. Although Baird's design was modified by others in 1920's, the viewer had to look through a glass lens at a small screen that gave off an orange glow.
>
> Zworykin's iconoscopic camera and Farnsworth's image dissector tube—similar to the modern CRT—led in 1936 to the development of modern TV. Regular broadcasts began in 1939, on the first network, NBC. Research stopped during WWII, but after that, sales grew, even though sets cost approximately $650, the equivalent of $4,000 today.
>
> Color broadcasts began in 1953; satellite broadcasting began in 1962; and home VCRs were introduced 1976. Key terms: television, history of television, NBC, color television, satellite broadcasting, video cassette recorders, Baird, Zworykin, Farnsworth.

EXERCISES

1. Choose a topic on which to write a report for this course. Make sure the topic is sufficiently focused that you will be able to cover it in some detail.
   a. Using Sheehy's *Guide to Reference Books*, plan a strategy for researching this topic.
      (1) Which guides, handbooks, dictionaries, and encyclopedias contain the background information you should read first?
      (2) Which basic reference books discuss your topics?
      (3) Which major indexes and abstract journals cover your topic?
   b. Write down the call numbers of the three indexes and abstract journals most relevant to your topic.
   c. Compare two abstract journals in your field on the basis of sponsoring organization, scope, clarity of listings, and timeliness. Also determine who writes the abstracts contained in each.
   d. Make up a preliminary bibliography of two books and five articles that relate to your topic.
   e. Using the bibliography from Exercise 1d, write down the call numbers of the books and journals (those that your library receives).
   f. Find one of the works listed and write a brief assessment of its value.
   g. Using a local industrial guide, make up a list of five persons who might have first-hand knowledge of your topic.
   h. Make up two sets of questions: one for an interview with one of the persons on your list from Exercise 1g, and one for a questionnaire to be sent to a large group of people.
   i. Photocopy the first two pages from an article listed in your bibliography from Exercise 1d. Paraphrase any three paragraphs, each on a separate note card. Also note at least two quotations, each on a separate card: one should be a complete sentence, and one an excerpt from a sentence.

2. Summarize the passage on pages 94 to 97, from a government report, *Facing America's Trash: What Next for Municipal Solid Waste?* ("MSW" is an abbreviation for Municipal Solid Waste. Table 3.5 is not included here.)

REFERENCES

Congress of the United States, Office of Technology Assessment. 1989. *Facing America's trash: What next for municipal solid waste?* Washington, D.C. Document OTA-0-424.

McComb, G. 1991. *Troubleshooting and repairing VCRs.* 2d ed. Blue Ridge Summit, Pa.: TAB/McGraw-Hill, Inc.

Office of Scientific and Technical Information, U.S. Department of Energy. 1990. *Energy Research Abstracts*, 15, 8, iii.

Parks, P. L. et al. 1986. Faculty and student perceptions of computer applications in nursing. *Journal of Professional Nursing* 2 (March–April): 105.

CHEMICAL COMPOSITION

Basic Chemical Composition

MSW consists mostly of water, various elements (e.g., carbon, hydrogen, oxygen, chlorine, and nitrogen), and incombustible materials (e.g., glass, metals, ceramics, minerals, clay, and dirt) (16). In addition, various trace metals and organic chemicals can be present, but little aggregated information exists on their concentrations in MSW prior to recycling, incineration, or landfilling.

One chemical of particular concern is chlorine because it can be involved in the formation of dioxins and other chlorinated organics, as well as hydrogen chloride, during incineration (ch. 6). The major sources of chlorine in MSW appear to be paper and plastics. In Baltimore County, Maryland, for example, paper was estimated to contribute 56 percent of the total chlorine in the combustible portion of MSW; in Brooklyn, New York, plastics contributed an estimated 52 percent (4).

Chlorine is used directly to make certain products, such as PVC plastics and insulation and textiles. Chlorine is also used to bleach pulp for paper-making. In the pulping process, chemicals remove roughly three-fourths of the lignin (which makes up about half of wood), and bleaching removes the rest. Elemental chlorine (as a gas) has been the preferred bleaching chemical because it is cheaper, effective in dissolving lignin while maintaining the strength of the pulp, and can achieve higher-brightness paper than alternative bleaches. The alternatives, which include hypochlorite, chlorine dioxide, peroxide, and oxygen, generally are less efficient and more expensive than chlorine gas.

Combustibility

Some components of MSW are combustible—organic materials such as paper, plastics, textiles, rubber, and wood. The organic fraction of MSW was estimated to be about 81 percent by weight in 1986 (10). It appears to be growing slowly, primarily because the portions of paper and plastics in MSW also are growing.

One measure of MSW that is related to combustibility is "higher heating value" (HHV), or the number of Btu of energy that could be produced per unit of MSW. In general, MSW can generate from 4,500 to 6,000 Btu per pound. The average Btu value of MSW may be increasing because both plastic and especially paper, which have increased over the last 10 years, have high Btu values (figure 3-3). Paper wastes comprise a large portion of MSW and thus contribute much of its average total HHV. Food and yard wastes both have low Btu values, while inorganic materials such as metals and glass have no Btu value.

However, MSW is not homogeneous, either in its Btu values or its composition, between different locations or even over short periods at the same location. For example, combustibility can vary drastically because the portion of yard wastes can more than double during certain seasons. Yard wastes have high moisture content and low Btu values, so the overall HHV of the MSW decreases during summer and fall, when large amounts of yard waste are generated. Moisture content is also important because it affects the stability of the combustion process (16) and combustion efficiency during "cold starts" of an incinerator (ch. 6). In addition, evaporating moisture during the initial stages of combustion requires the use of energy and thereby affects operating costs.

Removing particular materials from MSW prior to incineration (e.g., through source separation) can affect combustibility.[7] For example, removing yard wastes and inorganic recyclables such as glass and metals can reduce moisture and increase average HHV. In contrast, removing paper and plastics lowers HHV and increases moisture content. The net effect will depend on exactly what is removed.

Degradation

Some of the materials (e.g., paper and yard wastes) in MSW decompose or degrade, while others do not. In general, the rate of decomposition depends on local landfill conditions, such as temperature, moisture, oxygen levels, and pH (ch. 7). In theory, a large portion of MSW should eventually decompose because it tends to have a high level of degradable carbon. For example, one study estimated that degradable carbon comprised 34 to 59 percent of MSW (24). Another study estimated that

---

[7]The potential trade-offs between recycling and incineration of different materials are also discussed in chapters 1 and 6.

**Figure 3-3—Relative Btu Values per Pound
for Materials in MSW**

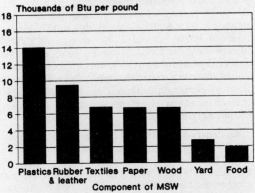

SOURCE: Franklin Associates, Ltd., *Waste Paper, The Future of a Resource 1980-2000*, prepared for the Solid Waste Council of the Paper Industry (Prairie Village, KS: December 1982).

paper products and textiles were composed of about 40 percent degradable carbon, while yard and food wastes were composed of less than 20 percent carbon (3).

The landfill excavation study, however, has revealed some interesting insights about decomposition. In these landfills, paper products in particular, but also food wastes, have not degraded rapidly; in fact, it appears that degradation in general may be slow (29). For example, newspapers that were still readable after years of burial were found in all of the studied landfills. Paper and food waste excavated from part of one landfill were in the same condition as similar materials buried 5 to 10 years earlier in another part of the landfill.

Toxic Substances and Household Hazardous Waste

When MSW is landfilled, incinerated, or recycled, some of the composite metals and organic chemicals have the potential to harm public health and the environment (chs. 5, 6, 7). These are often called toxic or potentially harmful substances, although their potential effects on health and the environment depend on rates of exposure and

dosage, sensitivity of exposed individuals, and other factors.

Toxic Substances in MSW

Many potentially harmful metals and organic chemicals are components of products and packaging that are used at residences and offices and then discarded as MSW. Available data focus on three metals—mercury, lead, and cadmium. For example, mercury is a component of most household batteries, as well as fluorescent light bulbs, thermometers, and mirrors. Sources of lead include solder in steel cans and electronic components, automobile batteries, paint pigments, ceramic glazes and inks, and plastics. About two-thirds of all lead in MSW (after recycling) is estimated to be from automobile batteries (11). Cadmium is found in metal coatings and platings; rechargeable household batteries; pigments in plastics, paints, and inks; and as a heat stabilizer in plastics. Nickel/cadmium batteries are the largest source, accounting for an estimated 52 percent after recycling, and plastics contribute about 28 percent.

The **noncombustible** portion of MSW is estimated to contain 98 percent of the lead and 64 percent of the cadmium (11). This suggests that separating noncombustible materials from MSW that is to be incinerated would be likely to reduce the amounts of these metals in emissions and ash (see ch. 6). Furthermore, because plastics account for an estimated 71 percent of the lead and 88 percent of the cadmium in the remaining **combustible** portion of MSW, efforts to manufacture plastic products without these metals also might help reduce amounts of these metals in emissions and ash. The toxicity issue is discussed in more detail in chapter 4.

Household Hazardous Wastes

Household hazardous wastes (HHW) are discarded products that contain potentially toxic substances, but that tend to be stored at residences for relatively long periods of time before being discarded.[8] Although there is no standardized definition of what products and materials comprise HHW, they generally include common household items such as cleaning products, automobile products, home maintenance products (e.g., paint, paint thinner, stain,

[8]The term "household hazardous wastes" is not used here in the legal sense of being a hazardous waste as defined in RCRA, although some of the substances in such wastes may be classified as hazardous in RCRA (see ch 8).

varnish, glue), personal care products, and yard maintenance products (e.g., pesticides, insecticides, herbicides). In most cases these items are not hazardous while in storage, or during use if properly handled, but they may release potentially toxic substances after they have been discarded.

More than 100 substances that are listed as RCRA hazardous wastes are present in household products (table 3-5). The substances include metals (e.g., mercury, lead, silver) and organic chemicals (e.g., trichloroethylene, benzene, toluene, parathion).

Several studies have looked at the amounts of HHW generated. In two communities, Marin County, California, and New Orleans, Louisiana, HHW from single-family dwellings was sorted and weighed (42). Between 0.35 and 0.40 percent of the total MSW was considered hazardous, and each household threw away an average of 50 to 60 grams of HHW each week.[9] Other studies in Albuquerque, New Mexico, and the Puget Sound area in Washington reached similar conclusions: in general HHW comprises less than 1 percent of MSW (25, 41). Data from Los Angeles County, California, Portland, Oregon, and several localities in Michigan indicated that the quantities of actual constituents of concern were even lower, less than 0.2 percent (20). This has led some analysts to conclude that placing HHW in landfills is not a problem (20). However, the extent to which HHW contributes to environmental problems at landfills is unclear. Given the total quantity of MSW generated each year, even the apparently low proportion of 0.2 percent would mean that about 300,000 tons of potentially toxic substances in HHW are discarded each year.[10] Yet, when spread among thousands of facilities, the potential impacts should be lessened.

Data from residences in several areas (Tucson and Phoenix, Arizona; Marin County, California; and New Orleans, Louisiana) have been compiled to indicate which HHW products were most commonly discarded; the data include containers but exclude automobile batteries (45). The largest category was household maintenance products, making up 37 percent by weight. Household batteries contributed

19 percent, cosmetics 12 percent, household cleaners 12 percent, automobile maintenance products 11 percent, and yard maintenance products 4 percent. About 80 percent of the automobile products was motor oil. Socioeconomic status appears to affect the types of HHW generated. Households in higher-income neighborhoods discarded more pesticides and yard products than did lower-income neighborhoods; cleaning materials were more common in middle-income neighborhoods; and automobile maintenance products were more common in lower-income neighborhoods (31, 45).

One study at a California landfill indicates similar trends (20). Two thousand fifty-six containers of HHW (whether empty or with residue) received at the landfill were sorted and counted. Of the six categories of containers, 40 percent had household and cleaning products; 30 percent automotive products; 16 percent personal products; 8 percent paint and related products; 3 percent insecticides, pesticides, and herbicides; and 4 percent were other products considered hazardous.

The effects of a one-day collection program for HHW in Marin County on subsequent generation of HHW raise an intriguing dilemma (31). Two months after the collection day was held, the amount of HHW in the normal MSW pickup was twice as high as it was before the collection day. This suggests that the educational effect of the collection day was short-lived or, as seems more likely, that people did not want to keep HHW around after they learned about it. If the latter proves true, regular collection days would be needed to keep HHW out of the normal MSW collection system. Chapter 8 discusses HHW programs in more detail.

Other Sources of Toxic Substances

Household products and materials in landfills and incinerators are not the only sources of potentially harmful chemicals in MSW. Under RCRA, businesses that generate less than 100 kilograms of hazardous wastes per month are allowed to deposit them in solid waste landfills (including municipal landfills) or have them burned in MSW incinerators (36, 37). These businesses are known as "very small

[9]These data refer to the weight of that portion of the waste that contains the hazardous ingredients, not including contaminated containers or other contaminated articles such as paint brushes and oil-soaked rags. Thus, they probably underestimate total amounts.

[10]Many hazardous household products also are emptied into sewer systems (40). When household cleaners are used, for instance, the product is washed down the drain and ends up in municipal sewage treatment plants.

quantity generators'' and include vehicle maintenance shops (which handle lead-acid car batteries and used motor oil), drycleaners, pesticide application services, and others (10,37). One study estimated that there are about 450,000 very small quantity generators in the country and that they generate about 197,000 tons of hazardous waste annually (1). How much of this waste is sent to MSW landfills and incinerators is unknown. Even if all of it is discarded at MSW landfills, it would represent much less than 1 percent of all landfilled waste; however, it does contain toxic substances, and about one-fourth of all MSW landfills accept such wastes (ch. 7).

In addition, some nonhazardous industrial wastes are discarded in MSW landfills (ch. 7). Although most nonhazardous wastes currently are managed ''on-site,'' pressure to send them to off-site landfills may increase in the future if regulations guiding on-site management become more stringent.

It also is important to note that some of the materials in MSW are not always handled by MSW management methods. For example, liquid cleansers may be washed down the drain and into the municipal sewage treatment system (40). Pesticides (e.g., from spraying lawns) can be carried by rain into storm drains, which generally discharge into surface waters. Pesticides also can be dumped on the ground or into sewers, or stored at home.

The Writing Process

The highly complex mental functions involved in writing are not completely understood. One thing, however, is certain: just as no two persons think alike, no two persons use the same process in writing. And the same person might well use different techniques on different occasions. A short memo calls for one technique; a formal report calls for another. And a large manual, written collaboratively with three other people, calls for a third. In one sense, therefore, the term "writing process" is misleading.

This chapter discusses the processes that work best for most people. First it discusses the basic three-stage approach to writing: prewriting, drafting, and revising. Then it discusses the common techniques used to test the effectiveness of documents. The chapter concludes with two sections on special aspects of the writing process: using a word processor and writing collaboratively.

Most of the information in this chapter is based on recent research into how professional writers work. The principal finding of this research is that, despite the considerable variations, most professional writers do their best when they treat writing as a three-part process:

- ☐ prewriting
- ☐ drafting
- ☐ revising

It would be a mistake, however, to think that the writing process proceeds smoothly from prewriting to drafting to revising. Many students, as well as professionals, think that there must be something wrong with them because the writing process never seems easy. For instance, at a certain point they think they have completed their research, and so they begin to draft. But right in the middle of the draft they realize they need some more information, and they end up going back to the library or the laboratory. Does this mean they did something wrong, that they didn't plan carefully enough? Probably not. For even the most experienced writers, it's two steps forward, one step back. And some days, it's one step forward, two steps back. Don't be discouraged if you find yourself revising the basic structure of the document when you thought you were polishing the final draft. The only writer who ought to be worried is the one who dashes off a quick draft and can't find anything to fix.

Many technical persons dislike writing and try to get it over with as quickly as possible. They treat writing as a two-part process: writing and typing. They start with a clean sheet of paper, write "I. Introduction" at the top, and hope the rest will follow. It doesn't. No wonder they dislike writing. A few geniuses can write like that, but the rest of us find that it doesn't work.

We end up staring at the blank sheet for a half-hour or so, unable to think of what to say. After another half-hour we have created a miserable little paragraph. We then spend the rest of the morning trying to turn that paragraph into coherent English. We change a little here and there, clean up the grammar and punctuation, and consult the thesaurus. Finally, it's time to go to lunch. We return, happy that we have at least created a good paragraph. Then we reread what we've written. We realize that all we have done is wasted a morning. We start to panic, which makes it all the more difficult to continue writing.

Some writers use a somewhat more effective and efficient process. They realize they need some sort of outline—a plan to follow. But they do only a half-hearted job on the outline, scribbling a few words on a pad. Outlining has bad associations for most people. Most of us at one time or another have had to hand in outlines to English teachers or history teachers who seemed more interested in correcting the format—the Roman numerals and the indentation—than in suggesting ways to help write the papers. And sometimes the outline assignment was a punishment for having written a poorly organized paper. "Take this home and outline it. Then you'll see what a mess it is!"

The writing process described in the following section makes writing easier because it breaks it down into smaller, more manageable tasks. Writing becomes less mysterious and more like the other kinds of technical tasks that we carry out all the time.

Prewriting

Prewriting is what you do before you actually start to write sentences and paragraphs. For experienced writers, prewriting is a crucial step. In fact, when they have completed prewriting, most writers consider themselves halfway home.

Prewriting consists of four steps:

1. analyzing your audience, purpose, and strategy
2. finding information
3. generating ideas
4. outlining

The first step—analyzing your audience, purpose, and strategy—was discussed in Chapter 3. The second step—finding information—was discussed in Chapter 4.

Analyzing Your Audience, Purpose, and Strategy

The first step in any writing assignment is to analyze your audience and purpose. Briefly, it involves profiling your readers—their background, experience, responsibilities, personal factors, and reasons for reading— and identifying your purpose in writing—to provide information, change your readers' attitudes, or motivate action. Once you are comfortable with your analysis of the audience and purpose, you can plan a strategy, which will consider what kind of document to write, as well as such matters as length, scope, structure, organization, and style.

Chapter 3 also discussed the importance of making sure your primary reader agrees with your strategy. It is a good idea to write to this person, briefly explaining your strategy, and give him or her an opportunity to think about what you are planning before replying to you.

Although it is important to keep audience and purpose in mind throughout the writing process, you must be absolutely sure at this stage that you have a clear understanding of the writing situation. You are committing yourself to a plan of attack, and if it doesn't work out, you will have to rewrite—or throw out—long sections of your draft.

Finding Information

The techniques of finding information were the subject of Chapter 4. The first task, once you have a topic to write about, is to determine how much you already know it. Then you supplement this information with primary and secondary research. Primary research—creating new information—consists of such techniques as conducting inspections and experiments, administering questionnaires, and writing inquiry letters. Secondary research consists of finding existing information by interviewing experts and by using the library.

It is worth mentioning again that the writing process is not a lock-step march. As you prepare for the next task—generating ideas—keep in mind that you will probably come back to create or find more information, and that you might also want to refine your analysis of the audience, purpose, and strategy.

Generating Ideas

How do you know when you have done enough research? There is no simple answer. Sometimes you will have a very clear idea of what information you need, and you are sure you have more than enough of the right kind. On some rare occasions you will have done all the research that there is to do; there are only a few articles available on the subject, and you have studied them all. More often, you will know you have done all the research you can because you are running out of time and you realize you have to start writing.

For one reason or another, it is time to generate ideas. By "ideas" I don't mean formal, page-length explanations that are ready to be placed into the document, although if you have these in mind at this point, all the better. By "ideas" I mean subtopics of the overall topic you are writing about. For instance, if you are going to write about a new technology that mixes water with oil to increase the efficiency and reduce pollution in power plants, your ideas probably will include the theory of the new technology, because almost certainly your readers will need to know it. However, you might also think of some other ideas — such as the history of this technology, its cost, its advantages and disadvantages — that might not make their way into the document. Many of these ideas might be little more than phrases or questions at this stage; logic tells you that you ought to know about them, and you remember reading about them, but you are not prepared to write about them yet. Generating ideas is a way to start mapping out what will be included in the document, where it will go, and what additional information you will need to get.

How do people generate ideas? However they can. Many people don't even know what process they use. And for some people, the best ideas come when they aren't consciously thinking about their topic. A billboard, a television commercial, a conversation with a neighbor — anything can trigger an idea. The brain works on a problem when we are driving or taking a shower or sleeping. Some creative people actually keep a pad and pen on their nightstand because they frequently wake up in the middle of the night with a great idea.

Even though we can't completely understand or control the process of generating ideas, we can help it along. Four different techniques — along with many variations of them — have proven useful for a lot of people:

- [ ] talking
- [ ] free writing
- [ ] brainstorming
- [ ] sketching

Talking

Chapter 4 recommended talking with someone as a way to determine what you already know about your topic. Talking with someone also works well at this stage as a way to generate and start to organize ideas.

You have done your primary and secondary research, but the information is spinning around in your head, and you realize you are starting to forget some of it. A real danger at this point is discouragement; there is so much to know about your topic, and the information doesn't fall into neat categories. How will you make sense of it — and how will you present it to your readers so that it makes sense to them?

Choose someone to work with. Another student is a good choice, because you will be able to return the favor. Ask the person to throw questions at you as you speak. If you start with your main idea — "Our company is considering instituting some sort of health-promotion program for its employees" — the person could ask "Why?" You respond, "Because our health insurance premiums are too high and because too many of our employees suffer preventable health problems." The person then asks, "How high are the premiums? What is the rate of increase?" or "What percentage of your employees get sick? What is an acceptable rate? Is the rate changing?"

A careful listener can help you tremendously just by asking the journalistic questions: Who? What? When? Where? Why? and How? However, there are other obvious questions that get at relationships and causes: What caused this? What will the effect of this be? What are your options? A careful listener doesn't have to be a detective, just someone trying to understand what you are saying and what it means.

If the person asks these questions, you will very quickly get a sense of how clearly you understand your topic, what additional information you will need to get, and what questions your readers will want answered in your document. You will also find yourself making new connections between ideas; the mere act of forcing yourself to put your ideas into sentences helps you synthesize the information.

Free Writing

You might have used free writing in one of your previous writing classes. Free writing consists of writing without any plans or restrictions. If your topic is the technology for mixing oil and water in power plants, you write whatever comes into your mind about it. You don't consult an outline, you don't stop to think about sentence construction, you don't consult a reference book. You just let the pen or the cursor move.

In most cases, the free-writing text never becomes a part of the final document. Your explanations might be only half-correct; they almost certainly will be unclearly expressed. And half of what you write might be merely phrases or questions that you think you might need to answer. But something valuable might be happening anyway: the mere act of trying to make sentences helps you understand what you do and do not understand. And one sentence might spark an important idea.

Figure 5.1 is an example of a brief free-writing text created in about five minutes by a student in a technical-writing course. She worked for a company considering a health-promotion program and had chosen that as her writing topic. This free-writing text is a mess: a series of incomplete thoughts, and more questions than answers. Yet the writer did start to work out some of the problems her company would face (keeping the cost down, motivating employees, possible liability for injuries during the program) and some possible solutions to research (such as asking the insurance company for assistance and finding out if there is an organization that oversees the different kinds of programs). By starting to think through the problems her company would face, the writer is of course starting to think about her own problems: what kind of information her readers will need. Not bad for five minutes' work.

Brainstorming

Brainstorming is another popular way to generate ideas for your document. Brainstorming involves spending 10 or 15 minutes listing ideas about your subject. List the ideas as quickly as you can, using short

> Insurance rates have gone up over 10% a year for last five years. Other problem: the loss of employees to preventable health problems. Check out health-promotion programs. How much do they cost? Are they administered by us or by a subcontractor? How about reduced-rate memberships at health clubs? Will we have further liability if someone gets hurt in the programs? How much would we save—in premiums and reduced health problems? Is there an organization that evaluates these programs? First we have to find out how many employees would be interested. Maybe the insurance company could help us with some of this information.

FIGURE 5.1   Free-Writing Text

phrases, not sentences. For instance, if you are describing global warming, the first five items in your brainstorming list might look like this:

    definition of g.w.
    role of CFCs
    political issues
    effect on smokestack industries
    international agreements to limit g.w.

As the word *brainstorming* suggests, you are not trying to impose any order on your thinking. You will probably skip around from one idea to another. Some of the ideas will be subsets of larger ideas. Don't worry about it; you will straighten that out later.

Like free writing, brainstorming has one purpose: to free your mind so that you can think of as many ideas as possible that relate to your subject. When you get to the next stage—constructing an outline—you will probably find that some of the items you listed do not in fact belong in the document. Just toss them out. The advantage of brainstorming is that for many people it is the most effective and efficient way to catalogue what *might* be important to the document. A more structured way of generating ideas, ironically, would miss more of them.

Brainstorming is actually a way of classifying everything you know into two categories: (1) information that probably belongs in the document and (2) information that does not belong. Once you have decided what material probably belongs, you can start to write the outline.

costs of replacing sick employees

necessary equipment

payback period?

history of health-promotion plans

how to measure interest?

increased productivity of workers?

how to measure success?

here or at a health club?

necessary publicity

converting space

start-up costs

reduce premiums

make it mandatory?

do interest surveys?

maintenance costs

national trends?

how to maintain interest

who would staff it?

FIGURE 5.2   Brainstorming List

Figure 5.2 is a brainstorming list created by the writer who is reporting to management about health-promotion programs. She has already gathered and skimmed all the information she thinks she will need from the professional journals.

The list contains a lot of good ideas that will certainly be discussed in the document, as well as some ideas (such as the history of health-promotion programs) that will probably be omitted.

Sketching

The three techniques for generating ideas that have been discussed so far—talking with someone, free writing, and brainstorming—are similar in that they are nonhierarchical. That is, they do not differentiate between the relative importance of ideas; minor ideas are mixed in with major ideas. Some people prefer a somewhat more structured way to generate ideas.

Sketching ideas—making a visual representation rather than a verbal one—can help you start to visualize the structure that the document will

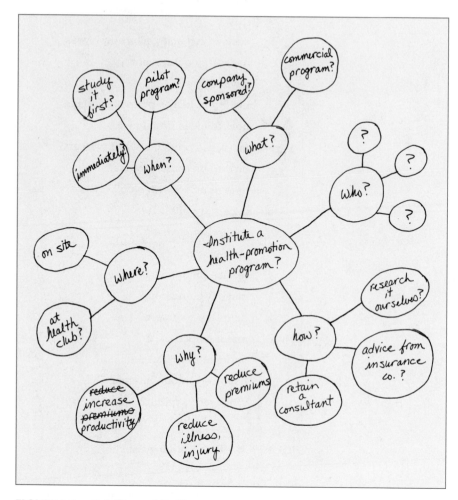

FIGURE 5.3   Satellite and Its Planets

take. Sketching has an additional benefit: it can help you see what information is weak or missing altogether.

Two kinds of sketches are used often in the professional world: the planet and its satellites, and the idea tree.

To create the planet and its satellites, start with a main idea or a main question. It is the planet; draw a circle around it in the middle of a sheet of paper. Less important or smaller ideas become satellites revolving around the planet. And the satellites can have their own satellites.

Figure 5.3 shows what a first version of the planet and its satellites might look like for the report on health-promotion programs. The writer has started with the journalistic questions. Notice that the writer doesn't yet know how to answer the question "Who?" She has to go back to her sources to find out who actually administers these kinds of programs.

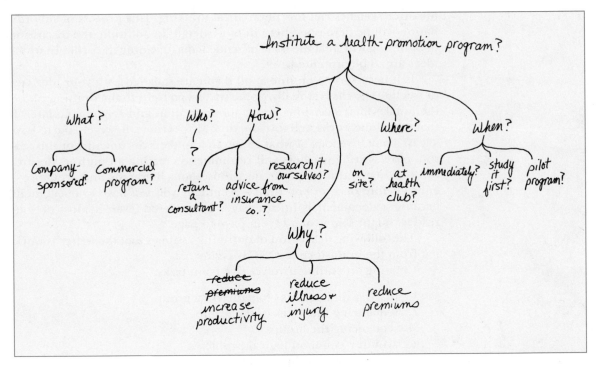

FIGURE 5.4   Idea Tree

The other popular format for sketching ideas is the idea tree. There is only one major difference between the planet format and the idea tree: the movement from larger to smaller ideas is from the center to the perimeter on the planet format and from top to bottom on the idea tree. Figure 5.4 shows an idea tree.

As you think more about your topic, you will extend and revise your sketch. You might want to rethink the sketch if, for example, one of the branches on your idea tree is substantially shorter or longer than the others. A short branch might signal that you need more research. However, it might just be a similar idea that is fine as it is. A long branch could signal that you went into much more detail on that one idea than you did on the other ideas. On the other hand, it could merely be a more complicated idea. Counting branches and satellites provides no sure answers.

Beginning to Outline

This section and the next one—developing an outline—explain how to create a structure for presenting your ideas. The previous section, on sketching, described part of what will be treated here. The satellite and the idea-tree sketches call for hierarchical thinking. This process of subordination will be developed here in more detail. In addition, the discussion of outlining will treat the issue of sequencing: deciding the order in which ideas are to be presented.

Is it necessary to outline at all if you are satisfied with your idea tree or satellite sketch? Not really. Some writers go right to the drafting phase, then make their decisions about subordination and sequencing later. If that technique works well for you, fine. Most writers, however, like to have a least a tentative idea of what comes first in their document. For this reason, most writing teachers will recommend creating an outline. Regardless of when you make these decisions about hierarchy and sequence — either before or after you draft — eventually you *will* have to make them, because a document is linear; pages are ordered, even if most of your readers might skip around from page to page.

The following discussion of outlining assumes that the writer is working from the brainstorming list in Figure 5.2.

Creating an outline involves five main tasks:

1. placing similar items together in a group
2. sequencing the items in the group
3. sequencing the groups
4. avoiding common logical problems
5. choosing an outline format

Placing Similar Items Together in a Group

Start by looking at the first item on your brainstorming list. For the writer working on the feasibility study on health-promotion programs, that item is "costs of replacing sick employees." She decides that this item should be part of the background for her report. She scans the brainstorming list, looking for another background item. Finally she comes upon "reduce premiums," "history of health-promotion plans" and "national trends."

She rewrites the items as follows:

Background

costs of replacing sick employees

increased productivity of workers?

reduce premiums

history of health-promotion programs

national trends

Looking at this list, she realizes that she has omitted an important item: a definition of health-promotion programs. She adds it to the list.

She then goes to the second item on the brainstorming list: "necessary equipment." She scans the list, looking for other items related to costs. She writes this list:

Costs

    necessary equipment

    necessary publicity

    converting space

    start-up costs

    maintenance costs

    who would staff it?

The writer realizes that this list contains two items that embrace the others: start-up costs and maintenance costs. She revises the list accordingly:

Costs

    start-up costs

        necessary equipment

        necessary publicity

        converting space

    maintenance costs

        who would staff it?

Looking at the maintenance costs, she realized that she has forgotten insurance premiums for the new program, so she adds it to the list. She also adds another maintenance expense: maintaining interest in the program.

    The writer follows the same procedure as she goes through her original brainstorming list: grouping related items and adding new items that she had omitted.

    Finally, she has created a series of small lists:

Background

    costs of replacing sick employees

    increased productivity of workers?

    reduce premiums

    history of health-promotion programs

    national trends

    definition of health-promotion programs

Costs

    start-up costs

        necessary equipment

necessary publicity

converting space

maintenance costs

who would staff it?

how to maintain interest

insurance premiums

Payback period?

Benefits

How to measure interest?

do interest surveys?

How to measure success?

The writer realizes that these lists are very rough. Some of the groups, such as the last one, have no subgroups. And the writer realizes that one of the items from her original brainstorming list—"make it mandatory?"—will be covered in the definition of health-promotion programs.

At this point, some writers will want to start drafting the document, even though the outline is incomplete. They like to see what they have to say, and the only sure way to do that is to start to write. Once they have started drafting, they feel better able to sequence items. They juggle paragraphs instead of juggling outline items.

Other writers feel more comfortable working from a more refined outline. For these writers, the next step is to sequence the items within each group.

Sequencing the Items in a Group

The writer looks at the background section of her outline:

Background

costs of replacing sick employees

increased productivity of workers?

reduce premiums

history of health-promotion programs

national trends

definition of health-promotion programs

She realizes that any report should begin with a definition of important terms. Next, she decides, should be a brief history of the programs. Three other items—"costs of replacing sick employees," "increased productivity," and "reduce premiums"—she decides could be worked into the history section, because they were the principal reasons to start the programs in the first place. However, she also realizes that they could be a part of the definition of health-promotion plans. She makes a note on the outline to check later to see where the information works best. The remaining item—"national trends"—follows naturally from the history section. The tentative sequence, then, looks like this:

Background

    definition of health-promotion programs

    history of health-promotion programs

        costs of replacing sick employees

        increased productivity of workers?

        reduce premiums

    national trends

Next, the writer turns to the extensive section on costs and benefits in the tentative outline:

Costs

    start-up costs

        necessary equipment

        necessary publicity

        converting space

    maintenance costs

        who would staff it?

        how to maintain interest

        insurance premiums

Payback period?

Benefits

She decides to work from major items to minor items, although some writers prefer to work in the opposite direction. Because "costs and benefits" is a common term in her field, she chooses the following sequence:

    costs

    benefits

    payback period

Now she starts to sequence subunits. Within the costs section, "start-up costs" should precede "maintenance costs" because the chronological sequence seems most appropriate here.

Costs

    start-up costs

    maintenance costs

Within the "start-up costs" section, the writer faces a decision. She can use a chronological pattern, which might call for publicity followed by space and equipment. Or she can use a pattern that calls for the more-important items followed by the less-important items. Because she is describing costs, she determines that her readers will want to read about the major costs first. Therefore, she decides on the following sequence:

Start-up costs

    converting space

    equipment

    publicity

Turning to the maintenance costs, the writer decides to continue with the "more important to less important" pattern. She refines her headings as she rewrites them. In addition, the writer realizes she should add "equipment maintenance" to the list:

Maintenance costs

    insurance premiums

    staffing

    equipment maintenance

    continued publicity

The writer decides to leave the "payback period" item where it is, at the end of the costs section, because it logically follows the discussion of start-up and maintenance costs. The complete costs section now looks like this:

Costs

    start-up costs

        converting space

        equipment

        publicity

    maintenance costs

        insurance premiums

        staffing

        equipment maintenance

        continued publicity

    payback period

The writer now turns to the two remaining groups in her outline: "How to measure interest?" and "How to measure success?" She cannot think of any ways to measure interest except through surveys. Realizing that a unit cannot be partitioned into a single unit, she decides to revise the item as follows:

> Using surveys to measure interest

Turning to the problem that her "How to measure success" group contains no subgroups, she turns back to the journal literature and identifies several accepted techniques used in industry:

> How to measure success
>
>> participation rates
>>
>> absenteeism rates
>>
>> long-term disability rates
>>
>> changes in productivity

The writer has started with the most objective factors and moved toward the more subjective ones because she feels her readers will be most interested in knowing first how to gather the most objective evidence.

Sequencing the Groups

Sequencing the groups involves the same logical processes as sequencing the items within each group. The first step is to determine the needs of the audience. In the outline being discussed here, the readers are expecting a traditional report structure: an executive summary and introduction followed by the background, discussion, and recommendations. The writer decides to sequence the groups she has outlined, then add the other elements. "Background," of course, comes first. "Measuring interest" is a logical second group, because unless the employees are interested, there is no point proceeding with the analysis. "Costs" is logical next, followed by "benefits" and "measuring success." With the traditional report elements added, the writer's sequence for the major items looks like this:

> executive summary
>
> introduction
>
> background
>
> discussion
>
>> using surveys to measure interest
>>
>> costs
>>
>> benefits
>>
>> measuring success
>
> recommendations

As this discussion shows, outlines are created and constantly refined as the writer tries to organize the material. Some writers prefer to create only a rough outline and then start drafting, stopping occasionally to revise and expand the outline. However, even those writers who like to create a very specific outline often end up changing it as they draft the document. Whatever technique helps you meet your audience's needs and achieve your purpose is the best technique for you.

Avoiding Common Logical Problems in Outlining

As you create and refine your outline, avoid two common logical problems:

- ☐ faulty coordination
- ☐ faulty subordination

Faulty coordination is the process of equating items that are not of equal value or not of the same level of generality. Look, for example, at the following listing:

common household tools

    screwdriver

    drill

    claw hammer

    ball-peen hammer

Although there are claw hammers and ball-peen hammers, this listing is not logical, because the other items are not similarly differentiated. For instance, there are electric drills and manual drills, as well as different kinds of screwdrivers. The listing should read:

common household tools

    screwdriver

    drill

    hammer

A related problem of coordination involves overlapping items, that is, creating more than one heading under which an item might be placed:

accounting services

    Automated Data Services

    Data Ready

    The Accountants

The Tax Service, Inc.

Because "The Tax Service, Inc." is an accounting service, it should be subordinated to "accounting services." However, if you wish to devote more attention to The Tax Service, Inc. than to the other companies, you could revise the list as follows:

The Tax Service, Inc.

three other accounting services

    Automated Data Services

    Data Ready

    The Accountants

Adding the phrase *three other* before "accounting services" prevents the overlapping.

Faulty subordination occurs when an item is made a subunit of a unit to which it does not belong:

power sources for lawnmowers

    manual

    gasoline

    electric

    riding mowers

In this excerpt, "riding mowers" is out of place because it is not a power source of a lawnmower. Whether it belongs in the outline at all is another question, but it certainly doesn't belong here.

A second kind of faulty subordination occurs when only one subunit is listed under a unit:

types of sound-reproduction systems

    records

        phonograph records

    tapes

        cassette

        open reel

        DAT

    compact discs

It is illogical to list "phonograph records" if there are no other kinds of records. To solve the problem of faulty subordination, incorporate the single subunit into the unit:

types of sound-reproduction systems

   phonograph records

   tapes

      cassette

      open reel

      DAT

   compact discs

Choosing an Outline Format

If you are creating an outline purely for your own use, you don't have to worry about its format, as long as you can understand it and work effectively from it. However, you might want to use one of the standard outline formats if your outline will be read by somebody else, or if your document will contain a table of contents, which is basically an outline of the document. (See Chapter 19 for a discussion of tables of contents.)

There are two common kinds of outlines:

- topic outlines
- sentence outlines

In a topic outline, the headings are phrases, such as "increased productivity." In a sentence outline, the headings are sentences, such as "A major advantage of health-promotion programs is that they increase worker productivity."

Relatively few writers use sentence outlines, because they know they will have to revise the sentences later. The main advantage of the sentence outline is that it keeps you honest: if you don't know what to say about an item, you will realize it clearly when you try to come up with a sentence.

If you will be submitting your outline to somebody else, you will have to make one more decision: whether to use a traditional alpha-numeric system or the decimal system.

The traditional system looks like this:

```
I.
    A.
        1.
        2.
    B.
        1.
            a.
            b.
                (1)
                (2)
                    (a)
                    (b)
        2.
    C. etc.
```

The decimal system, which is used in the military and often in the sciences, looks like this:

1.0
    1.1
        1.1.1
        1.1.2
    1.2

An advantage of the decimal system is that it is simple to use and to understand. You don't have to remember how to represent the different heading levels. In the same way, your readers can easily determine what level they are reading: a three-digit number is a third-level heading, and so forth.

Use whichever system helps you stay organized — or is required by your organization.

Developing an Outline

As you start to turn your rough ideas into an outline, you might not be thinking of patterns of development. Some topics just naturally lend themselves to a particular kind of development. For instance, when you are describing an event, such as the Chernobyl accident, you intuitively think in terms of chronology. When you want to explain why the accident happened, you intuitively choose a cause-effect pattern.

In many cases, however, no single pattern of development springs to mind. If you wish to describe a potential site for constructing a building, you could use a number of different patterns, such as spatial survey of the property from one end to the other, or a more-important-to-less-important pattern to direct your readers' attention first to the most essential features of the property. A single document is likely to contain a number of different patterns of development — chronological at one point, spatial at another.

Because you might have to make many decisions about patterns of development, it is useful to realize that some standard patterns usually work well in particular situations. Understanding these different patterns — and how to combine them to meet your specific needs — gives you more options as you put together your document.

The following eight patterns of development can be used effectively in technical writing:

- ☐ chronological (time)
- ☐ spatial
- ☐ classification/partition
- ☐ comparison/contrast

☐ general to specific
☐ more important to less important
☐ problems-methods-solutions
☐ cause-effect

■ *Chronological Pattern.* The chronological pattern works well when you want to describe a process, give instructions on how to perform a task, or explain how something happens.

The following outline is structured chronologically. The writer is outlining a brochure describing to a general audience the geological history of St. John, United States Virgin Islands.

 I. Introduction

 II. Stage One: Submarine Volcanism

  A. Logical Evidence: The Pressure Theory

  B. Empirical Evidence: Ram Head Drillings

III. Stage Two: Above-Water Volcanism

  A. Logical Evidence

   1. The Role of the Lowered Water Level

   2. The Role of Continental Drift

  B. Empirical Evidence: Modern Geological Data

IV. Stage Three: Depositing of Organic Marine Sediments

  A. The Role of the Ice Age

   1. Water Level

   2. Water Temperature

  B. The Role of "Drowned" Reefs

 V. Conclusion

Notice that on the A and B levels of this outline the reader continues the chronological pattern. "Logical evidence," the hypothetical explanations offered by earlier scientists, precedes the "empirical evidence," the measurable data gathered by modern scientists. In item IV, the Ice Age preceded and caused the drowned reefs, so it is discussed first.

■ *Spatial Pattern.* In the spatial pattern, items are organized according to their physical relationships to one another. The spatial pattern is useful in developing a description of a physical object.

A spatially organized outline follows. The writer is describing one type of computer keyboard.

1.0 The Keyboard

    1.1 The Main Section

        1.1.1 The Top Row

            1.1.1.1 The Function Keys

            1.1.1.2 The Cursor-Control Keys

        1.1.2 The Next Row

            1.1.2.1 The Escape Key

            1.1.2.2 The Number Keys

        1.1.3 The Letter Rows

            1.1.3.1 The Tab Key

            1.1.3.2 The Caps Lock Key

            1.1.3.3 The Letter Keys

            1.1.3.4 The Shift Key

        1.1.4 The Space Bar

    1.2 The Numeric Keyboard

        1.2.1 The Top Row: The ''Soft'' Function Keys

        1.2.2 The Middle Rows: The Number Keys

        1.2.3 The Bottom Row: The Arithmetic Keys

For this description, which in its finished form will include a labeled diagram of the keyboard, the writer has chosen to follow the physical shape of the item itself: the main section on the left, and the numeric keypad on the right. Within each of these sections, the writer works from the top to the bottom. Notice how he is careful to include the directional cues—such as "top row"—where appropriate in his headings.

■ *Classification/Partition Pattern.* Classification and partition are two related but different techniques. They are discussed in detail in Chapter 10.

    Classification is the basic process of outlining: placing items in categories based on a similar characteristic. Tennis rackets can be classified according to their price, their material (wood, aluminum, composite materials), the size of their heads, or the size of their grips.

    Partition is the process of dividing a single entity into its main parts. A component stereo system can be partitioned into its amplifier, tuner, speakers, cassette deck, and so on.

    Classification and partition are useful techniques for developing outlines about physical objects as well as more conceptual ideas. An outline about a bicycle or a nuclear submarine could be developed by partition; an outline about techniques for motivating employees could be developed by classification.

The following outline was written by a heating contractor who is writing to a rural fire department. He is proposing to install a heating system. He has partitioned the firehouse into major areas.

I. System for Offices, Dining Area, and Kitchen

    A. Description of Floor-Mounted Fan Heater

    B. Principal Advantages

        1. Individual Thermostats

        2. Quick Recovery Time

II. System for Bathrooms

    A. Description of Baseboard Heaters

    B. Principal Advantages

        1. Individual Thermostats

        2. Small Size

III. System for Garage

    A. Description of Ceiling-Mounted Fan Heater

    B. Principal Advantages

        1. High Power

        2. High Efficiency

This pattern looks like the one used to describe the keyboard, but it is different in an important way. The keyboard outline used space as the organizing principle: from left to right, from top to bottom. The firehouse outline uses another principle of organization—type of heating needs. In Item I, the offices, dining area, and kitchen are grouped together not because they are located in the same area but because they require the same kind of heat.

■ *Comparison/Contrast Pattern.* The comparison/contrast pattern is used frequently in one of the most common kinds of technical document: the feasibility report (see Chapter 22).

In writing a feasibility study, you are trying to determine whether an idea will work and whether it will be economical. Or you are trying to determine which one of several options is best for your organization. In either case, you will be comparing and contrasting items. For instance, if you want to see whether your company can afford to undertake an expensive marketing campaign, you are comparing and contrasting two options: staying with what you are doing now, and doing the marketing campaign. If you are trying to determine which of three brands of pasteurization machines to buy, you are comparing and contrasting the three.

The following excerpt from an outline was written by a civil engineer who was asked to determine which computer would best help the company analyze structural frames. He has already outlined the criteria the company needs to fulfill: reasonable cost, user friendliness, reliability, graphic capabilities, and so forth. In the excerpt shown here, he has created a very simple pattern.

1.0 Advantages and Disadvantages of the Four Computers

    1.1 Macintosh Plus

        1.1.1 Basic Profile

        1.1.2 Advantages

        1.1.3 Disadvantages

    1.2 Macintosh SE 020

        1.2.1 Basic Profile

        1.2.2 Advantages

        1.2.3 Disadvantages

    1.3 Macintosh SE 030

        1.3.1 Basic Profile

        1.3.2 Advantages

        1.3.3 Disadvantages

    1.4 Macintosh IIcx

        1.4.1 Basic Profile

        1.4.2 Advantages

        1.4.3 Disadvantages

In this excerpt, the writer is enabling his readers to compare and contrast the computers in two ways: (1) the advantages and disadvantages of each computer (that is, the relationship between 1.1.2 and 1.1.3) and (2) the advantages and disadvantages of all four of the computers (such as the relationship between 1.1 and 1.2).

Students are sometimes skeptical about an outline such as this one because it seems boring: the same pattern is repeated. But remember that your readers aren't looking to be entertained. They want information expressed clearly. A simple, straightforward pattern helps them find and understand that information.

For a more detailed discussion of comparison and contrast, see Chapter 10.

■ *General-to-Specific Pattern.* The movement from general to more specific information is common in sales literature and all kinds of reports intended for a multiple audience. The general information enables the manager or customer to understand the basics of the discussion without having to read all the details. The technical reader, too, appreciates the general information, which serves as an introductory overview.

Following is an outline for a sales brochure promoting a plastic coating used on institutional and industrial floors. The readers will be maintenance managers, who might not have heard of the product because it is fairly new. The purpose of the brochure is to persuade the readers to request a visit from the company's sales representative.

I. The Story of EDS Plastic Coating

    A. What Is EDS Plastic Coating?

    B. Why Is EDS the Industry Leader?

        1. Strength and Durability

        2. Long-Lasting Results

        3. Ease of Application

        4. Low Cost

    C. How to Find Out More about EDS Plastic Coating

II. How EDS Works

    A. How EDS Bonds on Different Floorings

        1. Wood

        2. Tile

            a. Flat

            b. Embossed

        3. Linoleum

            a. Flat

            b. Embossed

    B. How EDS Dries on the Floor

III. How to Apply EDS Plastic Coating

    A. Preparing the Surface

    B. Preparing the EDS Mixture

    C. Spreading EDS Plastic Coating

        1. Wetting the Applicators

            a. Spreading Pads

            b. Paint Brushes

    2. Determining Where to Begin

       a. Tiled Flooring

       b. Continuous Flooring

    3. Coordinating Several Workers

    4. Stopping for the Day

  D. Cleaning Up

    1. Spreading Pads

    2. Paint Brushes

Parts I and II of this outline are addressed to the person who would decide whether to seek more information about this product. The general information in Part I answers the three basic questions such a reader might have:

1. What is the product?
2. What are its important advantages?
3. How can I find out more about it?

Part II explains how the product works. Part III of the outline, which contains more specific information about how to use the product, is intended for the reader who actually has to supervise the use of the product. The pattern within this part is chronological, which is the obvious choice for a description of a process.

■ *More-Important-to-Less-Important Pattern.* Another basic pattern is the movement from more-important to less-important information. This pattern is effective even in describing events or processes that would seem to call for a chronological pattern. For example, suppose you were ready to write a report to a client after having performed an eight-step maintenance procedure on a piece of electronic equipment. A chronological pattern—focusing on what you did—would answer the following question: "What did I do, and what did I find?" A more-important-to-less-important pattern—focusing first on the problem areas and then on the no-problem areas—would answer the following question: "What were the most important findings of the procedure?" Most readers would probably be more interested, first, in knowing *what* you found than in *how* you found it.

The outline that follows is an example of the more-important-to-less-important pattern. The writer manages the engine-repair division of a company that provides various services to owners of small aircraft. The purpose of the report is to explain clearly what work was done and why.

I. Introduction

  A. Customer's Statement of Engine Irregularities

  B. Explanation of Test Procedures

II. Problem Areas Revealed by Tests

    A. Turbine Bearing Support

        1. Oversized Seal Ring Housing Defective

        2. Compressor Shaft Rusted

    B. Intermediate Case Assembly

        1. LP Compressor Packed with Carbon

        2. Oversized LP Compressor Rear Seal Ring Defective

III. Components That Tested Satisfactorily

    A. LP Turbine

    B. HP Turbine

    C. HP Compressor

In most technical writing, bad news is more important than good news, for the simple reason that breakdowns or problems have to be fixed. Therefore, the writer of this outline has described first the components that are not working properly.

■ *Problem-Methods-Solution Pattern.* This is a basic pattern for outlining a complete project: begin with the problem, discuss the methods you followed, and then finish with the results, conclusions, and recommendations.

The following outline was written by an electrical engineer working for a company that makes portable space heaters. He has been asked by his supervisor to determine the best way to improve the efficiency of the temperature controls used in their space heaters. After explaining the reasons for wanting to improve the efficiency, he describes the technology the company currently uses. Next he describes his research on the different methods of controlling temperature, concluding with the method he recommends.

I. The Need for Greater Efficiency in Our Heaters

    A. Financial Aspects

    B. Ethical Aspects

II. The Current Method of Temperature Control: The Thermostat

    A. Principle of Operation

    B. Advantages

    C. Disadvantages

III. Alternative Methods of Temperature Control

    A. Rheostat

        1. Principle of Operation

            2. Advantages

            3. Disadvantages

        B. Zero-Voltage Control

            1. Principle of Operation

            2. Advantages

            3. Disadvantages

    IV. Recommendation: Zero-Voltage Control

        A. Projected Developments in Semiconductor Technology

        B. Availability of Components

        C. Preliminary Design of Zero-Voltage Control System

        D. Schedule for Test Analysis

    V. References

Notice that the writer begins with a discussion of the problem (Part I), followed by a discussion of the existing type of control used on the company's products.

In Part III of this outline the writer has placed the discussion of the recommended alternative — zero-voltage control — after the discussions of the two other alternatives. The recommended alternative does not *have* to come last, but most readers will expect to see it in that position. This sequence *appears* to be the most logical — as if the writer had discarded the unsatisfactory alternatives until he finally hit upon the best one. In fact, we cannot know in what sequence he studied the different alternatives, or even whether he studied only one at a time. But the sequence of the outline gives the discussion a sense of forward momentum.

This pattern of development is particularly popular when the writer's company has been hired by the reader's company. In consulting arrangements of this sort, the reader expects to see a pattern that emphasizes the amount of work the writer performed.

■ *Cause-Effect.* Often in technical writing you will want to reason from an effect to a cause. Samples of the soup your company produces are showing an unusually large amount of salt, and you want to determine why. Sometimes you will reason from cause to effect: what would happen if your department's budget were increased only 2 percent instead of the requested 5 percent.

The following excerpt from an informal outline of an introduction was written by a student who worked part-time for a small dry-cleaning business. The problem he was investigating is that, because business has increased, the company needs to replace its automatic shirt-pressing machine.

The Current Problem

    Unacceptable Delays in Turnaround Time

    Excessive Overtime Expenses

    Inability to Find Trained Workers on Weekends

    Reliability Problems with Current Shirt-Pressing Machine

The Causes of the Problem

    Increased Clientele Because of Industrial Park

    Our Reputation for High-Quality Work

    Long Cycle Time of Our Current Shirt-Pressing Machine

    Age of Our Current Shirt-Pressing Machine

The rest of the writer's introduction discusses the purpose of the investigation and of the report, as well as the scope, organization, and major findings of the report.

In this excerpt, the writer wants to fill in the background section of his introduction. He starts with the current problem: the turn-around time is too long, workers are expensive and sometimes hard to find, and the current machine is breaking down. The next section outlines the causes of the problem: increased business, and the slowness and age of the current machine.

Drafting

Drafting is the process of turning an outline into sentences and paragraphs. Revision, the process of turning that draft into a document you are willing to submit to your readers, is discussed in the next section.

Some writers can produce a draft, fix a comma here and a word there, and have a professional-quality document. Other writers need to spend hours revising each page. For most people, some documents seem to write themselves, whereas others refuse to make any sense at all after several drafts.

It is impossible to offer firm guidelines on how to draft effectively. However, three principles seem to make the job easier for most people in most situations:

- draft quickly
- make it easy to expand and reorganize later
- start with the easiest ideas

Drafting Quickly

Perhaps the most useful piece of advice about drafting is to do it fast. Many students make the mistake of trying to draft slowly so that their first draft will be good enough to serve as a final draft. It probably won't be. Even Ernest Hemingway publicly lamented the fact that his first (and second and third) draft just wasn't very good.

Getting ideas down on paper is so complicated that you shouldn't expect to be able to say what you mean—and say it well—the first time around. Therefore, don't try to do too much. Try to turn the phrase from your outline into a few paragraphs. Often enough you will find that you aren't exactly sure what you had in mind when you wrote the outline. Or you will realize that you are missing some technical information.

Set a timer and draft for an hour without stopping. If you hit an item on your outline that you don't understand or that requires more research, just go on to the next item. Don't stop to revise your writing. Don't worry about clumsy sentence structure or odd spelling. Your goal is to create a big, rough draft that you can turn into a professional document later.

When it is time to stop drafting, many experienced writers like to stop in the middle of a paragraph, or even in the middle of a sentence. This way, when they start to draft again they can conclude the idea they were working on. If they had to start a new major section, they would be more vulnerable to writer's block, the mental paralysis that sets in when a writer stares at a blank page.

Making It Easy to Expand and Reorganize Later

You shouldn't worry about revising as you draft, but you should prepare for it. If you are working on paper, write clearly on one side only so you can cut and paste later. Skip lines so that you can add material easily.

Some people like to write on small pieces of paper, no bigger than 5 × 8 inch note cards, restricting themselves to one paragraph per card. They don't bother to create graceful transitions from one point to the next; they treat each point as separate. This way, when they revise they can lay the cards out on a table and try out different organizational patterns; they don't feel they are wasting the effort they expended earlier.

Starting with the Easiest Ideas

There is no rule that says you have to start writing at the start of the document. Most writers like to begin with a section from the middle, usually a technical section they are comfortable with. Postponing the introductory elements makes sense for another reason: because the introductory elements are based on the body—and have to flow smoothly into the body—you have to know what the body will say before you can introduce it. Until you have written the body, you can't be sure what it is going to say.

Revising

Revising is the process of making sure the document says what you want it to say—and that it says it professionally. As is the case with all other phases of the writing process, everyone uses a different technique for revising. But the important point is that you have to have a technique. You cannot hope to simply read through your document, waiting for the problems to leap off the page. Some of them might; most won't.

The following discussion of revising consists of two parts:

☐ learning how to look
☐ learning what to look for

Learning How to Look

All the points made here are merely common sense. Yet they work.

■ *Let It Sit.* This piece of advice, perhaps the most important, is the least followed.

You cannot revise your document effectively right after you have finished writing it. You can identify immediately some of the smaller writing problems—such as errors in spelling or grammar—but you cannot objectively assess the quality of what you have said: whether the information is clear, comprehensive, and coherent.

To revise your draft effectively, you have to set it aside for a time. Give yourself at least a night's rest. If possible, work on something else for a few days. This will give you time to "forget" the draft and approach it more as your readers will. By delaying your review, in a matter of a few minutes you will see problems that would have escaped your attention even if you had spent hours revising it immediately after completing the draft.

You're probably thinking that you don't have time to set it aside for a night. You've got other courses and other obligations, and you're trying to have something resembling a life. There is no easy solution to the time problem, and, by the way, it seems to get more difficult for people writing on the job. You just have to try to arrange your schedule so that you are done drafting a few days before the document is due.

■ *Read It Aloud.* Reading a document out loud gives you a better sense of what it sounds like than reading it silently. Any teacher of writing has seen it a hundred times. A student submits a paper whose first sentence is missing a necessary word. The teacher asks the student to look at the sentence and fix the problem, and she replies that she can't see any problem. When asked to read the sentence out loud, the student immediately "hears" the mistake. Big problems, too, such as illogical reasoning or missing evidence, also seem more obvious when you hear your document.

Actually read your draft out loud to identify possible areas for improvement. You will feel foolish doing this. Welcome to the club.

- *Get Help from Someone Else.* An obvious technique for distancing yourself from the draft is to get some help from another person. This will provide a more objective assessment of the writing. If possible, choose a person who fits the profile of your eventual readers; someone more knowledgeable about the subject than your audience will understand the document even if it isn't clear and therefore might not be sufficiently critical. After reading the draft, your colleague may be able to tell you its strong points, unclear passages, sections that need to be added, deleted, or revised, and so forth.

- *Use Checklists.* Before pilots take off, they run through a series of checklists; they can't count on remembering all the different safety checks to carry out. The same applies to writing.

  This text is full of checklists that summarize the major points you should look for in revising different kinds of documents. However, there is a big drawback with any checklist that anyone has written for you: it doesn't necessarily get at the things that you need most. For example, you might confuse *its* and *it's*. Don't count on a generic checklist to list this. Instead, keep a list of the points your instructors make about your writing and then customize the checklists you use.

- *Use Software.* There is a lot of software on the market that can help you spot different kinds of style problems, and some of it is very good. Like a checklist, however, a program is a generalized tool that cannot understand your context—the purpose and audience of your communication— and therefore cannot be relied upon to offer wise advice all the time. The role of computers in writing is discussed in detail later in this chapter.

Learning What to Look For

The key to effective revising is that you have to go through the document a number of times, looking for different kinds of problems each time. For this reason, people are often shocked to learn that, for many professional writers, revising takes much more time than drafting. Revising can eat up more than half of the total time devoted to a document. In looking for different kinds of problems to fix, most writers prefer a top-down pattern; that is, they look first for the most important and the largest problems and proceed to the less important, smaller ones. This way, they don't waste time fixing awkward sentences contained in paragraphs that will be thrown out.

Here are the major items to check for in revising:

- *Comprehensiveness.* In your first pass through the document, concentrate on the largest issue: comprehensiveness. Does the document discuss all the topics the readers will need discussed? Having completed the draft you might have a different perspective on the document; should anything be added? Go through the draft making sure you have included all the items from the outline; perhaps in writing the draft you omitted some.

- *Accuracy.* Having all the necessary information will do you—and your readers—no good if the information is inaccurate. Now is the time to go back to your sources to check to see that your explanations are accurate and that you have your equations and statistics right. Make sure too that your draft is straightforward and unbiased; remove any distortions or misrepresentations.

- *Organization.* Visualize your readers trying to use your document. Is the organization as clear and useful as it can be? In particular, look at the headings to see that the sequence is clear and logical. If you now think that a different organizational pattern will work better, make the changes.

- *Emphasis.* Looking at your headings will also give you a sense of the emphasis given to the different topics. If a relatively minor topic seems to be treated at great length, with numerous subheadings, check the draft itself. The problem might be merely that you created more headings at that point than you did in treating some of the other topics. But if your treatment is in fact excessive, mark passages to consider condensing.

- *Style.* Once you are satisfied that you have included the right information in the right order, revise for style. Have you used an appropriate level of vocabulary for your audience? Have you used consistent terminology throughout and provided a glossary—a list of definitions—if some of your readers will need it?

  Have you provided an appropriate mix of sentence types (simple, complex, compound, compound-complex) for your audience? Are the sentences grammatically correct? Have you avoided awkward constructions? See Chapter 6 for a detailed discussion of technical-writing style.

- *Spelling.* Readers take spelling very seriously. When they see misspellings, they tend to think that your command of the technical material might be similarly flawed.

In revising your document for spelling errors, remember that if you merely read the sentences and paragraphs, many of the errors will slip by. Most people will see the word *hte* as the word *the*, because that is the spelling the brain expects. Therefore, it is a good idea to read the words out of context. If you can stand it, read each word individually, starting at the right margin and working toward the left.

In checking the spelling on a computerized document, you can get a spell checker to do most of the work for you. Keep in mind, however, that a computer can only check whether you have spelled the word right; it cannot tell whether you have spelled the right word. The next section, on word processing, discusses spell checkers further.

Watching the Revision Process in Action

This section shows the process one student used to revise a job-application letter.

After studying the job ad and brainstorming, she produced this draft:

> 3404 Vista Street
> Philadelphia, PA 19136
>
> January 19, 19XX
>
> Ms. Susan Saunders, Director of Personnel
> Martin Marietta Aero and Naval Systems
> 103 Chesapeake Park Plaza
> Baltimore, MD 21220
>
> Dear Ms. Saunders:
>
> I am writing in response to your notice in the Philadelphia Inquirer. I would like to be considered for the position in Signal Processing. My academic training at Drexel University in Electrical Engineering, as well as my experience with RCA Advanced Technology Laboratory, would qualify me, I believe, for the available position.
>
> My education at Drexel University has provided me with a strong background in the areas of computer hardware and system design. Throughout my five years at Drexel University I have channeled my studies into the digital and computer oriented fields. I have gained much insight into the development and design of computer hardware and the digital signals which make all of this possible. The advanced engineering courses which I took served to take me beyond the basics to a better understanding of this field of engineering.

The enclosed résumé provides an overview of my education and experience. Could I meet with you at your convience to discuss my qualifications for this position? Please write to me at the above address or leave a message for me anytime at (215) 333-8440.

Yours truly,

Deborah Carter

Deborah Carter

In her first pass through the document, Deborah realized a big omission: she forgot the paragraph about her work experience. This is what she drafted:

Your ad states that a "background in VLSI design is perferred." While working at RCA I recieved much experience in the VLSI design area.

She added this paragraph after the introductory paragraph.

Next Deborah looked through her draft to make sure it was accurate. She found no problems.

Then she considered organization. Reviewing the advice she had read in several books about job-application letters, she decided that her education paragraph should precede her job-experience paragraph because it is best to begin with the stronger of the two. In switching the sequence of the paragraphs, she realized that her draft of the job-experience paragraph was inadequate: it provided too little information. She added material, making this her intermediate draft of the body of the letter:

I am writing in response to your notice in the Philadelphia Inquirer. I would like to be considered for the position in Signal Processing. My academic training at Drexel University in Electrical Engineering, as well as my experience with RCA Advanced Technology Laboratory, would qualify me, I believe, for the available position.

My education at Drexel University has provided me with a strong background in computer hardware and system design. Throughout my five years at Drexel I have concentrated on digital and computer oriented fields. I have gained much insight into the development and design of computer hardware and the digital signals which make all of this possible. The advanced engineering courses which I took served to take me beyond the basics to a better understanding of this field of engineering.

Your ad states that a "background in VLSI design is perferred." While working at RCA I recieved much experience in the VLSI design area. I was involved in a number of projects in which I assisted in the design and testing of VLSI standard cell integrated circuits. My position at RCA allowed me to become familiar with much of the hardware and software being used in the design area today.

The enclosed résumé provides an overview of my education and experience. Could I meet with you at your convience to discuss my qualifications for this position? Please write to me at the above address or leave a message for me anytime at (215) 333-8440.

Deborah concludes that this draft is better, but she is concerned about the lack of specific information in the middle two paragraphs. She goes back to her notes and substitutes specifics for generalizations:

My education at Drexel University has provided me with a strong background in computer hardware and system design. Throughout my five years at Drexel I have concentrated on digital and computer oriented fields, studying such subjects as system design, digital filters, and computer logic circuits. Because of my status as an honors student, I was permitted to take a graduate seminar in digital signal processing.

Your ad states that a "background in VLSI design is perferred." While working at RCA's Advanced Technology Laboratory, I was involved in the design and testing of two Gate Array (2700 and 5000 gates) VLSI standard cell integrated circuits. I also used CAD software to simulate, check, and evaluate several integrated circuit designs.

She prints out a triple-spaced copy of the letter and starts to revise it for style and spelling. Here are the revisions she made:

I am writing in response to your notice in the *January 17* Philadelphia

Inquirer. I would like to be considered for the position in

Signal Processing. My ~~academic training~~ *education* at Drexel University

in Electrical Engineering, as well as my experience with RCA

Advanced Technology Laboratory, would qualify me, I believe,

for the available position.

My education at Drexel ~~University~~ has provided me with a strong background in computer hardware and system design. Throughout my five years at Drexel I have concentrated on digital and computer oriented fields, studying such subjects as system design, digital filters, and computer logic circuits. Because ~~of my status as~~ *I am* an honors student, I was permitted to take a graduate seminar in digital signal processing.

Your ad states that a "background in VLSI design is ~~perferred~~ *preferred*." While working at RCA's Advanced Technology Laboratory, I was involved in the design and testing of two Gate Array (2700 and 5000 gates) VLSI standard cell integrated circuits. I also used CAD software to simulate, check, and evaluate several integrated circuit designs.

The enclosed résumé provides an overview of my education and experience. Could ~~I~~ *we* meet ~~with you~~ at your ~~convience~~ *convenience* to discuss my qualifications for this position? Please write to me at the above address or leave a message for me any time at (215) 333-8440.

As this example suggests, revising effectively takes some time and concentration. Deborah spent almost two hours revising the letter that took less than 20 minutes to draft.

Testing the Document

The discussion of revising has concentrated on two ways of assessing the quality of your document: reviewing it yourself and having someone else look at it. These techniques can be very effective.

In some situations, however, you might want to go further. You might want to actually test the quality of the document. Testing is appropriate if the document is extremely important—such as a corporate annual report—or if it is intended to teach the readers a specific body of information or the process for carrying out particular tasks.

There are two basic kinds of document testing:

- ☐ analyzing the document
- ☐ testing readers as they use the document

Analyzing the Document

Some of the common techniques of analyzing the document have been discussed already in the section on revising: reviewing the document numerous times and using checklists, style programs, and spell checkers.

Another way to test a document is to give it to an expert in the subject of the document. If, for instance, you have written an analysis of alternative fuels for automobiles, you could ask an automotive expert to review it for accuracy, comprehensiveness, and all the other aspects of effective technical writing. In the working world, important documents are routinely reviewed by technical experts, legal counsel, and marketing specialists before being released to the public.

Still another way to test a document is to apply a readability formula, which is a quasi-mathematical technique for assessing the relative difficulty of the writing. Readability formulas are discussed in the next chapter.

Testing Readers as They Use the Document

Analyzing a document can provide only limited information; it cannot tell you how well the document will work when it gets into the hands of the people who will be using it. To find out that information, you must test readers as they use the document.

Testing readers effectively entails several common-sense steps:

- ☐ determining what you want to test
- ☐ choosing representative test subjects
- ☐ making the tests realistic
- ☐ evaluating the data carefully

Determining What You Want to Test

The questions you want to answer are likely to vary from one kind of document to the next. A sales brochure might be tested to determine whether it makes a positive impression on potential customers. A legal contract might be tested to see if it is clear and understandable to the general reader. A set of instructions might be tested to see whether a general reader can carry out the task safely, easily, and effectively.

The first step in testing, therefore, is to make sure you and your coworkers agree on what you want to find out. Otherwise, you will not be able to decide on a way to design an effective test.

Choosing Representative Test Subjects

A basic principle of testing is that the test subjects—the people you are studying—must match the real users of the document as closely as possible. If the real users will be mostly men in their twenties, you should get men in their twenties to be your test subjects. Try to match the test subjects to the real users in terms of education, experience, and all the other factors discussed in the section on analyzing audience in Chapter 3.

Making the Tests Realistic

After getting realistic test subjects, make the tests themselves realistic. If you are testing a user's manual for a word-processing program, ask the test subject to use the draft to carry out some representative word-processing tasks on the appropriate computer. If you are testing a draft of a sales catalog, you might ask the reader to sit at a desk and find certain pieces of information in the catalog. In addition, at the conclusion of the test, you might ask the test subject to answer a questionnaire about the ease of use or the visual appeal of the catalog. See Chapter 4 for a discussion of questionnaires.

For documents that communicate information but that don't require that the reader actually do anything, you have to be more inventive. You can ask the reader to do the following:

- tell you what he or she is thinking while reading the document
- answer short-answer questions after reading the document
- write a paraphrase of a particular piece of information after reading the document
- indicate whether a new text accurately reflects what was said in a particular passage in the test document
- tell you what he or she thinks a passage means after reading a document

Gathering and Evaluating the Data

The data you collect from testing the documents must be evaluated as carefully as the data from any experiment. Basic principles of experimentation apply, such as having enough test subjects to ensure statistically valid results, and making sure that the data gathering does not unintentionally influence the test subjects' behavior.

If you interview the test subjects, ask each the same questions in the same way, and record the information accurately and precisely.

If you observe test subjects trying to carry out tasks, you must remain unobtrusive. In many organizations, special testing facilities enable the observer to observe the test subjects without their realizing it. The observer sits in an adjacent room and observes through a one-way mirror. Meanwhile, the activity in the testing room is being videotaped for further analysis. When the testing involves operating a computer, a keystroke program records every keystroke the test subject makes. This program produces a real-time record of what the test subject did, enabling the observer to evaluate the document's strengths and weaknesses.

Using a Word Processor

The following discussion does not cover the process of operating a word processor. Rather, it covers strategies of using the word processor effectively to improve the quality of your writing.

With a word processor, you can write faster and much more easily. You can make fewer errors. You can produce neater and more professional-looking writing. And you can collaborate more easily with other writers to produce longer, more ambitious documents.

Of course, there are some drawbacks. The instruction manuals for the hardware and software can be frustrating. And most users have lost some text because they forgot to save it onto a disk—or because the system "crashed." And for some people, using a word processor doesn't provide the same sense of the text that handwriting does. They miss the "feel" of the pen on paper.

In the following discussion of word processors and the writing process, the term *word processor* is used to refer to dedicated word processors (computers especially designed for word processing), microcomputers used with word-processing programs, and larger computers that the user accesses through remote terminals.

For most writers and most documents, word processors offer many advantages over pen and paper at every stage of the writing process—prewriting, drafting, and revising.

Prewriting

A number of programs exist that may help you think of topics to write about. These programs ask you to respond to a series of questions about your chosen topic, forcing you to think about the audience and purpose of the report and to indicate what kind of information you will need. There is no scientific proof at this point that using these programs is more effective than simply answering the same questions on paper. However, some professors report that as their students use these idea-generating programs several times they learn how to ask the right questions and therefore write better reports.

A second use of word processors is in taking notes. If your word processor is portable, you can type notes on it rather than writing them on index cards. When you want to use that information in your document, you can put it there easily, too, and you won't introduce any errors, because you'll simply move the entries into position.

Once you sit down to brainstorm and outline, the word processor will help you save time. You can create a long brainstorming list quickly and easily—and everything you write will be easy to read. Moving items from one position to another on the screen is almost effortless; therefore, you can easily try out different patterns as you classify items into groups. The same holds true when you sequence the groups. Creating and sequencing groups on paper can be cumbersome.

Drafting

The word processor is useful during the drafting stage because it helps you write faster. Knowing that you can easily move information from one place in the text to another encourages you to begin writing in the middle, on a technical point you are familiar with. When you can't think of what to write on one subject, just skip a few lines and go to the next subject on your outline. Shuffling your text later takes only a few seconds.

With a word processor, you can easily write your draft right on your outline. The advantage is that you are less likely to lose sight of your overall plan. Before you start to draft, make a copy of your outline. You'll be able to use it later to create a table of contents. On the other copy of the outline, start to draft. As you write, the portion of the outline below will scroll downwards to accommodate your draft. You don't have a separate outline and a separate draft: the outline becomes the draft.

Because word processors are relatively quiet and easy to type on, you will probably generate a much longer draft in a given time. You don't have the physical effort involved in writing by hand, and you don't have to return the carriage at the end of the line as you do on a typewriter. Producing a lot of writing quickly is exactly the point of drafting: you want to have material to revise later. Moreover, you can concentrate on what you are trying to say because making changes—large or small ones—requires so little effort.

The word processor can also help you break the habit of revising what you have just typed. Just turn the contrast knob to make the screen black. This technique, called invisible writing, encourages you to close your eyes or look at a printout of your outline or the keyboard. You will find that you don't stop typing so often.

A common feature on word processors—the search-and-replace function—also increases your speed during the drafting stage. The search-and-replace function lets you find any phrase, word, or characters and replace them with any other writing. For example, if you will need to use the word *potentiometer* a number of times in your document, you can simply type in *po* each time. Then, during the revising stage, you can instruct the word processor to change every *po* to *potentiometer*. This use of the search-and-replace function not only saves time as you draft; it also reduces the chances of misspelling, for you have to spell the word correctly only once.

Revising

Word processors make every kind of revising easier. Most obviously, your writing is legible, so you see what you've done without being distracted by sloppy handwriting. And because your writing is typed neatly, you have a more objective perspective on your work. You are looking at it as others will.

If the obvious typographical errors distract you, fix them so that you can concentrate on substantive changes. All word processors have easy-to-use add and delete functions so that you change *hte* to *the* in a few seconds.

You can also make major revisions to the structure and organization of the document easily. All word processors let you move text—anything from a single letter to whole paragraphs—simply and quickly. Therefore, you can try out different versions of the document without having to cut and paste pieces of paper. Most word processors also have a copy function, which lets you copy text, such as an introductory paragraph, and move the copy to some other location without moving the original text. With the copy function you can in effect create two different versions of the document simultaneously and decide which one works better.

A number of different editing programs are available that help you identify problems:

- A *spell checker* compares what you have typed with a dictionary, usually of 20,000 to 90,000 words. The program usually can check approximately 5,000 to 10,000 words per minute. The program will alert you when it sees a word that isn't in its dictionary. Although that word might be misspelled, it might be a correctly spelled word that isn't in the dictionary. You can add the word to your dictionary so that it will know in the future that the word is not misspelled. If the word is misspelled, the program will suggest a number of words that it "thinks" you might be trying to spell. You simply choose the one you want from the list—or look up the spelling in a dictionary and make the necessary changes on the keyboard.

  However, no spell-checker program can tell whether you have used the correct word; it can tell you only whether the word you have used is in its dictionary. Therefore, if you have typed "We need too dozen test tubes," the spell checker will not see a problem.

■ A *thesaurus* lists similar words for many common words. A thesaurus program has the same strengths and weaknesses as a printed thesaurus: if you know the word you are looking for but can't quite think of it, the thesaurus will help you remember it. But the terms listed might not be closely enough related to the original term to function as synonyms. Unless you are aware of the shades of difference, you might be tempted to substitute an inappropriate word. For example, the word *journal* is followed in *Roget's College Thesaurus* by the word *diary*. A personal journal is a diary, but a professional periodical certainly isn't.

■ A *word-usage program* measures the frequency of use of particular words and the lengths of words. All of us overuse some words, but without the word processor we have a difficult time determining which ones. Word length is useful to know, for technical terms frequently are long words. After we analyze our audience and purpose, we need to consider the amount of technical terminology to include in the document. Readability formulas, which rely on word length and sentence length, are discussed in Chapter 6.

■ A *style program* helps identify and fix potential stylistic problems. For instance, a style program counts such factors as sentence length, number of passive-voice constructions and expletives, and type of sentence (see Chapter 6). Many style programs identify abstract words and suggest more specific ones. Many point out sexist terms and provide non-sexist alternatives. Many point out fancy words, such as *demonstrate*, and suggest substitutes, such as *show*.

Keep in mind just what these different programs can do and what they cannot. They can point out your use of the passive voice, but they cannot tell you whether the passive voice is preferable to the active voice in the particular sentence. In addition, some of the advice offered by style programs is just plain ignorant or of dubious value.

These different style programs in a way make your job as a writer more challenging: by pointing out potential problems, they force you to decide about issues that you might not have considered otherwise. But the payoff is that a wise use of the programs will give you a better document.

Many of the functions performed by these style programs can be performed with the search function. For instance, if you know that you overuse expletives ("It is . . . ," "there is . . . ," "there are . . .") you can search for words such as *is* and *there*. Some uses of these words, of course, will not be expletives, but you can revise those that are inappropriate. Or if you realize you overuse nominalizations (noun forms of verbs, such as *installation* for *install*), you can search for the common suffixes (such as "*-tion*," "*-ance*," and "*-ment*") used in them. (You've probably already noticed that the word *nominalization* is itself a nominalization.) These stylistic questions are discussed in Chapter 6.

Although word processors can help you do much of the work involved in revision, they cannot replace a careful reading by another person. Revision programs will calculate sentence length more accurately than a person can, but they cannot identify unclear explanations, contradictions, inaccurate data, inappropriate words, and so forth. Use the revision programs, revise your document yourself, then get help from someone you trust.

Three other aspects of word processing in document production are discussed in other chapters in this text:

- Chapter 4, Finding and Using Information, discusses database searching, a technique for accessing information stored electronically. Although databases can be accessed through a personal computer, most students find it simpler and more effective to seek the assistance of the professional librarian at the college or university library.

- Chapter 11, Graphics, discusses some of the common types of graphics that can be produced with graphics software on many kinds of computers.

- Chapter 19, Formal Elements of a Report, discusses some of the formatting techniques that word processors make convenient.

Writing Collaboratively

This chapter has described writing as essentially a solitary activity. The only major exception has been the advice to seek out a colleague who can help identify problems during the revision stage. In fact, most small documents—such as letters, memos, and brief reports—are in fact conceived and written by a single person.

However, much writing on the job is collaborative. Collaborative writing can be defined as any writing in which more than one person participates during at least one of the stages of the writing process: prewriting, drafting, and revising.

Understanding the Role of Collaboration

Collaboration is much more popular in business and industry than most people would think. According to one major study (Ede and Lunsford 1989), almost nine of ten businesspersons report that they write collaboratively at least sometimes. In addition, collaborative writing is becoming more and more popular in colleges and universities for big documents such as senior design proposals and reports.

What are the advantages of collaboration? In four important ways, the process of working together can improve both the document and the organization that produces it:

■ *Collaboration draws on a greater knowledge base.* Two heads are better than one, and three are better yet. The more people involved in planning and writing the document, the larger the knowledge base. This translates to a document that can be more comprehensive and more accurate than a single-author text.

■ *Collaboration creates a better sense of how the audience will read the document.* The process of working collaboratively involves having different people read drafts. Each person who reads and comments on a draft acts as an audience, raising questions and suggesting improvements that one person might not think of.

■ *Collaboration improves communication among employees.* Collaboration breaks down physical, organizational, and emotional barriers. People who work together for a long time get to know each other and learn about each other's jobs, responsibilities, and frustrations. A shared goal—putting together a professional-quality document that will reflect positively on everyone involved—can enable people to transcend their own isolation.

■ *Collaboration improves the socialization of new employees.* Another major benefit of collaboration is that it helps acclimate new employees to an organization. They get to know people from different parts of the organization and from different levels. Working collaboratively helps new employees learn the ropes: how to get things done, which people to see, what forms to fill out, and so forth. In addition, collaboration teaches the values that permeate the organization, such as the importance of shared goals and the willingness to work hard and sacrifice for an important initiative.

Following are some common examples of collaborative writing in industry:

■ A manufacturer of computer software has completed a program and wants to write a user's manual. The manual begins with notes written by the systems analyst who devised the program. After conversations with the systems analyst, the technical writer drafts the manual and solicits comments and revisions from the systems analyst. Finally, a third person edits the manual.

■ An electronics company wants to respond to a government proposal to design a piece of friend-or-foe radar-detection equipment. The project manager creates a proposal-writing team consisting of several engineers in different areas represented by the product to be designed, a graphic artist, several writers and editors, and a contracts specialist.

■ Three biologists who have conducted experiments on the reaction of a strain of bacteria to a genetically engineered molecule decide to write up their work for publication in a professional journal. Each biologist writes a different section after a meeting to decide the best strategy for the article. Finally, one of the three revises the whole article.

As these examples suggest, collaborative writing can involve any number of people, can address any kind of audience, and can lead to any kind of document. Still, some basic guidelines can make almost any kind of collaborative writing project effective and efficient:

■ *The participants must agree to a goal, a strategy, and a set of procedures.* All members of the group must agree on a goal: what the document is intended to be and to do. In addition, they must agree on a strategy, covering their analysis of audience and purpose and questions such as length, organization, style, and level of vocabulary. Finally, the group must formally agree to their operating procedures, including such matters as scheduling meetings and communicating among members.

■ *The participants must try to work together in harmony.* Given the stresses of modern corporate life, tempers can flare and small disagreements can become unproductive turf disputes or personal grudges. The participants must try to be polite, listening carefully and respectfully to the other participants' ideas. They should try to seek agreement rather than dwell on differences of opinion. They should be willing to compromise, realizing that, in most cases, a document representing several people's ideas works better than one representing only one person's.

■ *Each participant must have a specific role.* For example, one person investigates and writes up a particular section of the document; another person investigates another area and checks the entire document for comprehensiveness.

■ *The group must have a coordinator.* One person must be the leader. That person schedules and chairs meetings of the group, provides motivation and technical support, and helps the group reach consensus when differences of opinion occur.

Although people often work together while drafting, most collaboration occurs during prewriting and revising. In fact, the term *brainstorming* was coined a half-century ago at Hughes Aircraft to refer to planning sessions attended by as many as a dozen people. At these brainstorming sessions one person was the leader, responsible for keeping the ideas flowing. He or she would encourage the less outspoken participants and keep the more vocal ones from dominating the session. In brainstorming, a group of people can create a much longer and more useful list of topics than a solitary person can. One person's idea gives another person an idea. A skillful leader can motivate everyone to contribute.

During the drafting stage, most people work alone. This does not mean that they don't talk with other people or occasionally call for assistance, only that they create the individual sentences and paragraphs by themselves.

During the revision stage, collaboration is again useful. The collaboration can be relatively informal or formal. Informal collaboration might consist of one person asking another's opinion on a particular passage. In formal collaboration, the group members exchange their work and write comments and suggestions, and they meet to talk through any problems. This process can be repeated any number of times, depending on the length and complexity of the document. Eventually, time limits or deadlines force the group members to collect all the work and put it together in a single document, a task one person usually oversees. This person makes sure the different segments are consistent in terms of format, style, pagination, and so forth, to make the document look as if it were written by one person, not by a group.

Using Technology to Enhance Collaboration

Two people can work together effectively using a pen and paper, but modern technology makes it possible to bridge the barriers of time and space. Three technologies are particularly useful in collaborative writing:

- *Word processing.* Word processors with the most basic software allow group members to revise their writing easily and effectively. If the group members are networked through a central computer, they can transmit and revise different pieces of writing. Even if the writers are working on separate word processors, they can exchange disks.

  Word processors save a lot of time and effort in putting the document together. Obviously, they make it unnecessary to retype the different sections. In addition, the principal writer can use functions such as search-and-replace to make sure that technical vocabulary is being used consistently and to create a list of key words or an index.

  More sophisticated word-processing software lets group members write comments on the text; these comments can be displayed on the text or kept hidden, depending on the reader's preference. Keeping the comments hidden enables a number of different people to read and think about the original text without being influenced by another reader's ideas.

- *Electronic mail.* Electronic mail (E-mail) enables a word processor to transmit text and graphics through a telephone line to another. In 1991, some ten billion E-mail messages were sent. The Congressional Office of Technology Assessment has estimated that by the year 2000 some 140 billion pieces of mail—two-thirds of the nation's total—will be sent by E-mail, not by the Postal Service.

  To use E-mail, one person types in a text—a brief message or a longer document—and presses a few keys to send it through the telephone wire to another person. When the recipient turns on his or her computer, it displays a notice that a message is waiting. The recipient then retrieves the text and either prints it out or types in a response and sends it back.

Much of the current E-mail traffic is internal: one person sending E-mail to another in the same company. Currently, E-mail works most effectively when both persons are using the same computer operating system (such as MS-DOS) and the same E-mail network (such as Western Union's Easylink or MCI's MCI Mail). However, the seven major E-mail networks are making themselves compatible, so that a person linked to one network can easily communicate with a person linked to another network.

- *Fax machines.* A fax (short for facsimile) machine is a device that electronically scans a piece of paper, converts the image to a digital format, and transmits the code through a telephone wire to another fax machine, which transforms it back to letters and numbers and prints hard copy. The process takes approximately a quarter-minute per page.

  In collaboration, fax machines can transmit text and graphics across the country almost instantly, saving the time involved in sending a photocopy. Faxing is also used as a convenient alternative to E-mail, although it is slower and more expensive, when there are problems of network incompatibility or when one person does not have access to a computer.

WRITER'S CHECKLIST

Prewriting

1. Have you analyzed your audience, purpose, and strategy for the document?

2. Have you researched the appropriate information for the document?

3. Have you generated ideas by talking with someone about the topic, free writing, brainstorming, or sketching?

4. In outlining, have you grouped similar items?

5. Have you sequenced the items in the group appropriately?

6. Have you sequenced the groups appropriately?

7. In developing the outline, have you used one (or more) of the common patterns, such as chronological, spatial, or general-to-specific?

8. Have you avoided faulty coordination and faulty subordination in creating the outline?

9. If your outline is to be read by someone else, have you chosen one of the common outline formats?

Revising

1. Have you studied the document carefully several times, looking each time for a different aspect of the writing, such as comprehensiveness, accuracy, emphasis, paragraphing, sentence structure, style, and spelling?

2. Have you given the document to someone else to help you find potential problems?

Testing the Document

1. If it is appropriate to test the document, have you used a style program and had someone else read it?

2. If it is appropriate to test readers as they use the document, have you
   a. determined what you want to test?
   b. chosen representative test subjects?
   c. made the tests realistic?
   d. gathered and evaluated the data carefully?

EXERCISES

1. Choose a topic you are familiar with and interested in (such as some aspect of your academic study, a neighborhood concern, or some issue of public policy). Generate ideas that might be relevant in a report on the topic. Finally, after analyzing your ideas, arrange them in an outline, discarding irrelevant ones and adding necessary ones.

2. The brainstorming list below was created by a manager in the personnel department of a city government. A recent study conducted by the department shows that turnover and absenteeism are high and that productivity is low. The most important factor contributing to these problems appears to be an ineffective method for evaluating job performance.

   Turn this brainstorming list into an outline for a report to the city manager, recommending that the city investigate alternative methods of job-performance evaluation. Where necessary, note information that has to be added to the outline.

   Then, turn the brainstorming list into an outline for a memo addressed to the city workers, explaining that alternative methods of job-performance evaluation will be studied and that the best one will be implemented. Where necessary, note information that has to be added to the outline.

   characteristics of current method of evaluation
   goal: decrease turnover
   ethical reasons to improve system
   current evaluations not performed on any schedule
   purposes of an evaluation system
   evaluations should be performed on a regular schedule
   economic reasons to improve system
   system should be useful to mgt and workers
   characteristics of a fair system
   federal govt prevents discrimination
   objectives of the system should be clear
   number of suits against city has increased
   goal: decrease absenteeism
   criteria of our system are subjective
   why institute a fair system?
   legal reasons to improve system
   criteria should be objective
   goal: reward good performance, discourage bad performance

3. The following portions of outlines contain logical flaws. In a sentence each, explain the flaws.

   a.
   I. Advantages of collegiate football
       A. Fosters school spirit
       B. Teaches sportsmanship
   II. Increases revenue for college
       A. From alumni gifts
       B. From media coverage

b.
A. Effects of new draft law
   1. On Army personnel
   2. On males
   3. On females

c.
I. Components of a personal computer
   A. Central processing unit
   B. External storage device
   C. Keyboard
   D. Magnetic tape
   E. Disks
   F. Diskettes

d.
A. Types of common screwdrivers
   1. Standard
   2. Phillips
   3. Ratchet-type
   4. Screw-holding tip
   5. Jeweler's
   6. Short-handled

e.
A. Types of health care facilities
   1. Hospitals
   2. Nursing homes
   3. Care at home
   4. Hospices

4. The writer of the following draft of an outline was part of a four-student team that designed and built a prototype of a human-powered hydrofoil boat. The outline was requested by the faculty advisor for the team. What changes would you make to the outline before submitting it?

1. Introduction
  Previous history of human-powered vehicles
    canoes
    rowboats
    pedal boats
    rowing shells
  problems encountered with these vehicles and their limits
    weight
    drag
    drive
  Improvements, areas of concern
    increased efficiency of transmission
      airfoil propeller
      flexible shaft
    reduce drag
      hydrofoils
      weight reduction

make attempt to reach our ultimate goal of 20 knots
    build prototype
    optimize prototype
2. Methods
Look at previous human powered boat vehicles
    try to avoid their problems
    Build prototype
Test prototype
Optimize prototype
Test final product
3. Conclusions
Problem areas
    drag
        foil size
        tubing support
        pontoons
    weight
        pontoons
        crank
        channel
future outlook
4. Recommendations
Investigate the following
    finned struts
    inflatable pontoons
    adjustable foils
    steering mechanism
    lightweight materials
final remarks

5. The following is a draft of a letter sent by a town supervisor to all town residents. Make any necessary revisions.

Dear Resident of the Town of Garden City:

I am pleased to announce that your Town will introduce Enhanced 911 Emergency Service. This forward step into modern technology will greatly enhance in the dispatch of police, fire and emergency medical services. I'm certain you can appreciate the difficulties encountered while attempting to locate the cite of an emergency where there is no systematic house numbering to guide said emergency personnel. As part of the implementation process, we undertook revisions to the address plan that affect many of your residential, commercial and other properties. Many structures have been assigned a new address. Others, however, will continue with their present address.

We consider the changes beneficial. Deliveries to your home or business will be made easier. More importantly, tehy will improve the ability of your police, fire and emergency medical services to respond to any emergency.

Effective immediately, your address at this location in Garden City will be:

Nielsen, Richard
950 Luptons Point
Garden City, NY 10982

Please use this address in all your future dealings. If it is a change in what you now consider your present address to be be, obtaining a supply of "change of address" from your POst Office to use in notifying your regular mailers might be a good idea.

IT IS CRITICAL, however, that any changes be verified with 1) Garden City Building Department (516 765-2019) and 2) New York Telephone, who supplies the data for our Enhanced 911 system. You may call New York Telephone toll free at 800 462-7220, MOnday thru Friday between the hours of 9 a.m. and 5 p.m. In order for the system to work optimally, New York Telephone must be informed of your new house number.

I close with the fervent wish that you may never require emergency services, but you can rest assured that they will be there if you should ever need them.

6. You and two or three other students in your class are to collaborate on the following project. Your local chamber of commerce, which is revising its tourism brochure, feels that your college or university should be described in more detail. It has asked that your team draft a 2,000-word description of the sites and activities on your campus that might attract tourists. (Three or four photographs may be used to supplement the description.) With your collaborators, brainstorm, do the necessary research, outline, draft, and revise the description.

REFERENCE

Ede, L., and A. Lunsford. 1989. *Singular texts—plural authors: Perspectives on collaborative writing*. Carbondale, Ill.: Southern Illinois University Press.

6

Writing Effective Sentences

Perhaps you have heard that the best technical writing style is no style at all. This means simply that the readers should not be aware of your presence. They should not notice that you have a wonderful vocabulary or that your sentences flow beautifully—even if those things are true. In the best technical writing, the writer fades into the background.

This is as it should be. Few people read technical writing for pleasure. Most readers either *must* read it as part of their work or want to keep abreast of new developments in the field. People read technical writing to gather information, not to appreciate the writer's flair. For this reason, experienced writers do not try to be fancy. The old saying has never been more appropriate: *Write to express, not to impress.*

This chapter discusses two aspects of style: sentence construction and word choice. Other aspects of style, including how to maintain coherence and how to write effective paragraphs, are discussed in the next chapter.

Before you start to write any document, you should find out your organization's stylistic guidelines.

Determining the Appropriate Stylistic Guidelines

Most successful writers agree that the key to effective writing is revision: coming back to a draft and adding, deleting, and changing. Time permitting, you might write four or five different drafts before you finally have to stop. In revising, you will make many stylistic changes in an attempt to get closer and closer to the exact meaning you wish to convey. Some stylistic matters, however, can be determined before you start to write. Learning the "house style" that your organization follows will cut down the time needed for revision.

An organization's stylistic preferences may be defined explicitly in a company style guide that describes everything from how to write numbers to how to write the complimentary close at the end of a letter. In some organizations, an outside style manual, such as the *United States Government Printing Office Style Manual*, is the "rule book." (Figure 6.1 shows a page from this manual.) In many organizations, however, the stylistic preferences are implicit; no style manual exists, but over the years a set of unwritten guidelines has evolved. The best way to learn the unwritten house style is to study some letters, memos, and reports in the files and to ask more-experienced coworkers for explanations. Secretaries, in particular, are often valuable sources of information.

As was discussed in Chapter 5, a word processor is a valuable tool in making stylistic revisions. A number of style programs exist that help isolate many of the concepts described in the following discussion. Even without specialized programs, however, you can perform some of the same techniques with the search function of any word-processing program.

12. NUMERALS

(See also Tabular Work; Leaderwork)

12.1. Most rules for the use of numerals are based on the general principle that the reader comprehends numerals more readily than numerical word expressions, particularly in technical, scientific, or statistical matter. However, for special reasons numbers are spelled out in indicated instances.

12.2. The following rules cover the most common conditions that require a choice between the use of numerals and words. Some of them, however, are based on typographic appearance rather than on the general principle stated above.

12.3. Arabic numerals are generally preferable to Roman numerals.

NUMBERS EXPRESSED IN FIGURES

12.4. A figure is used for a single number of *10* or more with the exception of the first word of the sentence. (See also rules 12.9, 12.23.)

50 ballots	24 horses	about 40 men
10 guns	nearly 10 miles	10 times as large

Numbers and numbers in series

★**12.5.** Figures are used in a group of 2 or more numbers, or for related numbers, any one of which is *10* or more. The sentence will be regarded as a unit for the use of figures.

Each of 15 major commodities (9 metal and 6 nonmetal) was in supply.
but Each of nine major commodities (five metal and four nonmetal) was in supply.

Petroleum came from 16 fields, of which 8 were discovered in 1956.
but Petroleum came from nine fields, of which eight were discovered in 1956.

That man has 3 suits, 2 pairs of shoes, and 12 pairs of socks.
but That man has three suits, two pairs of shoes, and four hats.

Of the 13 engine producers, 6 were farm equipment manufacturers, 6 were principally engaged in the production of other types of machinery, and 1 was not classified in the machinery industry.
but Only nine of these were among the large manufacturing companies, and only three were among the largest concerns.

There were three 6-room houses, five 4-room houses, and three 2-room cottages, and they were built by 20 men. (See rule 12.21.)
There were three six-room houses, five four-room houses, and three two-room cottages, and they were built by nine men.
Only 4 companies in the metals group appear on the list, whereas the 1947 census shows at least 4,400 establishments.
but If two columns of sums of money add or subtract one into the other and one carries points and ciphers, the other should also carry points and ciphers.
At the hearing, only one Senator and one Congressman testified.
There are four or five things which can be done.

FIGURE 6.1 From the *United States Government Printing Office Style Manual*

Structuring Effective Sentences

Good technical writing is characterized by clear, correct, and graceful sentences that convey information without calling attention to themselves. This section consists of seven guidelines for structuring effective sentences:

- □ use a variety of sentence types
- □ choose an appropriate sentence length
- □ use readability formulas cautiously
- □ focus on the "real" subject
- □ focus on the "real" verb
- □ express parallel elements in parallel structures
- □ avoid confusing modifiers

Use a Variety of Sentence Types

You should understand how to use different sentence structures to make your writing more effective.

There are four basic types of sentences:

- ■ simple (one independent clause)

    The technicians soon discovered the problem.

- ■ compound (two independent clauses, linked by a semicolon or by a comma and one of the seven common coordinating conjunctions: *and, or, nor, for, so, yet*, and *but*)

    The technicians soon discovered the problem, but they found it difficult to solve.

- ■ complex (one independent clause and at least one dependent clause, linked by a subordinating conjunction)

    Although the technicians soon discovered the problem, they found it difficult to solve.

- ■ compound-complex (at least two independent clauses and at least one dependent clause)

    Although the technicians soon discovered the problem, they found it difficult to solve, and they finally determined that they needed some additional parts.

Most commonly used in technical writing is the simple sentence, because it is clear and direct. However, a series of three or four simple sentences can bore and distract the reader. The main shortcoming of the simple sentence is that it can communicate only one basic idea, because it is made up of only one independent clause.

Compound and complex sentences communicate more sophisticated ideas. The compound sentence coordinates: the two halves of the sentence are roughly equivalent in importance. The complex sentence subordinates: one half of the sentence is less important than the other.

Compound-complex sentences are useful in communicating very complicated ideas, but their length and difficulty can make them unsuitable for many readers.

One common construction deserves special mention: the compound sentence linked by the coordinating conjunction *and*. In many cases, this construction is a weak way to link two thoughts that could be linked more clearly. For instance, the sentence "Enrollment is up, and the school is planning new course offerings" represents a weak use of the *and* conjunction, because it doesn't show the cause-effect relationship between the two clauses. A better link would be *so*: "Enrollment is up, so the school is planning new course offerings." A complex sentence would clarify the relationship even more: "Because enrollment is up, the school is planning new course offerings." Another example of the weak *and* conjunction is the sentence "The wires were inspected, and none was found to be damaged." In this case, a simple sentence would be more effective: "The inspection of the wires revealed that none was damaged" or "None of the wires inspected was damaged."

A number of software programs classify sentences according to type, so that during revision you can see the kinds of sentences you have written. On the basis of your analysis of audience and purpose, you can then decide whether your mixture of sentence types is appropriate.

Choose an Appropriate Sentence Length

Although it would be distracting and artificial to keep a count of the number of words in your sentences, sometimes sentence length can work against effective communication. In revising a draft, you might want to compute an average sentence length for a representative passage of writing. (Many software programs can do this for you.)

There are no firm guidelines covering appropriate sentence length. In general, however, a length of 15 to 20 words is effective for most technical writing. A succession of 10-word sentences would be choppy; a series of 35-word sentences would probably be too demanding. Try to avoid series of overly long and overly short sentences.

Avoid Overly Long Sentences

How long a sentence is too long? There is no simple answer, because the ease of reading a sentence depends on its length, vocabulary, and structure, the reader's knowledge of the topic, and the purpose of the communication.

Yet in many technical-writing applications, you will have a lot of information to convey, and your first draft is likely to include sentences such as the following:

> The construction of the new facility is scheduled to begin in March, but it might be delayed by one or even two months by winter weather conditions, which can make it impossible or nearly impossible to begin excavating the foundation.

This sentence, which contains 41 words, is too long, not because it is hard to understand, but because it is tiring to read. To make it more readable, divide it into two sentences.

> The construction of the new facility is scheduled to begin in March. However, construction might be delayed until April or even May by winter weather conditions, which can make it impossible or nearly impossible to begin excavating the foundation.

Sometimes an overly long sentence can be fixed by creating a list:

> To connect the CD player to the amplifier, first be sure that the power is off on both units, then insert the plugs firmly into the jacks (the red plug into the right-channel jack and the black plug into the left-channel jack), making sure that you leave a little slack in the connecting cord to allow for inadvertent shock or vibration.

> To connect the CD player to the amplifier, follow these steps:
>
> - Be sure that the power is off on both units.
> - Insert the plugs firmly into the jacks. The red plug goes into the right-channel jack, and the black plug into the left-channel jack.
> - Make sure that you leave a little slack in the connecting cord to allow for inadvertent shock or vibration.

As this revision suggests, sometimes the best way to communicate technical information is to switch to a more graphically oriented format—lists or graphics.

Avoid Overly Short Sentences

Just as sentences can be too long, they can also be too short, as in the following example.

> The fan does not oscillate. It is stationary. The blade is made of plastic. This is done to increase safety. Safety is especially important because this design does not include a guard around the blade. A person could be seriously injured by putting his or her hand into the turning blade.

The problem here is not that the word count of the sentences is too low, but rather that the sentences are choppy and contain too little information. The best way to revise a passage such as this is to combine sentences, as in the following example:

> The fan is stationary, not oscillating, with a plastic blade. Because the fan does not have a blade guard, the plastic blade is intended to prevent injury if a person accidentally touches it when it is moving.

One symptom of excessively short sentences is that they needlessly repeat key terms, as in the following example:

> Computronics, a medium-sized consulting firm, consists of many diverse groups. Each group handles and develops its own contracts.

To speed up this passage, combine the sentences:

> Computronics, a medium-sized consulting firm, consists of many diverse groups, each of which handles and develops its own contracts.

Here is another example, followed by a revision:

> I have experience working with various microprocessor-based systems. Some of these systems include the Z80, 6800 RCA 1802, and the AIM 6502.

> I have experience working with various microprocessor-based systems, including the Z80, 6800 RCA 1802, and the AIM 6502.

Use Readability Formulas Cautiously

Readability formulas are mathematical techniques used to determine how difficult a sample of writing is to read and understand. More than one hundred readability formulas exist; they are being used increasingly by government agencies and private businesses in an attempt to improve writing. In addition, many style programs calculate readability formulas. For these reasons, you should become familiar with the concept behind them.

Most readability formulas are based on the idea that short words and sentences are easier to understand than long ones. One of the more popular formulas, Robert Gunning's "Fog Index," works like this:

1. Find the average number of words per sentence, using a 100-word passage.
2. Find the number of "difficult words" in that same passage, that is, words of three or more syllables, except for proper names, combinations of simple words (such as *horsepower*), and verbs whose third syllable is *-es* or *-ed* (such as *contracted*).
3. Add the average sentence length and the number of difficult words.

4. Multiply this sum by 0.4.

Average number of words per sentence: 13.9
Number of difficult words: 16

$$29.9 \times 0.4 = 12$$

The result represents an approximate grade level. Here, the average twelfth-grade student could understand the writing.

Readability formulas are easy to use, and it is appealing to think you can be objective in assessing your writing, but unfortunately they have not been proven to work. They simply have not been shown to reflect accurately how difficult it is to read a piece of writing (Selzer 1981; Battison and Goswami 1981). Other factors, such as the format and organization of the passage, have a greater effect on readability than do word length and sentence length (Huckin 1983).

The problem with readability formulas is that they attempt to evaluate a portion of the communication — the words on paper — without considering the reader. It is possible to measure how well a specific person understands a specific writing sample. But words on paper don't communicate until somebody reads them, and everyone is different. A "difficult word" for a lawyer might not be difficult for a biologist, and vice versa. And someone who is interested in the subject being discussed will understand more than the reluctant reader will.

Good writing has to be well thought out and carefully structured. The sentences have to be clear, and the vocabulary has to be appropriate for the readers. *Then* the writing will be readable.

Focus on the "Real" Subject

The conceptual or "real" subject of the sentence should also be the sentence's grammatical subject, and it should appear prominently in technical writing. Don't bury the real subject in a prepositional phrase following a useless or "limp" grammatical subject. In the following examples, notice how the limp subjects disguise the real subjects. (The grammatical subjects are italicized.)

Weak   The *use* of this method would eliminate the problem of motor damage.
Strong   This *method* would eliminate the problem of motor damage.

Weak   The *presence* of a six-membered lactone ring was detected.
Strong   A six-membered lactone *ring* was detected.

Another way to make the subject of the sentence prominent is to reduce the number of grammatical expletives: *it is, there is,* and *there are.* In most cases, these constructions just waste space.

Weak   There is no alternative for us except to withdraw the product.
Strong   We have no alternative except to withdraw the product.

> *Weak*   It is hoped the testing of the evaluation copies of the software will help us make this decision.
>
> *Strong*   I hope the testing of the evaluation copies of the software will help us make this decision.

Often, as in this second example, the expletive "it is" is used along with the passive voice. See the discussion of the passive voice later in this chapter.

Keep in mind that expletives are not "errors"; they are conversational expressions that can sometimes help the reader understand the information in the sentence.

With the expletive

It is hard to say whether the recession will last more than a few months.

Without the expletive

Whether the recession will last more than a few months is hard to say.

The version without the expletive is a little more difficult to understand because the reader has to read and remember a long subject—"Whether the recession will last more than a few months"—before getting to the verb—"is." However, the sentence could be rewritten in other ways to make it easy to understand and to eliminate the expletive.

I don't know whether the recession will last for more than a few months.

Nobody really knows whether the recession will last more than a few months.

Using the search function of any word-processing program, you can find most weak subjects: usually they're right before the word *of*. Expletives are also easy to find.

Focus on the "Real" Verb

A "real" verb, like a "real" subject, should stand out in every sentence. Few stylistic problems weaken a sentence more than nominalizing verbs. Nominalizing the real verb involves converting it into a noun, then adding another verb, usually a weaker one, to clarify the meaning. "To install" becomes "to effect an installation," "to analyze" becomes "to conduct an analysis." Notice how nominalizing the real verbs makes the following sentences both awkward and unnecessarily long. (The nominalized verbs are italicized.)

> *Weak*   Each *preparation* of the solution is done twice.
> *Strong*   Each solution is prepared twice.

> *Weak*   An *investigation* of all possible alternatives was undertaken.
> *Strong*   All possible alternatives were investigated.

> *Weak*   *Consideration* should be given to an acquisition of the properties.
> *Strong*   We should consider acquiring the properties.

Some software programs search for the most common nominalizations. With any word-processing program you can catch most of the nominalizations if you search for character strings such as *tion, ment,* and *ance.* Your search for "of" will also expose many nominalized verbs.

Express Parallel Elements in Parallel Structures

A sentence is parallel if its coordinate elements are expressed in the same grammatical form: that is, its clauses are either passive or active, its verbs are either infinitives or participles, and so forth. By creating and sustaining a recognizable pattern for the reader, parallelism makes the sentence easier to follow.

Notice how faulty parallelism weakens the following sentences.

> *Nonparallel*   Our present system is costing us profits and reduces our productivity. (*nonparallel verbs*)
> *Parallel*   Our present system is costing us profits and reducing our productivity.

> *Nonparallel*   The dignitaries watched the launch, and the crew was applauded. (*non-parallel voice*)
> *Parallel*   The dignitaries watched the launch and applauded the crew.

> *Nonparallel*   The typist should follow the printed directions; do not change the originator's work. (*nonparallel mood*)
> *Parallel*   The typist should follow the printed directions and not change the originator's work.

A subtle form of faulty parallelism often occurs with the correlative constructions, such as *either . . . or, neither . . . nor,* and *not only . . . but also*:

> *Nonparallel*   The new refrigerant not only decreases energy costs but also spoilage losses.
> *Parallel*   The new refrigerant decreases not only energy costs but also spoilage losses.

In this example, "decreases" applies to both "energy costs" and "spoilage losses." Therefore, the first half of the correlative construction should follow "decreases." Note that if the sentence contains two different verbs, the first half of the correlative construction should precede the verb:

> The new refrigerant not only decreases energy costs but also prolongs product freshness.

When creating parallel constructions, make sure that parallel items in a series do not overlap, thus changing or confusing the meaning of the sentence:

> *Confusing*   The speakers will include partners of law firms, businesspeople, and civic leaders.
> *Clear*   The speakers will include businesspeople, civic leaders, and partners of law firms.

The problem with the original sentence is that "partners" appears to apply to "businesspeople" and "civic leaders." The revision solves the problem by rearranging the items so that "partners" cannot apply to the other two groups in the series.

Use Modifiers Effectively

Technical writing is full of modifiers—words, phrases, and clauses that describe other elements in the sentence. To make your meaning clear, you must use modifiers effectively. That is, you must clearly communicate to your readers whether a modifier provides necessary information about its referent (the word or phrase it refers to) or simply provides additional information. Further, you must make sure that the referent itself is always clearly identified.

Distinguish between Restrictive and Nonrestrictive Modifiers

A **restrictive modifier**, as the term implies, restricts the meaning of its referent: that is, it provides information necessary to identify the referent. In the following example, the restrictive modifiers are italicized.

> The aircraft *used in the exhibitions* are slightly modified.

> Please disregard the notice *you just received from us.*

In most cases, the restrictive modifier doesn't require a pronoun, such as *that* or *which*. If you choose to use a pronoun, however, use *that*: "The aircraft that are used in the exhibits are slightly modified." (If the pronoun refers to a person or persons, use *who*.) Notice that restrictive modifiers are not set off by commas.

A **nonrestrictive modifier** does not restrict the meaning of its referent: it provides information that is not necessary to identify the referent. In the following examples, the nonrestrictive modifiers are italicized.

> The Hubble telescope, *intended to answer fundamental questions about the origin of the universe*, was launched in 1990.

> When you arrive, go to the Registration Area, *which is located on the second floor.*

Like the restrictive modifier, the nonrestrictive modifier usually does not require a pronoun. If you use one, however, choose *which* (*who* or *whom* when referring to a person). Note that nonrestrictive modifiers are separated from the rest of the sentence by commas.

Avoid Misplaced Modifiers

The placement of the modifier often determines the meaning of the sentence. Notice, for instance, how the placement of *only* changes the meaning in the following sentences.

Only Turner received a cost-of-living increase last year.
(*Meaning:* Nobody else received one.)

Turner received only a cost-of-living increase last year.
(*Meaning:* He didn't receive a merit increase.)

Turner received a cost-of-living increase only last year.
(*Meaning:* He received a cost-of-living increase as recently as last year.)

Turner received a cost-of-living increase last year only.
(*Meaning:* He received a cost-of-living increase in no other year.)

Misplaced modifiers—those that appear to modify the wrong referent—are common in technical writing. The solution is, in general, to place the modifier as close as possible to its intended referent. Frequently, the misplaced modifier is a phrase or a clause:

Misplaced	The subject of the meeting is the future of geothermal energy in the downtown Webster Hotel.
Correct	The subject of the meeting in the downtown Webster Hotel is the future of geothermal energy.
Misplaced	Jumping around nervously in their cages, the researchers speculated on the health of the mice.
Correct	The researchers speculated on the health of the mice jumping around nervously in their cages.

A special kind of misplaced modifier is called a **squinting modifier**—one placed ambiguously between two potential referents so that the reader cannot tell which one is being modified:

Unclear	We decided immediately to purchase the new system.
Clear	We immediately decided to purchase the new system.
Clear	We decided to purchase the new system immediately.
Unclear	The men who worked on the assembly line reluctantly picked up their last paychecks.
Clear	The men who worked reluctantly on the assembly line picked up their last paychecks.
Clear	The men who worked on the assembly line picked up their last paychecks reluctantly.

Avoid Dangling Modifiers

A **dangling modifier** is one that has no referent in the sentence:

Searching for the correct answer to the problem, the instructions seemed unclear.

In this sentence, the person doing the searching has not been identified. To correct the problem, rewrite the sentence to put the clarifying information either *within* the modifier or *next* to the modifier:

As I was searching for the correct answer to the problem, the instructions seemed unclear.

Searching for the correct answer to the problem, I thought the instructions seemed unclear.

Sometimes you can correct a dangling modifier by switching from the indicative mood (a statement of fact) to the imperative mood (a request or command):

Dangling   To initiate the procedure, the BEGIN button should be pushed.
Correct   To initiate the procedure, push the BEGIN button.

In the imperative, the referent—in this case, *you*—is understood.

Choosing the Right Words and Phrases

Effective technical writing consists of the right words and phrases in the right places. The following section includes five guidelines:

- ☐ use active and passive voice appropriately
- ☐ use first, second, and third person appropriately
- ☐ use simple, clear words and phrases
- ☐ use positive expressions
- ☐ use nonsexist language

Use Active and Passive Voice Appropriately

There are two voices: active and passive. In the active voice, the subject of the sentence performs the action expressed by the verb. In the passive voice, the subject receives the action. (In the following examples, the subjects are italicized.)

Active   *Brushaw* drove the launch vehicle.
Passive   The launch *vehicle* was driven by Brushaw.

Active   Many *physicists* support the big-bang theory.
Passive   The big-bang *theory* is supported by many physicists.

In most cases, the active voice is preferable to the passive voice. The active-voice sentence more clearly emphasizes the actor. In addition, the active voice sentence is shorter, because it does not require a form of the *to be* verb and the past participle, as the passive-voice sentence does. In the second example, for instance, the verb is "support," rather than "is supported," and "by" is unnecessary.

The passive voice, however, is generally more appropriate in four cases:

- The actor is clear from the context.

  Example   Students are required to take both writing courses.

  The context makes it clear that the college requires that students take both writing courses.

- The actor is unknown.

  Example   The comet was first referred to in an ancient Egyptian text.

  We don't know *who* referred to the comet.

- The actor is less important than the action.

  Example   The documents were hand-delivered this morning.

  It doesn't matter *who* the messenger was.

- A reference to the actor is embarrassing, dangerous, or in some other way inappropriate.

  Example   Incorrect data were recorded for the flow rate.

  It might be inappropriate to say *who* recorded the incorrect data.

Many people who otherwise take little interest in grammar have strong feelings about the relative merits of active and passive. A generation ago, students were taught that the active voice is inappropriate because it emphasizes the person who does the work rather than the work itself and thus robs the writing of objectivity. In many cases, this idea is valid. Why write, "I analyzed the sample for traces of iodine" when you can say "The sample was analyzed for traces of iodine"? If there is no ambiguity about who did the analysis, or if it is not necessary to identify who did the analysis, a focus on the action being performed is appropriate.

Supporters of the active voice argue that the passive voice creates a double ambiguity. When you write, "The sample was analyzed for traces of iodine," your reader is not quite sure who did the analysis (you or someone else) or when it was done (as part of the project being described or some time previously). Even though a passive-voice sentence can contain all the information found in its active-voice counterpart, often the writer omits the actor.

The best approach to the active-passive problem is to recognize how the two voices differ and use them appropriately. In the following examples, the writer mixes active and passive voice for no good reason.

Awkward	He lifted the cage door, and a hungry mouse was seen.
Better	He lifted the cage door and saw a hungry mouse.
Awkward	The new catalyst produced good-quality foam, and a flatter mold was caused by the new chute-opening size.
Better	The new catalyst produced good-quality foam, and the new chute-opening size resulted in a flatter mold.

A number of style programs can help you find the passive voice in your writing. Beware, however, of the advice some style programs offer about voice. A number of programs suggest that the passive voice is undesirable, almost an error. This is simply not so. Use passive voice when it works better than the active voice; don't be bullied into changing to the active voice.

With any word-processing program, however, you can search for *is* and *was*, the forms of the verb *to be* that are most commonly used in passive voice expressions. In addition, searching for *ed* will isolate the past participles, which also appear in many passive-voice expressions.

Use First, Second, and Third Person Appropriately

Closely related to the question of voice is that of person. The term *person* refers to the different forms of the personal pronoun:

First Person	I worked . . . , we worked . . .
Second Person	You worked . . .
Third Person	He worked . . . , she worked . . . , it worked . . . , the machine worked . . . , they worked . . .

Organizations that prefer the active voice generally encourage first-person pronouns: "We analyzed the rate of flow." Organizations that prefer the passive voice often *prohibit* first-person pronouns: "The rate of flow was analyzed." (Use the search function on the word processor to find *I* and *we*.)

Another common question about person is whether to use the second or the third person in instructions. In some organizations, instructions — step-by-step procedures — are written in the second person: "You begin by locating the ON/OFF switch." The second person is concise and easy to understand. Other organizations prefer the more formal third person: "The operator begins by locating the ON/OFF switch." Perhaps the most popular version is the second person in the imperative: "Begin by locating the ON/OFF switch." In the imperative, the *you* is implicit. Regardless of the preferred style, however, be consistent in your use of the personal pronoun.

Use Simple, Clear Words and Phrases

Technical writing should be clear, because your purpose in writing is to help your readers understand information or to carry out tasks. The following section offers six guidelines:

- ☐ be specific
- ☐ avoid unnecessary jargon
- ☐ avoid wordy phrases
- ☐ avoid clichés
- ☐ avoid pompous words
- ☐ avoid long noun strings

Be Specific

Being specific involves using precise words, providing adequate detail, and avoiding ambiguity.

Wherever possible, use the most precise word you can. A Ford Taurus is an automobile, but it is also a vehicle, a machine, and a thing. In describing the Ford Taurus, the word *automobile* is better than *vehicle*, because the less specific word also refers to trains, hot-air balloons, and other means of transport. As the words become more abstract—from *machine* to *thing*, for instance—the chances for misunderstanding increase.

In addition to using the most precise words you can, be sure to provide enough detail. Remember that the reader probably knows less than you do. What might be perfectly clear to you might be too vague for the reader.

> *Vague*   An engine on the plane experienced some difficulties.

What engine? What plane? What difficulties?

> *Clear*   The left engine on the Jetson 411 unaccountably lost power during flight.

Avoid ambiguity. That is, don't let the reader wonder which of two meanings you are trying to convey.

> *Ambiguous*   After stirring by hand for ten seconds, add three drops of the iodine mixture to the solution.

Stir the iodine mixture or the solution?

> *Clear*   Stir the iodine mixture by hand for ten seconds. Then add three drops to the solution.
>
> *Clear*   Stir the solution by hand for ten seconds. Then add three drops of the iodine mixture.

What should you do if you don't have the specific data? You have two options: to approximate—and clearly tell the reader you are doing so—or to explain why the specific data are unavailable and indicate when they will become available.

> *Vague*   The leakage in the fuel system is much greater than we had anticipated.
>
> *Clear*   The leakage in the fuel system is much greater than we had anticipated; we estimate it to be at least five gallons per minute, rather than two.

Several style programs isolate common vague terms and suggest more precise alternatives.

Avoid Unnecessary Jargon

Jargon is shoptalk. To a banker, *CD* means certificate of deposit; to an audiophile it means compact disc. The term *CAFE* (Corporate Average Fuel Economy) is a meaningful acronym among auto executives. Although jargon is often held up to ridicule, it is a useful and natural kind of communication in its proper sphere. Two baseball pitchers would find it hard to talk to each other about their craft if they couldn't use terms such as *slider* and *curve*.

In one sense, the abuse of jargon is simply a needless corruption of the language. The best current example is the degree to which computer-science terminology has crept into everyday English. In offices, employees are frequently asked to provide "feedback" or told that the coffee machine is "down." To many people, such words seem dehumanizing; to others, they seem silly. What is a natural and clear expression to one person is often strange or confusing to another. Communication breaks down.

An additional danger of using jargon outside a very limited professional circle is that it sounds condescending to many people, as if the writer is showing off—displaying a level of expertise that excludes most readers. While the readers are concentrating on how much they dislike the writer, they are missing the message.

If some readers are offended by unnecessary jargon, others are intimidated. They feel somehow inadequate or stupid because they do not know what the writer is talking about. When the writer casually tosses in jargon, many readers really *can't* understand what is being said.

If you are addressing a technically knowledgeable audience, feel free to use appropriate jargon. However, an audience that includes managers or the general public will probably have trouble with specialized vocabulary. If your document has separate sections for different audiences—as in the case of a technical report with an executive summary—use jargon accordingly. A glossary (list of definitions) is useful if you suspect that managers will read the technical sections.

Avoid Wordy Phrases

Wordy phrases weaken technical writing by making it unnecessarily long. Sometimes writers deliberately choose phrases such as "demonstrates a tendency to" rather than "tends to." The long phrase rolls off the tongue easily and appears to carry the weight of scientific truth. But the humble "tends to" says the same thing—and says it better for having done so concisely. The sentence "We can do it" is the concise version of "We possess the capability to achieve it."

Some wordy phrases just pop into writers' minds. We are all so used to hearing "take into consideration" that we don't realize that "consider" gets us there faster. Replacing wordy phrases with concise ones is therefore more difficult than it might seem. Avoiding the temptation to write the long phrase is only half of the solution. The other half is to try to root out the long phrase that has infiltrated the prose unnoticed.

Following are a wordy sentence and a concise translation.

> *Wordy*　I am of the opinion that, in regard to profit achievement, the statistics pertaining to this month will appear to indicate an upward tendency.
>
> *Concise*　I think this month's statistics will show an increase in profits.

Wordy Phrase	Concise Phrase	Wordy Phrase	Concise Phrase
a majority of	most	it is often the case that	often
a number of	some, many	it is our opinion that	we think that
at an early date	soon	it is our understanding that	we understand that
at the conclusion of	after, following		
at the present time	now	it is our recommendation that	we recommend that
at this point in time	now		
based on the fact that	because	make reference to	refer to
despite the fact that	although	of the opinion that	think that
due to the fact that	because	on a daily basis	daily
during the course of	during	on the grounds that	because
during the time that	during, while	prior to	before
have the capability to	can	relative to	regarding, about
in connection with	about, concerning	so as to	to
in order to	to	subsequent to	after
in regard to	regarding, about	take into consideration	consider
in the event that	if	until such time as	until
in view of the fact that	because		

FIGURE 6.2   Wordy Phrases and Their Concise Equivalents

One kind of wordiness to avoid is unnecessary redundancy, as in *end result, any and all, each and every, completely eliminate,* and *very unique.* Be content to say something once. Use "The liquid is green," not "The liquid is green in color."

> *Redundant*   We initially began our investigative analysis with a sample that was spherical in shape and heavy in weight.
>
> *Better*   We began our analysis with a heavy, spherical sample.

Some of the most commonly used wordy phrases and their concise equivalents are listed in Figure 6.2.

Avoid Clichés

The English writer George Orwell once offered some good advice about writing: "Never use a metaphor, simile or other figure of speech which you are used to seeing in print." Rather than writing, "It's a whole new ball game," write, "The situation has changed completely." Don't write, "I am sure the new manager can cut the mustard"; write, "I am sure the new manager can do his job effectively." If someone suggests that you "go for it," don't. Why would you want to sound like Rocky?

Sometimes, writers further embarrass themselves by getting their clichés wrong: expressions become so timeworn that users forget what the words mean. The phrase "a new bag of worms" has found its way into print; the producer Sam Goldwyn was famous for such statements as "A verbal contract isn't worth the paper it's written on." And the phrase "I could care less" often is used when the writer means just the opposite. The best solution to this problem is, of course, not to use clichés.

Following are a cliché-filled sentence and a translation into plain English.

Trite	Afraid that we were between a rock and a hard place, we decided to throw caution to the wind with a grandstand play that would catch our competition with its pants down.
Plain	Afraid that we were in a dilemma, we decided on a risky and aggressive move that would surprise our competition.

Avoid Pompous Words

Writers sometimes try to impress their readers by using pompous words, such as *initiate* for *begin, perform* for *do,* and *prioritize* for *rank.* When asked why they use big words where small ones will do, writers say they want to make sure their readers know they have a strong vocabulary, that they are well educated. In technical writing, plain talk is best. If you know what you're talking about, be direct and simple. Even if you're not so sure of what you're talking about, say it plainly; big words won't fool anyone for more than a few seconds.

Following are two pompous sentences translated into plain English.

Pompous	The purchase of a minicomputer will enhance our record maintenance capabilities.
Plain	Buying a minicomputer will help us maintain our records.
Pompous	It is the belief of the Accounting Department that the predicament was precipitated by a computational inaccuracy.
Plain	The Accounting Department thinks a math error caused the problem.

Some of the most commonly used fancy words and their plain equivalents are listed in Figure 6.3.

Several style programs isolate fancy words and expressions. Of course, you can use any word-processing program to search for terms you tend to overuse.

In the long run, your readers will be impressed with your clarity and accuracy. Don't waste your time thinking up fancy words.

Avoid Long Noun Strings

A noun string is a phrase consisting of a series of nouns (or nouns and adjectives and adverbs), all of which modify the last noun. For example, in the noun string "parking-garage regulations," all of the words modify *regulations.* Noun strings save time, and as long as your readers understand

Fancy Word	Plain Word	Fancy Word	Plain Word
advise	tell	impact (verb)	affect
ascertain	learn, find out	initiate	begin
attempt (verb)	try	manifest (verb)	show
commence	start, begin	parameters	variables, conditions
demonstrate	show	perform	do
employ (verb)	use	prioritize	rank
endeavor (verb)	try	procure	get, buy
eventuate (verb)	happen	quantify	measure
evidence (verb)	show	terminate	end, stop
finalize	end, settle, agree, finish	utilize	use
furnish	provide, give		

FIGURE 6.3   Fancy Words and Their Plain Word Equivalents

them, there is no problem. A "passive-restraint system" is easier to write than "a restraint system that is passive," and it won't confuse readers.

Hyphens can help you clarify some noun-string expressions by linking the two words that go together. For example, in the phrase *hot-water line*, the hyphen links *hot* and *water*; together they modify *line*. In other words, it is a line that carries hot water. For more information on hyphens, see Appendix A.

However, sometimes noun strings are so long or so complex that hyphens won't ensure clarity, as in the following examples:

preregistration procedure instruction sheet update
operator-initiated default-prevention technique
user-defined reporting expression screen prompts description

Even relatively short noun strings can be confusing. Richard Pompian (1990) writes that a group of advanced students of data processing thought "management information systems" meant "systems of information *about* management" rather than "systems of information *for* management."

An additional danger is that sometimes these noun strings just sound silly and pompous. If you are writing about a smoke detector, there is no reason to call it a "smoke detection device."

Use Positive Constructions

The term *positive construction* has nothing to do with cheerfulness or an optimistic outlook on life. Rather, it indicates that the writer is describing what something is, not what it is not. In the sentence "I was sad to see this project completed," "sad" is a positive construction. The negative construction would be "not happy."

Here are a few other examples of positive and negative constructions:

Positive Construction	Negative Construction
most	not all
few	not many
on time	not late, not delayed
positive	not negative
negative	not positive
inefficient	not efficient
reject	cannot accept

Why should you try to use positive rather than negative constructions? Because readers understand positive constructions more quickly and more easily than negative constructions. And when several negative constructions are used in the same sentence, the reader has to work much harder to untangle the meaning. Consider the following examples:

Difficult   Because the team was not notified of the deadline in time, it was not able to prepare a satisfactory report.

Simpler   Because the team was notified of the deadline late, it produced an unsatisfactory report.

Difficult   Without an adequate population of krill, the entire food chain of the Antarctic region would be unable to sustain itself.

Simpler   If the krill population were too low, the entire food chain of the Antarctic region would be destroyed.

Use Nonsexist Language

Sexist language favors one sex at the expense of the other. Although sexist language can shortchange males—as in some writing about female-dominated professions such as nursing—in most cases it victimizes the female. Common examples include nouns such as *workman* and *chairman* and pronouns as used in the sentence "Each worker is responsible for his work area."

In some organizations, the problem of sexist language is still considered trivial; in internal memos and reports, sexist language is used freely. In most organizations, however, sexist language is a serious matter. Unfortunately, it is not easy to eliminate all gender bias from writing.

A number of male-gender words have no standard nongender substitutes, and sometimes there is simply no graceful way to get around the pronoun *he*. Over the years different organizations have created synthetic pronouns—such as *thon, tey,* and *hir*—but these pronouns have never caught on. Some writers use *he/she* or *s/he,* although other writers consider these constructions awkward.

However, many organizations have formulated guidelines in an attempt to reduce sexist language.

The relatively simple first step is to eliminate the male-gender words. *Chairman*, for instance, is being replaced by *chairperson* or *chair*. *Firemen* are *firefighters*, *policemen* are *police officers*.

Rewording a sentence to eliminate masculine pronouns is also effective.

> *Sexist* The operator should make sure he logs in.
> *Nonsexist* The operator should make sure to log in.12

In this revision, an infinitive replaces the *he* clause.

> *Nonsexist* Operators should make sure they log in.

In this revision, the masculine pronoun is eliminated through a switch from singular to plural.

Notice that sometimes the plural can be unclear:

> *Unclear* Operators are responsible for their operating manuals.

Does each operator have one operating manual or more than one?

> *Clear* Each operator is responsible for his or her operating manual.

In this revision, "his or her" clarifies the meaning. *He or she* and *his or her* are awkward, especially if overused, but they are at least clear.

If you use a word processor, search for *he, man,* and *men*, words and parts of words most commonly associated with sexist writing. Some style programs search out the most common sexist terms and suggest nonsexist alternatives.

In some quarters, the problem of sexist language is sidestepped. The writer simply claims innocence: "The use of the pronoun *he* does not in any way suggest a male bias." Many readers find this kind of approach equivalent to that of a man who enters a crowded elevator, announces that he knows it is rude to smoke, and then proceeds to light up a cheap cigar. Sexism in language is a serious matter because, like other forms of discrimination, it assigns a value to a person based on an irrelevant factor—n this case, gender—instead of the relevant factor—performance.

For a full discussion of nonsexist writing, see *The Handbook of Nonsexist Writing* (Miller and Swift 1988).

EXERCISES

1. The following sentences might be too long for some readers. Break each sentence into two or more sentences.

   a. In the event we get the contract, we must be ready by June 1 with the necessary personnel and equipment to get the job done, so with this in mind a staff meeting, which all group managers are expected to attend, is scheduled for February 12.

    b. Once we get the results of the stress tests on the 125-Z fiber-glass mix, we will have a better idea where we stand in terms of our time constraints, because if it isn't suitable we will really have to hurry to find and test a replacement by the Phase 1 deadline.

    c. Although we had a frank discussion with Becker's legal staff, we were not able to get them to discuss specifics on what they would be looking for in an out-of-court settlement, but they gave us a strong impression that they would rather not take the matter to court.

2. The following examples contain choppy, abrupt sentences. Combine sentences to create a smoother prose style.

    a. I need a figure on the surrender value of a policy. The policy number is A4399827. Can you get me this figure by tomorrow?

    b. There are advantages to having your tax return prepared by a professional. There are also disadvantages. One of the advantages is that it saves you time. One of the disadvantages is that it costs you money.

    c. We didn't get the results we anticipated. The program obviously contains an error. Please ask Paul Davis to go through the program.

    d. The supervisor is responsible for processing the outgoing mail. The supervisor is also responsible for maintaining and operating the equipment.

3. In the following sentences, the real subjects are buried in prepositional phrases or obscured by expletives. Revise the sentences so that the real subjects appear prominently.

    a. There has been a decrease in the number of students enrolled in two of our training sessions.

    b. There is a need for the new personnel to learn how to use the computer.

    c. It is on the basis of recent research that I recommend the new CAD system.

    d. The completion of the new causeway will enable 40,000 cars to cross the ravine each day.

    e. The use of in-store demonstrations has resulted in a dramatic increase in business.

4. In the following sentences, unnecessary nominalization has obscured the real verb. Revise the sentences to focus on the real verb.

    a. Pollution constitutes a threat to the Wilson Wildlife Preserve.

    b. Evaluation of the gumming tendency of the four tire types will be accomplished by comparing the amount of rubber that can be scraped from the tires.

    c. Reduction of the size of the tear-gas generator has already been completed.

    d. The construction of each unit will be performed by three men.

    e. This manual enables the pilot to perform a landing under any flying conditions.

5. The following examples contain errors of parallelism. Revise the sentences to eliminate these errors.

   a. The next two sections of the manual discuss how to analyze the data, the conclusions that can be drawn from your analysis, and how to decide what further steps are needed before establishing a journal list.
   b. With our new product line, you would not only expand your tax practice, but your other accounting areas as well.
   c. Sections 1 and 2 will introduce the entire system, while Sections 3 and 4 describe the automatic application and step-by-step instructions.
   d. The evaluator should indicate the quality of the job done, the amount of time in which the job was completed, and how much the job cost.

6. The following sentences contain punctuation or pronoun errors related to the modifiers. Revise the sentences to eliminate the errors.

   a. The Greeting Record button records the greeting which is stored on a microchip inside the machine.
   b. This problem that has been traced to manufacturing delays, has resulted in our losing four major contracts.
   c. Please get in touch with Tom Harvey who is updating the instructions.
   d. The American Airlines flight, which landed yesterday at O'Hare, carried six members of the board of directors.

7. The following sentences contain misplaced modifiers. Revise the sentences to eliminate the errors.

   a. Over the past three years it has been estimated that an average of eight hours per week are spent on this problem.
   b. These employees have only been out of school for one year.
   c. Information provided by this program is displayed at the close of the business day on the information board.
   d. The computer provides a printout for the management team that shows the likely effects of the action.

8. The following sentences contain dangling modifiers. Revise the sentences to eliminate the errors.

   a. By following these instructions, your computer should provide good service for many years.
   b. To examine the chemical homogeneity of the plaque sample, one plaque was cut into nine sections.
   c. In an effort to corroborate the earlier findings, a microcellular formulation was tested.
   d. The boats in production could be modified in time for the February debut by choosing this method.

9. In the following sentences, the passive voice is used inappropriately. Rewrite the sentences to remove the inappropriate usages.

    a. Most of the information you need will be gathered as you
        document the history of the journals.
    b. After the stretcher has been squared by yourself, put a tack into
        each side of each joint.
    c. When choosing multiple programs to record, be sure that the
        proper tape speed has been chosen.
    d. During this time I also cowrote a manual on the Roadway
        Management System. Frequent trips were also made to the field.

10. The following sentences are vague. Revise them by replacing the vague
    elements with specific information. Make up any reasonable details.

    a. The results won't be available for a while.
    b. The fire in the lab caused extensive damage.
    c. A soil analysis of the land beneath the new stadium revealed an
        interesting fact.

11. The following sentences contain wordy phrases. Revise the sentences
    to make them more concise.

    a. The instruction manual for the new copier is lacking in clarity
        and completeness.
    b. The analysis should require a period of three months.
    c. The second forecasting indicator used is that of energy demand.
    d. The software packages enable the user to create display
        information with a minimum of effort.
    e. We remain in communication with our sales staff on a daily basis.

12. The following sentences contain clichés. Revise the sentences to
    eliminate the clichés.

    a. We hope the new program will positively impact all our
        branches.
    b. I would like to thank each and every one of you.
    c. With our backs to the wall, we decided to drop back and punt.
    d. If we are to survive this difficult period, we are going to have to
        keep our ears to the ground and our noses to the grindstone.
    e. DataRight will be especially useful for those personnel tasked
        with maintaining the new system.

13. The following sentences contain pompous words. Revise the sentences
    to eliminate the pomposity.

    a. This state-of-the-art beverage-procurement module is to be
        utilized by the personnel associated with the Marketing
        Department.
    b. It is indeed a not unsupportable inference that we have been
        unsuccessful in our attempt to forward the proposal to the
        proper agency in advance of the mandated date by which such
        proposals must be in receipt.
    c. Our aspiration is the expedition of the research timetable.

14. The following sentences contain long noun strings that the general reader might find awkward or difficult to understand. Rewrite the sentences to eliminate the long noun strings.

 a. The corporate relations enhancement committee meeting has been scheduled for next Thursday.
 b. The research team discovered a glycerin-initiated, alkylene-oxide-based, long-chain polyether.
 c. We are considering purchasing a 1M RAM, 40M hard disk, 286 microprocessor laptop.

15. In the following sentences, convert the negative constructions to positive constructions.

 a. Williams was accused by management of making predictions that were not accurate.
 b. We must make sure that all our representatives do not act unprofessionally to potential clients.
 c. The shipment will not be delayed if Quality Control does not disapprove any of the latest revisions.

16. The following sentences contain sexist language. Revise the sentences to eliminate the sexism.

 a. Each doctor is asked to make sure he follows the standard procedure for handling Medicare forms.
 b. Policemen are required to live in the city in which they work.
 c. Two of the university's distinguished professors—Prof. Harry Larson and Ms. Anita Sebastian—have been elected to the editorial board of Modern Chemistry.

REFERENCES

Battison, R., and D. Goswami. 1981. Clear writing today. *Journal of Business Communication* 18, no. 4: 5–16.

Huckin, T. N. 1983. A cognitive approach to readability. In *New essays in technical and scientific communication: Research, theory, and practice.* Ed. P. Anderson and J. Brockmann. Farmingdale, NY: Baywood.

Miller, C., and K. Swift. 1988. *The handbook of nonsexist writing.* 2d ed. New York: Harper & Row.

Pompian, R. 1990. When "bad technique" is better: Appropriate uses of controversial grammatical constructions. In *Proceedings of the 37th International Technical Communication Conference.* Washington, D.C.: Society for Technical Communication, WE-106–109.

Selzer, J. 1981. Readability is a four-letter word. *Journal of Business Communication* 18, no. 4: 23–34.

Writing Coherent Prose

The discussion of readability formulas in the previous chapter mentioned recent research showing that the process of understanding text is much more complicated than we had once thought. Beginning early in this century, readability formulas were devised on the premise that the two most important factors determining readability were the length of the words and of the sentences. We now believe that other factors are more important.

Because understanding occurs within a person, not within a document, it is a process involved with the reader's own knowledge, attitude, and reading behavior. In reading a document, a person extracts new information and tries to associate it with some given or previously known information.

This simple idea suggests ways to make your writing easy to understand. It is the basis of coherent writing, that is, writing that hangs together, on the level of sentences, paragraphs, longer passages, and even whole documents.

This chapter will discuss the given/new strategy, a basic approach to writing coherent prose, and then show how that principle applies to both the large and small units of prose.

Understanding the Given/New Strategy

In 1974, Susan Haviland and Herbert Clark presented their given/new strategy, which has since been recognized as a major breakthrough in understanding how people acquire new information. Their theory is simple and sensible.

Every sentence that we read or hear contains both given and new information. Given information is what the listener or reader is supposed to know already. New information, obviously, is what the person is not expected to know already. When we listen or read, we try to understand the information by searching through our memory for information that matches the given information in the sentence. Once we find a match, we can attach the new information to it, thereby changing our original information.

For example, the sentence "The proposal we submitted last month won the contract" consists of two pieces of information: (1) we submitted a proposal last month and (2) that proposal won the contract. If you write this sentence, readers will first try to understand the first part—"we submitted a proposal last month"—by searching for a memory of the company's having submitting that proposal. Then, they will add the second part—"that proposal won the contract"—to the first part. Thereafter, their memories will reflect the updated information.

On the other hand, if you wrote that same sentence, "The proposal we submitted last month won the contract," but your reader did not know that the company had submitted a proposal last month, communication would break down, at least momentarily, as the reader tried (unsuccessfully) to remember the proposal you are writing about. That reader could continue to try to remember, give up in frustration, or create a new meaning: "We must have submitted a proposal last month, and it apparently won the contract."

Haviland and Clark tested their theory. They wanted to see the effects of different ways of presenting given and new information. They hypothesized that if the link between the given and the new information were clear and explicit, readers would have an easier time understanding it than if the link were not so clear and explicit. They confirmed their hypothesis. Readers found links such as those in Set 1, below, easier to understand than those in Set 2.

Set 1   We got some beer out of the trunk. The beer was warm.
Set 2   We checked the picnic supplies. The beer was warm.

Again, these results are common sense. In both sets, the first sentence provides the given information that the readers rely on in reading the second sentence. Because the word "beer" is used in both sentences in Set 1, the reader easily adds the new information to the old. However, in Set 2, the writer has made things a little more difficult for the readers. When they get to the "beer" in the second sentence, they have to infer that the beer must have been part of the "picnic supplies" mentioned in the previous sentence. Obviously, this linkage — "beer = picnic supplies" — takes more effort than "beer = beer."

What does the given/new strategy have to do with technical writing? It proves a simple, fundamental idea: when you want to transmit information to people, you will be most successful if you deliver the new information gradually, building on what they already know. This idea applies to all units of discourse, from clauses and sentences up to whole documents. The rest of the chapter will show how this idea works.

Using Structuring Units Effectively

The given/new strategy, originally devised to describe sentence-level coherence, offers a useful approach to understanding how larger sections of prose communicate.

The following sections cover five important kinds of structuring units that help you establish and maintain coherence in most kinds of technical writing:

- □ titles
- □ headings
- □ lists
- □ introductions
- □ conclusions

Titles

Almost every kind of technical document begins with a title. Even a letter is likely to have a subject line, which functions as a title. A good title is critical because it gives your readers their first opportunity to understand what you will be writing about and what you want to accomplish in the document. In other words, a good title communicates your subject and purpose.

Most writers don't draft a title until they complete the document. The reason is simple: until they finish the document, they cannot be sure what it says.

Precision is the key to a good title. If you are writing a feasibility study on the subject of offering free cholesterol screening at your company, make sure the title contains the key terms: "cholesterol screening" and "feasibility." The following title would be effective:

> Offering Free Cholesterol Screening at Thrall Associates:
> A Feasibility Study

If your document is an internal report discussing company business, you might not need to identify the organization; the following would be clear:

> Offering Free Cholesterol Screeni ng:
> A Feasibility Study

Of course, you could present the purpose before the subject:

> A Feasibility Study of Offering Free Cholesterol Screening

Do not substitute general terms, such as "health screening" for "cholesterol screening." Keep in mind that key terms from your title might be used in various kinds of indexes; the more precise your terms, the more useful your readers will find the title.

Your readers should be able to translate your title into a clear and meaningful sentence. For instance, the title "A Feasibility Study of Offering Free Cholesterol Screening" could be translated to the following sentence: "This document reports on whether it is feasible to offer free cholesterol screening for our employees."

However, notice what happens when the title is incomplete: "Free Cholesterol Screening." All the reader knows is that the document has something to do with free cholesterol screening. But is the writer recommending that free cholesterol screening be instituted? Or is the writer reporting on how well an existing program is working out?

Following are more examples of effective titles.

> Choosing a Laptop: A Recommendation

> An Analysis of the Kelly 1013 Packager

Open Sea Pollution-Control Devices: A Summary

A Forecast of Smoking Habits in the United States
in the Coming Decade

Headings

A heading is a lower-level title inside a document. A clear and informative heading is vital in technical writing because it announces the subject of the discussion that follows it. This announcement helps your readers understand what they are reading or, in some cases, helps them decide they don't want to bother. For the writer, a heading eliminates the need to announce the subject in a sentence such as "Let us now turn to the advantages of the mandatory enrollment process."

When you draft your headings, try to make them as informative as you can. Avoid the long noun strings, such as "Production Enhancement Proposal Analysis Techniques." (See Chapter 6 for a discussion of noun strings.) Instead, write what Edmond Weiss (1982) calls headlines: phrases that clarify the subject and theme of the discussion that follows. For instance, instead of writing "Production Enhancement Proposal Analysis Techniques," write something like one of the following:

Techniques for Analyzing the Proposal to Enhance Production

This title says, more clearly than the original, that the writer is going to describe the techniques.

The writer can improve the title even more by adding more information: some indication of how many techniques will be described:

Three Techniques for Analyzing the Proposal to Enhance Production

And the writer can go one step further, by indicating what he or she wishes to say about the three techniques:

Advantages and Disadvantages of the Three Techniques for
Analyzing the Proposal to Enhance Production

Don't worry if the heading seems a little long; clarity is more important than brevity.

What Are the Three Techniques for Analyzing the Proposal to
Enhance Production?

The question form is particularly appropriate when addressing less knowledgeable readers.

How to Analyze the Proposal to Enhance Production

The "how to" form is useful in instructional material, such as manuals.

Analyzing the Proposal to Enhance Production

The verbal form, with the *-ing* ending, is an effective way to suggest a process.

For information on how to format the headings, see Chapter 12, Document Design.

Lists

Many sentences in technical writing are long and complicated:

> We recommend that more work on heat-exchanger performance be done with a larger variety of different fuels at the same temperature, with similar fuels at different temperatures, and with special fuels such as diesel fuel and shale-oil-derived fuels.

Here readers cannot concentrate on the information because they must worry about remembering all the *with* phrases following *done*. If they could "see" how many phrases they had to remember, their job would be easier.

Revised as a list, the sentence is easier to follow:

> We recommend that more work on heat-exchanger performance be done:
>
> 1. with a larger variety of different fuels at the same temperature
> 2. with similar fuels at different temperatures
> 3. with special fuels such as diesel fuels and shale-oil-derived fuels

In this version, the placement of the words on the page reinforces the meaning. The readers can easily see that the sentence contains three items in a series. And the fact that each item begins at the same left margin helps, too.

If you don't have enough space to list the items vertically, or if you are not permitted to use a vertical list, arrange the items horizontally:

> We recommend that more work on heat-exchanger performance be done (1) with a larger variety of different fuels at the same temperature, (2) with similar fuels at different temperatures, and (3) with special fuels such as diesel fuels and shale-oil-derived fuels.

Make sure the items in the list are presented in a parallel structure. (See Chapter 5 for a discussion of parallelism.)

Nonparallel

Here is the sequence we plan to follow:

1. construction of the preliminary proposal
2. do library research
3. interview with the Bemco vice president
4. first draft
5. revision of the first draft
6. after we get your approval, typing of the final draft

Parallel

Here is the sequence we plan to follow:

1. write the preliminary proposal
2. do library research
3. interview the Bemco vice president
4. write the first draft
5. revise the first draft
6. type the final draft, after we receive your approval

In this example, the original version of the list is sloppy, a mixture of noun phrases (items 1, 3, 4, and 5), a verb phrase (item 2), and a participial phrase preceded by a dependent clause (item 6). The revision uses parallel verb phrases and deemphasizes the dependent clause in item 6 by placing it after the verb phrase.

Note that technical writing does not have to look "formal," with traditional sentences and paragraphs covering the whole page. Lists make your writing easier to read and to understand.

Introductions

Most documents begin with an introduction, a brief section intended to help readers understand what you are going to say, how you are going to say it, and why you are going to say it. To use the terminology from the given/new strategy, the introduction functions as the first piece of given information. Your readers will rely on it to understand the body of your document, the new information. If your readers finish reading your introduction but still do not know exactly what you are going to be writing about, and why it is important, they might lose interest or become confused or frustrated.

Even though no two introductions are the same, most answer a number of similar questions. As you read the explanations of these questions, keep several points in mind:

□ Few documents call for all these questions. Sometimes the questions are irrelevant or obvious and therefore do not need to be addressed.
□ There is no single "correct" sequence for addressing the questions.

☐ Sometimes it is convenient to combine the answers to several of the questions.

With those qualifications in mind, here is a list of the major questions addressed in introductions. A sample of a full introduction will follow the list.

■ *What is the subject of the document?* This question would seem to be the most basic. A reader who doesn't understand the answer to this one won't get much out of the document. It is a good idea to announce the subject explicitly even if you think your reader already knows it from earlier documents or discussions with you, and even though you think the title clearly indicates the subject. The following passage, from the introduction to a long technical report, clearly indicates its subject:

> The scientific community seems unable to prove that global warming is actually occurring. Why is there no consensus? What information is missing, and when will that information become available?

In introducing the subject of the document, you can simply come right out and announce it, using the word *subject*:

> The subject of this report is the scientific community's inability to prove that global warming is actually occurring. Why is there no consensus? What information is missing, and when will the information become available?

Sometimes the subject of the document is a problem that the writers have attempted to solve. In such a case, it makes sense to use the word *problem*:

> This report concerns a problem in the fuel-transmission system that has accounted for a 12 percent decrease in power in the 2300 series motors.

■ *What is the purpose of the document?* This question is often answered along with the question about the subject of the document. Readers want to know, early on, what you are trying to do in the document, what effect you wish to have on them. For more information about purpose in a document, see the discussion in Chapter 3. Following is a statement of purpose from an introduction to a feasibility report:

> This report has two main purposes: (1) to summarize the recent literature on the status of solar-powered vehicles and (2) to recommend the most fruitful directions for research over the next decade.

Notice that this statement of purpose also indicates clearly the subject of the document.

■ *What is the background of the subject?* The background consists of the information the readers will need to know to understand the discussion of the subject itself. That information usually consists of history, theory, or both. For instance, if you are trying to explain the significance of the fall of the Berlin Wall, the background would include the partitioning of Germany by the victorious forces after World War II and the history of the Cold War. If you are trying to explain to a child how a bird flies, the background would include some basic physics. If you are trying to explain why many people are concerned about the safety of nuclear energy, the background would include some history—Three Mile Island and Chernobyl—and a theoretical discussion of how nuclear accidents can occur.

Sometimes you might separate the background from the rest of the introduction. This separation is appropriate if the background is particularly long or if some of your readers are already familiar with it.

Following is the background section from an introduction to a memo:

> This leveling off of compact disc sales—only 28 percent increase over last year—comes as something of a disappointment, given the sales patterns over the last few years:
>
> | | |
> |---|---|
> | 1990 | 43.6 million units |
> | 1989 | 31.4 million units |
> | 1988 | 19.7 million units |
>
> We appear to be entering the mature sales phase of this product, the same pattern experienced by cassettes in their first few years.

■ *What is the scope of the document?* The scope of a document is determined by its boundaries. In other words, the scope indicates what is included in the document, and what is excluded. Following is an excerpt from a scope statement from the introduction of a manual:

> Part IV of the manual introduces the interuniversity E-mail network, including how to send mail to colleagues on the BITNET network and how to access colleagues in CSNET, ARPNET, and USENET. However, Part IV does <u>not</u> describe the basics of E-mail. For information on how to use E-mail on the IBM 3090 or the Prime, see Part II.

As this example indicates, when you write a scope statement, be clear about what you will not discuss.

■ *What is the organization of the document?* Closely related to scope is organization. If the scope statement answers the question "What is included?" the organization statement answers the question "What comes first, and why?" You want to explain your organizational plan because you want your readers to concentrate on your discussion; if they are wondering why you are discussing a particular subject at this point, they won't be paying full attention. Following is a statement of organization from the first paragraph of a letter:

> The first section of this letter describes the background of the relevant legislation, HR3008, because it is necessary to understand how the bill took shape in order to appreciate how the committee reached its decision. Then, the letter provides a detailed discussion of the majority viewpoint and, finally, the minority viewpoint.

■ *What are the key terms that will be used in the document?* If the document will use one or several key terms often, the introduction is a logical location for definitions. See Chapter 8 for a fuller discussion of where to place such definitions. The following definition is taken from the introduction to an employee's operations manual:

> At Winston, Inc., "flextime" refers to the policy that lets employees choose, within specified guidelines, the time at which they will begin and end work on normal days. The basic purposes of flextime are to give employees greater freedom to accommodate their other commitments, such as child-care responsibilities, as well as to increase the number of hours in the day during which we can be reached by our suppliers and customers in different time zones. For the specific guidelines on flextime, see Appendix C, page 119.

The following introduction to a research report was written by a materials engineer working for a company that uses large turbines in its manufacturing process. The report is titled "Nondestructive Testing Methods for Evaluating Plasma-coated Turbines: A Feasibility Report."

> The turbines in our Large Scale Manufacturing Division must be protected against corrosion. Every six months the exposed areas of the turbine are coated with plasma sprays and, after a visual inspection, are put back in service.
>
> The problem is that, because the turbine surface is so irregular, the plasma spray can leave the surface with a too-thin layer or even miss spots entirely and leave the surface vulnerable to corrosion. If corrosion occurs, the turbine has to be shut down, costing up to $12,000 in lost production and repairs. In 1990, we suffered seven forced shutdowns from corrosion.
>
> The purpose of this report is to present the findings of a research project to determine whether any of the currently available methods of nondestructive testing would be effective in helping us determine the quality of the plasma coating

before the turbines are put back into service. If such a method were feasible, we could dramatically reduce the number of corrosion-caused turbine shutdowns.

For this investigation, we studied two general classes of nondestructive testing methods: thermal and acoustic. For each class of method, we considered the technical effectiveness, ease of application, and cost.

The body of the report presents separate discussions of the two general classes of nondestructive testing methods. Detailed comparisons of the classes, according to criteria, are presented in the appendixes.

Conclusions

The word *conclusion* has two related but different meanings in writing. The first meaning is the one you are probably more familiar with from your earlier courses in writing: a conclusion is the last part of a document, the portion that summarizes the major points. The other meaning of conclusion is the inference or inferences that the writers draw from the results of an investigation. For example, the conclusion of your study of long-distance phone companies might be that MCI is the best one for your needs. Obviously, these two types of conclusions often overlap: the ending of a document is a logical place to put the inferences.

The discussion here considers the first kind of conclusion: how to end a document. For a discussion of the second kind of conclusion, see Chapter 19, Formal Elements of Reports.

The first point to understand is that sometimes technical documents don't conclude; they just stop. For example, the typical reference manual doesn't have a conclusion section. Nor does a parts catalog. If you can't think of how to conclude a technical document, ask yourself whether you ought to just stop writing.

In most cases, however, technical documents do have conclusions. The following section describes some of the typical questions addressed in conclusions.

■ *What are the main points established in the document?* After the reader has finished the body of the discussion, especially if the body is more than a few pages, you might want to summarize your important points, in case your readers have forgotten some of them. Here is an example of a conclusion, from the report on nondestructive testing methods for plasma coatings of turbines, that summarizes the key points.

Our analysis yielded the following main points. Thermal methods, including infrared camera and liquid crystals, would not be effective because, although they are very accurate, their responses are not unique to a particular material, thereby rendering their data useless. Ultrasonics appears to be the most promising area, with the point-contact mode preferable to the immersion mode because it is more convenient and less expensive.

- *What should be done next?* Even though a technical document ends, the project that it concerns might well continue. One way to conclude such a document is to offer suggestions—called recommendations— about what ought to be done next. For a further discussion of recom- mendations, see Chapter 19, Elements of Formal Reports.

  Here is a conclusion based on recommendations, from a journal article about an experimental study of the effects of paragraph length on comprehension.

  > The fact that the experimental data do not support the hypothesis—that shorter paragraphs are easier to understand— could be explained by the test methodology. Written free recall, while accurately reflecting the comprehension of the better writers, might not accurately reflect the comprehension of the less-capable writers. Therefore, the current experiment should be duplicated using recognition rather than free recall to determine whether the data-gathering methodology was a confounding variable.

- *How can you find out more information?* One common way to conclude technical documents is to help the readers determine how to get more information.

  Sometimes this is part of a sales strategy, as when a set of operat- ing instructions concludes with a statement such as the following: "We are confident that you will enjoy many years of fine service from your new television set. If you would like to learn more about our complete line of home electronics, please call us, toll free, at (800) 567-8453 between 9 A.M. and 5 P.M. Eastern time."

  Sometimes the conclusion is not a sales message. The following statement concludes a memo: "If you would like to be placed on the mailing list for future communications on the EMI Project, please place a check mark in the box below and return this copy to my office."

- *How can we help you in the future?* Many kinds of technical documents conclude with a polite offer to provide future services. A transmittal letter that accompanies a report to management might conclude, "If I can be of any further assistance with this or any other project, please do not hesitate to contact me." The following conclusion comes from a letter sent by a subcontractor to a client at the completion of a job.

  > Here at Huppman Electronics we pride ourselves on our 35 years of offering the finest professional services to the

electronics industry. We thank you for the opportunity to have served you, and hope that your trust in us has been fulfilled. If the need for our services arises again, we hope you will think of us.

Writing Coherent Paragraphs

There are two kinds of paragraphs: transitional paragraphs and body paragraphs.

Transitional paragraphs help the reader get from one major point to another. In most cases, they summarize the previous point and help the reader understand how it relates to the next one.

The following example of a transitional paragraph is taken from a manual explaining how to write television scripts. The writer has described six principles of writing for an episodic program, including introducing characters, pursuing the plot, and resolving the action at the end of the episode.

> The six basic principles of writing for episodic television, then, are the following:
>
> 1. reintroduce the characters
> 2. make the extra characters episode-specific
> 3. present that week's plot swiftly
> 4. make the characters react according to their personalities
> 5. resolve the plot neatly
> 6. provide a denouement that hints at further developments
>
> But how do you put these six principles into action? The following section provides specific how-to instructions.

Notice that the first sentence contains the word *then* to signal that it is introducing a summary. Note, too, that the final sentence of the transitional paragraph clearly indicates the relationships between what precedes it and what follows it.

Body paragraphs are the other kind. A body paragraph is the basic structural unit for communicating the technical information. A body paragraph could be defined as a group of sentences (or sometimes a single sentence) that is complete and self-sufficient but that also contributes to a larger discussion. The challenge of creating an effective paragraph in technical writing is to make sure, first, that all the sentences clearly and directly substantiate one main point, and, second, that the whole paragraph follows logically from the material that precedes it. Readers tend to pause between paragraphs (not between sentences) to digest the information given in one paragraph and link it with that given in the previous paragraphs. For this reason, the paragraph is the key unit of composition. Readers might forgive or at least overlook a slightly fuzzy sentence. But if they can't figure out what a paragraph says or why it appears where it does, communication is likely to break down.

Structure Paragraphs Clearly

Too often in technical writing, paragraphs seem to be written for the writer, not the reader. They start off with a number of details: about who worked on the project before and what equipment or procedure they used; about the ups and down of the project, the successes and setbacks; about specifications, dimensions, and computations. The paragraph winds its way down the page until, finally, the writer concludes: "No problems were found."

This structure—moving from the particular details to the general statement—accurately reflects the way the writer carried out the activity being described, but it makes the paragraph difficult to follow. As you put a paragraph together, focus on your readers' needs. Do they want to "experience" your writing, to regret your disappointments and celebrate your successes? Probably not. They just want to find out what you have to say.

The Topic Sentence

Help your readers. Put the point—the topic sentence—up front. Technical writing should be clear and easy to read, not full of suspense. If a paragraph describes a test you performed on a piece of equipment, include the result in your first sentence: "The point-to-point continuity test on Cabinet 3 revealed no problems." Then go on to explain the details. If the paragraph describes a complicated idea, start with an overview: "Mitosis occurs in five stages: (1) interphase, (2) prophase, (3) metaphase, (4) anaphase, and (5) telophase." Then describe each phase. In other words, put the "bottom line" on top.

Notice, for instance, how difficult the following paragraph is because the writer structured the discussion in the same order she performed her calculations:

> Our estimates are based on our generating power during eight months of the year and purchasing it the other four. Based on the 1990 purchased power rate of $0.034/KW (January through April cost data) inflating at 4 percent annually, and a constant coal cost of $45–$50, the projected 1992 savings resulting from a conversion to coal would be $225,000.

Putting the bottom line on top makes the paragraph much easier to read. Notice how the writer adds a numbered list after the topic sentence.

The projected 1992 savings resulting from a conversion to coal are $225,000. This estimate is based on three assumptions: (1) that we will be generating power during eight months of the year and purchasing it the other four, (2) that power rates inflate at 4 percent from the 1990 figure of $0.034/KW (January through April cost data), and (3) that coal costs remain constant at $45–$50.

The topic sentence in technical writing functions just as it does in any other kind of writing: it summarizes or forecasts the main point of the paragraph.

Why don't writers automatically begin with the topic sentence if that helps the readers? One reason is that it is easier to present the events in their natural order, without first having to decide what the single most important point is. Perhaps a more common reason is that writers feel uncomfortable "exposing" the topic sentence at the start of a paragraph. Most people who do technical writing were trained in science, engineering, or technology; they were taught that they must not reach a conclusion before obtaining sufficient evidence to back it up. Putting the topic sentence up top *looks* like stating a conclusion without "proving" it, even though the proof follows the topic sentence directly. Beginning the paragraph with the sentence "The committee concluded that human error caused the overflow" somehow seems risky. It isn't. The real risk is that you might frustrate or bore your readers by making them hunt for the topic sentence.

The Support

After the topic sentence comes the support. The purpose of the support is to make the topic sentence clear and convincing. Sometimes a few explanatory details can provide all the support needed. In the paragraph about estimated fuel savings presented earlier, for example, the writer simply fills in the assumptions used in making the calculation: the current energy rates, the inflation rate, and so forth. Sometimes, however, the support must carry a heavier load: it has to clarify a difficult thought or defend a controversial one.

Because every paragraph is unique, it is impossible to define the exact function of the support. In general, however, the support fulfills one of the following roles:

- [ ] to define a key term or idea included in the topic sentence
- [ ] to provide examples of illustrations of the situation described in the topic sentence
- [ ] to identify causes, factors that led to the situation
- [ ] to define effects, implications of the situation
- [ ] to defend the assertion made in the topic sentence

The techniques used in developing the support include those used in most nonfiction writing, including definition, comparison and contrast, classification and partition, and causal analysis. These techniques are described further in Part Two.

Keep Paragraphs to a Manageable Length

How long should a paragraph of technical writing be? In general, 75 to 125 words will provide enough length for a topic sentence and four or five supporting sentences. Long paragraphs are more difficult to read than short paragraphs simply because readers have to concentrate longer. Long, unbroken stretches of type can so intimidate some readers that they actually will skip them.

Don't let an arbitrary minimum guideline about length take precedence over your analysis of the audience and purpose. Often you will need to write very brief paragraphs. You might need only one or two sentences—to introduce a graphic, for example. A transitional paragraph also is likely to be quite short. If a brief paragraph fulfills its function, let it be. Do not combine two ideas in one paragraph in order to achieve a minimum word count.

While it is confusing to include more than one basic idea in a paragraph, often you will find it necessary to divide one idea into two or more paragraphs. A complex idea that would require 200 or 300 words probably should not be squeezed into one paragraph.

The following example shows how a writer addressing a general audience divided one long paragraph into two:

> High-tech companies have been moving their operations to the suburbs for two main reasons: cheaper, more modern space and a better labor pool. A new office complex in the suburbs will charge anywhere from half to two-thirds of the rent charged for the same square footage in the city. And that money goes a lot further, too. The new office complexes are bright and airy, with picture windows looking out on lush landscaping. New office space is already wired for computers; and exercise clubs, shopping centers, and even libraries are often on-site.

> The second major factor attracting high-tech companies to the suburbs is the availability of experienced labor. Office workers and middle managers are abundant; many suburbanites, especially women returning to the labor force after their children start school, are highly trained and willing to make the short trip to the office complex. In addition, the engineers and executives, who tend to live in the suburbs anyway, are happy to forgo the commuting, the city wage taxes, and the noise and stress of city life.

A strict approach to paragraphing would have required one paragraph, not two, because all the information presented supports the topic sentence that opens the first paragraph. Many readers, in fact, could easily understand a one-paragraph version. However, the writer found a logical place to create a second paragraph and thereby communicated better.

Another writer might have approached the problem differently, making each "reason for moving to the suburbs" a separate paragraph.

High-tech companies have been moving their operations to the suburbs for two main reasons: cheaper, more modern space and a better labor pool.

First, office space is a bargain in the suburbs. A new office complex will charge anywhere from half to two-thirds of the rent charged for the same square footage in the city. And that money goes a lot further, too. The new office complexes are bright and airy, with picture windows looking out on lush landscaping. New office space is already wired for computers; and exercise clubs, shopping centers, and even libraries are often on-site.

Second, experienced labor is plentiful. Office workers and middle managers are abundant; many suburbanites, especially women returning to the labor force after their children start school, are highly trained and willing to make the short trip to the office complex. In addition, the engineers and executives, who tend to live in the suburbs anyway, are happy to forgo the commuting, the city wage taxes, and the noise and stress of city life.

The original topic sentence becomes a transitional paragraph that leads clearly and logically into the two explanatory paragraphs.

Use Coherence Devices Within and Between Paragraphs

After you have blocked out the main structure of the paragraph—the topic sentence and the support—make sure the paragraph is coherent. In a coherent paragraph, thoughts are linked together logically and clearly. Parallel ideas are expressed in parallel grammatical constructions. Even if the paragraph already moves smoothly from sentence to sentence, emphasize the coherence by adding transitional words and phrases, repeating key words, and using demonstratives.

Maintaining coherence *between* paragraphs is the same process as maintaining coherence *within* paragraphs: place the transitional device as close as possible to the beginning of the second element. For example, the link between two sentences within a paragraph should be near the start of the second sentence:

The new embossing machine was found to be defective. However, the warranty on the machine will cover replacement costs.

The link between the two paragraphs should be near the start of the second paragraph:

The complete system would be too expensive for us to purchase

now . . . .

. . . .

> <u>In addition</u>, a more advanced system is expected on the market within six months . . . .
>
> . . . .
>
> . . . .

Transitional Words and Phrases

Transitional words and phrases help the reader understand a discussion by pointing out the direction the thoughts are following. Here is a list of the most common logical relationships between two thoughts and some of the common transitions that express those relationships:

Relationship	*Transitions*
addition	also, and, finally, first (second, etc.), furthermore, in addition, likewise, moreover, similarly
comparison	in the same way, likewise, similarly
contrast	although, but, however, in contrast, nevertheless, on the other hand, yet
illustration	for example, for instance, in other words, to illustrate
cause-effect	as a result, because, consequently, hence, so, therefore, thus
time or space	above, around, earlier, later, next, to the right (left, west, etc.), soon, then
summary or conclusion	at last, finally, in conclusion, to conclude, to summarize

In the following examples, the first versions contain no transitional words and phrases. Notice how much clearer the second versions are.

Weak	Neurons are not the only kind of cell in the brain. Blood cells supply oxygen and nutrients.
Improved	Neurons are not the only kind of cell in the brain. *For example,* blood cells supply oxygen and nutrients.
Weak	The project was originally expected to cost $300,000. The final cost was $450,000.
Improved	The project was originally expected to cost $300,000. *However,* the final cost was $450,000.
Weak	The manatee population of Florida has been stricken by an unknown disease. Marine biologists from across the nation have come to Florida to assist in manatee-disease research.
Improved	The manatee population of Florida has been stricken by an unknown disease. *As a result,* marine biologists from across the nation have come to Florida to assist in manatee-disease research.

Key Words

Repetition of key words—generally, nouns—helps the reader follow the discussion. Notice in the following example how the first version can be confusing.

> *Unclear*   For months the project leaders carefully planned their research. The cost of the work was estimated to be over $200,000.

What is the work, the planning or the research?

> *Clear*   For months the project leaders carefully planned their *research*. The cost of the *research* was estimated to be over $200,000.

Out of a misguided desire to be "interesting," some writers keep changing their important terms. *Plankton* becomes *miniature seaweed*, then *the ocean's fast food*. Leave this kind of word game to TV sportscasters; technical writing must be clear.

Demonstratives

In addition to transitional words and phrases and repetition of key phrases, demonstratives—*this, that, these,* and *those*—can help the writer maintain the coherence of a discussion by linking ideas securely. Demonstratives should in almost all cases serve as adjectives rather than as pronouns. In the following examples, notice that a demonstrative pronoun by itself can be confusing.

> *Unclear*   New screening techniques are being developed to combat viral infections. These are the subject of a new research effort in California.

What is being studied in California, new screening techniques or viral infections?

> *Clear*   New screening techniques are being developed to combat viral infections. *These techniques* are the subject of a new research effort in California.
>
> *Unclear*   The task force could not complete its study of the mine accident. This was the subject of a scathing editorial in the union newsletter.

What was the subject of the editorial, the mine accident or the task force's inability to complete its study of the accident?

> *Clear*   The task force failed to complete its study of the mine accident. *This failure* was the subject of a scathing editorial in the union newsletter.

Even when the context is clear, a demonstrative pronoun used without a noun refers the reader to an earlier idea and therefore interrupts the reader's progress.

> *Interruptive*   The law firm advised that the company initiate proceedings. This resulted in the company's search for a second legal opinion.
>
> *Fluid*   The law firm advised that the company initiate proceedings. *This advice* resulted in the company's search for a second legal opinion.

Transitional words and phrases, repetition of key words, and demonstratives cannot *give* your writing coherence: they can only help the reader to appreciate the coherence that already exists. Your job is, first, to make sure your writing is coherent and, second, to highlight that coherence.

Turning a "Writer's Paragraph" into a "Reader's Paragraph"

The best way to demonstrate what has been said in this discussion is to take a weak paragraph and improve it. The following paragraph is from a status report written by a branch manager of the utility company mentioned in Chapter 1. The paragraph explains how the writer decided on a method to increase the company's business within his particular branch.

There were two principal alternatives considered for improving Montana Branch. The first alternative was to drill and equip additional sources of supply with sufficient capacity to provide for the present and projected system deficiencies. The second alternative was to provide for said deficiencies through a combination of additional sources of supply and a storage facility. Unfortunately, ground-water studies which were conducted in the Southeast Montana area by the consulting firm of Smith and Jones indicated that although ground water is available within this general area of our system, it is limited as to quantity, and considerable separation between well sources is necessary in order to avoid interference between said sources. This being the case, it becomes necessary to utilize the sources that are available or that can be developed in the most efficient manner, which means operating them in conjunction with a storage facility. In this way, the sources only have to be capable of providing for the average demand on a maximum day, and the storage facility can be utilized to provide for the peaking requirements plus fire protection. Consequently, the second alternative as mentioned herein above was determined to be the more desirable alternative.

First, let's be fair. The paragraph has been taken out of its context—a 17-page report—and was never meant to stand alone on a page. Also, it was not written for the general reader, but for an executive of the water company—someone who, in this case, is technically knowledgeable in the writer's field. Still, an outsider's analysis of an essentially private communication can at least isolate the weaknesses.

The most important element of a paragraph is the topic sentence, so you look for it in the usual place: the beginning. At first glance, the topic sentence looks good. The statement that two principal alternatives were considered appears to introduce the rest of the paragraph. You assume that the paragraph will define the two alternatives.

The more you think about the topic sentence, however, the less good it looks. The problem is suggested by the word *principal*. Most complex decisions eventually come down to a choice between the two best alternatives; why bother labeling the two best alternatives "principal"? The writer almost sounds as if he is congratulating himself for weeding out the undesirable alternatives. (Further, he begins with the expletive "there are" and uses a passive construction that weakens the sentence.)

Structured as it is, the paragraph focuses on the process of choosing an alternative rather than on the choice itself. Perhaps the writer thinks his readers are curious about how he does his job; perhaps he wants to suggest that he's done his best, so that if it turns out that he made the wrong choice, at least he cannot be criticized for negligence; or perhaps he is unintentionally recreating the scientific method by describing how he examined the two alternatives and finally came to a decision.

For whatever reasons, the writer has built a lot of suspense into the paragraph. It proceeds like a horse race, with first one alternative gaining, then the other. Consequently, the reader ends up worrying about the outcome and can't concentrate on the specific reasons that one choice prevailed over the other. Quite likely, the only thing you will remember about the paragraph is the final emphatic statement: that the writer decided to go ahead with the second alternative. And given the complexity of the paragraph, it would be easy to forget what that alternative is.

Without too much trouble the paragraph could be made much easier to understand. A careful topic sentence—one that defines the choice of a technical solution to the problem—would be a constructive start. Then the writer could elaborate on the necessary details of the solution. Only after he has finished his discussion of the solution should he define and explain the alternative that was rejected. Using this structure, the writer would clearly emphasize what he *did* decide to do, putting what he decided *not* to do in a subordinate position.

Here is the writer's paragraph translated into a reader's paragraph:

> We found that the best way to improve the Montana Branch would be to add a storage facility to our existing supply sources. Currently, we can handle the average demand on a maximum day; the storage facility will enable us to meet peaking requirements and fire-protection needs. In conducting our investigation, we considered developing new supply sources with sufficient capacity to meet current and future needs. This alternative was rejected, however, when our consultants (Smith and Jones) did ground-water studies that revealed that insufficient ground water is available and that the new wells would have to be located too far apart if they were not to interfere with each other.

One clear advantage of the revision is that it is about half as long as the original. The structure of the first version is largely responsible for its length: the writer announced that two alternatives were considered; defined them; explained why the first was impractical and why it led to the second alternative; and finally, announced the choice of the second alternative. A direct structure eliminates all this zigzagging and clarifies the paragraph.

The only possible objection to the streamlined version is that it is *too* clear, that it leaves the writer vulnerable in case his decision turns out to have been wrong. But if the decision doesn't work out, the writer will be responsible no matter how he described it, and he will end up in more trouble if a supervisor has to investigate who made the decision. Good writing is the best bet under any circumstances.

EXERCISES

1. For each of the following titles, write a brief evaluation. How clearly does the title indicate the subject and purpose of the document? On the basis of your analysis, rewrite each title.

   a. Recommended Forecasting Techniques for Haldane Company
   b. Robotics in Japanese Manufacturing
   c. A Study of Disc Cameras
   d. Agriculture in the West: A Ten-Year View
   e. Synfuels—Fact or Hoax?

2. For each of the following headings, write a brief evaluation. How clearly does the heading indicate the subject of the text that will follow it? On the basis of your analysis, rewrite each title to make it more clear and informative. Invent any necessary details.

   a. Multigroup Processing Technique Review Board Report Findings
   b. The Great Depression of 1929
   c. Low-level Radiation and Animals
   d. Minimize Down Time
   e. Intensive Care Nursing

3. The information contained in the following sentences could be conveyed better in a list. Rewrite each sentence in the form of a list.

   a.
   The freezer system used now is inefficient in several ways: the chef cannot buy in bulk or take advantage of special sales, there is a high rate of spoilage because the temperature is not uniform, and the staff wastes time buying provisions every day.

   b.
   The causes of burnout can be studied from three areas: physiological—the roles of sleep, diet, and physical fatigue; psychological—the roles of guilt, fear, jealousy, and frustration; environmental—the roles of the physical surroundings at home and at work.

   c.
   There are many problems with the on-line registration system currently used at Dickerson. First, lists of closed sections cannot be updated as often as necessary. Second, students who want to

register in a closed section must be assigned to a special terminal. Third, the computer staff is not trained to handle the student problems. Fourth, the Computer Center's own terminals cannot be used on the system; therefore, the university has to rent 15 extra terminals to handle registration.

4. The following introduction is from a report entitled, "Autofocusing Lens System for the Visually Impaired: A Prototype Design." In a paragraph, evaluate the effectiveness of this introduction.

Introduction

According to the National Society to Prevent Blindness, over eleven million people are visually impaired. Of these cases of visual impairment, the low vision cases have a portion of their retina damaged. However, the rest of their retina is intact and usable. This problem can be corrected by using a telescope for enlarging a part of their central field of vision over the entire retina. As a result, the effect of the damaged retina is minimized and better resolution is obtained for the patient.

Various manually adjustable bioptic telescopes have been available on spectacles. However, the telescopes have to be constantly focused depending on the distance between the object and the telescope. To most older people with arthritis, manual focusing is a serious handicap. Thus the object of this project is to design an autofocusing system for the low-vision case of the visually impaired. The focusing system will be completely automatic as in the autofocusing cameras on the market.

This report describes my project to design and build an autofocusing lens system for the visually impaired. A simple explanation of the autofocusing system is as follows. A sensor array receives the same image that is seen by the telescope in front of the eye. When the image is not in focus, the array system detects this and produces various output signals. These signals are digitally processed or interpreted to obtain an error signal depending on the focus error. This error signal is used to drive a motor to focus the telescopes.

5. The following conclusion appears in a description of the preregistration process at a university. The description, written by students, is addressed to freshmen. In a paragraph, evaluate the effectiveness of the conclusion.

Conclusion

To summarize, the preregistration involves determining which courses you must take and which you would like to take, making up a schedule from the Course Selection Booklet, and filling out the Preregistration Form and sending it back to the Registrar's Office.

Two other points should be mentioned. First, don't be afraid to talk to teachers about their courses. They will be happy to give you a lot of good information. Second, make sure you get the forms in on time—there's a late fee of $20.

6. The following paragraph concludes a letter sent by an automobile dealer to his customers approximately one month after they receive delivery of a car. In a paragraph, evaluate the effectiveness of this conclusion.

Again, Mr. Horton, we offer our congratulations on your recent purchase of the Zephyr. By now, we are sure that you are as big a fan of the Zephyr as we here at Midtown Zephyr are. If any of your friends and neighbors admires your car and asks you where you purchased it, we would appreciate it if you told them, ''I got it at Midtown Zephyr. Why don't you go see them today?''

7. Provide a topic sentence for each of the following paragraphs.

a. _____

_____

_____

All service centers that provide gas and electric services in the tri-county area must register with the TUC, which is empowered to carry out unannounced inspections periodically. Additionally, all service centers must adhere to the TUC's Fair Deal Regulations, a set of standards that encompasses every phase of service-center operations. The Fair Deal Regulations are meant to guarantee that all centers adhere to the same standards of prompt, courteous, and safe work at a fair price.

b. _____

_____

_____

The reason for this difference is that a larger percentage of engineers working in small firms may be expected to hold high-level positions. In firms with fewer than 20 engineers, for example, the median income was $33,200. In firms of 20 to 200 engineers, the median income was $30,345. For the largest firms, the median was $27,600.

8. Develop the following topic sentences into full paragraphs.
   a. Job candidates should not automatically choose the company that offers the highest salary.
   b. Every college student should learn at least the fundamentals of computer science.
   c. The one college course I most regret not having taken is _____ .
   d. Sometimes two instructors offer contradictory advice about how to solve the same kind of problem.

9. In this exercise, several paragraphs have been grouped together into one long paragraph. Recreate the separate paragraphs by marking where the breaks would appear.

Books on Compact Discs

Compact discs, which have revolutionized the recorded-music industry, are about to do the same for the book-publishing industry. A number of reference books—such as trade directories and multivolume encyclopedias—are already available in compact disc format. And now trade publishers are working out the legal issues involved in publishing their books in compact disc format. How is the compact disc technology applied to books? The heart of the system is the same as that used for recorded music. Any kind of information that can be digitized—converted to the numbers 0 and 1—can be transferred to the 4.7-in. diameter compact discs. Words and pictures, of course, are digitized in the common personal computer. Instead of outputting the digitized information exclusively as sound, the new technology hooks up a compact disc player to a computer. The information stored on the disc is then output as words and pictures on the screen and as sound emitted through a speaker. Compact discs offer several important advantages over traditional delivery systems for printed information. First, compact discs can hold a tremendous amount of information. A 100-volume encyclopedia could fit on a single disc. The space storage advantages are considerable. Second, compact discs offer the accessing ease of an on-line system. If the user wants information on subatomic particles, he or she simply types in the phrase and the system finds every reference to the subject. On request, the citations or even the entries themselves can be printed out on paper. And third, information stored on compact discs can be updated much less expensively than paper information. The subscriber or purchaser simply receives an updated disc periodically.

10. In the following paragraphs, transitional words and phrases have been removed. Add an appropriate transition in each blank space. Where necessary, add punctuation.

a.

As you know, the current regulation requires the use of conduit for all cable extending more than 18″ from the cable tray to the piece of equipment. _____ conduit is becoming increasingly expensive: up 17 percent in the last year alone. _____ we would like to determine whether the NRC would grant us any flexibility in its conduit regulations. Could we _____ run cable without conduit for lengths up to 3′ in low-risk situations such as wall-mounted cable or low-traffic areas? We realize _____ that conduit will always remain necessary in high-risk situations. The cable specifications for the Unit Two report to the NRC are due in less than two months; _____ we would appreciate a quick reply to our request, as this matter will seriously affect our materials budget.

b.

Several characteristics of personal computers limit the kind of local area network (LAN) we can use. _____ personal computers are portable. Losing the ability to move the personal computers would be an unacceptable loss. This is especially true in light of our plans to reorganize the office next year. _____ the number of personal computers is increasing and is expected to continue to increase at least for another decade. The LAN must be easily expandable. _____ we do not have the time or expertise to maintain or repair a LAN. Any LAN we lease _____ would have to be virtually maintenance-free. _____ the choice of a LAN will require careful analysis.

11. The following paragraphs are poorly organized and developed. Rewrite the paragraphs, making sure to include in each a clear topic sentence, adequate support, and effective transitions.

a.

The cask containing the spent nuclear fuel is qualified to sustain a 30-foot fall to an unyielding base. If the cask is allowed to pierce a floor after a 30-foot fall, we can safely assume that the cask is not qualified to maintain its integrity. This could result in spent fuel being released from the cask, resulting in high radioactive exposure to personnel.

b.

The use of a single-phase test source results in partial energization of the polarizing and operating circuits of relay units other than the one under test. It is necessary that these "unfaulted" relay units be disabled to assure that only the unit being tested in providing information regarding operation or nonoperation. The disabling is accomplished by disengaging printed circuit cards from the card sockets in the relay logic unit. The following table lists the cards to be disengaged by card address in the logic unit under test with respect to STF position.

c.

The results indicate that the plaques are not chemically homogeneous: the percentage of NILOC varied among the various sections. This fact represents a problem with determinations such as were conducted in this study. Since only a portion of each plaque submitted was analyzed, differences in the amount of materials extracted could be solely a function of the portion of the plaque.

REFERENCES

Haviland, S. E., and H. H. Clark. 1974. What's new? Acquiring new information as a process in comprehension. *Journal of Verbal Learning and Verbal Behavior* 13, 512–521.

Weiss, E. H. 1982. *The writing system for engineers and scientists.* Englewood Cliffs, N.J.: Prentice-Hall.

PART

TWO

Techniques of Technical Writing

Definitions

The world of business and industry depends on clear and effective definitions. Without written definitions, the working world would be chaotic. For example, suppose that you learn at a job interview that the potential employer pays tuition and expenses for its employees' job-related education. That's good news, of course, if you are planning to continue your education. But until you study the employee-benefits manual, you will not know with any certainty just what the company will pay for. Who, for instance, is an *employee?* You would think an employee would be anybody who works for and is paid by the company. But you might find that for the purposes of the manual an employee is someone who has worked for the company in a full-time capacity (35 hours per week) for at least six uninterrupted months. A number of other terms would have to be defined in the description of tuition benefits. What, for example, is *tuition?* Does the company's definition include incidental laboratory or student fees? What is *job-related education?* Does a course on methods of dealing with stress qualify under the company's definition? What, in fact, constitutes *education?* All these terms and many others must be defined for the employees to understand their rights and responsibilities.

Definitions play a major role, therefore, in communicating policies and standards "for the record." When a company wants to purchase air-conditioning equipment, it might require that the supplier provide equipment certified by a professional organization of air-conditioner manufacturers. The organization's definitions of acceptable standards of safety, reliability, and efficiency will provide some assurance that the equipment is high in quality.

Definitions, of course, have many uses outside legal or contractual contexts. Two such uses occur very frequently. Definitions can help the writer clarify a description of a new technology or a new development in a technical field. When a new animal species is discovered, for instance, it is named and defined. When a new laboratory procedure is devised, it is defined and then described in an article printed in a technical journal. Definitions can also help a specialist communicate with a less knowledgeable audience. A manual that explains how to tune up a car will include definitions of parts and tools. A researcher at a manufacturing company will use definitions in describing a new product to the sales staff.

Definitions, then, are crucial in all kinds of technical writing, from brief letters and memos to technical reports, manuals, and journal articles. All kinds of readers, from the layperson to the expert, need effective definitions to carry out their jobs.

Definitions, like every other technique in technical writing, require thought and planning. Before you can write a definition to include in a document, you must carry out three steps:

1. Analyze the writing situation.
2. Determine the kind of definition that will be appropriate.
3. Decide where to place the definition.

Analyzing the Writing Situation

The first step in writing effective definitions is to analyze the writing situation: the audience and purpose of your document.

Unless you know whom you are addressing and how much that audience already knows about the subject, you cannot know which terms to define or the kind of definition to write. Physicists wouldn't need a definition of *entropy*, but a group of lawyers might. Builders know what a Molly bolt is, but some insurance agents don't. If you are aware of your audience's background and knowledge, you can easily devise effective informal definitions. For example, if you are describing a cassette deck to a group or readers who understand automobiles, you can use a familiar analogy: "The PAUSE button is the brake pedal of the cassette deck."

Keep in mind, too, your purpose in writing. If you want to give your readers only a basic understanding of a concept—say, time-sharing vacation resorts—a brief, informal definition will usually be sufficient. However, if you want your readers to understand an object, process, or concept thoroughly and be able to carry out tasks based on what you have written, then a more formal and elaborate definition is called for. For example, a definition of a "Class 2 Alert" written for operators at a nuclear power plant will have to be comprehensive, specific, and precise.

Understanding the Different Types of Definitions

The preceding analysis of the writing situation has suggested that definitions can be short or long, informal or formal. There are three basic types:

- ☐ parenthetical definitions
- ☐ sentence definitions
- ☐ extended definitions

Parenthetical Definition

A parenthetical definition is a brief clarification placed unobtrusively within a sentence. Sometimes a parenthetical definition is a mere word or phrase:

The crane is located on the starboard (right) side of the ship.

Summit Books announced its desire to create a new colophon (emblem or trademark).

> United Engineering is seeking to purchase the equity stock
> (common stock) of Minnesota Textiles.

A parenthetical definition can also take the form of a longer explanatory phrase or clause:

> Motorboating is permitted in the Jamesport Estuary, the portion of
> the bay that meets the mouth of the Jamesport River.
>
> The divers soon discovered the kentledge, the pig-iron ballast.
>
> Before the metal is plated, it is immersed in the pickle, the acid bath
> that removes scales and oxides from the surface.

Parenthetical definitions are not, of course, comprehensive. They serve mainly as quick and convenient ways of introducing new terms to the readers. Because parenthetical definition is particularly common in writing addressed to general readers, make sure that the definition itself is clear. You have gained nothing if your readers don't understand your clarification:

> Next, check for blight on the epicotyl, the stem portion above the
> cotyledons.

If your readers are botanists, this parenthetical definition will be clear (although it might be unnecessary, for if they know the meaning of *cotyledons*, they are likely to know *epicotyl*). However, if you are addressing the general reader, this definition will merely frustrate them.

Sentence Definition

A sentence definition is a one- or two-sentence clarification. It is more formal than the parenthetical definition. Usually, the sentence definition follows a standard pattern: the item to be defined (the *species*) is placed in a category of similar items (the *genus*) and then distinguished from the other items (by the *differentia*):

Species	= *Genus*	+ *Differentia*
A flip flop	is a circuit	containing active elements that can assume either one of two stable states at any given time.
An electrophorus	is an instrument	used to generate static electricity.
Hypnoanalysis	is a psychoana-lytical technique	in which hypnosis is used to elicit unconscious information from a patient.

Species		= *Genus*	+ *Differentia*
A Bunsen burner	is	a small laboratory burner	consisting of a vertical metal tube connected to a gas source.
An electron microscope	is	a microscope	that uses electrons rather than visible light to produce magnified images.

In many cases, a sentence definition will also include some sort of illustration. The definitions of electrophorus, Bunsen burner, and electron microscope, for example, would probably be accompanied by photographs or diagrams.

Sentence definitions are useful when your readers require a more formal or more informative clarification than parenthetical definition can provide. Sentence definitions are often used to establish a working definition for a particular document: "In this report, the term *electron microscope* will be used to refer to any microscope that uses electrons rather than visible light to produce magnified images." In sentence definitions, keep three points in mind.

- Be as specific as you can in writing the *genus* and the *differentia*. Remember, you are trying to distinguish the item from all other similar items. If you write, "A Bunsen burner is a burner that consists of a vertical metal tube connected to a gas source," the imprecise *genus*— "a burner"—defeats the purpose of your definition: there are many types of large-scale burners that use vertical metal tubes connected to gas sources. If you write, "Hypnoanalysis is a psychoanalytical technique used to elicit unconscious information from a patient," the imprecise *differentia*—"used to elicit..."—ruins the definition: there are many psychoanalytical techniques used to elicit a patient's unconscious information. If more than one *species* is described by your definition, you have to sharpen either the *genus* or the *differentia*, or both.

- Avoid writing circular definitions, that is, definitions that merely repeat the key words of the *species* (the ones being defined) in the *genus* or *differentia*. In "A required course is a course that is required," what does "required" mean? Required of whom, by whom? The word is never defined.

  Similarly, "A balloon mortgage is a mortgage that balloons" is useless. However, you can use *some* kinds of words in the *species* as well as in the *genus* and *differentia*; in the definition of *electron microscope* given earlier, for example, the word *microscope* is repeated. Here, *microscope* is not the "difficult" part of the species; readers know what a microscope is. The purpose of defining *electron microscope* is to clarify the *electron* part of the term.

■ Be sure the *genus* contains a noun or a noun phrase rather than a phrase beginning with *when, what,* or *where.*

Incorrect	A brazier is what is used to . . .
Correct	A brazier is a metal pan used to . . .
Incorrect	An electron microscope is when a microscope . . .
Correct	An electron microscope is a microscope that . . .
Incorrect	Hypnoanalysis is where hypnosis is used . . .
Correct	Hypnoanalysis is a psychoanalytical technique in which . . .

Extended Definition

An extended definition is a long (one- or several-paragraph), detailed clarification of an object, process, or idea. Often an extended definition begins with a sentence definition, which is then elaborated. For instance, the sentence definition "An electrophorus is an instrument used to generate static electricity" tells you the basic function of the device, but it leaves many questions unanswered: How does it work? What does it look like? An extended definition would answer these and other questions.

Extended definitions are useful, naturally, when you want to give your readers a reasonably complete understanding of the item. And the more complicated or more abstract the item being defined, the greater the need for an extended definition.

There is no one way to "extend" a definition. Your analysis of the audience and purpose of the communication will help to indicate which method to use. In fact, an extended definition will sometimes employ several different methods. Often, however, one of the following techniques will work effectively:

- □ illustration
- □ exemplification
- □ analysis
- □ principle of operation
- □ comparison and contrast
- □ analogy
- □ negation
- □ etymology

Illustration

Perhaps the most common way to write an extended definition in technical writing is to create some sort of graphic, such as a photograph, diagram, schematic, or flow chart, and then explain the graphic.

Graphics are particularly useful in defining physical objects. The simplest way to define a two-person saw, for instance, would be to draw a rough diagram. However, illustrations are also useful in defining concepts and ideas. A definition of the meteorological concept of temperature inversion, for instance, might include a diagram. So might definitions of lasers, compound fractures, and buying stocks on margin.

The following excerpt from an extended definition of parallelogram shows an effective combination of words and illustrations.

A parallelogram is a four-sided plane whose opposite sides are parallel with each other and equal in length. In a parallelogram, the opposite angles are the same, as shown in the sketch below.

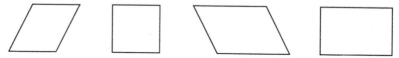

Exemplification

Exemplification, using an example (or examples) to clarify an object or idea, is particularly useful in making an abstract term easy to understand. The following paragraph is an extended definition of the psychological defense mechanism called *conversion* (Wilson 1964: 84).

A third mechanism of psychological defense, "conversion," is found in hysteria. Here the conflict is converted into the symptom of a physical illness. In a case of conversion made famous by Freud, a young woman went out for a long walk with her brother-in-law, with whom she had fallen in love. Later, on learning that her sister lay gravely ill, she hurried to her bedside. She arrived too late and her sister was dead. The young woman's grief was accompanied by sharp pain in her legs. The pain kept recurring without any apparent physical cause. Freud's explanation was that she felt guilty because she desired the husband for herself, and unconsciously converted her repressed feelings into an imaginary physical ailment. The pain struck her in the legs because she unconsciously connected her feelings for the husband with the walk they had taken together. The ailment symbolically represented both the unconscious wish and a penance for the feelings of guilt which it engendered.

Notice that the first two sentences in this paragraph are essentially a sentence definition that might be paraphrased as follows: "Conversion is a mechanism of psychological defense by which the conflict is converted into the symptoms of a physical illness."

This extended definition is effective because the writer has chosen a clear and interesting (and therefore memorable) example of the subject he is describing. No other examples are necessary in this case. If conversion were a more difficult concept to describe, an additional example might be useful.

Analysis

Analysis is the process of dividing a thing or idea into smaller parts so that the reader can more easily understand it.

A load-distributing hitch is a trailer hitch that distributes the hitch load to all axles of the tow vehicle and the trailer. The crucial component of the load-distributing hitch is the set of spring bars that attaches to the trailer. For a complete understanding of the load-distributing hitch, however, the following other components should be explained first:

1. the shank
2. the ball mount
3. the sway control
4. the frame bracket

The shank is the metal bar that is attached to the frame of the tow vehicle. . . .

This extended definition of a load-distributing hitch, which would be accompanied by a diagram, uses analysis as its method of development. The hitch is divided into its major components, each of which is then defined and described.

Principle of Operation

The principle of operation is an effective way to develop an extended definition, especially that of an object or process. The following extended definition of a thermal jet engine is based on the mechanism's principle of operation.

A thermal jet engine is a jet-propulsion device that uses air, along with the combustion of a fuel, to produce the propulsion. In operation, the thermal jet engine draws in air, increases its pressure, heats it by combustion of the fuel, and finally ejects the heated air (and the combustion gases). The increased velocity of the ejected mixture determines the thrust: the greater the difference in velocity between air entering and leaving the unit, the greater the thrust.

Note that this extended definition begins with a sentence definition.

Comparison and Contrast

Comparison and contrast is another useful technique for developing an extended definition. With this technique, the writer discusses similarities or differences between the item being defined and an item with which the readers are more familiar. The following definition of a bit brace begins by comparing and contrasting it to the more common power drill.

A bit brace is a manual tool used to drill holes. Cranked by hand, it can theoretically turn a bit to bore a hole in any material that a power drill can bore. Like a power drill, a bit brace can accept any number of different sizes and shapes of bits. The principal differences between a bit brace and a power drill are:

1. A bit brace drills much more slowly.
2. A bit brace is a manual tool, and so it can be used where no electricity is available.
3. A bit brace makes almost no noise in use.

The bit brace consists of the following parts. . . .

Analogy

An analogy is a specialized kind of comparison. In a traditional comparison, the item being defined is compared to a similar item. An electron microscope is compared to a common microscope, or a bit brace is compared to a power drill. In an analogy, the item being defined is compared to an item that in some ways differs completely, but that shares some essential characteristic or characteristics.

For instance, the central-processing unit of a computer is often compared to a brain. Obviously, these two items are very different—except that the function of the central-processing unit in the computer is similar to that of the brain in the body. Computer software is often compared to audiotapes or compact discs. Again, the differences are many, but the similarity is essential: the tape or disc contains the code that enables the stereo system to fulfill its function, just as the software contains the code that enables the computer to do its job.

The following excerpt from an extended definition of computer literacy shows how an analogy can clarify an unfamiliar concept.

> Computer literacy is the ability to use computers effectively. If you can operate a personal computer to do word processing or create a database, you certainly are computer literate. If you can operate a digital watch or program a VCR or use an automated teller machine at a bank, you also can be said to be computer literate. To use an analogy, computer literacy is like automotive literacy: if you know how to operate an automobile safely to get from one place to another, you possess automotive literacy. Just as you don't have to understand the principle of the internal combustion engine to drive a car, you don't have to understand the concepts of RAM and ROM to use a computer. . . .

Negation

A special kind of contrast is sometimes called *negation* or *negative* statement. Negation is the technique of clarifying a term by distinguishing it from a different term with which the reader might have confused it.

> An ambulatory patient is not a patient who must be moved by ambulance. On the contrary, an ambulatory patient is one who can walk without assistance from another person. . . .

Negation is rarely the only technique used in an extended definition. In fact, negation is used most often in a sentence or two at the start of a definition: for after you state what the item is *not*, you still have to define what it *is*.

Etymology

Etymology, the history of a word, is often a useful and interesting way to develop a definition.

> No-fault auto insurance, as the name suggests, is a type of insurance that ignores who or what caused the accident. When the accident damage is less than a specific amount set by the state, the people involved in the accident may not use legal means to determine who was at fault. What this means to the driver is that . . . .

> The word mortgage was originally a compound of mort (dead) and gage (pledge). The meaning of the word has not changed substantially since its origin in Old French. A mortgage is still a pledge that is "dead" upon either the payment of the loan or the forfeiture of the collateral and payment from the proceeds of its sale.

Etymology is a popular way to begin a definition of acronyms: abbreviations that are pronounced as words. For example, "SCUBA stands for self-contained underwater breathing apparatus. . . ." Or, "COBOL, which is the Common Business-oriented Language, was originally invented to. . . ."

Etymology, like negation, is rarely used alone in technical writing, but it is an effective way to introduce an extended definition.

The following extended definition of *nowcasting*, a weather forecasting technique, shows some of the common ways to elaborate a sentence definition. The definition is directed to a general audience. Marginal notes have been added.

sentence definition

> Nowcasting is a new word used to describe a short-term weather forecasting technique: describing the current weather and forecasting over the next 2 to 12 hours for a limited geographical area. Nowcasting relies on the most modern data-processing technology: computer systems that let the user manipulate and interpret tremendous amounts of data almost instantaneously.

comparison and contrast

principle of operation

> Traditional weather forecasting is more accurate for periods of 12 to 36 hours in the future than it is for the 2-to-12-hour period. The reason for this is that traditional forecasting relies on data in the atmosphere. Temperature, humidity, and wind are measured by satellites and by radiosondes carried by balloons. These data are then fed into mathematical models that measure motion, mass, and thermodynamics. The resulting forecasts are reasonably accurate, but they cannot focus on a limited geographic area.

principle of operation

exemplification

> By contrast, nowcasting supplements satellite data with numerous ground-based readings of clouds, temperature, and humidity. The advantage of using ground-based readings is that the nowcaster can make predictions on effects caused by the peculiarities of the local terrain. For example, the nowcaster can predict squall lines, fronts, mountain-valley wind patterns, and land-

sea breezes that are too small and local to be measured by the satellites or upper-atmosphere radiosondes. The resulting data are highly accurate for a limited geographic area—but only for a few hours.

The first paragraph contains what is essentially a sentence definition: "Nowcasting is a weather forecasting technique that describes the current weather and forecasts the weather for the next 2 to 12 hours for a limited geographical area." Also notice that the first sentence uses etymology in referring to the origin of the term *nowcasting*.

The two paragraphs that follow compare and contrast forecasting and nowcasting. Each paragraph describes the principle of operation of the technique, and then comments on its effectiveness.

Deciding Where to Place the Definitions

In many cases, the writer does not need to decide where to place a definition. In writing your first draft, for instance, you may realize that most of your readers will not be familiar with a term you want to use. If you can easily provide a parenthetical definition that will satisfy your readers' needs, simply do so.

Often, however, in assessing the writing situation before beginning the draft, you will conclude that one or more complicated terms will have to be introduced, and that your readers will need more detailed and comprehensive clarifications, perhaps sentence definitions and extended definitions. In these cases, you should plan—at least tentatively—where you are going to place them.

Definitions can be placed in four different locations:

- □ in the text
- □ in footnotes
- □ in a glossary
- □ in an appendix

The text itself can accommodate any of the three kinds of definitions. Parenthetical definitions, because they are brief and unobtrusive, are almost always included in the text; even if a reader already knows what the term means, the slight interruption will not be annoying. Sentence definitions are often placed within the text. If you want all your readers to see your definition or you suspect that many of them need the clarification, the text is the appropriate location. Keep in mind, however, that unnecessary sentence definitions can bother your readers. Extended definitions are rarely placed within the text, because of their length. The obvious exception, of course, is the extended definition of a term that is central to the discussion; a discussion of recent changes in workers' compensation insurance will likely begin with an extended definition of that kind of insurance.

Footnotes are a logical place for an occasional sentence definition or extended definition. The reader who needs the definition can find it easily at the bottom of the page; the reader who doesn't need it will ignore it. Footnotes are difficult to type, however, and they make the page look choppy. (Some word-processing software can set up footnotes automatically.) If you are going to need more than one footnote for every two or three pages, consider creating a glossary.

A glossary, an alphabetized list of definitions, can handle sentence definitions and extended definitions of fewer than three or four paragraphs. A glossary can be placed at the start of a document (such as after the executive summary in a report) or at the end, preceding the appendixes. A glossary is a convenient collection of definitions that otherwise might clutter up the document. For more information on how to set up a glossary, see Chapter 19.

An appendix is an appropriate place to put an extended definition, one of a page or longer. A definition of this length would be cumbersome in a glossary or in a footnote, and—unless it explains a crucial term—too distracting in the text. Because the definition is an appendix to the document, it will be listed in the table of contents. In addition, it can be referred to—with the appropriate page number—in the text.

WRITER'S CHECKLIST

This checklist covers parenthetical, sentence, and extended definitions.

1. Are all necessary terms defined?

2. Are the parenthetical definitions
   a. appropriate for the audience?
   b. clear?

3. Does each sentence definition
   a. contain a sufficiently specific *genus* and *differentia*?
   b. avoid circular definition?
   c. contain a noun in the *genus*?

4. Are the extended definitions developed logically and clearly?

5. Are the definitions placed in the most useful location for the readers?

1. Add parenthetical definitions of the italicized terms in the following sentences.
   a. Reluctantly, he decided to *drop* the physics course.
   b. Last week the computer was *down*.
   c. The culture was studied *in vitro*.
   d. The tire plant's managers hope they do not have to *lay off* any more employees.
   e. The word processor comes complete with a *printer*.

2. Write a sentence definition for each of the following terms:
   a. a catalyst
   b. a digital watch
   c. a job interview
   d. a cassette deck
   e. flextime

3. Write an extended definition of one of the following terms, or of a term used in your field of study:
   a. flextime
   b. binding arbitration
   c. energy
   d. an academic major (don't focus on any particular major; define what a major is)
   e. quality control

4. Revise any of the following sentence definitions that need revision.
   a. Dropping a course is when you leave the class.
   b. A thermometer measures temperature.
   c. The spark plugs are the things that ignite the air-gas mixture in a cylinder.
   d. Double parking is where you park next to another car.
   e. A strike is when the employees stop working.

5. Identify the techniques used to create the following extended definition:

Holography, from the Greek holos (entire) and gram (message), is a method of photography that produces images that appear to be three-dimensional. A holographic image seems to change as the viewer moves in relation to it. For example, as the viewer moves, one object on the image appears to move in front of another object. In addition, the distances between objects on the image also seem to change.

Holographs are produced by coherent light, that is, light of the same wavelength, with the waves in phase and of the same amplitude. This light is produced by laser. Stereoscopic images, on the other hand, are created by incoherent light— random wavelengths and amplitudes, out of phase. The incoherent light, which is natural light, is focused by a lens and records the pattern of brightness and color differences of the object being imaged.

How are holographic images created? The laser-produced light is divided as it passes through a beam splitter. One portion of the light, called the reference beam, is directed to the emulsion—the "film." The other portion, the object beam, is directed to the subject and then reflected back to the emulsion. The reference beam is coherent light, whereas the object beam becomes incoherent because it is reflected off the irregular surface of the subject. The resulting dissonance between the reference beam and the object beam is encoded; it records not only the brightness of the different parts of the subject but also the different distances from the laser. This encoding creates the three-dimensional effect of holography.

REFERENCE

Wilson, J. R. 1964. *The mind.* New York: Time, Inc.

Descriptions

Technical writing is filled with descriptions of objects, mechanisms, and processes. For our purpose here, a *description* is a verbal and visual representation.

What is an object? The range is enormous: from physical sites such as volcanoes and other kinds of natural phenomena to synthetic artifacts such as hammers. A tomato plant is an object, as is an automobile tire or a book.

A mechanism is a synthetic object consisting of a number of moving, identifiable parts that work together as a system. A compact disc player is a mechanism, as are a voltmeter, a lawn mower, a submarine, and a steel mill. A complex object, such as a submarine, is a mechanism because, despite its immense complexity, it can be described as a synthetic object consisting of several different "parts." Each of these parts—such as the heating, ventilating, and air-conditioning systems—is of course made up of dozens of different, smaller mechanisms.

A process is an activity that takes place over time: the earth was formed, steel is made, animals evolve, plants perform photosynthesis. Descriptions of processes should be distinguished from instructions: process descriptions explain how something happens; instructions tell how to do something. The readers of a process description want to understand the process, *but they don't want to perform the process.* A set of instructions, on the other hand, is a step-by-step guide intended to enable the readers to perform the process. A process description answers the question "How is wine made?" A set of instructions answers the question "How do I go about making wine?"

Instructions are discussed in Chapter 16.

Understanding the Role of Description

Descriptions of objects, mechanisms, and processes appear in virtually every kind of technical writing. A company studying the feasibility of renovating an old factory or building a new one includes a description of the existing facility. That old factory is a complex mechanism. A rational decision can be made only if the company understands it thoroughly.

A writer who wants to persuade his readers to authorize the purchase of some equipment includes a mechanism description in the proposal. An engineer who is trying to describe to her research-and-development department the features she would like incorporated in a piece of equipment now being designed includes a mechanism description in her report. An engineer who is trying to describe to the sales staff how a product works, so that they can advertise it effectively, includes a mechanism description. Mechanism descriptions are used frequently by companies communicating with their clients and customers; operating instructions, for instance, often include mechanism descriptions, as do advertising materials.

Process descriptions are common in technical writing because often readers who do not actually perform a process nevertheless need to understand it. If, for example, a new law limits the amount of heated water that a nuclear power plant may discharge into a river, the plant managers have to understand how water is discharged. If the plant violates the law, engineers have to devise alternative solutions—ways to reduce the quantity or temperature of the discharge—and describe them so their supervisors can decide which alternative to implement. Or consider a staff accountant who audits one of her company's branches and then writes up a report describing the process and its results. Her readers will review her report to determine whether she performed the audit properly and to learn her findings.

Notice that a description rarely appears as a separate document. Almost always, it is part of a larger report. For example, a maintenance manual for a boiler system might begin with a mechanism description, to help the reader understand how the system operates. However, the ability to write an effective description is so important in technical writing that it is discussed here as a separate technique.

Analyzing the Writing Situation

Before you begin to write a description as part of a longer document, consider carefully how the audience and purpose of the document should affect the way you describe the object, mechanism, or process.

What does the audience already know about the general subject? If, for example, you are to describe an electron microscope, you first have to know whether your readers understand what a microscope is. If you want to describe how industrial robots will affect car manufacturing, you first have to know whether your readers already understand the current process. In addition, you need to know whether they understand robotics.

The audience will determine your use of technical vocabulary as well as your sentence and paragraph length. Another audience-related factor is the use of graphics. Less knowledgeable readers need simple illustrations; they might have trouble understanding sophisticated schematics or decision charts.

What are you trying to accomplish in the description? If you want your readers to understand how a basic computer works, you will write a *general description*: a description that applies to several different varieties of computer. If, however, you want your readers to understand the intricacies of a particular computer, you will write a *particular description*. A general description of a typewriter will use classification to point out that there are manual, electric, and word-processing typewriters and then describe typical models of each. A particular description of a Smith Corona Model 2100 will treat that one typewriter in great detail. An understanding of your purpose will determine every aspect of the description, including length, amount of detail, and number and type of graphics included.

Understanding the Structure of Object and Mechanism Descriptions

Object and mechanism descriptions have the same basic structure. In the following discussion, the word *item* will refer to both objects and mechanisms.

Most descriptions of items have a four-part structure:

1. a title or section heading
2. a general introduction that tells the reader what the item is and what it does
3. a part-by-part description of the item
4. a conclusion that summarizes the description and tells how the parts work together

The Title or Section Heading

If the description of the item is to be a separate document, give it a title. If the description is to be part of a longer document, give it a section heading (see Chapter 7 for detailed discussions of titles and headings).

In either case, clearly indicate the subject and indicate whether the description is general or particular. For instance, the title of a general description might be "Description of a Minivan"; of a particular description, "Description of the 1991 Mazda MPV."

The General Introduction

The general introduction provides the basic information that your readers will need to understand the detailed description that follows. In writing the introduction, answer these five questions about the item:

□ What is it?
□ What is its function?
□ What does it look like?
□ How does it work?
□ What are its principal parts?

Of course, in some cases the answer would be obvious and therefore should be omitted. For instance, everyone knows the function of a computer printer.

■ *What is the item?* Generally, the best way to answer this question is to provide a sentence definition (see Chapter 8): "An electron microscope is a microscope that uses electrons rather than visible light to produce magnified images." Or "The Aleutian Islands are a chain of islands curving southwest from the tip of Alaska toward the tip of the Kormandorski Islands of the Soviet Union." Then elaborate if necessary.

■ *What is the function of the item?* State clearly what the item does: "Electron microscopes are used to magnify objects that are smaller than the wavelengths of visible light." Often, the function of a mechanism is implied in its sentence definition: "A hydrometer is a sealed, graduated tube, weighted at one end, that sinks in a fluid to a depth that indicates the specific gravity of the fluid." Of course, some objects have no "function." The Aleutian Islands, as valuable as they might be, have no function in the sense that microscopes do.

■ *What does the item look like?* Include a photograph or drawing if possible (see Chapter 11). If not, use an analogy or a comparison to a familiar item: "The cassette that encloses the tape is a plastic shell, about the size of a deck of cards." Mention the material, texture, color, and the like, if relevant.

Sometimes, an object is best pictured with both graphics and words. For example, a map showing the size and location of the Aleutian Islands would be useful. But a verbal picture would be useful, too: "The Aleutian Islands, a rugged string of numerous small islands, show signs of volcanic action. The islands are treeless and fog enshrouded. The United States military installations on the Aleutians are the main source of economic activity. . . ."

■ *How does the item work?* In a few sentences, define the operating principle of the item:

> A tire pressure gauge is essentially a calibrated rod fitted snugly within an open-ended metal cylinder. When the cylinder end is placed on the tire nozzle, the pressure of the air escaping from the tire into the cylinder pushes against the rod. The greater the pressure, the farther the rod is pushed.

Sometimes, objects do not "work"; they merely exist.

■ *What are the principal parts of the item?* Few objects or mechanisms described in technical writing are so simple that you could discuss all the parts. And, in fact, you rarely will need to describe them all. Some parts will be too complicated or too unimportant to be mentioned;

others will already be understood by your readers. A description of a bicycle would not mention the dozens of nuts and bolts that hold the mechanism together; it would focus on the chain, the pedals, the wheels, and the frame.

Similarly, a description of the Aleutian Islands would mention the four main groups of islands but not all the individual islands, some of which are very small.

At this point in the description, the principal parts should merely be mentioned; the detailed description comes in the next section. The parts can be named in a sentence or (if there are many parts) in a list. Name the parts in the order you will describe them.

The introduction is an excellent place to put a graphic aid that complements your listing of major parts. Your goal is to enable the readers to follow the part-by-part description easily. If you partition the Aleutian Islands into four main island groups, your map should identify those same four groups. See Chapter 11 for a discussion of graphics.

The information provided in the introduction generally follows this five-part pattern. However, don't feel you have to answer each question in order, or in separate sentences. If your readers' needs or the nature of the item suggests a different sequence—or different questions—adjust the introduction accordingly.

The Part-by-Part Description

The body of the description—the part-by-part description—is essentially like the introduction in that it treats each major part as a separate item. That is, in writing the body you define what each part is, then, if applicable, describe its function, operating principle, and appearance. The discussion of the appearance should include shape, dimension, material, and physical details such as texture and color (if essential). For some descriptions, other qualities, such as weight or hardness, might also be appropriate. If a part has any important subparts, describe them in the same way. A description of an item therefore resembles a map with a series of detailed insets. A description of a personal computer would include a keyboard as one of its parts. The description of the keyboard, in turn, would include the numeric keypad as one of its parts. And the description of the numeric keypad would include the function keys as *its* parts. This process of ever-increasing specificity would continue as required by the complexity of the item and the needs of the readers.

In partitioning the item into its parts, discuss them in a logical sequence. The most common structure reflects the way the item works or is used. In a stereo radio set, for instance, the "sound" begins at the radio receiver, travels into the amplifier, and then flows out through the speakers. Another common sequence is based on the physical structure of the item: from top to bottom, outside to inside, and so forth. A third sequence is to move from more-important to less-important parts. (Be careful when you use this sequence: Will your readers understand and agree with your assessment of importance?) Most descriptions could be organized in a number of different ways. Just make sure you consciously choose a pattern; otherwise you might puzzle and frustrate your readers.

Graphics should be used liberally throughout the part-by-part description. In general, try to create a graphic for each major part. Use photographs to show external surfaces, line drawings to emphasize particular items on the surface, and cutaways and exploded diagrams to show details beneath the surface. Other kinds of graphics, such as graphs and charts, are often useful supplements. If, for instance, you are describing the Aleutian Islands, you might create a table that shows total land area and habitable land area of the island groups.

The Conclusion

Descriptions generally do not require elaborate conclusions. A brief conclusion is necessary, however, if only to summarize the description and prevent the readers from placing excessive emphasis on the part discussed last in the part-by-part description.

A common technique for concluding descriptions of mechanisms and of some objects is to describe briefly how the parts function together. The conclusion of a description of a telephone, for example, might include a paragraph such as the following:

> When the phone is taken off the hook, a current flows through the carbon granules. The intensity of the speaker's voice causes a greater or lesser movement of the diaphragm and thus a greater or lesser intensity in the current flowing through the carbon granules. The phone receiving the call converts the electrical waves back to sound waves by means of an electromagnet and a diaphragm. The varying intensity of the current transmitted by the phone line alters the strength of the current in the electromagnet, which in turn changes the position of the diaphragm. The movement of the diaphragm reproduces the speaker's sound waves.

Following is a description of a modern long-distance running shoe (based on Kyle 1986). Marginal notes have been added.

The writer begins with a clear title.

The writer provides a brief background.

GENERAL DESCRIPTION OF A LONG-DISTANCE RUNNING SHOE

When track and field events became sanctioned sports in the modern world some hundred and fifty years ago, the running shoe was much like any other: a heavy, high-topped leather shoe with a leather or rubber sole. In the last decade, however, advances in technology have combined with increased competition among manufacturers to create long-distance running shoes that fulfill the two goals of all runners: decreased injuries and increased speed.

Introduction

The writer states the subject and kind of description.

This paper is a generalized description of a modern, high-tech shoe for long-distance running.

The modern distance running shoe has five major components:

The major components are listed.

1. the outsole
2. the heel wedge
3. the midsole
4. the insole
5. the shell

The writer refers to the graphic.

Figure 1 is an exploded diagram of the shoe.

Here the writer sketches the complete shoe, then shows its components in an exploded diagram. (See Chapter 11 for a discussion of graphics.)

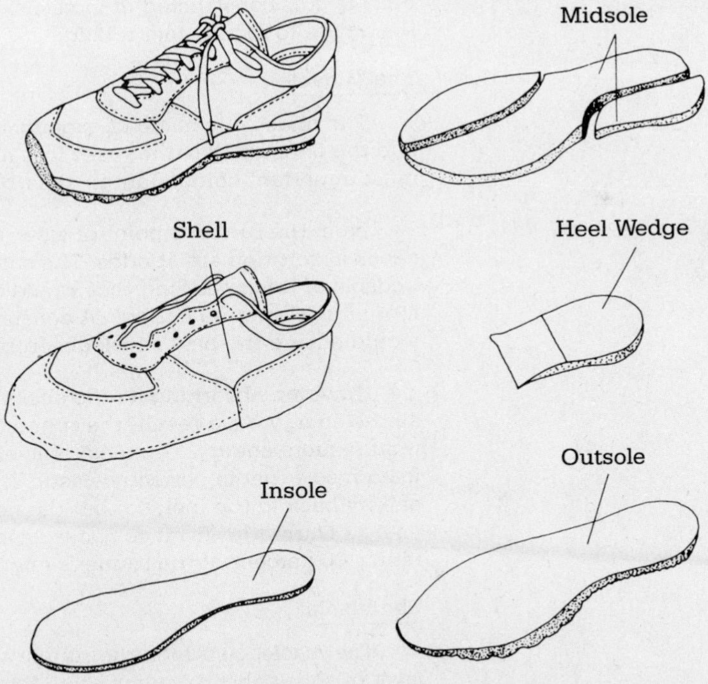

Figure 1. Exploded Diagram of a Long-Distance Running Shoe

The Components

The five principal components of the shoe will be discussed from bottom to top.

The Outsole

The outsole is made of a lightweight, rubberlike synthetic material. Its principal function is to absorb the runner's energy safely as the foot lands on the surface.

As the runner's foot approaches the surface, it supinates—rolls outward. As the foot lands, it pronates—rolls inward. Through tread design and increased stiffness on the innerside, the outsole helps reduce inward rolling.

Inward rolling is a major cause of foot, knee, and tendon injuries because of the magnitude of the force generated during running. The force on the foot as it touches the running surface can be up to three times the runner's weight. And the acceleration transmitted to the leg can be 10 times the force of gravity.

The Heel Wedge

The heel wedge is a flexible platform that absorbs shock. Its purpose is to prevent injury to the Achilles tendon. Like the outsole, it is constructed of increasingly stiff materials on the inner side to reduce foot rolling.

The Midsole

The midsole is made of expanded foam. Like the outsole and the heel wedge, it reduces foot rolling. But it also is the most important component in absorbing the shock.

From the runner's point of view, running efficiency and shock absorption are at odds. The safest shoe would have a midsole of thick padding that would crush uniformly as the foot hit the running surface. A constant rate of deceleration would ensure the best shock absorption.

However, absorbing all the shock would mean absorbing all the energy. As a result, the runner's next stride would require more energy. The most efficient shoe would have a foam insole that is perfectly elastic. It would return all the energy back to the foot, so that the next stride required less energy. Currently, distance shoes have midsoles designed to return 40 percent of the runner's energy back to the foot.

The Insole

The insole, on which the runner's foot rests, is another layer of shock-absorbing material. Its principal function, however, is to provide an arch support, a relatively new feature in running shoes.

The Shell

The shell is made of leather and synthetic materials such as nylon. It holds the soles on the runner's foot and provides ventilation. The shell accounts for about one-third of the nine ounces a modern shoe weighs.

Conclusion

The conclusion summarizes the major points of the description.

Today, scientific research on the way people run has led to great improvements in the design and manufacture of different kinds of running shoes. With the vast numbers of runners, more and more manufacturers have entered the market. The results are a lightweight, shock-absorbing running shoe that balances the needs of safety and increased speed.

Understanding the Structure of the Process Description

The structure of the process description is essentially the same as that of the object or mechanism description. The only real difference is that the process is partitioned into a reasonable (usually chronological) sequence of steps rather than parts. Most process descriptions contain the following four components:

1. a title or section heading
2. a general introduction that tells the reader what the process is and what it used for
3. a step-by-step description of the process
4. a conclusion that summarizes the description and tells how the steps work together

The Title or Section Heading

If the process description is to be a separate document, add a title. If the description is to be part of a longer document, add a section heading (see Chapter 7 for detailed discussions of titles and headings).

Whether you are writing a title or a section heading, clearly indicate the subject and whether the description is general or particular. For instance, the title of a general description might be "Description of the Process of Designing a New Production Car"; of a particular description, "Description of the Process of Designing the General Motors Saturn."

graphic: flow chart

The General Introduction

The general introduction gives your readers the basic information they need to understand the detailed description that follows. In writing the introduction, answer these six questions about the process:

- ☐ What is the process?
- ☐ What is its function?
- ☐ Where and when does it take place?
- ☐ Who or what performs it?
- ☐ How does it work?
- ☐ What are its principal steps?

Naturally, sometimes the answer would be obvious and therefore should be omitted. For example, there is no need to explain why skiers want to get the right-size skis and boots. And sometimes the question is inappropriate. For instance, there is no "function" to the process of creating the universe.

- ■ *What is the process?* Generally, the best way to answer this question is to provide a sentence definition (see Chapter 8): "Debugging is the process of identifying and eliminating any errors within the program." Then elaborate if necessary.

- ■ *What is the function of the process?* State clearly the function of the process: "The central purpose of performing a census is to obtain up-to-date population figures by which to revise legislative redistricting and revenue-sharing." Make sure the function is clear to your readers; if you are unsure whether they already know the function, state it anyway. Few things are more frustrating for your readers than not knowing why the process should be performed.

- ■ *Where and when does the process take place?* State clearly the location and occasion for the process: "The stream is stocked at the hatchery in the first week of March each year." You can generally add these details simply and easily. Again, omit these facts only if you are certain your readers already know them.

- ■ *Who or what performs the process?* Most processes are performed by people, by natural forces, by machinery, or by some combination of the three. In most cases, you do not need to state explicitly that, for example, the young trout are released into the stream by a person; the context makes that point clear. In fact, much of the description usually is written in the passive voice: "The water temperature is then measured." Do not assume, however, that your readers already know what agent performs the process, or even that they understand you when you have identified the agent. Someone who is not knowledgeable about computers, for instance, might not know whether a compiler is a person or a thing. The term *word processor* often refers ambiguously to a piece of equipment and to the person who operates it. Confusion at this early stage of the process description can ruin the effectiveness of the document.

■ *How does the process work?* In a few sentences, define the principle or theory of operation of the process:

> The four-treatment lawn-spray plan is based on the theory that the most effective way to promote a healthy lawn is to apply different treatments at crucial times during the growing season. The first two treatments—in spring and early summer—consist of quick-acting organic fertilizers and weed- and insect-control chemicals. The late-summer treatment contains postemergence weed-control and insect-control chemicals. The last treatment—in the fall—uses long-feeding fertilizers to encourage root growth over the winter.

■ *What are the principal steps of the process?* Name the principal steps of the process, in one or several sentences or in a list. Name the steps in the order you will describe them in the body. The principal steps in changing an automobile tire, for instance, include jacking up the car, replacing the old tire with the new one, and lowering the car back to the ground. Changing a tire also includes secondary steps—such as placing blocks behind the tires to prevent the car from moving once it is jacked up, and assembling the jack—that you should explain or refer to at the appropriate points in the description.

The information given in the introduction to a process description generally follows this pattern. However, don't feel you must answer each question in order, or in separate sentences. If your readers' needs or the nature of the process suggests a different sequence—or different questions—adjust the introduction.

As is the case with object and mechanism descriptions, process descriptions benefit from clear graphics in the introduction. Flow charts that identify the major steps of the process are particularly common.

The Step-by-Step Description

The body of the process description—the step-by-step description—is essentially like the introduction, in that it treats each major step as if it were a process. Of course, you do not repeat your answer to the question about who or what performs the action unless a new agent performs a particular step. You do answer the other principal questions—what the step is, what its function is, and when, where, and how it occurs. In addition, if the step has any important substeps that the reader will need to know in order to understand the process, you should explain them clearly.

Structure the step-by-step description chronologically: discuss the initial step first and then discuss each succeeding step in the order in which it occurs in the process. If the process is a closed system and hence has no "first" step—such as the cycle of evaporation and condensation—explain to your readers that the process is cyclical. Then simply begin with any principal step.

causality

Although the structure of the step-by-step description should be chronological, don't present the steps as if they have nothing to do with one another. In many cases, one step leads to another causally (see Chapter 10 for a discussion of cause and effect). In the operation of a four-cycle gasoline engine, for instance, each step sets up the conditions under which the next step can occur. In the compression cycle, the piston travels upward in the cylinder, compressing the mixture of fuel and air. In the power cycle, a spark ignites this compressed mixture. Your readers will find it easier to understand and remember your description if you clearly explain the causality in addition to the chronology.

A word about tense. Discuss steps in the present tense, unless, of course, you are writing about a process that occurred in the historical past. For example, a description of how the earth was formed would be written in the past tense: "The molten material condensed. . . ." However, a description of how steel is made would be written in the present tense: "The molten material is then poured into. . . ."

Whenever possible, use graphics within the step-by-step description to clarify each point. Additional flow charts are useful, but you will often want to include other kinds of graphics, such as photographs, drawings, and graphs. For instance, in the description of a four-cycle gasoline engine, you could illustrate the position of the valves and the activity occurring during each step. For example, you could show the explosion during the ignition step with arrows pushing the piston down within the cylinder.

The Conclusion

Process descriptions usually do not require long conclusions. If the description itself is brief—less than a few pages—a short paragraph summarizing the principal steps is all you need. For longer descriptions, a discussion of the implications of the process might be appropriate.

Following is the conclusion from a description of the four-cycle gasoline engine in operation:

> In the intake stroke, the piston moves down, drawing the air-fuel mixture into the cylinder from the carburetor. As the piston moves up, it compresses this mixture in the compression stroke. In the power stroke, a spark from the spark plug ignites the mixture, which burns rapidly, forcing the piston down. In the exhaust stroke, the piston moves up, expelling the burned gases.

Following is a particular description of the stages involved in the construction of the tunnel under the English Channel. The description is addressed to a general audience. Marginal notes have been added.

The writer begins with a clear title.	**TUNNELING UNDER THE ENGLISH CHANNEL**

Introduction

The writer provides the background: the purpose of the process and who is doing it.

Almost two hundred years ago, the idea of linking England and France by a tunnel under the English Channel was proposed to Napoleon. The plan was not pursued. In the intervening years the idea has been revived several times, and in 1875 a one-mile tunnel was dug at the foot of the White Cliffs of Dover. Fears of invasion from the continent led to repeated protests from the British military. In 1973, Britain and France agreed to a 32-mile rail link, but austerity measures forced Britain to cancel. The British and French have finally agreed to a plan for a tunnel that is expected to be operational by 1993.

convention is to use present tense

This paper describes the stages that have occurred—and will occur—in building the English Channel tunnel:

The writer lists the stages of the process.

parallel list

1. determining the objectives
2. determining the constraints
3. selecting a plan
4. constructing the original tunnel
5. improving the original tunnel

The tunnel will link Dover to Calais, as shown in Figure 1.

The Stages of the English Channel Tunnel Project

This overview clarifies the sequence of the stages.

Currently, the objectives and restraints have been determined, a plan has been selected, and the original tunnel has been constructed. The original tunnel, which will be a rail link, is expected to be supplemented with a roadway sometime early in the next century. These five stages will be discussed in their chronological sequence.

The writer provides a map of the area.

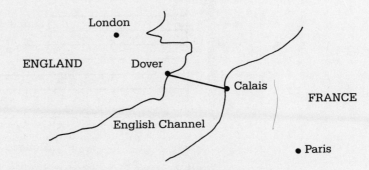

Figure 1 The Route of the English Channel Tunnel

Determining the Objectives

The main goal of the tunnel is to reduce the travel time between London and Paris. Figure 2 shows that the tunnel is expected to draw the capitals much closer. This will increase trade and tourism.

A second objective for both nations is to increase employment and thereby improve their governments' political fortunes. The tunnel project is expected to provide 60,000 jobs for some five or six years.

Determining the Constraints

In addition to the obvious technical questions that both parties considered, the main constraint was offered by Britain's conservative Prime Minister, Margaret Thatcher. Her government insisted that the project be financed privately. The approved project calls for a consortium of three British banks, three French ones, and ten construction companies.

Selecting a Plan

The competition came down to the proposals offered by four finalists. The British favored the Channel Expressway, which consisted of twin road and rail tunnels. However, objections based on its cost and the difficulties involved in ventilation doomed this idea.

The French backed Euroroute, which called for the construction of two artificial islands in the channel. The islands were to be linked by a tunnel and linked to the shores by bridges. Because of concerns over the enormous price—$15 billion—and the islands' vulnerability to terrorist attacks, Euroroute was abandoned.

A third proposal called for the construction of a bridge across the 23-mile waterway. However, questions about the

The writer discusses the unsuccessful proposals before the successful one.

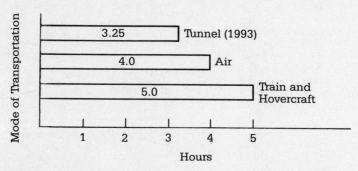

Figure 2 Travel Time Between London and Paris (Downtown to Downtown)

new composite material that was supposed to replace steel in the bridge hurt the plan.

The successful proposal is discussed in more detail than the others.

The proposal that won will cost $6 billion. It calls for constructing a 31-mile twin-tube rail tunnel. The two tubes will be linked by a service corridor. Specially designed rail cars will carry autos, buses, and trucks under the channel in about half an hour. Passengers will be able to stay in their vehicles or in passenger compartments. Trains will leave every three minutes and are expected to carry 4,000 vehicles an hour each way. Estimates are that some 67 million passengers will cross the channel each year—half by the tunnel. Currently, the annual passenger rate is 47 million.

Constructing the Original Tunnel

The tunnel was bored by 11 Tunnel-Boring Machines (TBMs), equipped with steel cutting disks that ate into the chalk that makes up the seabed. The chalk is the perfect material for a tunnel: impervious to water yet easy to cut. Lasers linked to microprocessors guided the cutters. Once the tunnel was cut, it was lined with 23-inch-thick prefabricated concrete joined by steel bands.

The service tunnel was completed in 1990; the rail tunnels are expected to be completed by 1993.

Improving the Original Tunnel

Both France and Britain were reluctant to give up the idea of a roadway. Therefore, one clause of the contract calls for the construction of a roadway by the year 2000. If the consortium fails to complete the roadway by that date, the governments may open up bidding to other companies.

Conclusion

The English Channel Tunnel project is the most ambitious engineering undertaking of the twentieth century. Since the Second World War, the value of the English Channel as a protection of the British from the Continent has all but disappeared. The tunnel is expected to strengthen and symbolize the economic interdependence of the British and the Europeans.

WRITER'S CHECKLIST

Descriptions of Objects and Mechanisms

1. Does the title or section heading identify the subject and indicate whether the description is general or particular?

2. Does the introduction to the object or mechanism description
   a. define the item?
   b. identify its function (where appropriate)?
   c. describe its appearance?
   d. describe its principle of operation (where appropriate)?
   e. list its principal parts?
   f. include a graphic identifying all the principal parts?

3. Does the part-by-part description
   a. answer, for each of the major parts, the questions listed in item 1?
   b. describe each part in the sequence in which it was listed in the introduction?
   c. include graphics for each of the major parts of the mechanism?

4. Does the conclusion
   a. summarize the major points made in the part-by-part description?
   b. include (where appropriate) a description of the item performing its function?

Process Descriptions

1. Does the title or section heading identify the subject and indicate whether the description is general or particular?

2. Does the introduction to the process description
   a. define the process?
   b. identify its function (where appropriate)?
   c. identify where and when the process takes place?
   d. identify who or what performs it?
   e. describe how the process works?
   f. list its principal steps?
   g. include a graphic identifying all the principal steps?

3. Does the step-by-step description
   a. answer, for each of the major steps, the questions listed in item 1?
   b. discuss the steps in chronological order or in some other logical sequence?
   c. make clear the causal relationships among the steps?
   d. include graphics for each of the principal steps?

4. Does the conclusion
   a. summarize the major points made in the step-by-step description?
   b. discuss, if appropriate, the importance or implications of the process?

EXERCISES

1. Write a description of one of the following items or of a piece of equipment used in your field. Be sure to specify your audience and indicate the type of description (general or particular) you are writing. Include appropriate graphics.
   a. a carburetor
   b. a locking bicycle rack
   c. a deadbolt lock
   d. a folding card table
   e. a lawn mower
   f. a photocopy machine
   g. a cooling tower
   h. a jet engine
   i. a telescope
   j. an ammeter
   k. a television set
   l. an automobile jack
   m. a stereo speaker
   n. a refrigerator
   o. a computer
   p. a cigarette lighter

2. In an essay, evaluate the effectiveness of one of the following descriptions.
   a.

DESCRIPTION OF A CAMERA

A camera is really just a box that does not admit light. On one end is the lens, on the other is the film. The lens focuses an image of what it sees on the film. As seen in Figure 1, this image is upside down.

How does the image become reversed again? The film is made of silver halide, which undergoes a chemical change when it is exposed to light—the more light, the greater the change. During the process of developing the film, the change is emphasized further. When a positive is made from this negative, the image comes out correct.

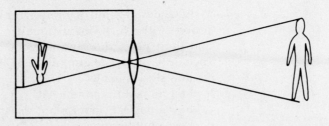

Figure 1

b.

COMPACT DISCS

A compact disc is about 12 cm in size, with a 15 mm hole in the center. See Figure 1. The disc itself is covered in plastic. Compact discs have music recorded on one side, although there is no reason not to put music on both sides. Today's compact discs can hold over seventy minutes of music.

The code on the disc consists of two components: pits and lands. Pits are the indentations etched into the disc by a laser when it is manufactured. A pit is only 0.5 μm wide; 20,000 pits would fit in an inch. Lands are the areas between the pits that are not etched. The pits go in a circle from the inside of the disc to the outside. If the pits were stretched out in a line, they would go for three miles.

When a disc is being played, it rotates clockwise from 200 to 500 rpm. The speed difference is caused by the fact that the disc travels at different speeds depending on what part of the disc the laser is reading.

Here is how the disc works. The laser in the player sends out a very thin beam of infrared light that hits the pits and lands on the disc. The light that hits the pits reflects back somewhat; the light that hits the lands reflects back almost completely. The photodetector sees this as a series of on and off flashes of light. Inside the player, this digital signal is turned back into analog, then sent to an amplifier and speakers.

Figure 1 Compact Disc

---

3. Rewrite one of the descriptions in Exercise 2.

4. Write a description of one of the following processes or a similar process with which you are familiar. Be sure to indicate your audience and the type of description (general or particular) you are writing. Include appropriate graphics.
   a. how steel is made
   b. how an audit is conducted
   c. how a nuclear power plant works
   d. how a bill becomes a law
   e. how a suspension bridge is constructed
   f. how a microscope operates

g. how we hear
h. how a dry battery operates
i. how a baseball player becomes a free agent
j. how cells reproduce

5. In an essay, evaluate the effectiveness of one of the following process descriptions.

a.

THE FOUR-STROKE POWER CYCLE

The power to drive an engine comes from the four-stroke cycle.

Intake Stroke: Inside the cylinder, the piston moves down, creating a vacuum that draws in the air/fuel mixture through the intake valve.

Compression Stroke: The intake valve shuts and the cylinder moves back up, compressing the air/fuel mixture in what is called the combustion chamber. The purpose of compression is to increase the power of the explosion that will occur in the next stroke.

Power Stroke: The spark across the electrodes of the spark plug ignites the air/fuel mixture, pushing the piston down. This in turn moves the wheels of the car.

Exhaust Stroke: The burned gases escape through the exhaust valve as the piston moves up again.

---

b.

COMPOSTING

Composting is a biological process by which organic wastes are turned into compost. In this process, microbial action reduces the volume of the waste by as much as half, resulting in a substance that can be used to condition soil.
Composting requires several steps:

- sorting
- size reduction
- digestion
- upgrading
- selling

First the waste is sorted, to remove inorganic materials such as glass, metal, and other nonbiodegradable things. Sorting is accomplished by mechanical means.

Next, the waste is shredded or pulverized to make it smaller. This step is necessary not only to make the waste easier to handle but also to assist in the biological activity in the next step.

Third, the waste is digested in a windrow, which is a long pile of the waste. Every few days, the pile is mixed to increase the digestion process. The digestion process takes about five weeks, although it may take up to nine weeks. The high heat produced during this stage kills most of the pathogenic organisms present in the waste. Windrows require a lot of space. A city of one million people would need about 250 acres just for the windrows. There are also mechanical systems to substitute for windrows; these require only about one sixth the space.

Several methods exist to upgrade the compost: drying, screening, and granulating.

Selling the compost is the big problem because it is expensive to truck it to farm areas, and because inorganic fertilizers are quite inexpensive.

Conclusion

However, as incineration becomes less common due to air-pollution standards, composting may become a more popular means of municipal waste treatment. Currently, only about two percent of the municipal waste is composted.

6. Rewrite one of the process descriptions in Exercise 5.

__REFERENCE__   Kyle, C. R. 1986. Athletic clothing. *Scientific American* 254, no. 3: 104–110.

10

Organizing Discussions

Chapter 5 discussed a number of common organizational patterns used in developing outlines, including the chronological, spatial, and general-to-specific.

The present chapter discusses several of these patterns in more detail and shows how they work not only in an outline but in extended discussions:

- ☐ more important to less important
- ☐ comparison and contrast
- ☐ classification and partition
- ☐ problems-methods-solution
- ☐ cause and effect

Keep in mind that these patterns are not molds into which you can pour your information. You will often have to modify and combine them to suit your readers and the special requirements of your information. You might find, for instance, that even though the overall pattern for your document is problems-methods-solution, one section calls for a comparison and contrast pattern, and one of *its* subsections requires a more-important-to-less-important pattern.

More-Important-to-Less-Important Pattern

When you have several points you wish to make about a particular subject, but the points don't naturally fall into a simple pattern such as chronology or spatial development, consider the more-important-to-less-important organization. For instance, you might want to explain why you chose to attend your college or university. Let's say the three factors were its location, its relatively low tuition and fees, and the quality of its program in your field. Chronology won't work because the factors don't exist in time, and spatial arrangement won't work because they don't exist in space. An effective strategy would be to start with the most important of the three, then treat the second-most important, and so on.

The more-important-to-less-important pattern recognizes that readers of technical writing want the bottom line—the most critical information—as quickly as possible. If they pick up a report that analyzes whether their company should adopt a new management structure, the first question they want answered is, Should we? Once they have learned the writers' main idea—that the company should or shouldn't adopt it or ought to study it more—they want to understand the writers' reasoning. If, for instance, the writers think that the company should not change management structures, they will offer reasons, such as that the current structure works fine, the proposed structure wouldn't work well, the changeover would be too disruptive and expensive, and so forth.

If the writers believe there is no serious problem with the current management structure, they should present that point first, because of the common-sense notion, "If it ain't broke, don't fix it." Why risk the trouble and expense of switching to a new system that is likely to have its own shortcomings?

You want to impress your readers, right away, with the quality of your reasoning. Therefore, don't begin with a minor point—such as that switching to a new management system would involve revising the organizational chart and the stationery—and hope to build up to the more compelling reasons. If you begin with a minor point, you might lose your credibility with the readers; they might toss the report aside, thinking that it's full of trivialities.

Be straightforward with your readers. If you have two very important points and three less-important points, present them that way: group the two big points together, and label them "Major Reasons to Retain Our Current Management Structure." Then present the less-important reasons as "Other Reasons to Retain Our Current Management Structure." Being straightforward is effective for two reasons: the material is easier to follow because it is clearly organized, and you look trustworthy.

Figure 10.1, an excerpt from an executive's memo in a company that sells equipment to manufacture semiconductors, shows the more-important-to-less-important organizational structure.

Comparison and Contrast

Much technical writing is organized by comparison and contrast—the same technique by which most of us learned as children. What is a zebra? A zebra looks like a horse (comparison), but it has stripes (contrast). In technical writing, comparison-and-contrast discussions are developed much more fully, of course, but the technique is essentially the same.

Comparison and contrast is a common writing technique in the working world because organizations constantly decide between alternatives. Should we purchase the model A computer, or the B, or the C? Should we hire person X or person Y? Should we carry out a comprehensive analysis of this potential site for a new plant, or would a more limited analysis be preferable? Probably the most common kind of report in technical writing is a feasibility report, in which the writer investigates a course of action and explains whether it is possible and practical. Almost always, this kind of report requires a substantial comparison-and-contrast section.

There are two major steps in comparing and contrasting items:

- ☐ establishing criteria
- ☐ organizing the discussion

A THREE-POINT PROGRAM TO IMPROVE SERVICE

What we have learned from the recent conference of semiconductor purchasers and from our own focus groups is that customers expect and demand better service than the industry currently provides. By this I don't mean returning phone calls, although that sort of common courtesy remains important. I mean something much more ambitious and difficult to attain: helping our customers do their job by anticipating and addressing their total needs. For this reason, I have formed a Customer Satisfaction Panel, chaired by Maureen Bedrich, whose job will be to develop policies that will enable us to improve the quality of the service we offer our customers.

I have asked the Panel to consider three major areas:

- improving the ease of use of our equipment
- improving preventive and corrective maintenance
- improving our compatibility with other vendors' products

Improving the Ease of Use of Our Equipment

The most important area to improve is user friendliness. When we deliver a new product to a customer, we have to sit down with them and explain how to integrate it into their own manufacturing process. This is time-consuming and costly not only for us, but for them as well. We must explore the option of automating this integration process. Automated process control, which is already common in such industries as machine-vision, would allow our customers to integrate our semiconductors much more efficiently. In addition, it would help them determine, through process control, whether our equipment is functioning according to specification.

Improving Preventive and Corrective Maintenance

The second most important area for study is improving preventive and corrective maintenance. Our customers will no longer tolerate down times approaching 10 percent; they require no more than 2–3 percent. In the semiconductor equipment field, preventive maintenance is critical because gases used in vapor-deposition systems periodically have to be removed from the inside of the equipment. Customers want to be able to plan for these stoppages to reduce costs. Currently, we have no means of helping them plan.

We also have to assist our customers with unscheduled maintenance. Basically, this means modularizing our equipment

FIGURE 10.1   More-Important-to-Less-Important Organizational Pattern

so that the faulty module can be pulled out of the equipment and replaced on the spot. This redesign will take many months, but I'm sure it will become a major selling point.

Improving Our Compatibility with Other Vendors' Products

Finally, we have to accept the fact that nobody in our industry is likely to control the market, and so we have to make our products more compatible with other manufacturers'. This involves a willingness to put our people on site to see what the customer's setup is and help them determine how to modify our product to fit in efficiently. We can no longer offer a "take-it-or-leave-it" product.

I hope you will extend every effort to work constructively with Maureen and her committee over the coming months to ensure that we improve the overall service we offer our customers.

FIGURE 10.1 *Continued*

Establishing Criteria

The first step in comparing and contrasting items is to establish the criteria: the standards or needs against which you will be analyzing the items. For instance, if you need to choose an elective course to take next semester, your only criterion might be that it must be offered at 10 o'clock on Mondays, Wednesdays, and Fridays.

Almost always, however, you will need to consider several criteria in carrying out the comparison and contrast. For example, you might need a three-credit science course that meets at 10 o'clock on Mondays, Wednesdays, and Fridays. In this case, you have three criteria: number of credits, general subject area, and schedule. When you are comparing and contrasting items for a study on the job, you are likely to have six, eight, or even more criteria. For a recommendation report on which kind of computer your company should buy, for example, your comparison and contrast would probably include all the following criteria:

1. availability of software
2. amount and type of storage capacity
3. ease of operation
4. reliability
5. availability of peripherals
6. accessibility of maintenance and service personnel
7. initial costs and maintenance costs

The more criteria you use in your comparison and contrast, the more precise your analysis will be. If you want to buy a used car but your only criterion is price, you don't have to ponder whether to buy the $1,000 car or the $2,000 car. However, when you add more and more criteria—such as age, reliability, and features—you will probably conclude that the more expensive car is "better"—according to the analysis.

Choose sufficient and appropriate criteria when you compare and contrast items. Don't overlook an important criterion. A beautifully written comparison and contrast of office computers is of little value if you have neglected price. If your company can spend no more than $35,000 and you carry out a detailed comparison and contrast of models that cost more than $50,000, you will have wasted everyone's time. Be careful, too, that the criteria you choose are sensible. If, for instance, your company has a large, empty storeroom whose temperature is just right for the computer, don't use size as one of your criteria: there is no reason to spend more money for a compact unit if a larger model would be perfectly acceptable.

Organizing the Discussion

Once you have chosen your criteria and carried out any necessary research, decide on a pattern to use in developing the passage. The two patterns from which to choose are called *whole-by-whole* and *part-by-part.*

In the whole-by-whole pattern, the first item is discussed, then the second, and so on:

Item 1
    Aspect A
    Aspect B
    Aspect C
Item 2
    Aspect A
    Aspect B
    Aspect C
Item 3
    Aspect A
    Aspect B
    Aspect C

The whole-by-whole pattern is effective if you want to focus on each item as a complete unit rather than on individual aspects of different items. For example, you are writing a feasibility report on purchasing an expensive piece of laboratory equipment. You have narrowed your choice to three different models, each of which fulfills your basic criteria. Each model has its advantages and disadvantages. Because your organization's decision on which model to buy will depend on an overall assessment rather than on any single aspect of the three models, you choose the whole-by-whole pattern, which gives your readers a good overview of the different models.

In the part-by-part pattern, each aspect is discussed separately:

Aspect A
    Item 1
    Item 2
    Item 3
Aspect B
    Item 1
    Item 2
    Item 3
Aspect C
    Item 1
    Item 2
    Item 3

The part-by-part pattern focuses on the individual aspects of the different items. Detailed comparisons and contrasts are more effective in the part-by-part pattern. For example, you are comparing and contrasting the three pieces of laboratory equipment. The one factor that distinguishes the three models is reliability: one model has a much better reliability record than the other two. The part-by-part pattern lets you create a section on reliability to highlight this aspect. Comparing and contrasting the three models in one place in the document rather than in three places makes your point more emphatic. You sacrifice a coherent overview of the different items for a clear, forceful comparison and contrast of an aspect of the three models.

Of course, you can have it both ways; if you want to use a part-by-part pattern to exphasize particular aspects, begin the discussion with a general description of the different items.

Once you have chosen the overall pattern—whole-by-whole or part-by-part—you have to decide how to organize the second-level items. That is, in a whole-by-whole passage, you have to sequence the "aspects"; in a part-by-part passage, you have to sequence the "items." For most documents, a more-important-to-less-important organizational pattern will work well because your readers want to get to the bottom line as soon as possible. For some documents, however, other patterns might work better. People who write for readers outside their own company often reverse the more-important-to-less-important pattern because they want to make sure their audience reads the whole discussion. This pattern is also popular with writers who are delivering bad news. If, for instance, you want to explain why you are recommending *not* going ahead with a popular course of action, the reverse sequence enables you to show your readers the problems with the popular plan before you present the plan you recommend. Otherwise, you might surprise the readers, and they might start to formulate objections in their minds before you have had a chance to explain your position.

Figure 10.2 is a comparison and contrast passage about dot-matrix printers and laser printers taken from a guide to buying personal computers. It is addressed to a general audience and employs the whole-by-whole organizational pattern.

Figure 10.3 shows the same passage, rewritten in the part-by-part pattern.

DOT-MATRIX AND LASER PRINTERS

There are two kinds of printers commonly used with PCs: dot-matrix and laser printers. Both kinds of printers are getting more sophisticated and less expensive every year. The following discussion will cover four aspects of the printers:

- operating principle
- print quality
- speed
- price

Dot-Matrix Printers

Operating Principle

A dot-matrix printer is an impact printer, that is, it works by hitting an ink ribbon, which leaves ink on the paper. Unlike a typewriter, in which metal letters hit the ribbon, in a dot-matrix printer little pins hit the ribbon. The computer tells the printer how to arrange the pins to form the letter or graphic you want. Less-expensive dot-matrix printers have a small number of pins, such as only 9; more expensive ones go up to 24.

Print Quality

Dot-matrix printers range from low-quality print to very high quality. A low-end 9-pin printer will print very poorly; you can read what it produces, but the print is faint and you can see the spaces between the little dots that make up the letters. A 24-pin printer can produce what is called "letter-quality" print, that is, print that looks as good as that produced by an expensive office typewriter.

The average dot-matrix printer has two or even three different printing modes. If you want the print quality to be high, you select high-quality, which instructs the printer to

---

FIGURE 10.2   Whole-by-Whole Comparison-and-Contrast Organizational Pattern

print each line several times. This repetition produces a darker, more professional print. If you don't care about quality, you choose draft mode, which instructs the printer to print each line only once.

One other aspect of print quality should be mentioned: printing of graphics. A dot-matrix printer, because of its pin technology, can print graphics, but in general the quality is only mediocre: up to perhaps 150 or 200 dots per square inch.

Speed

A dot-matrix printer in the draft mode can produce about two or three pages per minute. In the high-quality mode, the speed can be cut to one page or even less. In other words, the better you want the document to look, the longer it will take to print.

Price

Dot-matrix printers are relatively inexpensive. A low-end printer can cost less than $150; a high-end printer, which can produce letter-quality print, costs about $550.

The other expense involved with a dot-matrix printer is the ribbon, which ranges from about $5 to $15 and lasts about 500 double-spaced pages on the medium-quality print option.

Laser Printers

Operating Principle

In a laser printer, a rotating drum receives an electric charge in a pattern that corresponds to the letters or graphics you want to print. As the blank paper passes over the drum, the electric charge attracts a powdery substance called toner, which adheres to the paper only in those places that have received the electric charge. A laser beam is then directed onto the paper, burning the toner into the paper.

Print Quality

A laser printer produces much better quality print than a dot-matrix. Whereas a dot-matrix produces about 150–200 dots per square inch, the laser printer produces 600. What this means is that, for most readers, the text looks as good as typeset text, that is, printed books. And graphics are substantially sharper than those produced on a dot-matrix printer.

Speed

A low-end laser printer will produce about four pages per minute; a high-end, eight or more. In other words, a laser

FIGURE 10.2   *Continued*

. printer is faster than most dot-matrix printers even when they are printing in the draft mode, their fastest speed.

Price

A basic laser printer costs less than $1,000, almost twice as much as a high-end dot matrix. A sophisticated one costs close to $2,000.

Laser printers don't use ribbons, but toner cartridges. A new cartridge costs about $70 and lasts for about 2,000 pages. The cost per page, then, is about equal for laser printers and dot-matrix printers.

Conclusion

With computer printers, you get what you pay for. If all you need is a readable copy of your text and graphics, a $200 dot-matrix printer will do the job. If you need higher-quality text and graphics, or fast output, you will have spent close to $1,000 for a laser printer.

FIGURE 10.2   *Continued*

DOT-MATRIX AND LASER PRINTERS

Here is the same passage as in Figure 10.2 presented in the part-by-part pattern.

There are two kinds of printers used commonly with PCs: dot-matrix and laser printers. Both kinds of printers are getting more sophisticated and less expensive every year. The following discussion will cover four aspects of the printers:

- operating principle
- print quality
- speed
- price

Operating Principle

Dot-Matrix Printers

A dot-matrix printer is an impact printer, that is, it works by hitting an ink ribbon, which leaves ink on the paper. Unlike a typewriter, in which metal letters hit the ribbon, in a dot-matrix printer little pins hit the ribbon. The computer tells the printer how to arrange the pins to form the letter or symbol you want. Less-expensive dot-matrix printers have a small number of pins, such as only 9; more expensive ones go up to 24.

FIGURE 10.3   Part-by-Part Comparison-and-Contrast Organizational Pattern

Laser Printer

In a laser printer, a rotating drum receives an electric charge in a pattern that corresponds to the letters or graphics you want to print. As the blank paper passes over the drum, the electric charge attracts a powdery substance called toner, which adheres to the paper only in those places that have received the electric charge. A laser beam is then directed onto the paper, burning the toner into the paper.

Print Quality

Dot-Matrix Printers

Dot-matrix printers range from low-quality print to very high quality. A low-end 9-pin printer will print very poorly; you can read what it produces, but the print is faint and you can see the spaces between the little dots that make up the letters. A 24-pin printer can produce what is called "letter-quality" print, that is, print that looks as good as that produced by an expensive office typewriter.

The average dot-matrix printer has two or even three different printing modes. If you want the print quality to be high, you select high-quality, which instructs the printer to print each line several times. This repetition produces a darker, more professional print. If you don't care about quality, you choose draft mode, which instructs the printer to print each line only once.

One other aspect of print quality should be mentioned: printing of graphics. A dot-matrix printer, because of its pin technology, can print graphics, but in general the quality is only mediocre: up to perhaps 150 or 200 dots per square inch.

Laser Printer

A laser printer produces much better quality print than a dot-matrix. Whereas a dot-matrix produces about 150–200 dots per square inch, the laser printer produces 600. What this means is that, for most readers, the text looks as good as typeset text, that is, printed books. And graphics are substantially sharper than those produced on a dot-matrix printer.

FIGURE 10.3   *Continued*

Speed

Dot-Matrix Printers

A dot-matrix printer in the draft mode can produce about two or three pages per minute. In the high-quality mode, the speed can be cut to one page or even less. In other words, the better you want the document to look, the longer it will take to print.

Laser Printers

A low-end laser printer will produce about four pages per minute; a high-end, eight or more. A laser printer is therefore faster than most dot matrix printers even when they are printing in the draft mode, their fastest speed.

Price

Dot-Matrix Printers

In general, dot-matrix printers are inexpensive. A low-end printer can cost less than $150; a high-end printer, which can produce letter-quality print, costs about $550.

The other expense involved with a dot-matrix printer is the ribbon, which ranges from about $5 to $15 and lasts about 500 double-spaced pages on the medium-quality print option.

Laser Printers

A basic laser printer costs less than $1,000, almost twice as much as a high-end dot matrix. A sophisticated one costs close to $2,000.

Laser printers don't use ribbons, but toner cartridges. A new cartridge costs about $70 and lasts for about 2,000 pages. The cost per page, then, is about equal for laser printers and dot-matrix printers.

Conclusion

With computer printers, you get what you pay for. If all you need is a readable copy of your text and graphics, a $200 dot-matrix printer will do the job. If you need higher-quality text and graphics, or fast output, you will have to spend close to $1,000 for a laser printer.

FIGURE 10.3   *Continued*

Classification and Partition

Classification is the process of putting items into different categories that share certain characteristics. For instance, all the students at a university could be classified by gender (males and females), age (17-years old, 18, 19, and so forth), major (nursing, finance, sociology), and any number of other characteristics.

You can create categories within categories. For instance, within the category of students majoring in business at your college or university, you can create subcategories: male business majors and female business majors. If necessary, you could keep creating subcategories until you run out of items to categorize: male business majors born in the United States and male business majors not born in the United States.

Partition is the process of breaking down a unit into its components. A stereo system could be partitioned into the following components: turntable, cassette deck, CD player, tuner, amplifier, and speakers. Each component is separate; the only relation among them is that together they form a stereo system. Similarly, a stereo system that lacks any of these components would be incomplete. Each component can of course be partitioned further. Partition underlies descriptions of objects, mechanisms, and processes, as explained in Chapter 9.

The most important concept to keep in mind when you are classifying or partitioning is that often you are establishing a variety of levels. As you know from outlining, you must avoid confusing major and minor categories (faulty coordination). An item that is out of place can destroy any system that you employ. Consider the following attempt to classify keyboard instruments:

> pianos
> organs
> harpsichords
> synthesizers
> concert grand pianos

Clearly, "concert grand pianos" is a subcategory of pianos and does not belong in this system of classification. The following partition of a small sailboat is similarly faulty:

> hull
> sails
> rudder
> centerboard
> mast
> mainsail
> jib

"Mainsail" and "jib" are subcategories of sails and represent a level of partition beyond that defined by the rest of this list.

The following guidelines should help you prevent this error:

☐ Choose a basis consistent with your audience and purpose.
☐ Use only one basis at a time.
☐ Avoid overlap.
☐ Be inclusive.
☐ Arrange the categories in a logical sequence.

■ *Choose a basis of classification or partition that is consistent with your audience and purpose.* Never lose sight of your audience—their backgrounds, skills, and reasons for reading—and your purpose. If you were writing a brochure for hikers in a particular state park, you would probably classify the local snakes according to whether they are poisonous or nonpoisonous, not according to size, color, or any other basis. If you were writing a manual describing do-it-yourself maintenance for a particular car, you would most likely partition the car into its component parts so that the owner could locate specific discussions.

■ *Use only one basis of classification or partition at a time.* If you are classifying fans according to movement, do not include another basis of classification:

> oscillating
> stationary
> inexpensive

"Inexpensive" is inappropriate here because it has a different basis of classification—price—from the other categories. Also, do not mix classification and partition. If you are partitioning a fan into its major components, do not include a stray basis of classification:

> motor
> blade
> casing
> metal parts

"Metal parts" is inappropriate here because motor, blade, and casing are partitions based on function; "metal parts," as opposed to "plastic parts," might be a useful classification, but it does not belong in this partition.

■ *Avoid overlap.* Make sure that no single item could logically be placed in more than one category of your classification or, in partition, that no listed component includes another listed component. Overlapping generally results from changing the basis of classification or the level at which you are partitioning a unit. In the following classification of bicycles, for instance, the writer introduces a new basis of classification that results in overlapping categories:

> mountain bikes
> racing bikes
> touring bikes
> ten-speed bikes

The first three categories here have use as their basis; the fourth category, number of speeds. The fourth category is illogical, because a particular ten-speed bike could be a mountain bike, a touring bike, or a racing bike. In the following partition of a guitar, the writer changes the level of the focus:

> body
> bridge
> neck
> frets

The first three items here are basic parts of a guitar. "Frets," as part of the neck, represents a further level of partition.

■ *Be inclusive.* When classifying or partitioning, be sure to include all the categories. For example, a classification of music according to type would be incomplete if it included popular and classical music but not jazz. A partition of an automobile by major systems would be incomplete if it included the electrical, fuel, and drive systems but not the cooling system.

If your purpose or audience requires that you omit a category, tell your readers what you are doing. If, for instance, you are writing in a classification passage about recent sales statistics of General Motors, Ford, and Chrysler, don't use "American car manufacturers" as a classifying tag for the three companies. Although all three belong to the larger class "American car manufacturers," there are a number of other American car manufacturers. Rather, classify General Motors, Ford, and Chrysler as "the big three" or refer to them by name.

■ *Arrange the categories in a logical sequence.* After establishing your categories and subcategories of classification or partition, arrange them according to some reasonable plan: time (first to last), space (top to bottom), importance (most to least), and so on. For example, a classification of cassette decks might begin with the simplest kind and proceed to the most sophisticated. See Chapter 5 for a discussion of patterns of development.

Several examples of classification and partition are shown in Figures 10.4 through 10.6.

The first example, Figure 10.4, is taken from a description of background-sound systems used in industry. This classification is based on who controls the music being broadcast. The various music systems could easily be classified according to other bases: the source of the music (FM radio and nonradio); cost (expensive and inexpensive systems); physical origin (on the job site or off).

The various background-sound systems available for industrial use can be classified according to whether the client has any control over the content of the programs. Piped-in phone line systems and FM radio systems are examples of those services that are not controlled by the client. In the piped-in phone line system, a special phone hook-up brings in the music, which is programmed entirely by the vendor. The FM radio system uses a special receiver to edit out news and other nonmusic broadcasts. Sound systems that can be controlled by the client include on-the-premises systems and customized external systems. An on-the-premises system is any arrangement in which a person or persons at the job site selects and broadcasts recorded music. A customized external system lets the client choose from a number of different programs or, for an extra fee, stipulate what the vendor is to broadcast.

FIGURE 10.4   Classification

Figure 10.5, a classification of nondestructive testing techniques, shows how the writer uses classification and subclassification effectively. She is introducing nondestructive testing to a technical audience.

Notice that the writer has classified nondestructive testing by technology. (One method, ultrasonic attenuation, is subclassified.) Another basis for classification would be the sensitivity of the techniques: very sensitive and less sensitive.

Figure 10.6, a description of a record turntable (Isganitis 1977), is an example of partition. For more examples of partition, see Chapter 9, which covers descriptions of objects, mechanisms, and processes.

Nondestructive testing of structures permits early detection of stresses that can cause fatigue and ultimately structural damage. The least sensitive tests isolate macrocracks. More sensitive tests identify microcracks. The most sensitive tests identify slight stresses. All sensitivities of testing are useful because some structures can tolerate large amounts of stress—or even cracks—before their structural integrity is threatened.

There are four techniques for nondestructive testing:

1. body-wave reflection
2. surface-wave reflection
3. ultrasonic attenuation
4. acoustic emission

Body-Wave Reflection

In this technique, a transducer sends an ultrasonic pulse through the test material. When the pulse strikes a crack, part of the pulse's energy is reflected back to the transducer. Body-wave reflection cannot isolate stresses: the pulse is sensitive only to relatively large cracks.

Surface-Wave Reflection

The transducer generates an ultrasonic pulse that travels along the surface of the test material. Cracks reflect a portion of the pulse's energy back to the transducer. Like body-wave reflection, surface-wave reflection picks up only macrocracks. Because cracks often begin on interior surfaces of materials, surface-wave reflection is a poor predictor of serious failures.

Ultrasonic Attenuation

The transducer generates an ultrasonic pulse either through or along the surface of the test material. When the pulse strikes cracks or the slight plastic deformations associated with stress, part of the pulse's energy is scattered. Thus, the amount of the pulse's energy decreases. Ultrasonic attenuation is a highly sensitive method of nondestructive acoustic testing.

Two methods of ultrasonic attenuation exist. One technique reflects the pulse back to the transducer. In the other technique, a second transducer can be set up to receive the pulses sent through or along the surface of the material.

Acoustic Emission

When a test specimen is subjected to a great amount of stress, it begins to emit waves, some of which are in the ultrasonic range. A transducer attached to the surface of the test specimen records these waves. Current technologies make it possible to interpret these waves accurately for impending fatigue and cracks.

FIGURE 10.5 Classification and Subclassification

GENERAL DESCRIPTION OF A RECORD TURNTABLE

Introduction

The music that comes out of the speakers of a record player is encoded in the grooves of the phonograph record. The first step in getting the music out of the record and into the air is performed by the record turntable, the component that "reads" the grooves of the record and passes the encoded signal on to be amplified and finally transmitted.

Figure 1 shows the four basic components of the record turntable.

The operating principle of the turntable is simple. The motor and the drive system rotate the platter, on which the record rests. As the record rotates, the tonearm picks up the recorded signals and transmits them to the amplifier.

The Components

The Motor

The function of the motor is to provide a source of power to turn the platter at an accurate and consistent speed. Speed accuracy is important in preventing "wow and flutter," the short-term fluctuations that give the recorded music a harsh and muddy sound.

Although several different kinds of motors are used in turntables, the most common kind is the AC synchronous motor, which operates at a fixed speed determined by the frequency of the AC voltage from the power line. A good turntable will have a speed accuracy better than ± 0.10 percent.

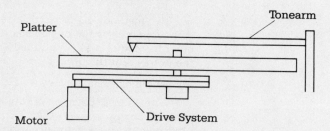

Figure 1   Basic Components of a Turntable

FIGURE 10.6   Partition

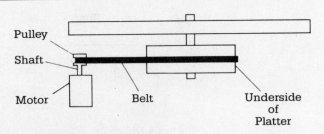

Figure 2 Belt-Drive System

The Drive System

The drive system transmits the power from the motor to the platter. A good drive system will not introduce any mechanical vibrations that cause "rumble," a humming sound.

Three basic drive systems are used in turntables:

1. belt-drive
2. idler-rim drive
3. direct-drive

In the belt-drive system (see Figure 2), a pulley is mounted on the end of the motor shaft. This pulley drives a flexible rubber belt coupled to the underside of the platter. The advantage of the belt-drive system is that the belt acts as a mechanical damper, reducing the rumble and the wow and flutter. However, the belt eventually deteriorates, varying the speed at which the platter turns.

In the idler-rim system (see Figure 3), the motor shaft is linked to the platter by a hard rubber wheel. Idler-rim drive is a sturdy system, but because the platter is directly linked to the motor, rumble is relatively high.

In the direct-drive system (see Figure 4), the platter rests directly on the motor shaft. The direct-drive is powered by a DC servocontrolled motor, which can be operated at the exact

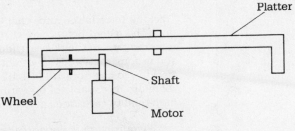

Figure 3 Idler-Rim Drive System

FIGURE 10.6 *Continued*

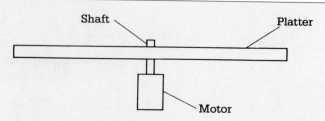

Figure 4 Direct-Drive System

platter speed required: 33 1/3 or 45 RPM. Direct-drive requires no linkage, and thus it produces the least rumble and wow and flutter. Direct-drive is a sturdy and reliable system, but it is expensive.

The Platter

The platter is the machined metallic disc (sometimes covered with a soft material such as felt) on which the record rests while it is being played. There is little variation in the quality of different platters.

The Tonearm

The tonearm houses the cartridge, the removable case that contains the stylus (the "needle") and the electronic circuitry, which converts the mechanical vibrations of the record grooves into electrical signals. The cartridge is the crucial component of the turntable because it actually touches the grooves of the record, thus affecting not only the sound quality but also the record wear.

The design of every kind of tonearm is unique. However, the operating characteristics of a tonearm can be measured. Four factors are generally considered in describing the performance of a tonearm:

1. tracking force
2. pivot friction
3. tracking error
4. tonearm resonance

Tracking force is the force that the stylus and tonearm exert upon the record being played. The amount of tracking force required by a particular cartridge depends on the quality of its stylus construction. High-quality cartridges require low tracking forces from 1/2 to 1 1/2 grams. A tonearm should therefore be capable of operating at low tracking forces so that high-quality cartridges may be used.

Pivot friction is horizontal and vertical friction in the pivot

FIGURE 10.6 *Continued*

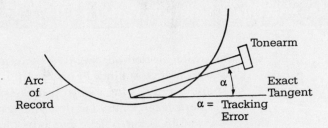

Figure 5 Illustration of Tracking Error (Top View)

of the tonearm. Excessive pivot friction requires higher tracking forces to prevent the stylus from jumping the record groove. This in turn increases record and stylus wear.

Tracking error (or tangent error) results when the cartridge is not held exactly tangent to the record circumference, as shown in Figure 5. The tracking error figure for a tonearm is a measure of the error in the angle of the cartridge with respect to the exact tangent. Depending on the arm geometry, tracking error will vary from point to point along the record radius. When tracking error occurs, the stylus does not ride properly in the record groove, distorting the sound and damaging the record grooves. The maximum value of tracking error for a high-quality tonearm is approximately 5 degrees at the outer edge of a 12-inch record.

Tonearm resonance occurs when the tonearm encounters a signal (or vibration) of a frequency equal to its own resonant frequency. At this resonant frequency, the arm undergoes abnormally large vibrations, resulting in mistracking of the record groove by the stylus. Resonance is a common problem in many mechanical devices and is not easily avoided. However, proper design can limit the range of frequencies over which resonance will occur. For tonearms, the ideal range for resonant frequencies is between 10 and 15 Hz, because it is below the audible frequency range (20 to 20,000 Hz).

Conclusion

Turntables can cost from about $30 to more than ten times that amount. They range from single-play models, which require that the operator lift the tonearm and place it on the record, to sophisticated systems that the operator can program to search out particular songs without playing the entire side of the record. However, the basic principle of all turntables is the same: a motor produces power that rotates the platter. The cartridge picks up and deciphers the signals embedded in the grooves of the record that is resting on the platter. The signal is then sent along to the other components—the preamplifier, amplifier, and speakers—to produce audible sound.

FIGURE 10.6   *Continued*

Problem-Methods-Solution Pattern

The problem-methods-solution pattern is an excellent way to organize a discussion of most kinds of projects. It is easy to write—and easy to read—because it reflects the logical processes used in carrying out the project. This pattern can be used to describe past work or future work; a proposal, for example, is a problem-methods-solution discussion intended to persuade your readers to let you carry out a project.

The three components of this pattern are simple to identify:

☐ *problem:* what was not working (or not working as effectively as it should), or what opportunity exists for improving current processes
☐ *methods:* the procedure you performed either to confirm your analysis of the problem or to figure out how to solve it
☐ *solution:* whether your analysis of the problem was correct, or what you discovered or devised to solve the problem

For example, the problem might be that you know you need to buy a bicycle, but there are a number of different types of bikes on the market, and you don't know the advantages and disadvantages of the different types. So you decide to do some research. Your methods include talking to friends who know about bikes, visiting bike shops, and reading books or journal articles about the subject. Based on this research, you devise a solution: to put an ad in the school paper requesting to hear from anyone who wants to sell a used 18-speed mountain bike.

As you write a problems-methods-solution passage, be clear and specific. Don't say that our energy expenditures are getting out of hand. Instead, say that the energy usage has increased 7 percent in the last year, and that the utility costs have risen 11 percent. Then calculate the total increase in energy costs.

In describing your methods, make sure your readers understand not only exactly what you did, but also why you did it that way. Most technical problems can be approached using several different methods; therefore, you might have to justify your choices. If you chose to read about bicycles, why those particular books and journals? The fact that they were available isn't good enough; they have to be current and authoritative.

Finally, when you describe the solution, don't overstate. Be very careful about saying, "This decision will enable us to increase our market share from 7 to 10 percent within 12 months." The world is too complicated for that sort of self-assurance. Instead, be more careful: "This decision promises to increase our market share significantly: an increase from 7 to 10 percent or even 11 percent is quite possible." This way, your document won't come back to haunt you if things don't turn out as well as you had anticipated.

A problems-methods-solution discussion can be sequenced in a number of ways. The most common way, naturally, is to start with the problem, follow with the methods, and conclude with the solution. However, sometimes you might want to present the problem first, then go directly to the solution, leaving the methods for last. This sequence deemphasizes the methods, which is appropriate if your readers either already know them well or have little need to understand them. In some articles the methods come first; the writers' strategy is to interest the readers in what they did in the lab or in the field, then go back and present the problem and solution. Similarly, some writers start with the solution, then discuss the problem and methods.

Sequencing is relatively unimportant as long as you (1) provide some kind of preliminary summary to give your readers an overview and (2) provide headings or some other format devices to enable your readers to find the information they want.

The problems-methods-solution discussion in Figure 10.7 is based on an article (Moore 1990) by a police officer writing in a futurist journal. His subject is recent attempts to deal with the problems that result when police officers pursue a suspect who is fleeing by car. Notice how the first heading introduces the problem; the second, the methods; and the third, the solution.

Cause and Effect

Technical writing often involves cause-and-effect reasoning. Sometimes you will reason forward: cause to effect. If we raise the price of a particular product we manufacture (cause), what will happen to our sales (effect)? The government has a regulation that we may not use a particular chemical in our production process (cause); what will we have to do to keep the production process running smoothly (effect)? Sometimes you will reason backward: from effect to cause. Productivity went down by 6 percent in the last quarter (effect); what factors led to this decrease (causes)? The federal government has decided that used-car dealers are not required to tell potential customers about the cars' defects (effect); why did the federal government reach this decision (causes)?

Cause-and-effect reasoning, therefore, provides a way to answer the following two questions:

What will be the effect(s) of X?
What caused X?

Establishing causal links requires great care. The following discussion of creating causal links is based on the theory of philosopher Stephen Toulmin (1980).

What's Wrong with Police Pursuit Methods?

Throughout this century, law-enforcement agencies have accepted the idea that when a criminal or a criminal suspect is attempting to flee in a car, it is the department's responsibility to try to capture him or her. Each year, police pursuits result in the deaths of hundreds of criminals and suspects as they lose control of their cars. These deaths are regrettable, but the response, by both the police and the general public, has been that this is the price you pay when you try to elude the law. However, there is another statistic that must not be overlooked. Even though police pursuits result in the capture of more than 1,000 suspects or criminals each year in the United States, more than 100 innocent civilians and police officers are killed in police pursuits; the suspect's car or the pursuit vehicle loses control and hits pedestrians or other cars. In recent years, these tragedies have been the subject of court action, and some departments and police officers have been convicted of civil and criminal charges resulting from injuries and deaths, sometimes even when the suspect's car caused the accident. It is time to ask ourselves whether chasing the bad guys is an outmoded procedure that often causes more problems than it solves.

We should be working toward a comprehensive solution: a set of high-tech tools that will keep the suspect from trying to flee by car. If we devote our resources to this plan, we should realize our goal within five years.

What We Should Be Doing about the Problem

We should be advocating and sponsoring research in two broad areas: preventing the suspect from getting in the car in the first place, and using safer methods of police pursuit. Technology exists, or is within grasp, to assist in both areas.

First, preventing the criminal or suspect from getting in the car. Detroit has prototype systems that call for the driver to perform simple motor skills; these systems prevent most drunk drivers from starting their cars. Convicted drunk drivers could be forced to wear electronic identification tags, just as some other sorts of criminals do, to alert the police when they try to start their cars. Similar electronic systems could be devised that stop car thieves, or at least slow them down. Several states already allow an electronic tagging system, hidden within the car, that the police can activate when the car is reported stolen. The thief cannot hear the electronic beeper and doesn't even know that the police are tracking him; he is usually arrested while he sits watching TV in his living room.

FIGURE 10.7   Problem-Methods-Solution Organizational Pattern

Second, using more sophisticated pursuit methods. When we pursue by car, we are engaged in a fairly equal contest with the suspect or criminal. We should try to change the odds. Some departments are experimenting with military-style helicopters equipped with massive searchlights and infrared tracking devices and heat-sensitive cameras that can identify suspects on the ground after they have abandoned a car and try to flee on foot. Helicopters would also enable the police to shoot a beeper device that attaches to the car, making it easier and safer to pursue it. Perhaps the most promising technology already exists: if all production cars sold in the United States were equipped with a radio receiver designed to respond to a certain frequency transmission, the police car or helicopter could disable the fleeing car by simply zapping it with a radio wave. The radio wave would shut down the car's electrical system.

Police organizations across the nation have tremendous political clout. For too long we have let the insurance industry push for research and development in these areas. It is time that we step up our own lobbying and research efforts. After all, our officers, as well as innocent civilians, are dying every day.

Conclusion

In researching these potential solutions, we must of course consider the issues of privacy, human rights, and individual freedoms. Yet we cannot risk continuing to chase criminals and suspects at high speeds through streets filled with civilians. We should borrow and adapt technology devised by the military services, and assist the auto makers in developing high-tech devices that make it harder to elude the police and easier for us to stop the suspect or criminal. If we take these steps, within several years we can present a comprehensive, cost-effective set of measures that will reduce dramatically the number of pursuit injuries and fatalities.

FIGURE 10.7   *Continued*

The Three Elements of a Causal Discussion

There are three elements to a persuasive causal discussion:

- ☐ the assertion
- ☐ the facts
- ☐ the reasoning

The assertion is the claim you wish to prove or make compelling. Here are some examples of assertions:

Our company should institute flextime.

Our company should not institute flextime.

The search for alternative fuels has been hindered by myopic government policy.

By the end of next year, we will have the new rolling machine in full production.

The key to success in our industry is a commitment to the customer.

The assertion is the conclusion you want your readers to accept—and, if appropriate, act on—after they have finished reading your discussion.

The facts are the information you want your audience to consider as they read your discussion. For a discussion of the merits of flextime, the facts might include the following:

The turnover rate among our female employees is double that of our male employees.

Replacing a staff-level employee costs us about one half the employee's annual salary; replacing a professional-level employee, a whole year's salary.

At exit interviews, over 40 percent of our female employees under the age of 38 state that they quit because they cannot work and also be home for their school-age children.

Other companies have found that flextime significantly decreases the turnover rate among female employees under the age of 38.

Other companies have found that flextime has additional benefits and introduces no significant problems.

The reasoning is the way you derived the assertion from the facts. In the discussion of flextime, the reasoning involves three links:

Flextime appears to have reduced the problem of high turnover among younger female employees at other companies.

Our company is similar to other companies.

If flextime has proven helpful at other companies, it is likely to prove helpful at our company.

For your discussion to be effective, your readers of course have to understand your facts and your reasoning. In addition, however, they must accept that your facts are accurate and relevant to the subject, and that you have not left out any relevant ones. And finally, they must accept that your reasoning is logical.

It is relatively simple to make sure your facts are accurate, but how do you make sure they are also sufficient and complete? This is not an easy question, because standards vary from field to field. In general, however, common sense is the best guide. For instance, if you state that twice as many female employees as male employees under the age of 38 quit last year, but the facts are that two women and one man quit, you don't have a persuasive case. If, however, 43 women and 21 men quit, that would seem to be more significant. Another example: if your reader believes, from talking to his neighbor about flextime, that it causes new problems—decreased car-pooling opportunities and increased utility bills, for example—but you haven't addressed these potential problems, that reader might doubt your assertion. You have to meet the skeptical or hostile reader's possible objections to your case by qualifying your assertion or conceding its drawbacks.

And how do you convince your readers that your reasoning is logical? If there were an easy answer to this question, the world would be much more peaceful. But again, you have to use your common sense. If your argument depends on the idea that your company is like the other companies that have found flextime to be successful, you have to decide whether that point needs to be made explicitly. If all the examples you can cite are from the same industry as yours, from companies of about the same size, with employees of approximately the same background, it might be necessary only to list some of those companies. On the other hand, if the companies you can cite are not apparently similar to yours, you might want to explain why you think that a plan that worked well for them is likely to work well for you, despite your differences.

Logical Fallacies

As you create causal discussions, be careful to avoid three basic logical errors:

☐ inadequate sampling
☐ *post hoc* reasoning
☐ begging the question

■ *Inadequate sampling.* Don't draw conclusions about causality unless you have an adequate sample of evidence. It would be illogical to draw the following conclusion:

> The new Gull is an unreliable car. Two of my friends own Gulls, and both have had reliability problems.

The writer would have to study a much larger sample and compare the findings with those for other cars in the Gull's class before reaching any valid conclusions. However, sometimes incomplete data are all you have to work with. It is proper in these cases to qualify your conclusions:

> The Martin Company's Collision Avoidance System has performed according to specification in extensive laboratory tests. However, the system cannot be considered effective until it has performed satisfactorily in the field. At this time, Martin's system must be considered very promising.

■ Post hoc *reasoning*. The complete phrase—*post hoc, ergo propter hoc*—means "After this, therefore, because of this." The fact that A precedes B does not mean that A caused B. The following statement is illogical:

> There must be something wrong with the new circuit breaker in the office. Ever since we had it installed, the air conditioners haven't worked right.

The air conditioners' malfunctioning *might* be caused by a problem in the circuit breaker; on the other hand, it might be caused by inadequate wiring or by problems in the air conditioners themselves.

■ *Begging the question*. To beg the question is to assume the truth of what you are trying to prove. The following statement is illogical:

> The Matrix 411 printer is the best on the market because it is better than the other printers.

The writer here states that the Matrix 411 is the best because it is the best; he does not show why he thinks it is the best.

Figure 10.8, from a discussion of why mammals survived but dinosaurs did not (Jastrow 1977: 104–105), illustrates a very effective use of cause-and-effect reasoning. Notice that Figure 10.8, addressed to a general audience, uses the three elements of a causal discussion in each of his paragraphs: the assertion, the facts, and the reasoning. In the first paragraph, for instance, the writer asserts that one reason ancestral mammals were more intelligent than the dinosaurs is that their environment demanded greater intelligence. The facts are that the mammals were outnumbered by their predators and had to search for food at night, when especially sharp sensory perception is required. The reasoning, explained in the final paragraph, is the general theory of natural selection: that only those mammals with great intelligence survived and reproduced, and that this fact improved the intelligence of these mammals.

Why were the ancestral mammals brainier than the dinosaurs? Probably because they were the underdogs during the rule of the reptiles, and the pressures under which they lived put a high value on intelligence in the struggle for survival. These little animals must have lived in a state of constant anxiety—keeping out of sight during the day, searching for food at night under difficult conditions, and always outnumbered by their enemies. They were Lilliputians in the land of Brobdingnag. Small and physically vulnerable, they had to live by their wits.

The nocturnal habits of the early mammal may have contributed to his relatively large brain size in another way. The ruling reptiles, active during the day, depended mainly on a keen sense of sight; but the mammal, who moved about in the dark much of the time, must have depended as much on the sense of smell, and on hearing as well. Probably the noses and ears of the early mammals were very sensitive, as they are in modern mammals such as the dog. Dogs live in a world of smells and sounds, and, accordingly, a dog's brain has large brain centers devoted solely to the interpretation of these signals. In the early mammals, the parts of the brain concerned with the interpretation of strange smells and sounds also must have been quite large in comparison to their size in the brain of the dinosaur.

Intelligence is a more complex trait than muscular strength, or speed, or other purely physical qualities. How does a trait as subtle as this evolve in a group of animals? Probably the increased intelligence of the early mammals evolved in the same way as their coats of fur and other bodily changes. In each generation, the mammals slightly more intelligent than the rest were more likely to survive, while those less intelligent were likely to become the victims of the rapacious dinosaurs. From generation to generation, these circumstances increased the number of the more intelligent and decreased the number of the less intelligent, so that the average intelligence of the entire population steadily improved, and their brains grew in size. Again the changes were imperceptible from one generation to the next, but over the course of many millions of generations the pressures of a hostile world created an alert and relatively large-brained animal.

FIGURE 10.8   **Cause-and-Effect Organizational Pattern**

WRITER'S CHECKLIST

The following checklist covers five organizational patterns:

More Important to Less Important

1. Have you clearly indicated that you are using the more-important-to-less-important organizational pattern?

2. Has your discussion made clear why the first point is the most important, the second the second-most important, and so forth?

Comparison and Contrast

1. Have you included the necessary criteria?

2. Have you chosen a structure—whole-by-whole or part-by-part—that is appropriate for your audience and purpose?

Classification and Partition

1. Have you used only one basis at a time?

2. Have you chosen a basis consistent with the audience and purpose of the document?

3. Have you avoided overlap?

4. Have you included all the appropriate categories?

5. Have you arranged the categories in a logical sequence?

Problem-Methods-Solution

1. Have you described the problem, methods, and solution clearly and specifically?

2. If appropriate, have you justified your methods?

3. Have you avoided overstating your solution?

4. Have you sequenced the discussion in a way that is consistent with the audience and purpose of the document?

Cause and Effect

1. Have you clearly expressed your assertion, your facts, and your reasoning?

2. Have you sequenced your assertion, your facts, and your reasoning appropriately for your audience and purpose?

3. Have you avoided inadequate sampling?

4. Have you avoided *post hoc* reasoning?

5. Have you avoided begging the question?

EXERCISES

1. Using the more-important-to-less-important organizational pattern, write a memo or letter on one of the following topics. Be sure to indicate your audience and purpose.

   a. the reasons you chose your college or university
   b. the reasons you chose your major
   c. the reasons you should (or should not) be required to study a foreign language (or learn a computer programming language)
   d. the three changes you would like to see most at your school
   e. the reasons that recycling should (or should not) be mandatory

2. Rewrite the following more-important-to-less-important paragraph to improve the clarity of its organization.

   Compact discs will have several advantages over the new digital audio tape. First, the listener can instantly access a particular track, and even a particular passage within a track, simply by pushing a button. Digital audio tape, like analog audio tape, has to be wound on the spool, which takes time. Second, with a compact disc, nothing touches the encoded music, so they will last indefinitely and will not degrade, with proper care. With digital tape, however, a tape head touches the tape, so it is inevitable that some of the encoded music will eventually be lost and the tape will be subject to jamming and stretching. Third, compact discs are sturdier than digital audio tape because it is not as vulnerable to temperature extremes.

3. Using a comparison-and-contrast organizational pattern, write a memo or letter on one of the following topics. Be sure to indicate the audience and purpose for your paper.

   a. lecture classes and recitation classes
   b. the album reviews in *Stereo Review* and *Rolling Stone*
   c. manual transmission and automatic transmission automobiles
   d. black-and-white and color photography

4. Rewrite the following comparison-and-contrast paragraph so that it contains a clear topic sentence.

   Both the "Gorilla" and the "Hunter" welding shields can be worn without a hard hat and allow the welder to move freely in a congested area. Neither provides head protection. The "Gorilla" is a leather hood. It is uncomfortable because it fogs up easily and absorbs unpleasant odors. Most welders prefer the "Hunter" because it does not fog up and because, being made of plastic, it does not absorb odors. Inspectors have found that welders wear it even when not required to do so. The result has been that sometimes it is difficult to find them. No problem is found in locating the "Gorilla" shields.

5. Using the problem-methods-solution pattern, write a letter or a memo on one of the following topics. Be sure to indicate the audience and purpose for your paper.

  a. how you solved a recent problem related to your education
  b. how you went about making a recent major purchase, such as a car, a personal computer, or a bicycle
  c. how you would propose reducing the time required to register for classes or to change your schedule
  d. how you would propose increasing the ties between your college or university and local business and industry.

6. Rewrite the following problems-methods-solution passage to improve its clarity and persuasiveness.

When the Hindenburg exploded over the fields of New Jersey, the future of hydrogen as a fuel went up in the smoke. Yet the dirigible tragedy, ironically, highlights some of the reasons that hydrogen should be considered again as a fuel. When the dirigible exploded, the hydrogen gas rose clear of the airship, and 62 of the 97 passengers survived. Modern fuels, such as gasoline or jet fuel, are of course responsible for many thousands of deaths due to explosions each year.

The reasons to consider hydrogen as a fuel are simple and obvious. First, it is in virtually limitless supply because it is a key element of water. Second, it produces no pollution. After hydrogen is burned, all that is left over is water. And third, it can be transported like natural gas and converted to electricity, but much more efficiently. What, then, are the problems? First, it is economically impractical to make it, transport it, and burn it. What is needed is a major research and development effort to eliminate the technical and economic barriers to the use of hydrogen as a fuel.

Some of this work is already underway. The best way to create hydrogen is to use solar energy to power the electrolyzers that split water. Other fuels exist, naturally, but there is no point using up fossil fuels to create a clean fuel. A recent Princeton study suggests that improvements in photovoltaic cells might make it economical to use solar energy as the fuel by the year 2000. In addition, super-strong composite materials are being developed that will make it possible to safely store the hydrogen in automobile fuel tanks and other small-scale applications. Finally, more powerful fuel cells are being developed to convert the hydrogen into electricity five times as efficiently as the same cells only two years ago. Within one or two years, these cells will have power-to-weight ratios sufficient to power helicopters and airplanes.

It is predicted that by the year 2000, hydrogen costing approximately $1.50 will produce as much energy as a gallon of gasoline. This highly competitive cost will spur the development of hydrogen as a fuel to power large and small machines, from ovens to cars and airlines. Already, the Deutsche Airbus company of Germany has announced plans to convert one of its jumbo jets to hydrogen. In the United States, Lockheed is actively investigating the fuel. The promise is one more fuel source—along with solar, wind, geothermal, and hydroelectric power—to ease the transition from fossil fuels.

7. Using classification, write a memo or letter on one of the following topics. Be sure to indicate your audience and purpose.
   a. foreign cars
   b. smoke alarms
   c. college courses
   d. personal computers
   e. insulation used in buildings
   f. cameras

8. Using partition, write a memo or letter about a piece of equipment or machinery you are familiar with, or about one of the following topics. Be sure to indicate your audience and purpose.
   a. a student organization on your campus.
   b. a tape cassette
   c. a portable radio
   d. a bicycle
   e. an electric clock

9. Rewrite the following classification paragraph to improve its structure, development, and writing style.

   There are three basic types of metal bicycle pedals, used by the serious cyclist. First is the double-sided type. When you are using toe clips, the double-sided type is just carrying unnecessary weight. Second is the quill pattern, which is like the double-side type except that one side is cut away to give the cyclist greater ground clearance and reduce the weight. The third type of metal bicycle pedal is the platform type, which spreads the load on to a greater part of the foot. All three types are available in steel and alloys, and some manufacturers are producing them in plastic for the cyclist on a budget. Also, all three types can be outfitted with toe clips and straps.

10. Using causal reasoning—either forward or backward—write a letter or a memo on one of the following topics. Be sure to indicate your audience and purpose.
    a. the number of women serving in the military
    b. the price of gasoline
    c. the emphasis on achieving high grades in college
    d. the prospects for employment in your field

11. Write a memo to your instructor, analyzing the following causal discussion (Rogers 1979). The subject is a method for determining how to plan a site for a solar-energy home. What are the strengths and weaknesses of the passage?

    The best way to evaluate the microclimate of a site is to observe it in detail for a full year. Most designers cannot afford to spend the time to do this, but on the other hand, if the habitat is to be occupied for generations perhaps such effort is entirely justified. . . .

I have been following the development of a greenhouse that was sited with great care. The structure is built into a south facing hillside. The site study continued for a full round of seasons. [The owner] decided on a rather unique approach to environmental accommodation. Basically, the idea was that if the given environment was too harsh, one could create an environment, in this case by building a big barn of a greenhouse, inside which a habitat that was compatible with the created environment could be constructed. One would step out one's door and have two seasons available. The greenhouse season, with growing plants and a chaise lounge, or the bright sun, snow, and sparkling cold of a typical New Mexico winter at 7,000 feet. The whole project was to be primarily solar heated.

[The owner] spent the best part of a year testing promising sites on a large tract of land. This involved a lot of winter days with a good book on sunny south slopes, and a lot of below freezing nights in a sleeping bag. She did her work so well that the inner dwelling is simply delineated, rather than isolated, from the primary greenhouse. Insulation was not required. Backup heat for the living area is provided by a wood stove, and heat stratification (warmer air will hug the ceiling area) keeps the sleeping loft comfortable after the fire goes out. The structure is about 65 × 20 feet, and the budget involved about $6,000 in cash and a lot of help from friends and persons interested in the project. A large part of the success of the habitat has resulted from the loving care that produced a quality site study before the first batter board was set.

12. In one paragraph each, describe the logical flaws in the following items.

    a. The election couldn't have been fair—I don't know anyone who voted for the winner.

    b. Environmentalists ought to be ignored because we don't need any more cranks in this world.

    c. Since the introduction of cola drinks at the start of this century, cancer has become the second greatest killer in the United States. Cola drinks should be outlawed.

REFERENCES

Isganitis, E. 1977. General description of a record turntable. Unpublished document.

Jastrow, R. 1977. *Until the sun dies.* New York: Norton.

Moore, R. E. 1990. Police pursuits: High-tech ways to reduce the risks. *Futurist* 24, no. 2 (July–August):26–28.

Rogers, B. T. 1979. Microclimate: "Don't fight the site," In *The solar age resource book* ed. Martin McPhillips, 1–5. New York: Everest House.

Toulmin, S. 1980. *The uses of argument.* London: Cambridge University Press.

Graphics

Graphics are the "pictures" of technical writing: maps, photographs, diagrams, charts, graphs, and tables. Few technical documents contain only text; graphics offer several benefits that sentences and paragraphs alone cannot provide.

- *Graphics are visually appealing.* Watch people skim a document. They almost automatically stop at the graphics and begin to study them. Readers are intrigued by graphics; that in itself can increase the effectiveness of your communication.

- *Graphics are easy to understand and remember.* Try to convey in words what a simple hammer looks like. It's not easy to describe the head and how it fits on the handle. But in 10 seconds you could draw a simple diagram that would clearly show the hammer's design.

- *Graphics let you emphasize particular information.* If you want to show your readers the location of the San Andreas fault, you can take a standard map of the United States and shade the fault area. If you want to show the details of the derailleur mechanism on a bicycle, you can present a diagram of the bicycle with a close-up of the derailleur so that your readers can see both the details of the mechanism and its location.

- *Graphics can save space.* Consider the following paragraph:

  > In the Boise area, some 90 percent of the population aged 18–24 watches movies or tapes on a VCR. They watch an average of 2.86 tapes a week. Of people aged 25–34, the percentage is 86, and the number of movies or tapes is 2.45. Among 35–49 year olds, the percentage is 82, and the number of movies or tapes is 2.19. Among the 50–64 age group, the percentage is 67, and the number of movies and tapes watched is 2.5. Finally, among those people 65 years old or older, the percentage is 48, and the number of movies and tapes watched weekly is 2.71.

  Presented as a paragraph, this information is uneconomical, not to mention boring and hard to understand and remember. Presented as a table, however, the information is more concise.

Age	Percent watching tapes/movies	Number of tapes watched per week
18–24	90	2.86
25–34	86	2.45
35–49	82	2.19
50–64	67	2.50
65 +	48	2.71

- *Graphics are almost indispensable in demonstrating relationships, which form the basis of most technical writing.* For example, if you wanted to show the number of nuclear power plants completed each year over the last decade, a line graph would be much easier to understand than a paragraph full of numbers. Graphics can also show the relationships among several variables over time, such as the numbers of four-cylinder, six-cylinder, and eight-cylinder cars manufactured in the United States during each of the last five years.

As you draft your document, think of opportunities to use graphics to clarify, emphasize, summarize, and organize information. McGuire and Brighton (1990) recommend that you be alert to certain words and phrases that often signal the opportunity to create a graphic:

categories	fields	process
components	functions	relates to
composed of	if and then	routines
configured	layers	sequence
consists of	numbers	shares
defines	phases	structured
features	procedures	summary of

If you find yourself writing a sentence such as, "The three categories of input modules are...," consider creating a graphic to reinforce your idea.

Characteristics of Effective Graphics

A graphic, like a paragraph, must be clear and understandable alone on the page, and it must be meaningfully related to the larger discussion. When you are writing a paragraph, it is easy to remember these two tasks: having to provide a transition to the paragraph that follows reminds you of the need for overall coherence. When you are creating a graphic, however, it is easy to put the rest of the document out of your mind temporarily as you concentrate on what you "draw."

The tendency, therefore, is to forget to relate the graphic to the text. Although some graphics merely illustrate points made in the text, others need to be explained to the readers. Graphics illustrate facts, but they can't explain causes, results, and implications. The text must do that.

Effective graphics are

☐ related directly to the writing situation
☐ labeled completely
☐ placed in an appropriate location
☐ integrated with the text

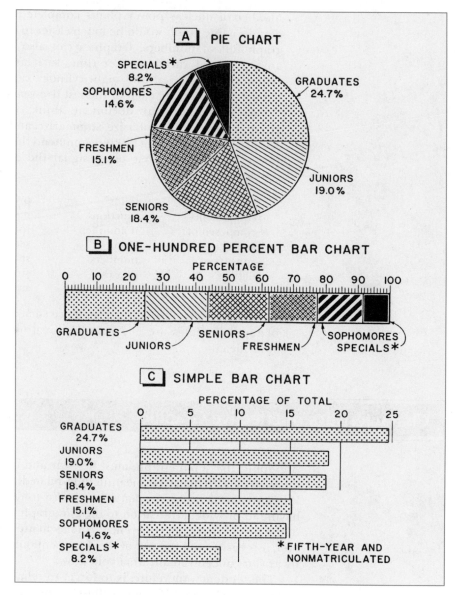

FIGURE 11.1   Different Graphics Communicating the Same Information

■ *A graphic should be related directly to the writing situation.* The first question to answer once you have decided to convey some body of information as a graphic is "What type of graphic would be most appropriate?" Some kinds of graphics effectively show deviations from a numerical norm, some effectively show statistical trends, and so forth. Some types are very easy to understand. Other, more sophisticated types contain considerable information, and the reader will need help to understand them. Think about your writing situation: the audience and purpose of the document. Can your readers handle a sophisticated graphic? Or will your purpose in writing be better served by a simpler one? A pie chart might be perfectly appropriate in a general discussion of how the federal government spends its money. However, so simple a graphic would probably be inappropriate in a technical article addressed to economists.

Figure 11.1 (Schmid and Schmid 1979, 147) shows how the same information—in this case, categories of students by class standing—can be communicated using different kinds of graphics. For many readers, the pie chart would be the simplest to understand; the 100 percent bar graph, probably the most difficult.

Figure 11.2 (Schmid and Schmid 1979, 8–9) shows that when you work with complicated numerical information, you often have to decide what facts to emphasize. For example, if you are researching the subject of military preparedness, you might find a table, such as the one at the start of Figure 11.2, that contains numerous statistics about the number of military vehicles disabled over a six-month period, and the proportion disabled due to lack of spare parts. The eight bar graphs and line graphs following the table show some of the ways you can emphasize different facts through different kinds of graphs.

- [ ] Graph A focuses on the total number of vehicles disabled but also shows the relative proportions of the two categories of causes.
- [ ] Graph B focuses on the relationship between the two categories of causes.
- [ ] Graph C focuses on the importance of the lack of spare parts by converting those statistics into percentages.
- [ ] Graph D provides the same information as Graph C but adds the range in percentage of disabilities due to lack of spare parts. The top of each bar represents the percentage of disabilities in the service with the most spare-parts problems; the bottom of each bar represents the percentage of disabilities in the service with the fewest spare-parts problems.
- [ ] Graph E focuses on each of the services, showing the total number of disabled vehicles in one month and the proportion disabled due to spare parts. The small graph at the bottom is a blowup of the two small bars for engineers and medical.
- [ ] Graph F presents the same information as Graph E in percentages.
- [ ] Graph G focuses on percentages of disabled vehicles due to lack of spare parts.
- [ ] Graph H, like Graph D, presents ranges, in percentages, for each of the seven services.

| General-Purpose Vehicle Disabled 4 Days or More | | | | | | | | | | | |
| | May | | June | | July | | August | | September | | October | |
Service	Total	Due to Lack of Parts	Total	Due to Lack of Parts	Total	Due to Lack of Parts	Total	Due to Lack of Parts	Total	Due to Lack of Parts	Total	Due to Lack of Parts
All services	922	625	1271	774	856	533	981	675	1247	679	1486	682
Chemical	98	48	78	43	78	23	39	17	91	51	95	51
Engineers	136	116	136	97	29	29	19	18	29	29	21	17
Medical	41	25	81	44	4	2	6	5	5	2	9	6
Ordnance	252	174	368	221	265	193	353	268	606	218	654	164
Quartermaster	116	63	76	41	61	21	61	23	59	31	58	28
Signal	69	44	80	34	28	15	31	23	39	34	46	30
Transportation	210	155	452	294	391	250	472	321	418	314	603	386

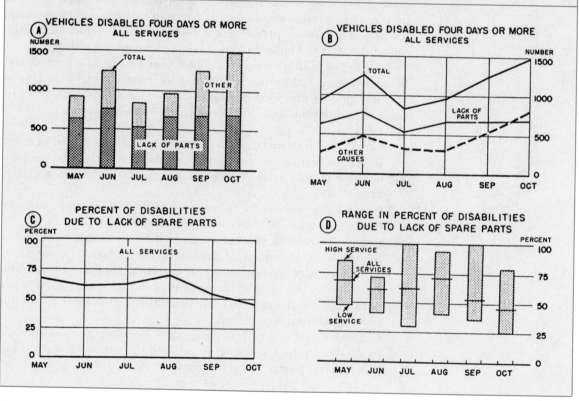

FIGURE 11.2   Different Graphics Emphasizing Different Points

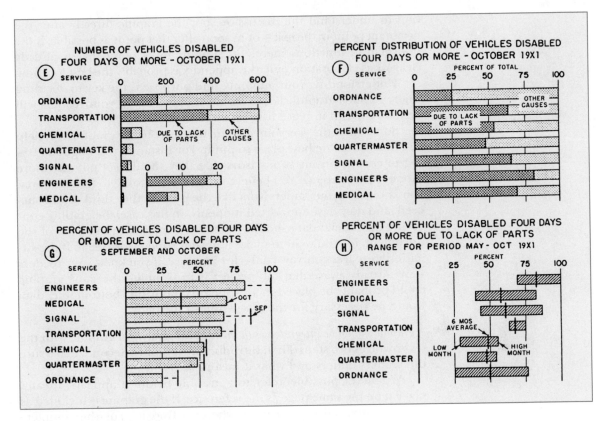

FIGURE 11.2 *Continued*

As Figure 11.2 makes clear, it is your job as the writer to figure out what aspect of the information you wish to emphasize. In addition, you must keep in mind that some graphics are more challenging than others. Some of these graphs, such as A, C, E, and F, would be perfectly clear to the general reader; the others, however, might not.

- *A graphic should be labeled completely.* Every graphic (except a brief, informal one) should have a clear and informative title. The columns of a table, and the axes of a graph, should be labeled fully, complete with the units of measurement. The lines on a line graph should also be labeled. Your readers should not have to guess whether you are using meters or yards as your unit of measure, or whether you are including in your table statistics from last year or just this year. If you did not discover or generate the information in the graphic, you must cite the source of the information.

- *A graphic should be placed in an appropriate location.* Graphics can be placed in any number of locations. If your readers need the information to understand the discussion, put the graphic directly after the pertinent point in the text—or as soon after that point as possible. If the information functions merely as support for a point that is already clear, or as elaboration of it, the appendix is probably the best location.

Understanding your audience and purpose is the key to deciding where to put a graphic. For example, if only a few of your readers will be interested in it, place it in an appendix.

Sometimes the way your readers will react to the graphic will help you determine the best place to put it. For instance, you might be trying to convince your supervisors that they should *not* put any more money into a project. You believe the project is going to fail. However, you know that your supervisors like the project: they think it will succeed, and they have supported it openly. In this case, the graphic conveying the crucial data should probably be placed in the body of the document, for you want to make sure your supervisors see it. If the situation were less controversial—for instance, if you knew your readers were already aware that the project is in trouble—the same graphic could probably be placed in an appendix. Only the "bottom line" data would be necessary for the discussion in the body.

■ *A graphic should be integrated with the text.* Integrating a graphic with the text involves two steps. First, introduce the graphic. Second, make sure the readers understand what it means.

Whenever possible, refer to a graphic before it appears. Ideally, place it on the same page as the reference. If the graphic is included as an appendix, tell your readers where to find it: "For the complete details of the operating characteristics, see Appendix B, page 24."

Writers often fail to explain clearly the meaning of the graphic. When introducing a graphic, ask yourself whether a mere paraphrase of its title will be clear: "Figure 2 is a comparison of the costs of the three major types of coal gasification plants." If you want the graphic to make a point, don't just hope your readers will see what that point is. State it explicitly: "Figure 2 shows that a high-sulfur bituminous coal gasification plant is presently more expensive than either a low-sulfur bituminous or anthracite plant, but more than half of its cost is cleanup equipment. If these expenses could be eliminated, high-sulfur bituminous would be the least expensive of the three types of plants."

Graphics and Computers

Recent advances in computerized graphics packages have made it simple to create a number of graphics easily. These packages let you do much more than create traditional tables, pie charts, bar graphs, and line graphs; they let you create all sorts of line drawings—from flow charts to maps, blue-prints, and diagrams—and print them out in color for increased emphasis.

Today's architecture students routinely create three-dimensional images using computed-assisted design (CAD) software. And more-advanced students use animation to create four-dimensional images, that is, three-dimensional images that change over time. Over the next few years, the hardware and software are expected to improve dramatically.

The great advantage of graphics software is that it enables you to cre-ate professional-quality graphics quickly and easily. After you input the data, the computer does all the work, displaying the information in differ-ent forms. You choose the form you think works best for your audience and purpose, then add any necessary textual labels.

One additional benefit is that graphics software lets you control the entire writing process more completely than ever before; you can create text and graphics with the same software program. This means that you can create whatever kind of graphic you want—and put it exactly where you want it. Instead of having to give the text to the typist and then fit the graphic in at the end of the text, as often happens today, you can try out a number of organizational patterns. Using a device called a scanner, you can even digitize a photograph and then manipulate it as you would any other graphic image.

One word of caution. Just as a style program offers advice that cannot substitute for your own judgment about a sentence, a graphics program creates images that might or might not accomplish what you want them to. You have to think about what you want the graphic to do and decide whether it succeeds. Sometimes, the computer-generated graphic is overly elaborate and actually obscures the point you are trying to communicate. You might have to tone it down.

Figure 11.3 shows two graphics created on the computer by under-graduate students. The graphic on the right shows a prototype for a device

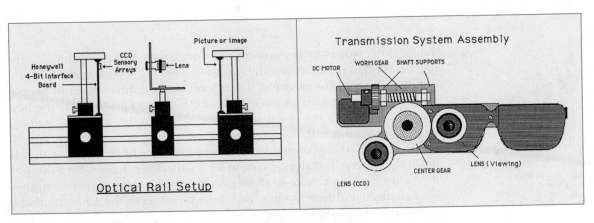

FIGURE 11.3   Graphics Created on Computers by Students

to improve the vision of people with a particular kind of eye disease. The graphic on the left shows some of the equipment used by the students in devising the prototype. Each graphic took the student less than an hour; creating drawings as professional looking as these by hand would have taken substantially more time.

Visit the computer center at your college or university to test out some of the many excellent graphics software packages that you can use to create professional-looking graphics for your documents not only in this course but also in most of your others. Also, consult the bibliography (Appendix D) for a list of books about computers and technical writing.

Types of Graphics

There are dozens of different types of graphics. Many organizations employ graphic artists to devise informative and attractive visuals. This discussion, however, will concentrate on the basic graphics that can be constructed without special training or equipment.

The graphics used in technical documents can be classified into two basic categories: tables and figures. Tables are lists of data—usually numbers—arranged in columns. Figures are everything else: graphs, charts, diagrams, photographs, and the like. Generally, tables and figures have their own sets of numbers: the first table in a document is Table 1; the first figure is Figure 1. In documents of more than one chapter (as in this book), the graphics are usually numbered chapter by chapter. Figure 3.2, for example, would be the second figure in Chapter 3.

Tables

Tables easily convey large amounts of information, especially quantitative data, and often provide the only means of showing several variables for a number of items. For example, if you want to show the number of people employed in six industries in 10 states, a table would probably be the best graphic. Tables lack the visual appeal of figures, but they can handle much more information with complete accuracy.

In general, tables should be structured so that the reader's eye travels down the column, not along a row, to see a parameter change. For example, if you are creating a table to show that the populations of three counties have changed at different rates over the last two years, the table should be structured like Table 1.

Table 1
Population Changes (1988–1990) of A, B, and C Counties

| Date | Population | | |
	County A	County B	County C
1988	35,912	46,983	53,572
1989	34,124	47,912	53,954
1990	29,876	47,321	62,876

This structure would help your readers see that the population of County A has fallen over the three-year period, whereas that of County C increased dramatically in 1990. Notice that the title of this table reflects a vertical reading of the data.

If, however, you wanted to compare the populations of the three counties year by year, the following structure would be more appropriate:

Table 1
1988–1990 Population of A, B, and C Counties

| County | Population | | |
	1988	1989	1990
A	35,912	34,124	29,876
B	46,983	47,912	47,321
C	53,572	53,954	62,876

This structure enables your readers to glance down the date columns to compare the populations of the three counties. Notice how the new title of the table reflects the new arrangement of data.

Figure 11.4, on page 284, shows the standard parts of a table. Notice that it identifies the table with both a number ("Table 1") and a substantive title. The title should encompass the items being compared, as well as the criteria of comparison.

Mallard Population in Rangeley, 1986–1990

Grain Sales by the United States to the Soviet Union in 1990

The Growth of the Robotics Industry in Japan, 1989–1991

Multiple Births in the Industrialized Nations in 1989

Note that most tables are numbered and titled above the data. The number and title are centered horizontally.

If all the data in the table are expressed in the same unit, indicate that unit under the title:

Farm Size in the Midwestern States
(in Hectares)

Table 1 Title (subtitle)			
Stub heading	Column Heading 1	Column Heading 2	Column Heading 3
Stub category 1			
Item A . . .	data[a]	data	data
Item B . . .	data	data	data
Item C . . .	data	data[b]	data
Item D	data	data	data
Stub category 2			
Item E . . .	data	data	data[c]
Item F . . .	data	data	data
Item G . . .	data	data	data

Notes: [a]Footnote
      [b]Footnote
      [c]Footnote
Source:

FIGURE 11.4   Parts of a Table

If the data in the different columns are expressed in different units, indicate the unit in the column heading.

Population | Per Capita Income
(in millions) | (in thousands of U.S. dollars)

Provide footnotes for any information that needs to be clarified. Also at the bottom of the table, below any footnotes, provide complete bibliographic information for the source of your information (if you did not generate it yourself).

The stub is the left-hand column, in which you list the items being compared in the table. Arrange the items in the stub in some logical order: big to small, important to unimportant, alphabetical, chronological, and so forth. If the items fall into several categories, you can include the names of the categories in the stub:

Sunbelt States
   Arizona
   California
   New Mexico

Snowbelt States
   Connecticut
   New York
   Vermont

If the items in the stub are not grouped in logical categories, skip a line every four or five items to help the reader follow the rows across the table. Leader dots, a row of dots that links the stub and the next column, are also useful.

The columns are the heart of the table. Within the columns, arrange the data as clearly and logically as you can. Line up the numbers consistently.

```
 3,147
   365
46,803
```

In general, don't change units. If you use meters for one of your quantities, don't use feet for another. If, however, the quantities are so dissimilar that your readers would have a difficult time understanding them as expressed in the same units, inconsistent units might be more effective.

```
 3 hr
12 min
 4 sec
```

This listing would probably be easier for most readers to understand than a listing in which all quantities were expressed in hours, minutes, or seconds.

Use leader dots if a column contains a "blank" spot: a place where there are no appropriate data:

```
 3,147
  . . .
46,803
```

Table 6 Test Results for Valves #1 and #2				
Valve Readings	Maximum Bypass Cv	Minimum Recirc. Flow (GPM)	Pilot Threads Exposed	Main ΔP at Rated Flow (psid)
Valve #1				
Initial	43.1	955	+3	4.5
Final	43.1	955	+3	. . .
Valve #2				
Initial	48.1	930	+3	4.5
Final	48.1	950	+2	. . .

FIGURE 11.5   Table

Make sure, however, that you don't substitute leader dots for a quantity of zero.

```
  3,147
      0
 46,803
```

Figure 11.5 is an example of an effective table. Figure 11.6 shows a table that communicates mostly verbal information. Figure 11.7 (McComb 1991, 133) shows how a table can be used to communicate a maintenance schedule.

Figures

Every graphic that is not a table is a figure. The discussion that follows covers the principal types of figures: bar graphs, line graphs, charts, diagrams, and photographs.

Bar Graphs

Bar graphs provide a simple, effective way of representing different quantities so that they can be compared at a glance. In bar graphs, the longer the bar, the greater the quantity. The bars can be drawn horizontally or vertically. Horizontal bars are generally preferred for showing different items at any given moment (quantities of different products sold during a single year), whereas vertical bars show how the same item varies over time (month-by-month sales of a single product). These distinctions are not ironclad, however; as long as the axes are labeled carefully, your readers should have no trouble understanding you.

Presentation Schedule for the Computer Stem

Time	Location	Speaker	Topic
9:00	Windsor Room	Mark Stevens	Word processing for Dyslexic Students
10:00	Cresthaven Room	Tracy Collins	LANs in the Composition Classroom
11:00	Barclay Room	Ahmed Singh	Using File Servers Effectively
12:00	Harrington Ballroom	Wallace Marx	The Effects of Word Processors on Revision Strategies

FIGURE 11.6   A Table That Communicates Verbal Information

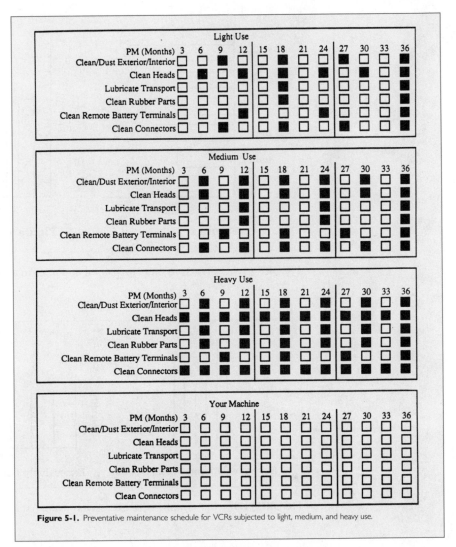

Figure 5-1. Preventative maintenance schedule for VCRs subjected to light, medium, and heavy use.

FIGURE 11.7   A Table Used to Communicate a Maintenance Schedule

Figure 11.8 shows the structure of typical horizontal and vertical bar graphs. When you construct bar graphs, follow four basic guidelines.

- *Make the proportions fair.* For all bar graphs, number the axes at regular intervals. Using a ruler or graph paper makes your job easier.

    For vertical bar graphs, choose intervals that will make your vertical axis about three-quarters the length of the horizontal axis. If your vertical axis is much longer than that, the differences between the height of the bars will be exaggerated. If the horizontal axis is too long, the differences will be unfairly flattened. Figure 11.9 shows two poorly proportioned graphs.

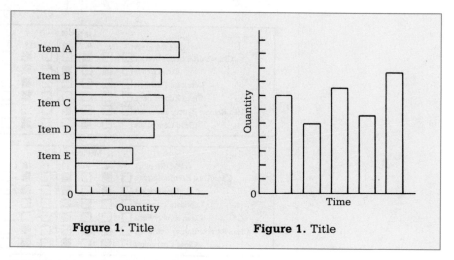

FIGURE 11.8   **Structure of a Horizontal and a Vertical Bar Graph**

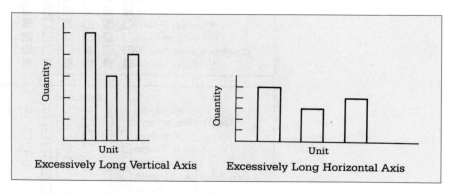

FIGURE 11.9   **Bar Graphs with Excessively Long Axes**

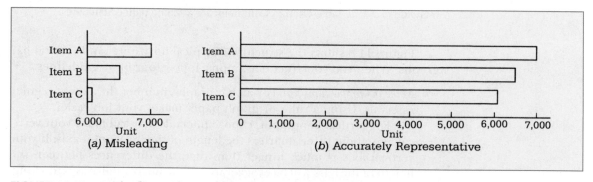

FIGURE 11.10   **Misleading and Accurately Representative Bar Graphs**

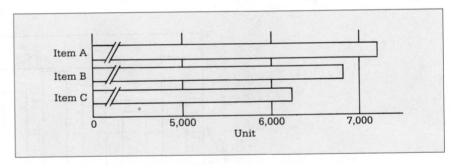

FIGURE 11.11   A Bar Graph with the Quantity Axis Clearly Broken

Make all bars equally wide, and use the same amount of space between them. The space between the bars should be about half the width of the bars themselves.

■ *If at all possible, begin the quantity scale at zero.* This will ensure that the bars accurately represent the quantities. Notice in Figure 11.10 how misleading a graph is if the scale doesn't begin at zero. Version *a* is certainly more dramatic than version *b*. In version *a*, the difference in the lengths of the bars suggests that Item A is much greater than Item B, and that Item B is much greater than Item C. Such misrepresentation is unethical.

If it is not practical to start the quantity scale at zero, break the quantity axis clearly, as in Figure 11.11.

■ *Use tick marks or grid lines to signal the amounts.* Ticks are the little marks drawn to the axis:

Grid lines are ticks extended through the bars:

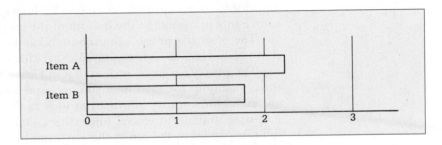

Grid lines are usually necessary only if you have several bars, some of which would be too far away from tick marks for readers to gauge the quantity easily.

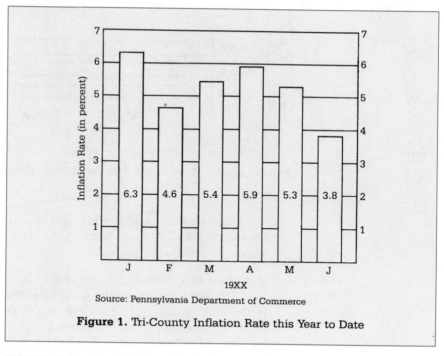

Source: Pennsylvania Department of Commerce

Figure 1. Tri-County Inflation Rate this Year to Date

FIGURE 11.12   Bar Graph

■ *Arrange the bars in a logical sequence.* In a vertical bar graph, use chronology for the sequence when you can. For a horizontal bar graph, arrange the bars in descending-size order beginning at the top of the graph, unless some other logical sequence seems more appropriate.

Figure 11.12 shows an effective bar graph.

Notice that most figures are titled underneath. Unlike tables, which are generally read from top to bottom, figures are usually read from the bottom up. If the graph displays information that you have gathered from an outside source, provide complete bibliographic information for that source in a brief note at the bottom of the graph.

The basic bar graph can accommodate many different communication needs. Here are a few common variations.

The *grouped bar graph,* such as that in Figure 11.13, lets you show two or three quantities for each item you are representing. Grouped bar graphs are useful for showing information such as the numbers of full-time and part-time students at several universities. One kind of bar represents the full-time students; the other, the part-time. To distinguish the bars from each other, use hatching (striping) or shading, and label one set of bars or provide a key. Leave at least one bar's width between sets of bars.

Another way to show this kind of information is through the *subdivided bar graph,* shown in Figure 11.14. A subdivided bar graph adds Aspect I to Aspect II, just as wooden blocks are placed on one another.

Related to the subdivided bar graph is the *100-percent bar graph*, which lets you show the relative proportions of the elements that make up several items. Figure 11.15 shows a 100-percent bar graph. This kind of graph is useful in portraying, for example, the proportion of full-scholarship, partial-scholarship, and no-scholarship students at a number of colleges.

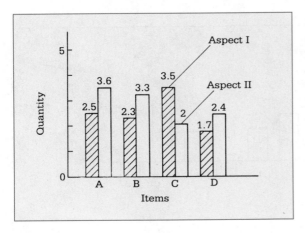

FIGURE 11.13   Grouped Bar Graph

FIGURE 11.14   Subdivided Bar Graph

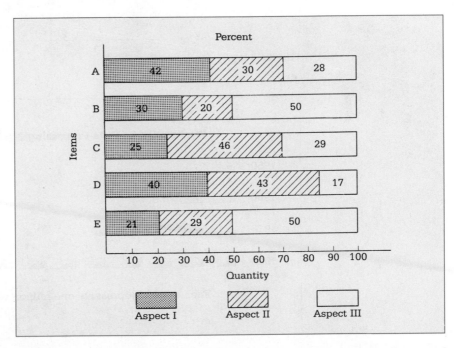

FIGURE 11.15   100-Percent Bar Graph

The *deviation bar graph*, shown in Figure 11.16, lets you know how various quantities deviate from a norm. Deviation bar graphs are often used when the information contains both positive and negative values, as with profits and losses. Bars on the positive side of the norm line represent profits; on the negative side, losses.

Pictographs are simple graphs in which the bars are replaced by series of symbols that represent the items (see Figure 11.17). Pictographs are gener-

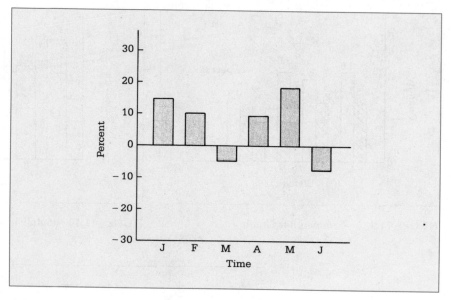

FIGURE 11.16   Deviation Bar Graph

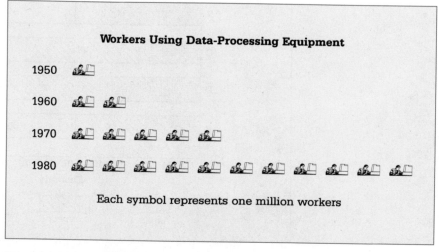

FIGURE 11.17   Pictograph

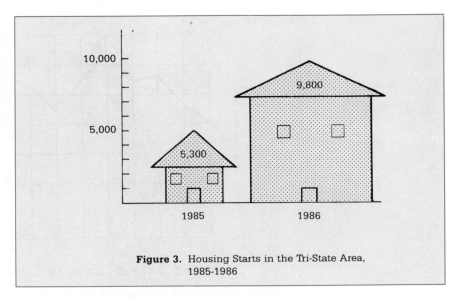

Figure 3. Housing Starts in the Tri-State Area, 1985-1986

FIGURE 11.18   Misleading Pictograph

ally used only to enliven statistical information for the general reader. The quantity scale is usually replaced by a statement that indicates the numerical value of each symbol. Hundreds of predrawn symbols and pictures, called clip art, are available; they can be inserted onto a graphic easily.

Pictographs are arranged horizontally rather than vertically: symbols sitting on top of each other would look foolish.

A related kind of error is to use a vertical format and make each symbol tall enough to represent the quantity desired. As Figure 11.18 shows, the larger symbol, if it is drawn proportionally, looks many times larger than it should: the reader sees the total area of the symbol rather than its height.

Line Graphs

Line graphs are like vertical bar graphs, except that in line graphs the quantities are represented not by bars but by points linked by a line. This line traces a pattern that in a bar graph would be formed by the highest point of each bar. Line graphs are used almost exclusively to show how the quantity of an item changes over time. Line graphs might portray the month-by-month sales or production figures for a product or the annual rainfall for a region over a given number of years. A line graph focuses the reader's attention on the change in quantity, whereas a bar graph emphasizes the actual quantities themselves. Figure 11.19 shows a typical line graph.

An advantage of the line graph for demonstrating change is that it can accommodate much more data than a bar graph. Because you can plot three or four lines on the same graph, you can compare trends conveniently. Figure 11.20 shows a multiple-line graph. However, if the lines

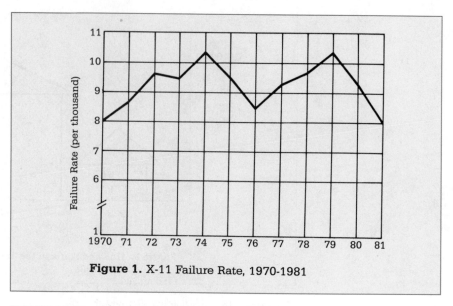

Figure 1. X-11 Failure Rate, 1970-1981

FIGURE 11.19   Line Graph with a Truncated Axis

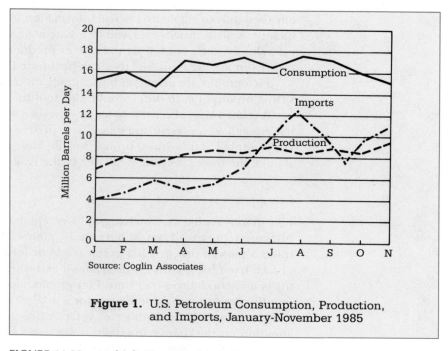

Source: Coglin Associates

Figure 1. U.S. Petroleum Consumption, Production, and Imports, January-November 1985

FIGURE 11.20   Multiple-Line Graph

intersect each other often, the graph will be unclear. If this is the case, draw separate graphs.

The principles of constructing a line graph are similar to those used for a vertical bar graph. The vertical axis, which charts the quantity, should begin at zero; if it is impractical to begin at zero because of space restrictions, clearly indicate a break in the axis, as in Figure 11.19. Where precision is required, use grid lines — horizontal, vertical, or both — rather than tick marks.

When you plan the line graph, be sure that the vertical and horizontal axes are reasonably proportioned. Otherwise, you might inadvertently mislead the reader. Figure 11.21 (Spear 1969, 56) shows how contracting and expanding the two axes change the impression created by a line graph.

Two common variations on the line graph — the *stratum graph* and the *ratio graph* — deserve special mention. A stratum graph shows an overall change while partitioning that change into its parts. Figure 11.22 shows a stratum graph. For the last year covered by the graph, 1988, the gross farm income was approximately $130 billion. Of that sum, approximately $45 billion was net farm income and $85 billion was production expenses.

A *ratio graph* is a line graph used to emphasize percentages of change rather than the change in real numbers. A ratio graph charts data that could not be represented fairly on a standard line graph. For example, you might wish to compare the month-by-month sales of a large corporation with those of a small one. You would have great trouble making a vertical axis that would accommodate a company with sales of $20,000 a month and one with sales of $20,000,000. Even if you had a giant piece of paper, the graph could not reflect a true relation between the companies. If both companies increased their sales at the same rate (such as 2 percent per month), the small company's line would appear relatively flat, whereas the big company's line would shoot upward, just because of the large quantities involved.

To solve this problem, the ratio graph compresses the vertical axis more and more as the quantities increase. Figure 11.23 shows how ratio graphs work. This graph measures two trends: the increase in the gross domestic product per capita and the decrease in average weekly hours worked. The vertical axis has the annotation "Index, 1950 = 100." This means that if the average worker's productivity in 1950 were defined as 100 units, the average worker's productivity, according to the line, in 1965 was approximately 140 units; in 1986 it was approximately 195 units. Why is the vertical axis not regular? The compressed scale accounts for inflation. If a regular scale had been used, the productivity line would have shot upward at a much greater slope, giving an inaccurate impression of how productive the average worker is.

Charts

Whereas tables and graphs present statistical information, most charts convey relationships that are more abstract, such as causality or hierarchy.

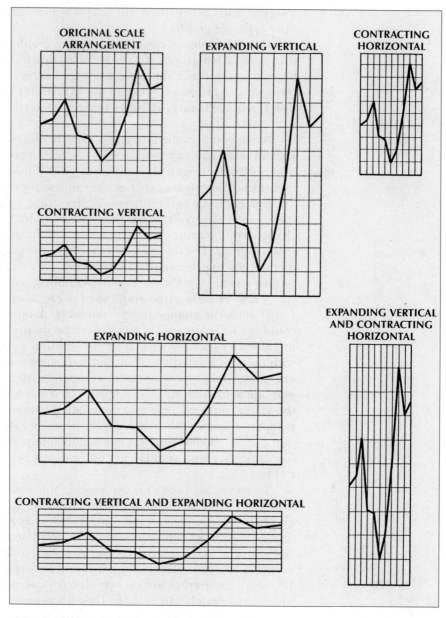

FIGURE 11.21   The Effects of Expanding and Contracting the Axes

(The pie chart, which is really just a circular rendition of the 100-percent bar graph, is the major exception.) Many forms of tables and graphs are well known and fairly standard. By contrast, only a few kinds of charts— such as the organization chart and flow chart—follow established patterns. Most charts reflect original concepts and are created to meet specific communication needs.

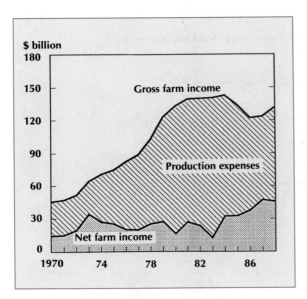

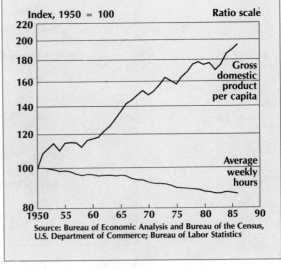

FIGURE 11.22   Stratum Graph (from *1990 Agricultural Chartbook*)

FIGURE 11.23   Ratio Graph (from *Productivity and the Economy: A Chartbook*, 51 1988)

The *pie chart* is a simple but limited design used for showing the relative size of the parts of a whole. Pie charts can be instantly recognized and understood by the untrained reader: everyone remembers the perennial "where-your-tax-dollar-goes" pie chart. The circular design effectively shows the relative size of as many as five or six parts of the whole, but it cannot easily handle more parts because, as the slices get smaller, judging their sizes becomes more difficult. (Very small quantities that would make a pie chart unclear can be grouped under the heading "Miscellaneous" and explained in a footnote. This "miscellaneous" section, sometimes called "other," appears after the other sections as you work in a clockwise direction.)

To create a pie chart, begin with the largest slice at the top of the pie and work clockwise in decreasing-size order, unless you have a good reason for arranging the slices in a different order. Label the slices (horizontally, not radially) inside the slice, if space permits. It is customary to include the percentage that each slice represents. Sometimes, the absolute quantity is added. To emphasize one of the slices—for example, to introduce a discussion of the item represented by that slice—separate it from the pie.

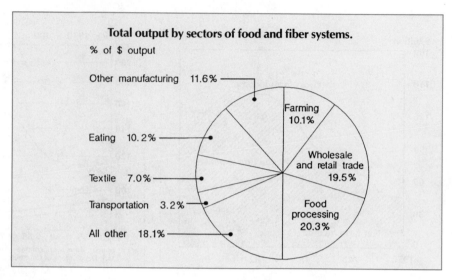

FIGURE 11.24 Pie Chart

Make sure your math is accurate as you convert percentages into degrees in dividing the circle. A percentage circle guide—a template with the circle already converted into percentages—is a useful tool. Graphics software packages can create sophisticated pie charts.

Figure 11.24 shows a typical pie chart. The farming category, although not the largest slice, begins the chart because the farming output is the subject of the chart.

A *flow chart*, as its name suggests, traces the stages of a procedure or a process. A flow chart might be used, for example, to show the steps involved in transforming lumber into paper or in synthesizing an antibody. Flow charts are useful, too, for summarizing instructions. The basic flow chart portrays stages with labeled rectangles or circles. To make it visually more interesting, use pictorial symbols instead of geometric shapes. If the process involves quantities (for example, paper manufacturing might "waste" 30 percent of the lumber), they can be listed or merely suggested by the thickness of the line used to connect the stages. Flow charts can portray open systems (those that have a "start" and a "finish") or closed systems (those that end where they began). A special kind of flow chart, called a decision chart (in which the flow follows different routes depending on yes/no answers to questions), is used frequently in computer science.

Figure 11.25 (Energy Information Administration 1989, 4) shows an open-system flow chart. Figure 11.26 (Michels, 1979, 28) shows a closed-system flow chart.

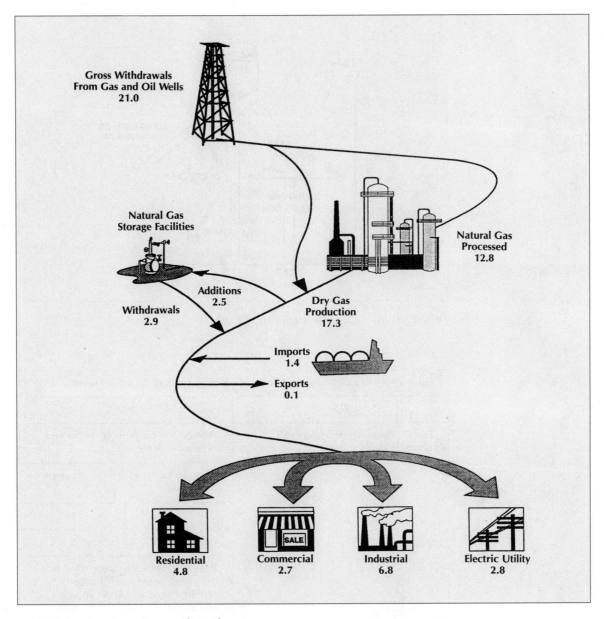

FIGURE 11.25   Open-System Flow Chart

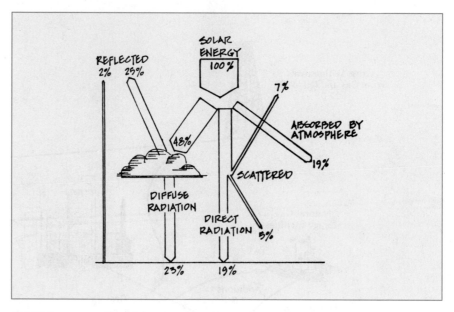

FIGURE 11.26   Closed-System Flow Chart

Do you want to replace all occurrences of the first word with the second word?		
YES	NO	
Press Y. The computer replaces all occurrences of the first word with the second word.	Press N. The cursor moves to the next occurrence of the word.	
	Do you want to replace that occurrence of the word?	
	YES	NO
	Press SHIFT/F2.	Press F2.
	Press F4 to move the cursor to the next occurrence of the word. Then press SHIFT/F2 or F2, as you did above, to replace that occurrence of the word or to leave it as is.	
Do you want to use the search function again?		
YES	NO	
Press F8.	Press F10	

FIGURE 11.27   Instructional Flow Chart

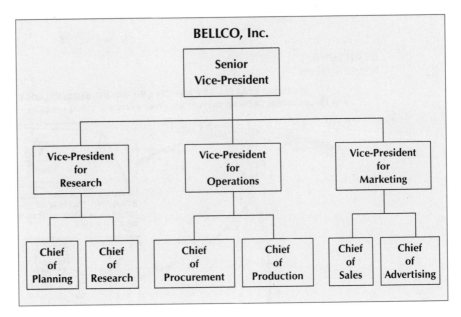

FIGURE 11.28   Organization Chart

A related kind of flow chart is used often in instructional materials to show the reader which of two or more paths to follow. Figure 11.27, from a software manual, is part of a longer discussion of how to use the search-and-replace function. At this point in the explanation, the user has already told the computer which word to search for and which word to replace it with. Now the computer prompts the user to decide whether or not *all* instances of the first word should be replaced.

An *organization chart* is a type of flow chart that portrays the flow of authority and responsibility in an organization. In most cases, the positions are represented by rectangles. The more important positions can be emphasized through the size of the boxes, the width of the lines that form the boxes, the typeface, or color. If space permits, the boxes themselves can include brief descriptions of the positions, duties, or responsibilities. Figure 11.28 is a typical organization chart. Unlike most other figures, organization charts are generally titled *above* the chart because they are read from the top down.

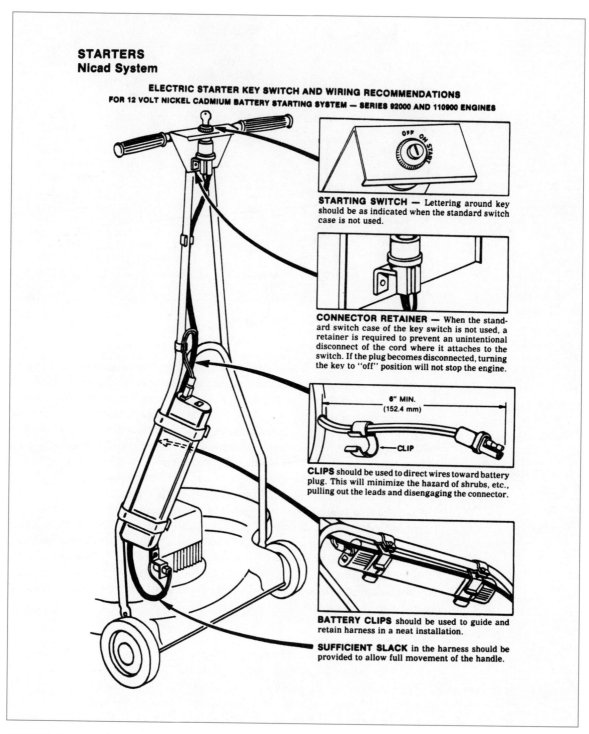

STARTERS
Nicad System

ELECTRIC STARTER KEY SWITCH AND WIRING RECOMMENDATIONS
FOR 12 VOLT NICKEL CADMIUM BATTERY STARTING SYSTEM — SERIES 92000 AND 110900 ENGINES

STARTING SWITCH — Lettering around key should be as indicated when the standard switch case is not used.

CONNECTOR RETAINER — When the standard switch case of the key switch is not used, a retainer is required to prevent an unintentional disconnect of the cord where it attaches to the switch. If the plug becomes disconnected, turning the key to "off" position will not stop the engine.

6" MIN. (152.4 mm)
CLIP

CLIPS should be used to direct wires toward battery plug. This will minimize the hazard of shrubs, etc., pulling out the leads and disengaging the connector.

BATTERY CLIPS should be used to guide and retain harness in a neat installation.

SUFFICIENT SLACK in the harness should be provided to allow full movement of the handle.

FIGURE 11.29   Diagram (from *Briggs & Stratton Service and Repair Instructions*)

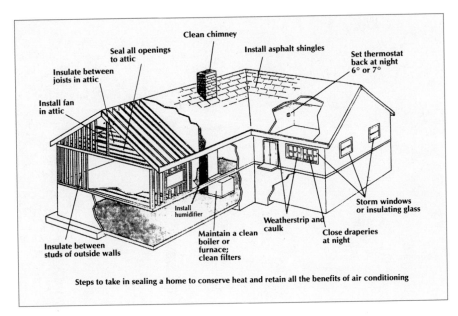

FIGURE 11.30   Cutaway Diagram

Diagrams

To portray physical objects, diagrams are often the most effective graphic. They can emphasize important external parts to help the reader locate them, as shown in Figure 11.29. *Cutaway diagrams* (Figure 11.30) let the reader "remove" a part of the surface to expose what is underneath. *Exploded diagrams* (Figure 11.31) separate components while maintaining their physical relationship. Some kinds of information cannot be communicated other than by a diagram. Figure 11.32 (Schmid and Schmid 1979, 25) displays several common kinds of optical illusions.

Diagrams can also be used to communicate abstract, conceptual information, as shown in the block diagram in Figure 11.33. Notice how much clearer the block diagram is than the prose version of the same information. Figure 11.34 shows that pictorial elements can be used in a diagram. Figure 11.35 shows that a simple diagram can clarify a difficult concept. Figure 11.36 (*1990 Agricultural Chartbook* 1990, 48) shows a popular kind of diagram: the map.

Photographs

Photographs are unmatched for realistically reproducing some kinds of images. If you want to show the different kinds of tire-tread wear caused by various alignment problems, a photograph is best. If you want your readers to recognize a new product such as a lawnmower, you will probably include a photograph. And recent advances in specialized kinds of photography—especially in internal medicine and biology—are expanding the possibilities of the art.

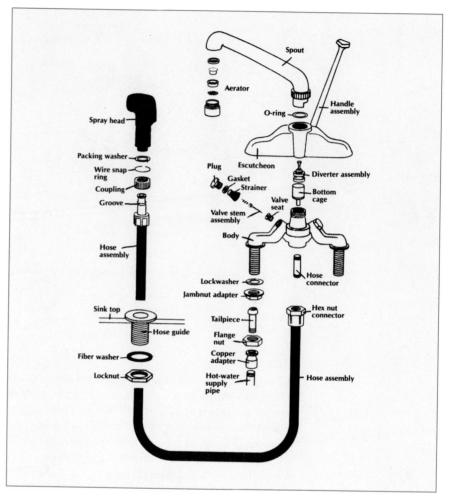

FIGURE 11.31   Exploded Diagram

Ironically, however, sometimes a photograph can provide too much information. In an advertising brochure for an automobile, a glossy photograph of the dashboard might be very effective, but if you are creating an owner's manual and you want to show how to find the trip odometer, a diagram will probably work better. And sometimes a photograph can provide too little information; the item you want to show can be inside the mechanism or obscured by some other component.

Modern photographic equipment—especially "idiot-proof" automatic 35-millimeter cameras—is so sophisticated that almost anyone can take a reasonably good snapshot. However, for publication purposes, photography is still best left to the professionals. Yet on occasion you will want to use existing photographs in a document. On these occasions, it is useful to know the basics of using photographs.

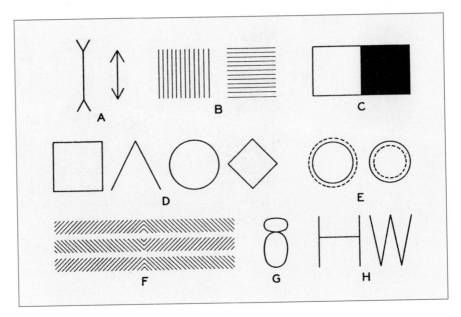

FIGURE 11.32   Optical Illusions.

A. The two vertical lines are of the same length, although the one at the left seems to be longer. **B.** Both shaded sections are identical in height and width. **C.** The white portion of the rectangle appears larger than the black portion, yet actually the two are identical in size. **D.** At their widest point the four geometric forms are exactly equal. **E.** The diameters of the two circles represented by full lines are identical. **F.** The three crosshatched bars are parallel and of identical width from end to end. The bending impression is an optical illusion resulting from this particular type of hatching. **G.** The upper and lower parts of this symbol are of the same width. **H.** The distances between the vertical lines of the letter *H* and the two upper points of the *W* are equal.

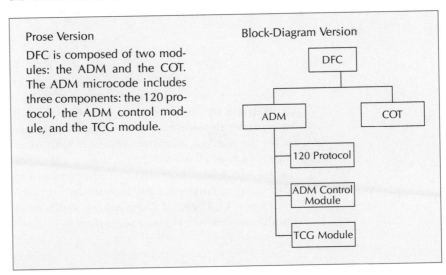

FIGURE 11.33   Block Diagram

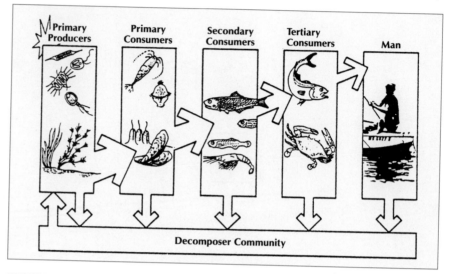

FIGURE 11.34   Diagram Containing Pictorial Elements

Converting a Binary Number into a Decimal Number

Binary Number:	1	1	1	0
	8	4	2	1
	×1	×1	×1	×0
	8	4	2	0

Total: 8 + 4 + 2 + 0 = 14 Decimal Number

FIGURE 11.35   Diagram Used to Clarify a Difficult Concept

If necessary, indicate the angle from which the photograph was taken. Your reader shouldn't have to wonder whether you were standing above or below the subject. If appropriate, include in the picture some common object, such as a coin or a ruler, to give a sense of scale. Eliminate extraneous background clutter that can distract your reader. And if appropriate, label components or important points.

Figure 11.37 (NCR Corporation 1987, 6–19) shows a labeled photograph, supplemented by a diagram.

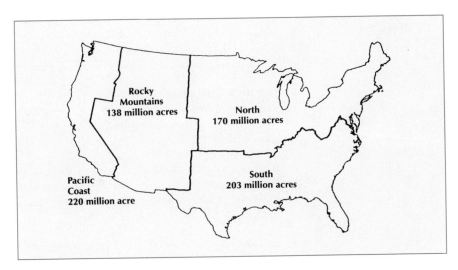

FIGURE 11.36   Forest Land Areas of the United States

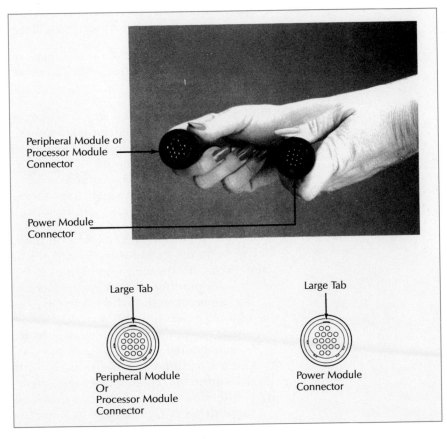

FIGURE 11.37   Labeled Photograph

EXERCISES

1. Create an organizational chart for some organization you belong to or are familiar with: a department at work, a fraternity or sorority, a campus organization, or student government.

2. Create a flow chart for some process you are familiar with, such as registering for courses, applying for a summer job, studying for a test, preparing a paper, or some task at work. Your audience is someone who will be carrying out the process you visualize.

3. Create a pie chart to show some information such as your expenses for the semester or the makeup (by major) of your technical-writing class. Write a few sentences that explain the significance of the information.

4. Define an audience and purpose and then draw a diagram of some object you are familiar with, such as a tennis racquet, a stereo speaker, or a weight bench.

5. Define an audience and purpose and then draw a diagram of some concept, such as the effects of smoking or the advantages of belonging to the student chapter of a professional organization.

6. Find out from the admissions department of your college or university the numbers of students from the different states, or from the different counties in your state. Communicate this information using a map.

7. Find and photocopy a bar graph, a line graph, a pictograph, a pie chart, a flow chart, and a diagram. For each of these graphics, write a brief discussion that responds to the following questions:
   a. Is the graphic necessary?
   b. Is it professional in appearance?
   c. Does it conform to the guidelines for that kind of graphic?
   d. Is it effectively integrated into the discussion?

8. The table below provides statistics on the money spent on alcoholic beverages for the period 1985 through the first half of 1989. Study the table, then perform the following tasks.
   a. Create two different graphics, each of which emphasizes the difference between the actual dollar expenditures and the inflation-adjusted expenditures.
   b. Create two different graphics, each of which emphasizes the trend in actual dollar expenditures.
   c. Create two different graphics, each of which emphasizes the trend in location of alcoholic beverage expenditures.
   d. Create two different graphics, each of which emphasizes the proportions of expenditures at home and away from home.

Change in Alcoholic Beverages Personal Consumption Expenditures:
1985–1988, January–June 1989
(percent)

Expenditures	Annual Change			Six Months Jan–June 1989	Annual Compound Change 1985–89
	1985–86	1986–87	1987–88		
Current dollars					
All alcoholic beverages	5.9	2.9	2.2	4.7	3.7
At-home	4.2	1.7	2.0	4.3	2.6
Away from home	8.7	4.9	2.6	5.4	5.4
Inflation-adjusted					
All alcoholic beverages	0.9	*	– 1.4	1.0	– 0.1
At-home	– 0.3	– 0.4	– 0.4	1.5	– 0.1
Away from home	1.9	0.8	3.1	– 0.1	– 0.2

*Less than 0.1 percent.
SOURCE: U.S. Department of Commerce, Bureau of Economic Analysis.

9. For each of the following graphics, write a paragraph evaluating its effectiveness and describing how you would revise it.

   a.

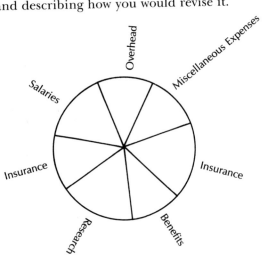

Expenses at Hillway Corporation

b. *Engineering and Liberal Arts Graduate Enrollment*

	1985	1986	1987
Civil Engineering	236	231	253
Chemical Engineering	126	134	142
Comparative Literature	97	86	74
Electrical Engineering	317	326	401
English	714	623	592
Fine Arts	112	96	72
Foreign Languages	608	584	566
Materials Engineering	213	227	241
Mechanical Engineering	196	203	201
Other	46	42	51
Philosophy	211	142	151
Religion	86	91	72

c.

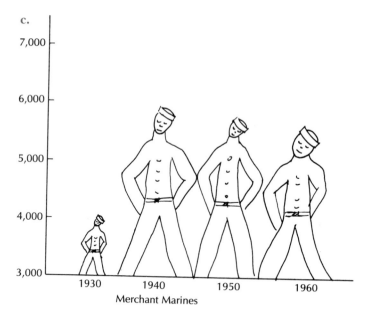

Merchant Marines

d.

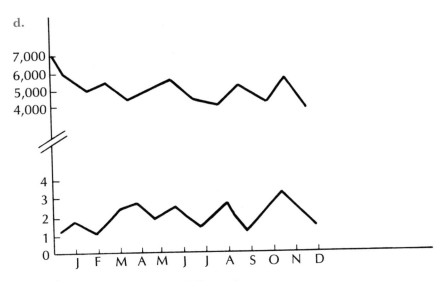

Sales: Hollins, Inc., and Gems, Inc.

e.

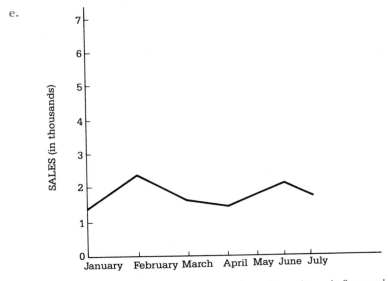

10. In each of the following exercises, translate the written information to at least two different kinds of graphics. For each exercise, which kinds work best? If one kind works well for one audience but not so well for another audience, be prepared to explain.

   a. Following are the profit and loss figures for Pauley, Inc., in early 1989: January, a profit of 6.3 percent; February, a profit of 4.7 percent; March, a loss of 0.3 percent; April, a loss of 2.3 percent; May, a profit of 0.6 percent.

   b. The prime interest rate had a major effect on our sales. In January, the rate was 9.5 percent. It went up a full point in February, and another half point in March. In April, it leveled off, and it dropped two full points each in May and June. Our sales figures were as follows for the Crusader 1: January, 5,700; February, 4,900; March, 4,650; April, 4,720; May, 6,200; June, 8,425.

c. Following is a list of our new products, showing for each the profit on the suggested retail price, the factory where produced, the date of introduction, and the suggested retail price.

The Timberline	*The Four Season*
Profit 28%	Profit 32%
Milwaukee	Milwaukee
March 1990	October 1989
$235.00	$185.00
The Family Excursion	*The Day Tripper*
Profit 19%	Profit 17%
Brooklyn	Brooklyn
October 1989	May 1989
$165.00	$135.00

d. This year, our student body can be broken down as follows: 45 percent from the tristate area; 15 percent from foreign countries; 30 percent from the other Middle Atlantic states; and 10 percent from the other states.

e. In January of this year we sold 50,000 units of the BG-1, of which 20,000 were purchased by the army. In February, the army purchased 15,000 of our 60,000 units sold. In March, it purchased 12,000 of the 65,000 we sold.

f. The normal rainfall figures for this region are as follows: January, 1.5 in.; February, 1.7 in.; March, 1.9 in.; April, 2.1 in.; May, 1.8 in.; June, 1.2 in.; July, 0.9 in.; August, 0.7 in.; September, 1.3 in.; October, 1.1 in.; November, 1.0 in.; December, 1.2 in. The following rainfall was recorded in this region: January, 2.3 in.; February, 2.6 in.; March, 2.9 in.; April, 2.0 in.; May, 1.6 in.; June, 0.7 in.; July, 0.1 in.; August, 0.4 in.; September 1.3 in.; October, 1.2 in.; November, 1.4 in.; December, 1.8 in.

g. You can access the Rightfile from programs written in six different languages. Rightfile classifies these languages as two groups (A and B) and provides separate files for each group. The A group includes C, PL/1, and Fortran. The B group includes COBOL and RPGII. The object module for the A group is GRTP. The object module for the B group is GRAP.

REFERENCES

Briggs & Stratton Corporation. 1984. *Briggs & Stratton service and repair instructions.* Part #270962.

Energy Information Administration, Office of Oil and Gas, 1989. *Natural Gas Annual 1989.* Washington, D.C.: U.S. Department of Energy.

McComb, G. 1991. *Troubleshooting and repairing VCRs.* 2d ed. Blue Ridge Summit, Pa.: TAB/McGraw-Hill, Inc.

McGuire, M. and P. Brighton. 1990. "Translating Text into Graphics." Session at the 37th International Technical Communication Conference.

NCR Corporation. 1987. *Installing your NCR 9200 System.* Dayton, Oh.: NCR Corporation.

Productivity and the economy: A chartbook. 1988. Washington, D.C.: U.S. Department of Labor, Bureau of Labor Statistics. Bulletin 2298.

1990 Agricultural chartbook. 1990. Washington, D.C.: U.S. Department of Agriculture.

Schmid, C. F. and S. E. Schmid. 1979. *Handbook of graphic presentation,* 2d ed. New York: John Wiley and Sons.

Spear, M. E. 1969. *Practical charting techniques.* New York: McGraw-Hill.

12

Page Design

As you look through a magazine, you read some ads and you skip others. It takes you only a fraction of a second to decide whether to spend the time reading the ad. Obviously, you aren't making a completely rational decision; you are merely responding to what your eye sees. If the ad looks interesting—if it offers an intriguing photograph or a creative use of type—you are likely to read it, even if you have no real interest in the product. If the ad looks boring or cluttered, you might skip it because it doesn't seem worth the effort.

Although most kinds of technical writing differ greatly from magazine ads, appearance is a critical element in both. A document communicates through words and graphics, but also through its page design. "Page design" is a broad term including all of the aspects the appearance of the page: margins, line length, style and size of the type, use of color, and so forth.

An effectively designed page is attractive to look at and easy to read. You want the reader to want to read your document, even when required to. And you want your reader to find it easy and pleasant to read; he or she shouldn't be bored or confused or fatigued by the mere act of reading it.

In addition, an effectively designed page emphasizes what is important. For example, a warning such as "Wear safety goggles when drilling!" might appear like this in a set of instructions:

**WEAR SAFETY GOGGLES
WHEN DRILLING!**

The large, boldface type, in capital letters, emphasizes the warning, as do the box and the stop sign.

This chapter is only a brief introduction to page design. Although you don't need to be an expert, you need to understand enough about it to increase the chances that your readers will read your documents sympathetically and understand your message.

If you work on a computer, you will have a big advantage. Most word-processing programs let you test different designs. You can effortlessly change features—the spacing from single to double, for example, or the type style from roman to boldface—and print out a sample for evaluation. Word-processing programs also suggest appropriate design elements through their default values; that is, when you open up a program, it will already be set for a certain line length, margin size, and so forth.

Through desktop publishing, the computer has revolutionized the production of all kinds of documents, from brochures and newsletters to complete books. This software provides many options for varying the design of all the verbal and visual elements of a document. As a result, much of the work that used to be done by outside printers is now being done in-house by writers, graphic artists, and designers sitting in front of computers. Already, job ads for technical writers are asking for desktop-publishing experience. As the years go by, employers will ask all professional-level employees for a more sophisticated understanding of how to design pages effectively.

This chapter will discuss the following three aspects of page design:

☐ using white space effectively
☐ choosing type fonts, families, and sizes
☐ designing titles, headings, and lists

Using White Space Effectively

White space is the part of the paper that does not have writing or graphics on it: the margins on all four sides of the page, the space between two lines of text, and the space between text and graphics. This section will cover four important uses of white space:

☐ margins
☐ columns
☐ leading
☐ justification

Margins

Most people are surprised to learn that about half the area of a typical page is devoted to margins. Why so much? Margins serve two main purposes:

■ They provide enough space so the document can be bound, or so the reader can easily hold the page without covering up the text.

■ They provide a neat frame. Pages on which the type extends all the way out to the end of the paper look sloppy.

For typed or word-processed pages printed on one side and attached with a staple or a paper clip, a margin of approximately 1½ inches on all four sides is appropriate. For documents printed on both sides and bound, such as manuals and books, a slightly more complicated formula is used. Figure 12.1 shows the common proportions used for left-hand and right-hand pages. The actual sizes of the margins are determined by a number of factors:

■ The size of the page. In general, the smaller the page, the smaller the margins.

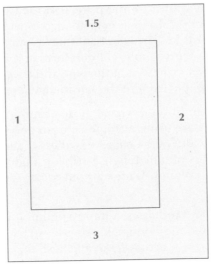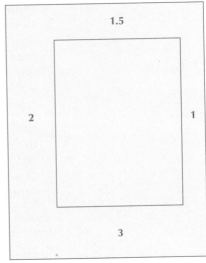

left-hand page   right-hand page

FIGURE 12.1 **Proportions of Margins for a Document That Is Bound Like a Book**

- The amount of text that needs to fit on the page. For an article in a journal that will be mailed, the margins tend to be small, to reduce postage costs.

- The technicality of the text. In general, the more technical the text, the larger the margins. This strategy makes the pages look less intimidating.

- The background of the readers. In general, the less knowledgeable the readers, the larger the margins.

Columns

The number of text columns to use on a page is a function of the size of the paper. For traditional 8½ × 11 inch paper, with standard-size typing or word processing, one column is used most often. For published manuals and books, however, multiple columns are more common; two columns is the most popular design, followed by three columns. To separate the columns, you can use white space alone or a vertical line with white space on both sides. In either case, the gap between two columns of text should be large enough — about a half-inch — so that the reader doesn't unintentionally keep reading into the next column.

Using a multi-column layout rather than a single-column layout can have several advantages:

- More words will fit on the page, and the lines will be easier to read. Research has shown that, in general, a short line—of perhaps 50 to 60 characters—is less tiring to read than a longer line, especially for long documents. With a multi-column format, you can choose a line length appropriate for the size of the type and the paper, as well as degree of difficulty of the text and the knowledge level of the readers.

- Graphics can fit on the page more economically. As Figure 12.2 suggests, if you are using a single-column design, a graphic will take up the whole width of the page, regardless of the size of the graphic itself. With a multi-column design, a graphic might be able to fit in the width of a single column. Wider graphics can spread across two or more columns.

As Figure 12.3 indicates, a multi-column format also permits a clear, simple, and economical design for step-by-step instructions. You can place the graphic directly across from the text that it refers to, rather than underneath it.

Leading

Leading (pronounced *ledding*) refers to the white space between lines, or between one line and a graphic. If the leading is too great, the page looks diffuse; the text loses its coherence, and the reader tires quickly. If the leading is too small, the page looks crowded and is difficult to read; the eye

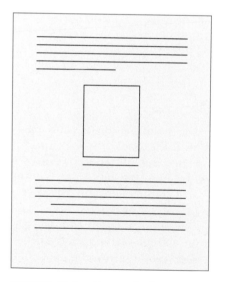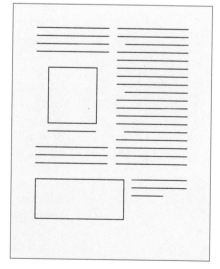

FIGURE 12.2   Economical Use of Space in a Multi-Column Layout

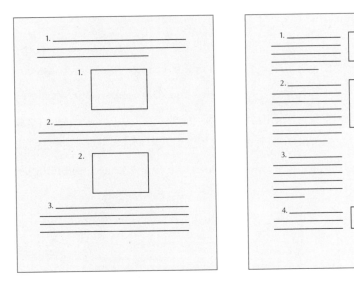

FIGURE 12.3   Simple Integration of Text and Graphics
in a Multi-Column Layout

sees the lines above and below the one being read, causing fatigue. Figure 12.4 shows the same text with three different amounts of leading.

Usually, leading is determined by the kind of document you are writing. Memos are single spaced; other documents, such as reports and proposals, are double spaced. Leading is also important in separating one section of text from another. Breaks between single-spaced paragraphs are usually two spaces. In double-spaced text and one-and-a-half spaced text, there is no extra leading, but the first line of each paragraph is indented, as in hand-written documents.

For breaks between sections, the leading is usually greater than between paragraphs. For instance, if you are typing a letter single spaced (double spaced between paragraphs), and you begin a new section with a heading, you would triple space before the heading. In other words, the leading *between* two sections should be greater than the leading *within* a section. Figure 12.5 (*1990 U.S. Industrial Outlook* 1990, 40-1) shows the effective use of leading (along with italics and boldface type) to distinguish one section from another.

Notice in Figure 12.5 the large amount of white space between the abstract and the introduction, and between the end of the introduction and the start of the materials-and-methods section.

Leading is used, too, to set off graphics from text, and to set off block quotations. If the text is single spaced, the leading would be double spaced; if the text is one-and-a-half or double spaced, the leading would be triple spaced.

a. Excessive Leading

Aronomink Systems has been contracted by Cecil Electric

Cooperative, Inc. (CECI), to design a solid-waste management

system for the Cecil County plant, Units 1 and 2, to be built at

Cranston, Maryland. The system will consist of two 600 MW

pulverized coal-burning units fitted with high-efficiency

electrostatic precipitators and limestone reagent FGD systems.

The coal will contain an estimated 3 percent sulfur and

10 percent ash. The station will output approximately 64 TPH

(DWB) of FGD sludge and 24 TPH fly ash at 100 percent load.

b. Appropriate Leading

Aronomink Systems has been contracted by Cecil Electric
Cooperative, Inc. (CECI), to design a solid-waste management
system for the Cecil County plant, Units 1 and 2, to be built at
Cranston, Maryland. The system will consist of two 600 MW
pulverized coal-burning units fitted with high-efficiency
electrostatic precipitators and limestone reagent FGD systems.
The coal will contain an estimated 3 percent sulfur and
10 percent ash. The station will output approximately 64 TPH
(DWB) of FGD sludge and 24 TPH fly ash at 100 percent load.

c. Inadequate Leading

Aronomink Systems has been contracted by Cecil Electric
Cooperative, Inc. (CECI), to design a solid-waste management
system for the Cecil County plant, Units 1 and 2, to be built at
Cranston, Maryland. The system will consist of two 600 MW
pulverized coal-burning units fitted with high-efficiency
electrostatic precipitators and limestone reagent FGD systems.
The coal will contain an estimated 3 percent sulfur and
10 percent ash. The station will output approximately 64 TPH
(DWB) of FGD sludge and 24 TPH fly ash at 100 percent load.

FIGURE 12.4   Degrees of Leading

40

Personal Consumer Durables

Shipments of selected personal consumer durables are forecast to increase 0.5 percent in 1990, in constant dollars, compared with a decline of 1.3 percent in 1989. Small gains are seen for jewelry, musical instruments, boat building and repairing, and sporting goods. Small declines are forecast for dolls, toys, and games; lawn and garden equipment; bicycles; and motorcycles.

The personal consumer durables industries covered in this chapter are jewelry (SICs 3911 and 3961); musical instruments (SIC 3931); dolls, toys, and games (SICs 3942 and 3944); lawn and garden equipment (SIC 3524); motorcycles, bicycles, and parts (SIC 3751); boat building and repairing (SIC 3732); and sporting and athletic goods (SIC 3949).

Following a 1 percent decline in 1989, constant-dollar shipments of personal consumer durables will rise an estimated 0.5 percent in 1990. The expected growth reflects improvement in jewelry, musical instruments, and sporting goods. Shipments of dolls, toys, and games; lawn and garden equipment; bicycles; and motorcycles are expected to decline.

In 1989, constant-dollar shipments by the boat building and repairing industry declined 7 percent, mainly because of temporarily rising interest rates. Similarly, the lawn and garden equipment industry was affected by the slowdown in residential housing starts and excessive retail inventories in 1989.

The major factor affecting the remaining personal consumer goods industries is the general economy. The present economic recovery, now in its eighth year, has eliminated the pent-up demand for many durable goods, resulting in saturated markets in many cases. Also threatening future shipments is the high consumer debt load. Consumer installment debt in mid-1989 was about 15.9 percent of personal income, compared with less than 12 percent in 1982, the bottom of the last recession. Any decision by consumers to lighten their debt load could cut consumer spending.

Another important factor affecting all personal consumer durables is foreign trade. If the value of the U.S. dollar relative to other currencies continues to rebound from the lows set in mid-1988, imports are likely to increase and provide stiff competition for many of these industries. Likewise, export markets gained in recent years would face increased foreign price competition. If these industries are to maintain their market shares, both here and abroad, they must price aggressively and strive for continued improvements in product quality and service.—*John M. Harris, Office of Consumer Goods, (202) 377-1178, September 1990.*

JEWELRY

Continued growth in personal consumer expenditures, plus increased exports, resulted in product shipments of jewelry reaching an estimated $5.6 billion (current dollars) in 1989, up about 4 percent from 1988. Shipments of precious metal

Trends and Forecasts: Jewelry, Precious Metal (SIC 3911)
(in millions of dollars except as noted)

Item	1987[1]	1988[2]	1989[3]	1990[4]	1987-88	1988-89	1989-90
						Percent Change	
Industry Data							
Value of shipments[5]	4,266	4,530	4,675	–	6.2	3.2	–
Value of shipments (1987$)	4,266	4,394	4,526	4,570	3.0	3.0	1.0
Total employment (000)	36.1	36.3	37.2	36.6	0.6	2.5	-1.6
Production workers (000)	25.4	26.0	27.1	26.5	2.4	4.2	-2.2
Average hourly earnings ($)	7.74	8.02	8.16	–	3.6	1.7	–
Product Data							
Value of shipments[6]	3,946	4,189	4,325	–	6.2	3.2	–
Value of shipments (1987$)	3,946	4,064	4,187	4,228	3.0	3.0	1.0
Trade Data							
Value of imports (ITA)[7]	2,048	2,156	2,698	–	5.3	25.1	–
Import/new supply ratio[8]	0.342	0.340	0.384	–	-0.6	12.9	–
Value of exports (ITA)[9]	240	344	472	–	43.3	37.2	–
Export/shipments ratio	0.061	0.082	0.109	–	34.4	32.9	–

[1]Industry and product data are preliminary. Trade data are adjusted to conform to the 1987 SIC.
[2]Estimated, except for exports and imports.
[3]Estimated.
[4]Forecast.
[5]Value of all products and services sold by establishments in the jewelry, precious metal industry.
[6]Value of products classified in the jewelry, precious metal industry produced by all industries.
[7]Import data were developed by the author.
[8]New supply is imports plus corresponding product shipments.
[9]Export data were developed by the author.
SOURCE: U.S. Department of Commerce: Bureau of the Census; International Trade Administration (ITA). Estimates and forecasts by ITA.

FIGURE 12.5   Leading Used to Distinguish One Section from Another

Justification

Justification refers to the alignment of words along the left and right margins. In technical writing, text is almost always left-justified; that is, except for paragraph indentations, the lines begin along a uniform left margin. Often, however, text is not right-justified, but "ragged right"; the lines end on an irregular right border.

Ragged right is most common in typewritten and word-processed text (even though most word-processing programs can right-justify text). Right-justified text is seen most often in typeset, formal documents.

Should you right-justify your assignments for technical writing? If you are using a standard word-processing program rather than a sophisticated desktop publishing system, probably not. Right-justification can actually make the text harder to read, because the software will insert irregular horizontal white space between words to push the line out to the right margin. As a result, spaces between words can vary, by as much as several hundred percent. Because a big space suggests a break between sentences, not a break between words, the reader can become confused, frustrated, and fatigued.

> This passage, for example, was typed on a word processor that was set to right-justify the lines. Note that the spacing between the words is irregular. Look, especially, at the space before "between" in line 2. That's just the way it turned out with this combination of words. With other combinations, the bigger spaces would have occurred in other places, but there almost certainly would have been some big spaces.

> What you are reading now, however, was typed on the same word processor, set to ragged right. Notice that the spacing between the words is regular throughout this excerpt. This regularity is easier on the eyes.

Some word-processing programs and typesetting systems will automatically hyphenate words that would not fit on the line. Hyphenation slows down the reader and can be distracting. Ragged-right text does not require hyphenation, the way right-justified text sometimes does.

Choosing Type Fonts, Families, and Sizes

When we talk about type fonts, families, and sizes, we are referring primarily to design elements used in word-processing, desktop publishing, and more sophisticated typesetting. Standard typewriters, even the best ones, offer very few options beyond roman type, underlining, and, in some cases, italics.

This section introduces three aspects of type: fonts, families, and sizes.

Type Fonts

A font is a set of letters, punctuation marks, and special characters of a particular style. When the first movable, reusable metal type was invented in the western world in the fifteenth century, the fonts resembled ornate hand lettering.

This paragraph, for example, was typed on a word processor using what the software designer calls the ITC Zapf Chancery Medium Italic font. You are not likely to see this style of font in any technical writing, because it is too ornate, and too hard to read.

This paragraph was typed with the Times font. It looks like the kind of type used by newspapers such as the <u>New York Times</u> in the nineteenth century.

This paragraph was typed with the Helvetica font, which has a very modern, high-tech look. The Helvetica font is an example of a sans-serif font; that is, a font in which there are no small horizontal lines at the ends of the major strokes in most of the letters. For instance, look at the letter <u>N</u> in <u>New York Times</u> in the paragraph above. The three little horizontal lines are the serifs. Sans serif fonts, of which there are many, are used generally for short documents; they are harder on the eyes than serif fonts because each letter is less distinct from every other one than in a serif type.

Most of the time, you will be using a standard font, such as Courier, that your software includes and that your printer can reproduce. The important points to remember are that different fonts create different impressions—from highly formal and ceremonial to very modern and high-tech—and that some fonts cause less fatigue than others.

Type Families

Each type font is actually a family of fonts, consisting of a number of variations on the basic style. Figure 12.6, for example, is the Helvetica font family.

You should know the various options you have within a particular family. In a complex document, you are likely to have three or four different levels of headings. Using some of the different members of a font family lets you create an attractive and functional set of headings.

With a typewriter, you have few options: uppercase and lowercase, and underlining. Here is a sample, arranged from more emphatic to less emphatic:

<u>SUMMARY</u>
SUMMARY
<u>Summary</u>
Summary

Helvetica Light
Helvetica Light Italic
Helvetica Regular
Helvetica Regular Italic
Helvetica Bold
Helvetica Bold Italic
Helvetica Heavy
Helvetica Heavy Italic
Helvetica Regular Condensed
Helvetica Regular Condensed Italic

FIGURE 12.6   Helvetica Family of Type

As you can see, your choices are limited. The visual contrast between SUMMARY and Summary is quite weak.

However, if you use a word processor and have access to some other members of a family, such as boldface, you can easily create striking contrasts without a jarring effect.

SUMMARY

SUMMARY

SUMMARY

SUMMARY

Type Sizes

The most obvious size difference lies between uppercase and lowercase letters within a font. To make your technical writing easy to read, use uppercase and lowercase letters just as you would in any other kind of writing. Some writers mistakenly think that a text printed in all uppercase letters is easier to read. It isn't. In fact, lowercase letters are easier to read because the individual variations from one letter to another are greater for lowercase than for uppercase. In addition, using all uppercase letters removes the visual cues that signal new sentences.

Using different sizes of the same font can create other visual cues. Sizes are measured in a unit called a point. Figure 12.7 shows the basic range of type sizes.

In most technical writing, you will be writing the basic text in 10 point or 12 point.

FIGURE 12.7 Helvetica Medium in 11 Different Sizes

This paragraph is printed in 10 point. This size is easy to read, provided it is reproduced on a letter-quality impact printer or laser printer. On most impact printers, however, the resolution isn't high enough for this small a size.

This paragraph is printed in 12 point. If you are using a dot-matrix printer, 12 point is the best size to use because it balances readability and economy.

This paragraph is printed in 14 point. This size is appropriate for titles or headings.

Some of the other sizes are used occasionally in technical communications. For footnotes, you might want to drop down to 8 or 9 point. And for text in slides or transparencies, 18 point or even 24 point might be the most appropriate.

Don't go overboard with size variations. You don't want your document to look like a sweepstakes advertisement in the mail. In technical writing, you should never call attention to your presentation of information; the information itself must remain the focus.

Designing Titles, Headings, and Lists

Chapter 7 discussed the strategy of creating effective titles, headings, and lists. The present section explains how to integrate these three elements into the page design. The principle is that you want titles, headings, and lists to stand out visually. Titles and headings should stand out because they announce and communicate new information; your reader needs to realize that you are beginning a new idea. Lists, too, should be visually emphatic, because the reader will find it easier to read, comprehend, and remember the information if it is clearly arranged on the page.

Designing Titles

Because a title is the most important heading in a document, you should display it clearly and prominently. If it is on a cover page or a title page, you might present it in boldface in a large size, such as 14 or 18 point. If it also appears at the top of the first page, you might make it slightly bigger than the rest of the text—perhaps 14 point for a document printed in 12 point—but not as big as on the cover page or title page.

Titles are usually centered on a page horizontally. For more information on title pages for reports, see Chapter 19.

Designing Headings

The principle of designing headings is similar to that of designing titles, with a few exceptions. In general, your readers should be able to see that you are beginning a new idea. For this reason, you will probably use boldface, italics, or underlining, as well as some variations in size, to set off headings.

Indentation can also help your readers. In general, the more important the level, the closer to the left margin it appears. First-level headings, therefore, usually begin at the left margin. Second-level headings are often indented five spaces; third-level headings, ten spaces.

When you indent a heading, indent the text that follows it. This strategy accomplishes two goals. First, the indented heading remains clear and emphatic; it does not get swallowed up by the text that follows it. And

Text Formatting: Technical Reference ☐ Programming Support

STRINGS

A *string* is a named group of *characters*, *not including* a newline character, that may be interpolated by name at any point.

Strings are used to store text, or series of escape sequences, and to read that data back in as needed.

Naming strings

The guidelines for macro naming apply equally to strings. Since request, macro, diversion, and string names share the *same* name list, it is critical that a string not be assigned with the name of any macro unless the intention is to overwrite that macro.

Alias, rename, remove

Like macros, strings may be aliased with `:alias`, renamed with `.rn`, and removed with `.rm`.

Creating strings

Strings are created with the `.ds` request and appended to with the `.as` request.

```
:ds Client "John
:as Client " Doe
```

long strings

Just as macros can be defined to return the values of strings and registers as they were at the time of creation or at the time of invocation, so can strings. Strings, like macros, are created in *copy mode*, so string definitions may contain concealed references to registers or other strings. The example below, creates a string which always reports the current indent.

```
:ds indent "The indent is \\n(.i units
```

Reading strings

Strings are read by using the `\*` escape sequence followed by a delimited string name.

one-character names

The string *x* is interpolated at any desired point with `\*x`.

two-character names

The string *xx* is interpolated at any desired point with `\*(xx`.

any name

Any string, whether its name has one character or fourteen, may be interpolated at any desired point with `\*{name}`.

Example

Using the string which was created above, the following input

```
Dear \*{Client},
```

yields:

Dear John Doe,

5 - 18

FIGURE 12.8   Creative Use of Margins in Designing Headings

Summary

In this example, the writer has skipped a line between the heading and the text that follows it. The heading stands out clearly.

Summary

In this example, the writer has not skipped a line between the heading and the text that follows it. The heading stands out, but not as emphatically.

Summary. In this example, the writer has begun the text on the same line as the heading. This style makes the heading stand out the least.

FIGURE 12.9   The Effect of Different Leading on the Impact of Headings

second, the indented text appears appropriately subordinate. In some documents, writers even decrease the type size for lower-level text. For instance, in a document that is printed in 12 point, all third-level and fourth-level text is printed in 10 point.

On occasion, writers will use margins creatively to highlight their headings. In Figure 12.8 (Maloney 1987), the writer places several levels of headings in the left margin to make them stand out visually.

In designing headings, you also need to use leading well. A noticeable distance between a heading and the text increases the impact of the heading. Consider the three examples in Figure 12.9.

Designing Lists

Chapter 7 explained the strategy of creating lists, discussed parallelism, and pointed out that vertical lists are easier to read and comprehend than horizontal lists. The present discussion explains the design and punctuation of lists.

In most cases, the items in the list are indented. The amount of indentation depends on the length of the items. Single words or short phrases might be indented so that the list appears centered; longer items might be indented only two or three characters.

Each listed item is preceded by a number, a letter, or a symbol (usually a bullet, a large, emphatic dot). Numbered lists are used to suggest sequence (as in the numbered steps in a set of instructions) or priority (the first item is the most important). Sometimes writers will number a list to emphasize

The new facility will offer three advantages:

- lower leasing costs
- easier commuting distance
- a greater pool of potential workers

The new facility will offer three advantages:

- The leasing costs will be lower.
- The commuting distance for most of the employees will be shorter.
- The pool of potential workers will be larger.

The new facility will offer three advantages:

- lower leasing costs. The lease will cost $1,800 per month; currently we pay $2,300.
- easier commuting distance. According to a recent questionnaire, our workers now spend an average of 18 minutes to travel to work. At the new location, the average would drop to 14 minutes.
- a greater pool of potential workers. In the last decade, the population has begun to shift westward, to the area of the new facility. We would be able to increase our potential work force, especially in the semiskilled and managerial categories.

The new facility will offer three advantages:

- Lower leasing costs.
- Easier commuting distance. According to a recent questionnaire, our workers now spend an average of 18 minutes to travel to work. At the new location, the average would drop to 14 minutes.
- A greater pool of potential workers. In the last decade, the population has begun to shift westward, to the area of the new facility. We would be able to increase our potential work force, especially in the semiskilled and managerial categories.

FIGURE 12.10   Punctuation of Different Kinds of Lists

the total number of items (as in the well publicized "Seven Warning Signals of Cancer" from the American Cancer Society). For other kinds of lists, bullets are more common. Letters are sometimes used for second-level items when numbers are used on the first level.

Punctuation of lists varies widely. Like most stylistic issues in technical writing, there is no universal standard, so you should find out how your organization treats this question. In general, however, you should punctuate the listed items as follows:

- If the items are fragments, such as "decreases costs," use a lowercase letter at the start and do not use a period or a comma at the end.

- If the items are complete sentences, use a capital letter at the start and a period at the end.

- If an item is a fragment followed by a complete sentence, begin the fragment with either a lowercase or uppercase letter—be consistent within the document—and end it with a period. Then begin the complete sentence with an uppercase letter and end it with a period.

- If the list consists of two different kinds of items—some are fragments and some are fragments combined with complete sentences—punctuate all the items the same way: with uppercase letters and periods.

Figure 12.10 shows these four possibilities. Note two things in this figure:

- The last item in a list of fragments is not followed by a period. The horizontal white space to the right of the last listed item, as well as the vertical white space that separates the list from the next line, clearly indicates the end of the list. Some writers, however, prefer a period after the last item.

- When an item extends to the second line, the first letter of the second line and all subsequent lines—these lines are called turnovers—is aligned under the first letter of the first line. In other words, the bullet or the number at the start of the list extends out to the left of the text. This format, called hanging indentation, highlights the bullet or the number, and thereby helps the reader "see" the organization of the passage. Figure 12.11 (*Information Network* 1987) shows several levels of hanging indentation used effectively.

Figure 12.12 (*Parts Invoice Writer* 1986) shows many of the design elements discussed in this chapter. The figure shows two pages from a user's guide in which the left-hand page is devoted to text and the right-hand page is devoted to graphics that clarify, exemplify, emphasize, or elaborate on the text. Notice in Figure 12.12 that the text page uses different size type, boldface, three kinds of rectangular markers, and lists with indented turnovers. The bulk of the page is double column, to save space. The graphics page, which has a primarily vertical orientation, corresponds to the three main points of the text page.

Appendix C: Planning checklist for EDI

This list provides an overview of the tasks that your company should consider to move from a traditional system of sending business documents on paper to an electronic system of sending business documents by EDI.

1. Establish your business objectives for using EDI.

2. Assign a project leader.

3. Establish managerial and technical contacts in all affected departments.

4. Set up a project team, defining each member's area of responsibility.

5. Establish contacts within your exchange partners' organizations.

6. Plan changes to current business procedures.

   • Plan changes in paper flow.

   • Plan changes in computer systems.

   • Identify tasks to be done.

   • Estimate the resources that you will need for each task.

   • Develop implementation schedules, assigning a responsible person and a date for each item.

Appendix C: Planning checklist for EDI   C-1

FIGURE 12.11   **Hanging Indentation** (Courtesy of International Business Machines Corporation)

A FEW OF THE BENEFITS

With Parts Invoice Writer, you have complete control over the way you price your parts. "Flexible pricing" is the key to Parts Invoice Writer's power and flexibility. With flexible pricing, you can calculate prices at the franchise level (one formula for all parts), at the customer level (a special formula for each customer), and at the part level (a formula for each part).

In addition, you can override established pricing formulas for an entire invoice and for individual line items on an invoice.

Improve Gross Profits, Minimize Effort through Flexible Pricing

Flexible pricing assists your Parts Department in realizing an improved gross profit with minimum effort. Flexible pricing controls are easy to maintain, yet flexible enough for the most sophisticated pricing strategy. After you set up flexible pricing at your dealership, the parts counter person no longer has to look for prices in price books or for discounts in customer listings. And no more computing discounted prices manually!

Special features include minimum and maximum price checks and point-of-sale overrides with a detailed audit of sale price deviations.

Cancelling Material Issues and Counter Tickets Is Easy

Any material issue or counter ticket with open status may be cancelled. When you cancel, the system:

- Reduces System/36 allocated inventory.

- Changes the document status to "cancelled."

- For counter tickets: posts an entry to account 8888 and prints a copy of the void invoice.

Additional Payment Options Available at Counter Ticket Closing

When closing a counter ticket, the parts counter person, parts manager, cashier, or accounting office staff can:

- Change the payment method.

- Adjust the total payment amounts, entering disputed amounts to expense accounts.

- Modify freight charges.

- Split the payment among three different terms, accounts, and control numbers.

FIGURE 12.12   Effective Design

Improve gross profits, minimize efforts:

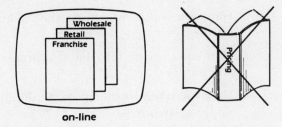

on-line

Cancelling material issues and counter tickets is easy:

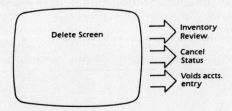

Additional payment options:

"Yes, I can adjust your method of payment. How would you like that payment split?"

"Yes, you're right. You were overcharged for that part. I can credit your account."

"Yes, the freight charge is wrong. I will adjust it."

FIGURE 12.12   *Continued*

WRITER'S CHECKLIST

1. Have you chosen appropriate margins, considering the size of the page, the amount of text to fit on the page, the level of difficulty of the text, and the background of the readers?

2. Have you chosen an appropriate number of columns, considering the size of the page, the amount of text to fit on the page, and the dimensions of the graphics?

3. Have you chosen an appropriate amount of leading?

4. Have you chosen an appropriate alignment: ragged right or right-justified?

5. Have you chosen a font that is appropriate for your subject, audience, and purpose?

6. Have you taken advantage of the different styles within the family of fonts you are using?

7. Have you taken advantage of different type sizes?

8. In designing your title, have you used prominent type styles and type sizes, and have you centered it?

9. In designing your headings, have you worked out a logical, consistent style for each hierarchical level?

10. In designing your lists, have you punctuated consistently, chosen an appropriate amount of indentation, and indented turnovers?

EXERCISES

1. Photocopy a page from a book or magazine and write a memo to your instructor describing and evaluating its design.

2. The article (Swart) on pages 335 to 337 appeared in the August 1990 issue of the journal *Personnel*. Write a memo to your instructor describing and evaluating the page design.

3. The description on pages 338 to 339 appears in a government document, *EPA Workshop on Radioactively Contaminated Sites* (EPA 520/1-90-009, March 1990). Revise the page design, rewriting any passages necessary to improve it.

An Overlooked Cost of Employee Smoking

J. Carroll Swart

Smoking restrictions in the workplace can pay off in many ways, including reductions in maintenance costs.

Although health concerns have been a primary reason for implementing smoking restrictions in the workplace, an important and often overlooked issue is the effect of smoking on an employer's routine and nonroutine maintenance costs. According to one researcher, William Weis of Seattle University, those costs are $1,250 more annually for a smoking employee than for a nonsmoking employee. In addition, various employers have reported reductions in maintenance costs after implementing strong restrictions on smoking:

• When a West Coast insurance company adopted a policy that permitted smoking only in a designated area in the lunchroom, the company's cleaning service voluntarily dropped its cleaning charge by 10% per month.

• An electronic components wholesaler banned smoking in the work-

RICHARD OSAKA

Smoking Costs

place and reduced its cleaning costs by more than half.

• A motel chain that now provides only nonsmoking rooms reduced its cleaning staff after adopting its no-smoking policy. Moreover, it claims that smoking rooms needed painting five times more often than did non-smoking rooms.

To companies searching for ways to reduce costs, the effects of employee smoking and routine maintenance costs is a relevant issue. Of equal if not more importance is the matter of employee smoking and nonroutine maintenance costs. Do furniture, carpeting, and draperies require special cleaning, and do the cleaning cycles occur frequently because of cigarette smoking? Do interior windows require frequent cleaning because of tobacco smoke? Do furniture and office furnishings experience shorter lives because of cigarette burns and cigarette smoke pollution? Do office walls and ceilings require painting because of cigarette smoking? Do display cases, office walls, ceilings, and ceiling lights require frequent washing because of cigarette smoking? Do computers and other electronic equipment, sensitive to smoke, require frequent servicing because of damage caused by cigarette smoking? In short, does smoking in offices damage sensitive electronic equipment, accelerate the depreciation of furniture and office furnishings, and make it necessary to frequently redecorate?

To find answers to these questions, I mailed a questionnaire to HR directors at 2,000 companies and asked them about their smoking policies and maintenance costs. The companies included banks, data-processing firms, savings and loan associations, utilities, and insurance carriers. They were selected by software programs on the basis of three criteria: a minimum of 500 shareholders, $5 million in assets, and the filing of financial data with the Securities and Exchange Commission on a quarterly basis.

Each company that received the questionnaire was instructed to consider all locations and facilities and indicate on a scale the degree that

---

EXHIBIT 1
Degrees of Restrictions on Employee Smoking—A Scale*

Policy A
It is the policy of the company to hire nonsmokers only. Smoking is prohibited off the job as well as on the job.

Policy B
Smoking is prohibited in all areas on company premises.

Policy C
Smoking is prohibited in all areas in company buildings.

Policy D
Smoking is prohibited in all areas in company buildings, with few exceptions. Smoking is permitted in the smoking section of the cafeteria (or room with a similar function in your company, if there is no cafeteria); in specially designated smoking rooms (smoking lounges); and in private offices, which may be designated "smoking permitted" or "no smoking" by the occupant.

Policy E
Smoking is prohibited in all common areas except those designated "smoking permitted.*
Smoking is permitted in specially designated smoking rooms (smoking lounges).
In open offices and in shared workspace areas where smokers and nonsmokers work together, where smokers' and nonsmokers' preferences are in conflict, employees and management will endeavor to find a satisfactory compromise. On failure to find a compromise, the preferences of the nonsmoker will prevail.
Private offices may be designated "smoking permitted" or "no smoking" by the occupant.

Policy F
It is the policy of the company to respect the preferences of both smokers and nonsmokers in company buildings. Where smokers' and nonsmokers' preferences are in conflict, employees and management will endeavor to find a satisfactory compromise. On failure to reach a compromise, the preferences of the nonsmoker will prevail.

Policy G
The company places no restrictions on employee smoking (the company does not have a smoking policy).

* "Smoking permitted" is synonymous with "designated smoking area." The latter term is increasing in usage.
A basic assumption is that all companies have policies prohibiting smoking in areas where there are safety and fire hazards and where sensitive equipment may be damaged. In reference to the scale above, the term "smoking policy" refers to a written statement or statements that place restrictions on smoking and intend to accommodate health concerns.

---

mainly reflects the company's posture on smoking. The scale, shown in Exhibit 1, lists six types of policies that restrict smoking (Policies A through F) and one that does not (Policy G).

Clearing the Smoke
The questionnaire mailing produced a response rate of 30% (608 companies). Among all responding companies, 68% had policies that restricted smoking (Policies A through F), whereas 32% had no restrictions on smoking (Policy G). Of the companies that restricted smoking, 84% had policies similar to Policies D,E, and F, whereas 16% had policies similar to Policies A, B, and C.

Overall, 23.3% of the surveyed companies that had smoking restrictions

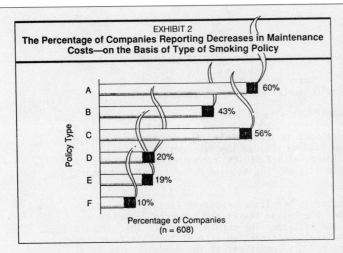

EXHIBIT 2
The Percentage of Companies Reporting Decreases in Maintenance Costs—on the Basis of Type of Smoking Policy

Policy Type

A — 60%
B — 43%
C — 56%
D — 20%
E — 19%
F — 10%

Percentage of Companies
(n = 608)

First, a smoking policy that prohibits smoking in all areas on a company's premises or in all areas of the company's buildings will probably reduce both routine and nonroutine maintenance costs. Policies that impose lesser degrees of restrictions on smoking are unlikely to produce beneficial results from a maintenance-cost standpoint.

Second, to maximize the benefits of strong smoking restrictions from a maintenance-cost perspective, an employer may want to consider reducing its custodial staff. Or, in instances in which outside suppliers of janitorial services are used, the employer may want to negotiate a more favorable contract with those suppliers. ◆

J. CARROLL SWART *is professor of management at the College of Business, Ball State University in Muncie, Indiana. He is the author of* A Flexible Approach to Working Hours *(AMACOM, 1978). This is his fifth contribution to* Personnel.

If you want to make photocopies or obtain reprints of this or other articles in Personnel, **please call (518) 891-5510 or write to AMA, P.O. Box 319, Saranac Lake, N.Y. 12983.**

in effect for one year or more reported reductions in maintenance costs. Moreover, maintenance-cost reductions were greatest at companies that placed strong restrictions on employee smoking (Policies A, B, and C). In contrast, companies that established less-restrictive policies on smoking (Policies D, E, and F) less often experienced decreases in maintenance costs (see Exhibit 2).

Some Recommendations

For employers that want to implement or change their smoking policy, two concluding remarks are appropriate.

27. RADIUM CHEMICAL COMPANY SITE SUMMARY

Shawn W. Googins, CHP
Environmental Protection Agency

The Radium Chemical Company (RCC) site is located in a light industrial and residential neighborhood in the Borough of Queens, New York. The site is at 60-06 27th Avenue, Woodside, and is immediately adjacent to the Brooklyn Queens Expressway, a major highway through the New York City area. It includes a 10,120-ft^2 building, with 7,850 ft^2 contaminated above New York State (NYS) limits for unrestricted release, and 4,000 ft^2 of surrounding land.

The RCC leased radium sources to hospitals, research facilities, and industrial firms throughout the United States and prepared radioluminous paints containing radium and tritium. The company moved to this site in 1955, abandoning another radioactively contaminated site at 235 East 44th Street, Manhattan, New York. During RCC's operations at the Woodside site it is estimated that approximately 1 Ci of radium sources was lost from the facility during shipment. Some of the lost sources were later recovered from the streets of New York City. The facility became an Environmental Protection Agency (EPA) removal action in 1985, at the request of the New York State Department of Environmental Conservation after the New York State Court declared the company abandoned.

Preliminary investigations indicate that contaminants include Ra-226, tritium, Sr-90 and various chemicals. The Ra-226 consists of about 120 Ci of sources previously used for cancer therapy, well logging, and research. Approximately 8 Ci of this are in the form of Ra-Be neutron sources. There is an undetermined quantity of powdered radium in paints, salts, solutions, and watch dials. There is an unknown quantity of tritium as tritiated water and in watch dials. Air sampling and smear surveys show that tritium is not a significant problem at the site; levels are either nondetectable or below NYS limits for unrestricted use. There is 50 mCi of Sr-90 present in eye applicator sources. Chemicals of concern at the site include hexanes, lacquers, hydroxymethylcellulose, 200 to 500 lb of mercury, ether, phosphoric acid, miscellaneous acids, bases, solvents, and approximately 300 lab pack containers.

Radiological contamination was found throughout the building interior, on the rooftop, and in the soil and storm drains. Alpha levels and exposure rates were particularly high in the glove box room, the vault, and the shipping area.

Radon levels range from 1 to 500 pCi/L inside the building; outside ambient air readings are between 0.5 and 0.8 pCi/L. Soil on the facility grounds ranges from 0.9 to 37 pCi/g, with 58 percent of the readings below 5 pCi/g and 15 percent greater than 15 pCi/g. Sediment from the storm drains ranges from 200 to 400 pCi/g.

RADIOLOGICAL CONDITIONS

Location	Exposure Rate mR/h	Alpha Levels dpm/100cm^2
Exterior:		
Walls	0.02 to 4.0	
Rooftop	0.1 to 50	general < 33
		hot spots 600
Interior:		
Shipping area	0.1 to 50	50 to 1,200
Repair room	3.0 to 25	200 to 7,200
Work shop	0.5 to 300	100 to 99,000
Glove box room	0.5 to 50	200 to 480,000
Vault	100 to 5000	52,000
Offices	0.03 to 0.2	

Upon taking over site management at the request of the State, EPA established site security measures and developed contingency plans in cooperation with local officials. Security measures included exclusion of the owner and employees from the site and installation of a perimeter fence and a CCTV surveillance system. Contingency planning with State and local officials included conducting a dose assessment for potential accident scenarios, review of State and local emergency plans, and providing instruction to local ambulance, fire, police, and hospital emergency room personnel.

Trailers have been brought on site to provide work areas, offices, laboratory facilities, and storage space. The existing exhaust system for the building has been secured and capped; a HEPA ventilation system has been installed, tested, and is in operation. Shielding and remote manipulators are used where necessary for source handling. A remote video system has been installed in the high radiation areas. Site radiological work practices include airlocks, frisking, protective clothing, respiratory protection, step off and sticky pads, and a Radiation Work Permit system. All activities involving radiation exposures are preplanned to keep exposures as low as reasonably achievable.

Environmental monitoring at the site includes air monitoring (of radon, particulates, and exhaust system), surface contamination surveys, and exposure rate surveys. The onsite laboratory is equipped with a gross alpha/beta system, liquid scintillation, and gamma spectroscopy (Hyperpure or Intrinsic germanium).

REFERENCES

Googins, S. W. 1990. Radium Chemical Company site summary. In *EPA workshop on radioactively contaminated sites.* Washington, D.C.: United States Environmental Protection Agency. EPA 520/1-90-009 (March).

Information network electronic data interchange implementation guide. 1987. Tampa, Fla.: IBM Information Services, C/1.

1990 U.S. Industrial Outlook. 1990. Washington, D.C.: United States Department of Commerce.

Parts invoice writer: Parts user's guide. 1986. N.p.: Volkswagen of America, Inc., 1/8–1/9.

Maloney, M. C., ed. 1987. Text formatting: Technical reference. Toronto: SoftQuad Inc. 5–18.

Swart, J. C. 1990. An overlooked cost of employee smoking. *Personnel* 67, 8 (August), 53–55.

PART

THREE

Common Applications

Memos

The memorandum is the workhorse of technical writing. It is how persons and groups communicate in an organization. Each day, the average employee may receive a half-dozen memos and send out another half-dozen. Whereas most memos convey routine news addressed to several readers, many organizations are turning to the memo for brief technical reports.

Most organizations have distribution routes—lists of employees who automatically receive copies of memos that pertain to their area of specialization or responsibility. A technician, for example, might be on several different distribution routes: one for all employees, one for the technical staff, and one for those involved with a particular project.

This chapter concentrates not on the "FYI" memo—the simple communication addressed "For Your Information"—but on the more substantive brief technical reports written in memo form, such as directives, responses to inquiries, trip reports, and field reports.

Like all technical writing, memos should be clear, accurate, comprehensive, concise, accessible, and correct. But they need not be ceremonious. Memo writing is technical writing with its sleeves rolled up.

The printed forms on which memos are written include a subject heading—a place for the writer to define the subject. If you use this space efficiently, you can define both your subject and purpose and thus begin to communicate immediately.

In the body of the memo, headings help your readers understand your message. For example, the simple heading *Results* enables them to decide whether or not to read the paragraph. (See Chapters 7 and 12 for a further discussion of headings.)

Most organizations have at least two sizes of memo forms: a full page and a half page. Often, writers are reluctant to put down only two or three lines on a full page, so they start to pad the message. The smaller page encourages conciseness.

The following discussion describes in detail how to write effective memos.

Writing Effective Memos

Readers expect memos to have certain structural features and to convey certain kinds of information, such as directives and trip reports.

However, memo writing is essentially like any other technical-writing process. You have to first understand your audience and purpose. Then you have to gather your information, create some sort of outline, write a draft, and finally revise it. Making the memo look like a memo—adding the structural features that your readers will expect—is relatively simple. You can build the structure into your outline, or you can shape the draft at some later stage. It doesn't matter "when" you make it look like a memo, as long as the finished product meets your reader's needs and expectations.

Structuring the Memo

The memo is made up of two components: the identifying information and the body.

The Identifying Information

The top of a memo should identify the writing situation as efficiently as possible. The writer names himself or herself, the audience, and the subject, ideally with some indication of purpose. Basically, a memo looks like a streamlined letter. It has no salutation ("Dear Mr. Smith") or complimentary close ("Sincerely yours"). Most writers put their initials or signature next to their typed names or at the end of the memo. The inside address — the mailing address of the reader — is replaced by a department name or an office number, generally listed after the person's name. Sometimes no "address" at all is given.

Almost all memos have five elements at the top: the logo or a brief letterhead of the organization and the "To," "From," "Subject," and "Date" headings. Some organizations have a "copies" or "cc" (carbon copy) heading as well. "Memo," "Memorandum," or "Interoffice" might be printed on the forms.

Organizations sometimes have preferences about the ways to fill out the headings. Some prefer full names of the writer and reader; others want only the first initials and the last names. Some prefer job titles; some not. If your organization does not object, include your job title and your reader's. In this way, the memo will be informative for a reader who refers to it after you or your reader has moved on to a new position as well as for readers elsewhere in the organization who might not know you and your reader. List the names of persons receiving copies of the memo (generally photocopies, not carbons) either alphabetically or in descending order of organizational rank. In listing the date, write out the month (March 4, 19XX or 4 March 19XX); do not use the all-numeral format (3/4/XX). Foreign-born people could be confused by the numerals, because in many countries the first numeral identifies the day, not the month.

The subject heading — the title of the memo — deserves special mention. Don't be *too* concise. Avoid naming only the subject, such as "Tower Load Test"; rather, specify what about the test you wish to address. For instance, "Tower Load Test Results" or "Results of Tower Load Test" would be much more informative than "Tower Load Test," which does not tell the reader whether the memo is about the date, the location, the methods, the results, or any number of other factors related to the test. (See Chapters 7 and 12 for more information on titles.)

The following examples show several common styles of filling in the identifying information of a memo.

AMRO   **MEMO**

To: B. Pabst
From: J. Alonso
Subject: MIXER RECOMMENDATION FOR PHILLIPS
Date: 11 June 19XX

NORTHERN PETROLEUM COMPANY   INTERNAL CORRESPONDENCE

Date: January 3, 19XX
To: William Weeks, Director of Operations
From: Helen Cho, Chemical Engineering Dept.
Subject: Trip Report—Conference on Improved Procedures
          for Chemical Analysis Laboratory

HARSON ELECTRONICS   MEMORANDUM

To: John Rosser, Accounting   From: Andrew Miller,
                                                Technical Services

                                      Subject: Budget Revision
                                                for FY 19XX

cc: Dr. William De Leon,   Date: March 11, 19XX
       President
       John Grimes, Comptroller

                    INTEROFFICE
To: C. Cleveland   cc: B. Aaron
From: H. Rainbow   K. Lau
Subject: Shipment Date of Blueprints J. Manuputra
          to Collier   W. Williams
Date: 2 October 19XX

Type the second and all subsequent pages of memos on plain paper. Include the following information in the upper left-hand corner.

1. the name of the recipient
2. the date of the memo
3. the page number

You might even define the communication as a memo and repeat the primary names. A typical second page of such a memo begins like this:

```
Memo to: J. Alders        April 6, 19XX
   From: R. Rossini       Page 2
```

The Body of the Memo

The average memo has a very brief "body." A message about the office closing early during heavy snow, for example, requires only one or two sentences. However, memos that convey complex technical information, and those that approach one page or longer (memos are always single-spaced), are most effective when they follow a basic structure. This structure gives you the same sense of direction that a full-scale outline does as you plan a formal report. Most substantive memos should include four parts:

1. purpose statement
2. summary
3. discussion
4. action

As with any kind of document, the organization of the memo does not necessarily reflect the writing sequence. After brainstorming and outlining the memo, start with the discussion—the most technical section. Until you have written the discussion, you cannot write the summary or the purpose statement, because both these parts depend on the discussion. For most writers, the most effective sequence of composition will be discussion, action, summary, and purpose.

Make sure to highlight the structure of the memo. Use headings and lists (see Chapters 7 and 12.)

Headings make the memo easier to read, easier *not* to read, and easier to refer to later. Headings are not subtle; they define clearly what the discussion is about and thus improve the reader's comprehension. In addition, headings represent a courtesy to executives who can read the purpose and summary and stop if they need no further details. Finally, headings enable readers to isolate quickly the information they need after the memo has been filed away for a while. Rather than having to reread a three-page discussion, they can turn directly to the summary, for instance, or to a subsection of the discussion.

Lists help your reader understand the memo. If, for instance, you are making three points, you can list them consistently in the different sections of the memo. Point number one under the "Summary" heading will correspond to point number one under the "Recommendations" heading, and so forth.

Purpose Statement

Memos are reproduced very freely. The majority of those you receive might be only marginally relevant to you. Many readers, after starting to read their incoming memos, ask, "Why is the writer telling me this?" The first sentence of the body—the purpose statement—should answer that question. Following are a few examples of purpose statements.

> I want to tell you about a problem we're having with the pressure on the main pump, because I think you might be able to help us.

> The purpose of this memo is to request authorization to travel to the Brownsville plant Monday to meet with the other quality inspectors.

> This memo presents the results of the internal audit of the Phoenix branch that you authorized March 13, 19XX.

> I want to congratulate you on the quarterly record of your division.

> This memo confirms our phone call of Tuesday, June 10, 19XX.

Notice that the best purpose statements are concise and direct. Make sure your purpose statement has a verb that clearly communicates what you want the memo to accomplish, such as *to request*, *to explain*, or *to authorize*. (See Chapters 3 and 5 for a more detailed discussion of purpose.) Some students of logical argument object to a direct statement of purpose— especially when the writer is asking for something, as in the example about requesting travel authorization. Rather than beginning with a direct statement of purpose, an argument for such a request would open with the reasons that the trip is necessary, the trip's potential benefits, and so forth. Then the writer would conclude the memo with the actual request: "For these reasons, I am requesting authorization to. . . ." Although some readers would rather have the reasons presented first, far more would prefer to know immediately why you have written. There are two basic problems with the indirect argument structure:

- ☐ Some readers will toss the memo aside if you seem to be rambling on about the Brownsville plant without getting to the point.
- ☐ Some readers will suspect that you are trying to manipulate them into doing something they don't want to do.

The purpose statement sacrifices subtlety for directness.

Summary

Along with the purpose statement, the summary forms the core of the memo. It has three main purposes:

- ☐ It helps all the readers follow the subsequent discussion.
- ☐ It enables executive readers to skip the rest of the memo if they so desire.
- ☐ It conveniently reminds readers of the main points.

Following are some examples of summaries:

> The proposed revision of our bookkeeping system would reduce its errors by 80 percent and increase its speed by 20 percent. The revision would take two months and cost approximately $4,000. The payback period would be less than one year.

> The conference was of great value. The lectures on new coolants suggested techniques that might be useful in our Omega line, and I met three potential customers who have since written inquiry letters.

> The analysis shows that lateral stress was the cause of the failure. We are now trying to determine why the beam could not sustain a lateral stress weaker than that it was rated for.

> In March, we completed Phase II (Design) on schedule. At this point, we anticipate no delays that will jeopardize our projected completion date.

The summary should reflect the length and complexity of the memo. It might range in length from one simple sentence to a long and technical paragraph. If possible, the summary should reflect the structure of the memo. For example, the discussion following the first sample summary should explain, first, the proposed revision of the bookkeeping system and, second, its two advantages: fewer errors and increased speed. Next should come the discussion of the costs and finally discussion of the payback period.

Discussion

The discussion elaborates on the summary. It is the most detailed, technical portion of the memo. Generally, the discussion begins with a background paragraph. Even if you think the reader will be familiar with the background, include a brief recap, just to be safe. Also, the background will be valuable to a reader who refers to the memo later.

Each background discussion is, of course, unique; however, some basic guidelines are useful. If the memo defines a problem—for example, a flaw detected in a product line—the background might discuss how the problem was discovered or present the basic facts about the product line: what the product is, how long it has been produced, and in what quantities. If the memo reports the results of a field trip, the background might discuss why the trip was undertaken, what its objectives were, who participated, and so forth.

Following is a background paragraph from a memo requesting authorization to have a piece of equipment retooled.

> **Background**
>
> The stamping machine, a Curtiss Model 6143, is used in the sheet-metal shop. We bought it in 1976 and it is in excellent condition. However, since we switched the size of the tin rolls last year, the stamping machine no longer performs up to specifications.

After the background comes the detailed discussion. Here you give your readers a clear and complete idea of what you have to say. You might divide the detailed discussion into the subsections of a more formal report: materials, equipment, methods, results, conclusions, and recommendations. Or you might give it headings that pertain specifically to the subject you are discussing. You might include small tables or figures but should attach larger ones to the back.

The discussion section of the memo can be developed according to any of the basic patterns for structuring technical documents:

- chronological
- spatial
- classification/partition
- comparison/contrast
- general to specific
- more important to less important
- problem-methods-solution
- cause/effect

These patterns are discussed in detail in Chapter 5.

Following is the detailed discussion section from a memo written by a salesperson working for the "XYZ Company," which makes electronic typewriters. The XYZ salesperson met an IBM salesperson by chance one day, and they talked about the XYZ typewriters. The XYZ salesperson is writing to his supervisor, telling her what he learned from the IBM salesperson and also what XYZ's research and development (R&D) department told him in response to the comments of the IBM salesperson.

> **DISCUSSION**
>
> **Salesperson's Comments:**
>
> In our conversation, he talked about the strengths of our machines and then mentioned two problems: excessive ribbon consumption and excessive training time.
>
> In general, he had high praise for the XYZ machines. In particular, he liked the idea of the rotary and linear stepping motor. Also, he liked having all the options within the

confines of the typewriter. He said that although he knows we have some reliability problems, the machines worked well while he was training on them.

The <u>major problem</u> with the XYZ machines, he said, is excessive ribbon consumption. According to his customers who have XYZ machines, the $5 cartridge lasts only about two days. This adds up to about $650 a year, about a third the cost of our basic Model A machine.

The <u>minor problem</u> with the machines, he said, is that most customers are used to the IBM programming language. Since our language is very different, customers are spending more time learning our system than they had anticipated. He didn't offer any specifics on training-time differences.

R&D's Response:

I relayed these comments to Susan Brown in R&D. Here is what she told me.

<u>Ribbon Consumption:</u> A recent 20 percent price reduction in the 4.2″ cartridge should help. In addition, in a few days our 4.9″ correctable cartridge—with a 40 percent increase in character output—should be ready for shipment. R&D is fully aware of the ribbon-consumption problem, and will work on further improvements, such as thinner ribbon, increased diameter, and new cartridges.

<u>Training Time:</u> New software is being developed that should reduce the training time.

If I can answer any questions about the IBM salesman's comments, please call me at x1234.

The basic pattern of this discussion is chronological: the writer describes first his discussion with the IBM salesperson and then the response from R&D. Within each of these two subsections the basic pattern is more important to less important: The ribbon-consumption problem is more serious than the training-time problem, so ribbon consumption is discussed first.

Action

Many memos present information that will eventually be used in major projects or policies. These memos will be included in the files on these projects or policies. Some reports, however, require follow-up action more immediately, by either the writer or the readers. For example, a memo addressing a group of supervisors might define what the writer is going to do about a problem discussed in the memo. A supervisor might use the action component to delegate tasks for other employees. In writing the action component of a memo, be sure to define clearly *who* is to do *what* and *when*. Following are two examples of action components.

Action:

I would appreciate it if you would work on the following tasks and have your results ready for the meeting on Monday, June 9.

1. Henderson to recalculate the flow rate.
2. Smith to set up meeting with the regional EPA representative for some time during the week of February 13.
3. Falvey to ask Armitra in Houston for his advice.

Action:

To follow up these leads, I will do the following this week:

1. Send the promotional package to the three companies.
2. Ask Customer Relations to work up a sample design to show the three companies.
3. Request interviews with the appropriate personnel at the three companies.

Notice in the first example that although the writer is the supervisor of his readers, he uses a polite tone in this introductory sentence.

Understanding Common Types of Memos

Each memo is written by a specific writer to a specific audience for a specific purpose. Every memo is unique. In one memo, for instance, you might be trying to persuade a group of readers that even though your idea for improving the public image of your company will cost some money in the short run, the campaign is necessary and will more than pay for itself in the long run. There is no magic formula for writing this kind of memo; you have to find some way to convince your readers that your idea makes sense.

Even though no two memos are the same, there are four broad categories of memos according to the functions they fulfill: the directive, the response to an inquiry, the trip report, and the field/lab report.

Notice as you read about each type of memo how the purpose-summary-discussion-action strategy is tailored to the occasion. Pay particular attention to the headings, lists, and indentation used to highlight structure.

The Directive

In a directive memo, you define a policy or procedure you want your readers to follow. If possible, explain the reason for the directive; otherwise, it might seem like an irrational order rather than a thoughtful request. For short memos, to prevent the appearance of bluntness, the explanation should precede the directive. For longer memos, the actual directive might precede the detailed explanation. This strategy will ensure that the readers will not overlook the directive. Of course, the body of the memo should begin with a polite explanatory note, such as the following:

> The purpose of this memo is to establish a uniform policy for dealing with customers who fall more than 60 days behind in their accounts. The policy is defined below under the heading "Policy Statement." Following the statement is our rationale.

Figure 13.1 provides an example of a directive.

Notice in Figure 13.1 that the directive is stated as a request, not an order. Unless your readers have ignored previous requests, a polite tone works best.

Also note that in spite of its brevity and simplicity, this example follows, without headings, the purpose-summary-discussion-action structure. The subject line identifies the purpose, the first paragraph is a combination of summary and discussion (extensive discussion is hardly necessary in this situation), and the second paragraph dictates the specific action to be taken.

Quimby   Interoffice

Date: March 19, 19XX
To: All supervisors and sales personnel
From: D. Bartown, Engineering
Subject: Avoiding customer exposure to sensitive
           information outside Conference Room B.

It has come to our attention that customers meeting in Conference Room B have been allowed to use the secretary's phone directly outside the room. This practice presents a problem: the proposals that the secretary is typing are in full view of the customers. Proprietary information such as pricing can be jeopardized unintentionally.

In the future, would you please escort any customers or non-Quimby personnel needing to use a phone to the one outside the Estimating Department? Your cooperation in this matter will be greatly appreciated.

FIGURE 13.1   Directive

National Insurance Company   **MEMO**

TO: J. M. Sosry, Vice President
FROM: G. Lancasey, Accounting
SUBJECT: National's Compliance with the Federal Pay Standards
DATE: February 2, 19XX

Purpose: This memo responds to your request for an assessment of our compliance with the Federal Non-Inflationary Pay and Price Behavior Standards.

Summary:
1. We are in compliance except for a few minor violations.
2. Legal Affairs feels we are exercising "good faith," a measure of compliance with the Standards.
3. Data Processing is currently studying the costs and benefits of implementing data processing of the computations.

The following discussion elaborates on these three points.

Discussion:
1. We are in compliance with the Standards except for the following details related to fringe benefits.
   a. The fringe benefits of terminated individuals have not yet been eliminated from our calculations. The salaries have been eliminated.
   b. The fringe benefits associated with promotional increases have not yet been eliminated from our calculations.
   c. The fringe benefits of employees paid $9,800 or less have not yet been eliminated from our calculations.

FIGURE 13.2   Response to an Inquiry   *(See page 354.)*

The Response to an Inquiry

Often you might be asked by a colleague to provide information that cannot be communicated on the phone because of its complexity or importance. In responding to such an inquiry, the purpose-summary-discussion-action strategy is particularly useful. The purpose of the memo is simple: to provide the reader with the information he or she requested. The sum-

Memo to J. M. Sosry
February 2, 19XX
Page 2

2. I met with Joe Brighton of Legal Affairs last Thursday to discuss the question of compliance. Joe is aware of our minor violations. His research, including several calls to Washington, suggests that the Standards define "good faith" efforts to comply for various-size corporations, and that we are well within these guidelines.

3. I talked with Ted Ashton of Data Processing last Friday. They have been studying the costs/benefits of implementing data processing for the calculations. Their results won't be complete until next week, but Ted predicts that it will take up to two months to write the program internally. He is talking to representatives of computer service companies this week, but he doubts if they can provide an economical solution.

As things stand now—doing the calculations manually—we will need three months to catch up, and even then we will always be about two weeks behind.

Action: I have asked Ted Ashton to send you the results of the cost/benefits study when they are in. I have also asked Joe Brighton to keep you informed of any new developments with the Standards.

If I can help further, please let me know.

FIGURE 13.2   *Continued*

mary states the major points of the subsequent discussion and calls the reader's attention to any parts of it that might be of special importance. The action section (if it is necessary) defines any relevant steps that you or some other personnel are taking or will take. Figure 13.2 provides an example of a response to an inquiry.

Notice in this sample memo how the numbered items in the summary section correspond to the numbered items in the discussion section. This parallelism helps readers find the discussion they want quickly.

Dynacol Corporation

INTEROFFICE MEMORANDUM

To: G. Granby, R&D
From: P. Rabin, Technical Services
Subject: Trip Report—Computer Dynamics, Inc.
Date: September 20, 19XX

Purpose:
This memo presents my impressions of the Computer Dynamics technical seminar of September 18. The purpose of the seminar was to introduce their new PQ-500 line of computers.

Summary:
In general, I was not impressed with the new line. The only hardware that might be of interest to us is their graphics terminal, which I'd like to talk to you about.

Discussion:
Computer Dynamics offers several models in its 500 series, ranging in price from $11,000 to $45,000. The top model has a 200 Mb memory. Although it's very fast at matrix operations, this feature would be of little value to us. The other models offer nothing new.

I was disturbed by some of the answers offered by the Computer Dynamics representatives, which everyone agreed included misinformation.

The most interesting item was the graphics terminal. It is user-friendly. Integrating their terminal with our system could cost $4,000 and some 4–5 person-months. But I think that we want to go in the direction of graphics terminals, and this one looks very good.

Recommendation:
I'd like to talk to you, when you get a chance, about our plans for the addition of graphics terminals. I think we should have McKinley and Rossiter take a look at what's available. Give me a call (x3442) and we'll talk.

FIGURE 13.3   Trip Report   *(See page 356.)*

The Trip Report

A trip report is a record of a business trip written after the employee returns to the office. Most often, a trip report takes the form of a memo. The key to writing a good trip report is to remember that your reader is less interested in an hour-by-hour narrative of what happened than in a carefully structured discussion of what was important. If, for instance, you attended a professional conference, don't list all the presentations—simply attach the agenda or program if you think your reader will be interested. Communicate the important information you learned—or describe the important questions that didn't get answered. If you traveled to meet a client (or a potential client), focus on what your reader is interested in: how to follow up on the trip and maintain a good business relationship with the client.

In most cases, the standard purpose-summary-discussion-action structure is appropriate for this type of memo. Briefly mention the purpose of the trip—even if your reader might already know its purpose. By doing this, you will be providing a complete record for future reference. In the action section, list either the pertinent actions you have taken since the trip or what you recommend that your reader do. Figure 13.3 provides an example of a typical trip report.

Notice in this example that the writer and reader appear to be relatively equal in rank: the informal tone of the "Recommendation" section suggests that they have worked together before. Despite this familiarity, however, the memo is clearly organized to make it easy to read and refer to later, or to pass on to another employee who might follow up on it.

Field and Lab Reports

Many organizations use memos to report on inspection and maintenance procedures. These memos, known as field or lab reports, include the same information that high-school lab reports do—the problem, methods, results, and conclusions—but they deemphasize the methods and can include a recommendations section.

A typical field or lab report, therefore, has the following structure:

1. purpose of the memo
2. problem leading to the decision to perform the procedure
3. summary
4. results
5. conclusions
6. recommendations
7. methods

Sometimes several sections are combined. Purpose and problem often are discussed together, as are results and conclusions.

Lobate Construction   **MEMO**

To:	C. Amalli	cc:	A. Beren
From:	W. Kabor		S. Dworkin
Subject:	Inspection of Chemopump		N. Mancini
	after Run #9		
Date:	6 January 19XX		

Purpose:
This memo presents the findings of my visual inspection of
the Chemopump after it was run for 30 days on Kentucky #10
coal and requests authorization to carry out follow-up
procedures.

Problem:
The inspection was designed to determine if the new
Chemopump is compatible with Kentucky #10, our lowest-
grade coal. In preparation for the 30-day test run, the
following three modifications were made:

1. New front bearing housing buffer plates of tungsten
   carbide were installed.
2. The pump casting volute liner was coated with tungsten
   carbide.
3. New bearings were installed.

Summary:
A number of small problems with the pump were observed,
but nothing serious or surprising. Normal break-in accounts
for the wear. The pump accepted the Kentucky #10 well.

Findings:
The following problems were observed:

1. The outer lip of the front-end bell was chipped along
   two-thirds of its circumference.

FIGURE 13.4   Lab Report   *(See page 358.)*

The lab report shown in Figure 13.4 illustrates some of the possible
variations on this standard report structure.

Notice the following points about this example:

☐ A single word—*visual*—constitutes the discussion of the inspection
procedure in the purpose section. Nothing else needs to be said.

Memo to C. Amalli
6 January 19XX
Page 2

    2.   Opposite the pump discharge, the volute liner received a slight wear groove along one-third of its circumference.

    3.   The impeller was not free-rotating.

    4.   The holes in the front-end bell were filled with insulating mud.

The following components showed no wear:

1. The 5½″ impeller.
2. The suction neck liner.
3. The discharge neck liner.

Conclusions:

The problems can be attributed to normal break-in for a new Chemopump. The Kentucky #10 coal does not appear to have caused any extraordinary problems. The new Chemopump seems to be operating well.

Recommendations:

I would like authorization to modify the pump as follows:

1. Replace the front-end bell with a tungsten carbide-coated front-end bell.
2. Replace the bearings on the impeller.
3. Install insulation plugs in the holes in the front-end bell.

I recommend that the pump be reinspected after another 30-day run on Kentucky #10.

If you have any questions, please call me at x241.

FIGURE 13.4 *Continued*

☐ In the "Summary" section, the writer lists the "bad news"—the problems—before the "good news." This is a logical order, because the bad news means more to the readers.

☐ The last sentence allows the reader to get in touch with the writer to ask questions or authorize the recommended modifications.

WRITER'S CHECKLIST

The following checklist covers the basic formal elements included in most memo reports.

1. Does the identifying information
   a. include the names and (if appropriate) the job positions of both you and your readers?
   b. include a sufficiently informative subject heading?
   c. include the date?

2. Does the purpose statement clearly tell the readers why you are asking them to read the memo?

3. Does the summary
   a. briefly state the major points developed in the body of the memo?
   b. reflect the structure of the memo?

4. Does the discussion section
   a. include a background paragraph?
   b. include headings to clarify the structure and content?

5. Does the action section clearly and politely identify tasks that you or your readers will carry out?

EXERCISES

1. As the manager of Lewis Auto Parts Store, you have noticed that some of your salespeople are smoking in the showroom. You have received several complaints from customers. Write a memo defining a new policy: salespeople may smoke in the employees' lounge but not in the showroom.

2. There are 20 secretaries in the six departments at your office. Although they are free to take their lunch hours whenever they wish, sometimes several departments have no secretarial coverage between 1:00 and 1:30 P.M. Write a memo to the secretaries, explaining why this lack of coverage is undesirable and asking for their cooperation in staggering their lunch hours.

3. You are a senior with an important position in a school organization, such as a technical society or the campus newspaper. The faculty adviser to the organization has asked you to explain, for your successors, how to carry out the responsibilities of the position. Write a memo in response to the request.

4. The boss at the company where you last worked has phoned you, asking for your opinion on how to improve the working conditions and productivity. Using your own experiences, write a memo responding to the boss's inquiry.

5. If you have attended a lecture or presentation in your area of academic concentration, write up a trip report memo to an appropriate instructor assessing its quality.

6. Write up a recent lab or field project in the form of a memo to the instructor of the course.

7. The following memos could be improved in tone, substance, and structure. Revise them, adding any reasonable details.

    a.

KLINE MEDICAL PRODUCTS

Date: 1 September 19XX
To: Mike Framson
From: Fran Sturdiven
Subject: Device Master Records

The safety and efficiency of a medical device depends on the adequacy of its design and the entire manufacturing process. To ensure that safety and effectiveness are manufactured into a device, all design and manufacturing requirements must be properly defined and documented. This documentation package is called by the FDA a "Device Master Record."

The FDA's specific definition of a "Device Master Record" has already been distributed.

Paragraph 3.2 of the definition requires that a company define the "compilation of records" that makes up a "Device Master Record." But we have no such index or reference for our records.

Paragraph 6.4 says that any changes in the DMR must be authorized in writing by the signature of a designated individual. We have no such procedure.

These problems are to be solved by 15 September 19XX.

---

    b.

City of Oakland
Office of the Mayor

To: Department Heads
From: Mayor Christine Rawlins
Subject: "Tolerance"

As you are surely aware, an unwritten policy of "tolerance" for public officials and police officers was in existence until recently at the Oakland Department of Law Enforcement. That policy was of grave concern to me and, as you know, resulted in a major public controversy and necessary personnel adjustments.

It is the unaltered policy of this administration that no department shall at any time adopt, utilize, adapt, or endorse any policy of favoritism in any aspect of its enforcement responsibility. The informing principle must be that no one in Oakland is above the law.

I fully expect that each department is operating under that principle, but let this serve as a reminder that this is our policy and there shall be no deviation from that policy.

Please convey this information to the necessary personnel in your department.

c.

Viking National Bank

Memorandum

To: George Delmore, Expense Recording
From: David Derahl, Internal Audit
Subject: Escheat Procedures
Date: Dec. 6, 19XX

Undeliverable checks received by the originating department should be voided immediately. The originating department should obtain a copy of the voided check and forward the original to Expense Recording. Efforts to determine a valid mailing address should be initiated by the originating department. If the check is less than seven months old and remains undeliverable, the liability should be recorded in the originating department's operating expense account.

This memo should define our recommended policy on escheat procedures. I hope it answers your question.

d.

Wayne Consulting Engineers, Inc.

February 13, 19XX
To: Morris Dansette, Vice President
From: Holly Ryder, Architecture Department
Subject: Lettering Machine

The Architecture department has witnessed various demonstrations of lettering machines. We have concluded that the Kroy 360 printer is the machine that would best suit Wayne's needs. The Kroy would be centrally located so all disciplines would have access. The Kroy is available in all of Wayne's standard fonts including Bodoni and Univers. The especially designed tape for architects and engineers is heat pressed and doesn't come off in blueprint or photocopy machines. The Kroy had memory capabilities of 32,000 characters.

A demonstration by the manufacturer's rep should be attended by people who will frequently use it, or one person per department. The cost of the Kroy is $1700. They will take our existing Kroy as a trade-in for $300. We can also trade in our tape supply and put it toward our first font purchase. Once ordered it takes 3 days to deliver. Total balance is due after 30 days. The Kroy comes with a one-year warranty.

If you have any other questions, please call me at x3229.

e.

Diversified Chemicals, Inc.

MEMO

Date: August 27, 19XX
To: R. Martins
From: J. Speletz
Subject: Charles Research Conference on Corrosion

The subject of the conference was high-temperature dry corrosion. Some of the topics discussed were

1 - thin film formation and growth on metal surfaces. The lectures focused on the study of oxidation and corrosion by spectroscopy.

2 - the use of microscopy to study the microstructure of thick film formation on metals and alloys. The speakers were from the University of Colorado and MIT.

3 - one of the most interesting topics was hot corrosion and erosion. The speakers were from Penn State and Westinghouse.

4 - future research directions for high-temperature dry corrosion were discussed from five viewpoints.
  1 - university research
  2 - government research
  3 - industry research
  4 - European industry research
  5 - European government research

5 - corrosion of ceramics, especially the oxidation of Si_3N_4. One paper dealt with the formation of Si ALON, which could be an inexpensive substitute for Si_3N_4. This topic should be pursued.

f.

Korvon Laboratories—Memo

To:	Ralph Eric
From:	Walt Kavon
Subject:	"Computers in the Laboratory"
Date:	May 1, 19XX

The seminar on "Computers in the Laboratory" was held in New York on April 22, 19XX. Approximately thirty managers of labs of various sizes attended.

The leader of the seminar, Mr. Daniel Moore, presented a program that included the following topics:

Modern analytical instrumentation
Maintaining quality in quality assurance
Harnessing the power of computers
Capital investments: justifying costs to management

The subjects of minicomputers versus terminals and how to increase reliability and reduce costs were discussed by several computer manufacturers' representatives.

My major criticism of the session was the ineffective leadership of Dr. Moore. Frequently, he read long passages from published articles. Often, he was very disorganized. I did enjoy, however, meeting the other lab managers.

g.

Technical Maintenance, Inc.   **Memo**

TO:	Rich Abelson
FROM:	Tom Donovan
SUBJECT:	Dialysis Equipment
DATE:	10/24/XX

The Clinic that sent us the dialysis equipment (two MC-311's) reported that it could not regulate the temperature precisely enough.

I found that in both 311's, the heater element did not turn off. The temperature control circuit has an internal trim potentiometer that required adjustment. It is working correctly now.

I checked out the temperature control system's independent backup alarm system that will alarm and shut down the system if the temperature reaches 40°C. It is working properly.

The equipment has been returned to the client. After phoning them, I learned that they have had no more problems with it.

Letters

Whether it is mailed or faxed, the letter is the basic means of communication between two organizations: close to 100 million business letters are written each working day. Although telephone use is constantly increasing, letters remain the basic link, because they provide documentary records. Often, phone conversations and transactions are immediately written up as letters, to become a part of the files of both organizations. The increasing use of electronic mail will not change this fact. Therefore, even as a new employee, you can expect to write letters regularly. And as you advance to positions of greater responsibility, you will write even more letters, for you will be representing your organization more often.

More than any other kind of technical writing, letters represent the dual nature of the working world. On the one hand, a letter must be every bit as accurate as a legal contract. On the other hand, it communicates to people. Your reader will form an opinion of your organization from your letter. Regarding your reader as a person—while representing your organization effectively—is the challenge of writing good business letters.

Writing Effective Letters

The previous chapter mentioned that readers expect to see traditional features in memos. The same holds true for letters. In fact, letter format has remained virtually unchanged for hundreds of years. For this reason, you will want to learn these formats, as well as the various kinds of letters, such as inquiries and sales messages.

Yet writing a letter is much like writing any other technical document. First you have to analyze your audience and determine your purpose. Then you have to gather your information, create some sort of outline, write a draft, and finally revise it. Making the letter look like a letter is the easy part. You can build the structure into your outline, or you can shape the draft at some later stage. It doesn't matter at what stage you make it look like a letter, provided the finished product meets your reader's needs and expectations.

Projecting the "You Attitude"

Like any other type of technical writing, the letter should be clear, concise, comprehensive, accessible, correct, and accurate. It must convey information in a logical order. It should not contain small talk: the first paragraph should get directly to the point without wasting words. And to enable the reader to locate information quickly and easily, it should use topic sentences at the start of the paragraphs. Often, letters use headings and indentation just as reports do. In fact, some writers use the term *letter report* to describe a technical letter of more than, say, two or three pages. In substance, it is a report; in form, it is a letter, containing all the letter's traditional elements.

Moreover, because it is a communication from one person to another, a letter must also convey a courteous, positive tone. The key is the "you attitude." This term means looking at the situation from your reader's point of view and adjusting the content, structure, and tone to meet the person's needs. The "you attitude" is largely common sense. If, for example, you are writing to a supplier who has failed to deliver some merchandise on the agreed-upon date, the "you attitude" dictates that you not discuss problems you are having with other suppliers—those problems don't concern your reader. Rather, you should concentrate on explaining clearly and politely that your reader has violated your agreement and that not having the merchandise is costing you money. Then you should propose ways to expedite the shipment.

Looking at things from the other person's point of view would be simple if both parties always saw things the same way. They don't, of course. Sometimes the context of the letter is a dispute. Nevertheless, good letter writers always maintain a polite tone. Civilized behavior is good business, as well as a good way to live.

Following are examples of thoughtless sentences, each followed by an improved version that exhibits the "you attitude."

Egotistical   Only our award-winning research and development department could have devised this revolutionary new sump pump.

Better   Our new sump pump features significant innovations that you may appreciate.

Blunt   You wrote to the wrong department. We don't handle complaints.

Better   Your letter has been forwarded to the Customer Service Division.

Accusing   You must have dropped the engine. The housing is badly cracked.

Better   The badly cracked housing suggests that your engine must have fallen onto a hard surface from some height.

Sarcastic   You'll need two months to deliver these parts? Who do you think you are, the Post Office?

Better   Surely you would find a two-month delay for the delivery of parts unacceptable in your business. That's how I feel too.

Belligerent   I'm sure you have a boss, and I doubt if he'd like to hear about how you've mishandled our account.

Better   I would prefer to settle our account with you rather than having to bring it to your supervisor's attention.

Condescending	Haven't you ever dealt with a major corporation before? A 60-day payment period happens to be standard.
Better	We had assumed that you honored the standard 60-day payment period.
Overstated	Your air-filter bags are awful. They're all torn. We want our money back.
Better	Nineteen of the 100 air-filter bags we purchased are torn. We would therefore like you to refund the purchase price of the 19 bags: $190.00

After you have drafted a letter, look back through it. Put yourself in your reader's place. How would you react if you received it? A calm, respectful, polite tone always makes the best impression and therefore increases your chances of achieving your goal.

Avoiding Letter Clichés

Related to the "you attitude" is the issue of letter clichés. Over the decades, a set of words and phrases has come to be associated with letters, phrases such as "as per your request." For some reason, many people think that these phrases are required. They're not. They make the letter sound stilted and insincere. If you would feel awkward or uncomfortable saying these clichés to a friend, avoid them in your letters.

Figure 14.1 (page 369) lists some of the common letter clichés and their more natural equivalents.

Following are two versions of the same letter: one written in letter clichés, the other in plain language.

Dear Mr. Smith:

Referring to your letter regarding the problem encountered with your new Eskimo Snowmobile. Our Customer Service Department has just tendered its report.

It is their conclusion that the malfunction is caused by water being present in the fuel line. It is our unalterable conclusion that you must have purchased some bad gasoline. We trust you are cognizant of the fact that while we guarantee our snowmobiles for a period of not less than one year against defects in workmanship and materials, responsibility cannot be assumed for inadequate care. We wish to advise, for the reason mentioned hereinabove, that we cannot grant your request to repair the snowmobile free of charge.

Permit me to say, however, that the writer would be pleased to see that the fuel line is flushed at cost, $30. Your Eskimo would then give you many years of trouble-free service.

Enclosed please find an authorization card. Should we receive it, we shall endeavor to perform the above-mentioned repair and deliver your snowmobile forthwith.

Sincerely yours,

Dear Mr. Smith:

Thank you for writing to us about the problem with your new Eskimo Snowmobile.

Our Customer Service Department has found water in the fuel line. Apparently some of the gasoline was bad. While we guarantee our snowmobiles for one year against defects in workmanship and materials, we cannot assume responsibility for problems caused by bad gasoline. We cannot, therefore, grant your request to repair the snowmobile free of charge.

However, no serious harm was done to the snowmobile. We would be happy to flush the fuel line at cost, $30. Your Eskimo would then give you many years of trouble-free service.

If you will authorize us to do this work, we will have your snowmobile back to you within four working days. Just fill out the enclosed authorization card and drop it in the mail.

Sincerely yours,

The second version of this letter not only avoids the clichés but also shows a much better understanding of the "you attitude." Rather than building the letter around the violation of the warranty, as the first writer does, the second writer presents the message as good news: the snowmobile is not ruined, and it can be returned in less than a week for a low cost.

Understanding the Elements of the Letter

Almost every letter has a heading, inside address, salutation, body, complimentary close, signature, and reference initials. In addition, some letters contain one or more of the following notations: attention, subject, enclosure, and copy.

For short, simple letters, you can compose the elements in the sequence in which they will appear. For more complex letters, however, you ought to use the strategy discussed with all the other technical-writing

Letter Clichés	Natural Equivalents
attached please find	attached is
cognizant of	aware that
enclosed please find	enclosed is
endeavor (verb)	try
herewith ("We herewith submit...")	(None. *Herewith* doesn't say anything. Skip it.)
hereinabove	previously, already
in receipt of ("We are in receipt of...")	"We have received..."
permit me to say	(None. Permission granted. Just say it.)
pursuant to our agreement	as we agreed
referring to your ("Referring to your letter of March 19, the shipment of pianos...")	"As you wrote in your letter of March 19, the..." or subordinate the reference at the end of your sentence.
same (as a pronoun: "Payment for same is requested...")	(Use the noun instead: "Payment for the merchandise is requested...")
wish to advise ("We wish to advise that...")	(The phrase doesn't say anything. Just say what you want to say.)
the writer ("The writer believes that...")	"I believe..."

FIGURE 14.1   Letter Clichés

applications: start with the body and continue to the end. Then go back and add the preliminary elements, ending with the first paragraph.

In the following paragraphs, the elements of the letter are discussed in the order they would ordinarily appear. Six common types of letters will be discussed in detail later in this chapter.

Heading

The typical organization has its own stationery, with its name, address, phone number, fax number, and perhaps a logo—the letterhead—printed at the top. The letterhead and the date the letter will be sent (typed two lines below the letterhead) make up the heading. When typing on blank paper use your address (without your name) and date as the heading. Type only the first page of any letter on letterhead stationery. Type the second and all subsequent pages on blank paper, with the name of the recipient, the page number, and the date in the upper left-hand corner. For example:

Mr. John Cummings
Page 2
July 3, 19XX

Do not number the first page of any letter.

Inside Address

The inside address is your reader's name, position, organization, and business address. If your reader has a courtesy title, such as *Dr., Professor,* or — for public officials — *Honorable,* use it. If not, use *Mr.* or *Ms.* (unless you know the reader prefers *Mrs.* or *Miss*). If your reader's position can fit conveniently on the same line as his or her name, add it after a comma; otherwise, place it on the line below. Spell the name of the organization the way it does: for example, International Business Machines calls itself IBM. Include the complete mailing address: the street, city, state, and zip code.

Attention Line

Sometimes you will be unable to address the letter to a particular person. If you don't know (and cannot easily find out) the person's first name or don't know the person's name at all, use the attention line:

Attention: Technical Director

Use the attention line if you want to make sure that the organization you are writing to responds even if the person you write to is unavailable. In this case, put the name of the organization or of one of its divisions on the first line of the inside address:

Operations Department
Haverford Electronics
117 County Line Road
Haverford, MA 01765

Attention: Charles Fulbright, Director

Subject Line

On the subject line, put either a project number (for example, "Subject: Project 31402") or a brief phrase defining the subject of the letter (for example, "Subject: Price Quotation for the R13 Submersible Pump").

Operations Department
Haverford Electronics
117 County Line Road
Haverford, MA 01765

Attention: Charles Fulbright, Director

Subject: Purchase Order #41763

Salutation

If you have no attention line or subject line, put the salutation two lines below the inside address. The traditional salutation is *Dear* followed by the reader's courtesy title and last name. Use a colon after the name, not a comma. If you are fairly well acquainted with your reader, use *Dear* followed by the first name. When you do not know the reader's name, use a general salutation, such as *Dear Technical Director:* or *Dear Sir or Madam:*

When you are addressing a group of people, use a salutation like *Ladies and Gentlemen:* or *Gentlemen:* (if all the readers are male) or *Ladies:* (if all the readers are female).

Or you can tailor the salutation to your readers, using a phrase like *Dear Members of the Restoration Committee:* or *Dear Members of Theta Chi Fraternity:*

This same strategy is useful for sales letters without individual inside addresses, as in the salutation *Dear Homeowner:* or *Dear Customer:*

Body

The body is the substance of the letter. Although you might write only a few words, generally you will have three or more paragraphs. The first paragraph introduces the subject of the letter, the second elaborates the message, and the third concludes it.

Complimentary Close

After the body of the letter, include one of the traditional closing expressions: *Sincerely, Sincerely yours, Yours sincerely, Yours very truly, Very truly yours.* Capitalize only the first word in the complimentary close, and follow all such expressions by a comma. Today, all the phrases have lost their particular meanings and connotations. They are interchangeable.

Signature

Type your full name on the fourth line below the complimentary close. Sign the letter, in ink, above the typewritten name. Most organizations prefer that you add, beneath your typed name, your position title. For example:

Very truly yours,

Chester Hall

Chester Hall
Personnel Manager

Reference Line

If someone else types your letters, the reference line identifies, usually by initials, both you and the typist. It appears a few spaces below the signature line, along the left margin. Generally, the writer's initials—which always come first—are capitalized, and the typist's initials are lowercase. For example, if Marjorie Connor wrote a letter that Alice Wagner typed, the standard reference notation would be *MC/aw*.

Enclosure Line

If the envelope contains any documents other than the letter itself, identify the number of enclosures:

For one enclosure	Enclosure
Or	Enclosure (1)
For more than one enclosure	Enclosures (2)
	Enclosures (3)

In determining the number of enclosures, count only the separate items, not the number of pages. A three-page memo and a ten-page report constitute only two enclosures. Some writers like to identify each enclosure by stating its title.

Copy Line

If you want the reader to know that other people are receiving a copy, use the symbol *cc* (for "carbon copy") or *xc* (for "Xerox copy," a photocopy) followed by the names of the other recipients (listed either alphabetically or according to organizational rank). If you do not want your reader to know about other copies, type *bcc* ("blind carbon copy") on the copies only—not on the original.

Learning the Format of the Letter

There are three popular formats used for letters: modified block, modified block with paragraph indentations, and full block. Figures 14.2, 14.3, and 14.4 show diagrams of letters written in these three formats.

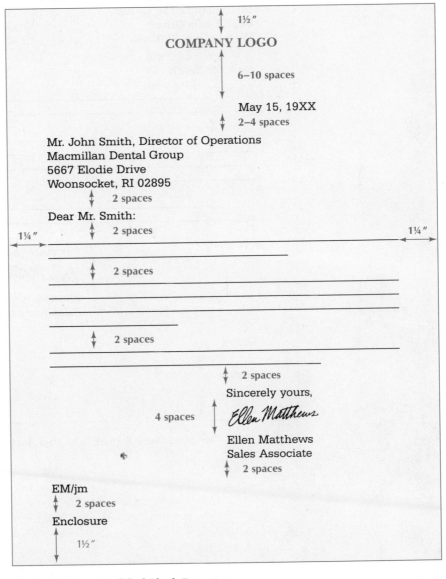

FIGURE 14.2   Modified Block Format

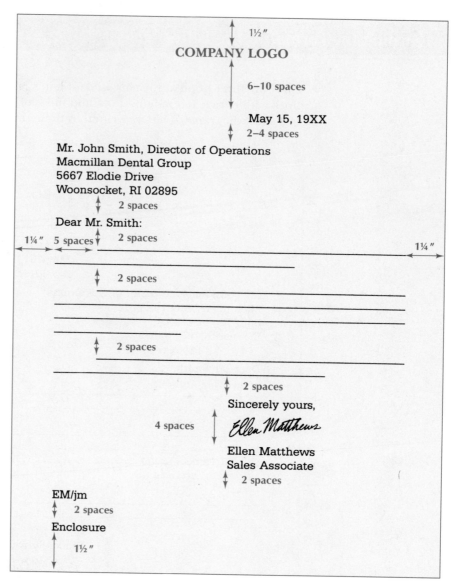

FIGURE 14.3 Modified Block Format, with Paragraph Indentations

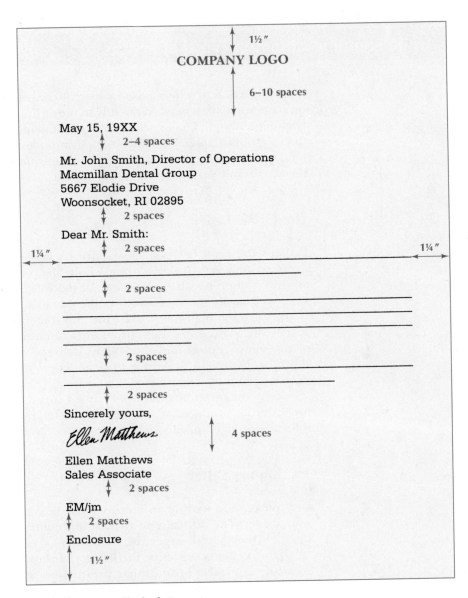

↕ 1½"

COMPANY LOGO

↕ 6–10 spaces

May 15, 19XX

↕ 2–4 spaces

Mr. John Smith, Director of Operations
Macmillan Dental Group
5667 Elodie Drive
Woonsocket, RI 02895

↕ 2 spaces

Dear Mr. Smith:

1¼" ← → ↕ 2 spaces → 1¼"

↕ 2 spaces

↕ 2 spaces

↕ 2 spaces

Sincerely yours,

Ellen Matthews

↕ 4 spaces

Ellen Matthews
Sales Associate

↕ 2 spaces

EM/jm

↕ 2 spaces

Enclosure

↕ 1½"

FIGURE 14.4 **Full Block Format**

Understanding Common Types of Letters

Dozens of kinds of letters exist for specific occasions. This chapter focuses on six types written most frequently in the technical world: order, inquiry, response to inquiry, sales, claim, and adjustment. The transmittal letter is discussed in Chapter 17. The job-application letter is discussed in Chapter 15. For a more detailed discussion of business letters, consult one of the several full-length books on the subject.

The Order Letter

Perhaps the most basic form of business correspondence is the order letter, written to a manufacturer, wholesaler, or retailer. When writing an order letter, be sure to include all the information your reader will need to identify the merchandise: the quantity, model number, dimensions, capacity, material, price, and any other pertinent details. Also specify the terms of payment (if other than payment in full upon receipt of the merchandise) and method of delivery. A typical order letter is shown in Figure 14.5. (Notice that the writer uses an informal table to describe the parts he orders.)

Many organizations have preprinted forms, called purchase orders, for ordering products or services. A purchase order calls for the same information that appears in an order letter.

The Inquiry Letter

Your purpose in writing an inquiry letter is to obtain information from your reader. The difficulty of writing an inquiry letter is determined by whether the reader is expecting your letter.

If the reader is expecting the letter, your task is easy. For example, if a company that makes institutional furniture has advertised that it will send its 48-page, full-color brochure to prospective clients, you need write merely a one-sentence letter: "Would you please send me the brochure advertised in *Higher Education Today*, May 13, 19XX?" The manufacturer knows why you're writing and, naturally, wants to receive letters such as yours, so no explanation is necessary. You may also ask a technical question, or set of questions, about any product or service for sale. Your inquiry letter might begin, "We are considering purchasing your new X-15 work stations for an office staff of 35 and would like some further information. Would you please answer the following questions?" The detail about the size of the potential order is not necessary, but it does make the inquiry seem serious and the potential sale substantial. An inquiry letter of this kind will get a prompt and gracious reply.

WAGNER AIRCRAFT

116 North Miller Road
Akron, OH 44313

September 4, 19XX

Franklin Aerospace Parts
623 Manufacturer's Blvd.
Bethpage, NY 11741

Attention: Mr. Frank DeFazio

Gentlemen:

Please send us the following parts by parcel post. All page numbers refer to your 19XX catalog.

Quantity	Model No.	Catalog Page	Description	Price
2	36113-NP	42	Seal fins	$ 34.95
1	03112-Bx	12	Turbine bearing support	19.75
5	90135-QN	102	Turbine disc	47.50
1	63152-Bx	75	Turbine bearing housing	16.15
			Total Price:	$118.35

Yours very truly,

Christopher O'Hanlon

Christopher O'Hanlon
Purchasing Agent

FIGURE 14.5   Order Letter

If your reader is not expecting your letter, your task is more difficult. If the reader will not directly benefit from supplying information, you must ask a favor. Only careful, persuasive writing will make your reader *want* to respond despite the absence of the profit motive.

In the first paragraph of this kind of letter, state why you are writing to *this* person or organization, rather than any other organization that could supply the same or similar information. You might use subtle

14 Hawthorne Ave.
Belleview, TX 75234

November 2, 19XX

Dr. Andrew Shakir
Director of Technical Services
Orion Corporation
721 West Douglas Avenue
Maryville, TN 31409

Dear Dr. Shakir:

I am writing to you because of Orion's reputation as a leader in the manufacture of adjustable X-ray tables. I am a graduate student in biomedical engineering at the University of Texas working on an analysis of diagnostic equipment. Would you be able to answer a few questions about your Microspot 311?

1. Can the Microspot 311 be used with lead oxide cassettes, or does it accept only lead-free cassettes?
2. Are standard generators compatible with the Microspot 311?
3. What would you say is the greatest advantage, for the operator, in using the Microspot 311? For the patient?

My project is due January 15. I would greatly appreciate your assistance in answering these questions. Of course, I would be happy to send you a copy of my report when it is completed.

Yours very truly,

Albert K. Stern

Albert K. Stern

FIGURE 14.6   Inquiry Letter

(See page 380.)

flattery at this point—for example, "I was hoping that, as the leader in solid-state electronics, you might be able to furnish some information about. . . ." Then you might explain why you want the information. Obviously, a company will not furnish information to a competitor. You have to show that your interests are not commercial—for instance, "I will be using this information in a senior project in agronomy at Illinois State University. I am trying to devise a. . . ." If you need the information by a certain

ORION

721 West Douglas Avenue
Maryville, TN 31409

(615) 619-8132
TECHNICAL SERVICES

November 7, 19XX

Mr. Albert K. Stern
14 Hawthorne Ave.
Belleview, TX 75234

Dear Mr. Stern:

I would be pleased to answer your questions about the Microspot 311. We think it is the best unit on the market.

1. The 311 can handle lead oxide or lead-free cassettes.
2. At the moment, the 311 is fully compatible only with our Duramatic generator. However, special wiring kits are available to make the 311 compatible with our earlier generator models—the Olympus and the Saturn. We are currently working on other wiring kits.
3. For the operator, the 311 increases the effectiveness of the radiological procedure while at the same time cutting down the amount of film used. For the patient, it cuts down the number of repeat exposures and total dose.

I am enclosing our brochure on the Microspot 311. If you would like copies, please let me know. I would be happy to receive a copy of your analysis when it is complete. Good luck!

Sincerely yours,

Andrew Shakir

Andrew Shakir, M.D.
Director of Technical Services

AS/le

Enclosure
cc — Robert Anderson, Executive Vice President

FIGURE 14.7   Response to an Inquiry   *(See page 380.)*

date, this might be a good place to mention it: "The project is to be completed by April 15, 19XX."

The inquiry letter should contain a numbered list of specific questions. Readers are understandably annoyed by thoughtless requests to send "everything you have" on a topic. They much prefer a set of carefully thought-out technical questions showing that the writer has already done substantial research. "Is your Model 311 compatible with Norwood's Model

B12?" is obviously much easier to respond to than "Would you please tell me about your Model 311?" If your questions can be answered in a small space, simply leave room for your reader's reply after each question or in the margin.

Because you are asking someone to do something for you, it is natural to offer something in return. In many cases, all you can offer are the results of your research. If possible, say you would be happy to send a copy of the report you are working on. And finally, express your appreciation. Don't say, "Thank you for sending me this information." Such a statement is presumptuous, because it assumes that the reader is both willing and able to meet your request. A statement such as the following is more effective: "I would greatly appreciate any help you could give me in answering these questions." Finally, if the answers will be brief, enclose a stamped self-addressed envelope for your reader's reply.

You should of course write a brief thank-you note to someone who has responded to your inquiry letter.

Figure 14.6 provides an example of a letter of inquiry.

The Response to an Inquiry

If you ever receive an inquiry letter, keep the following suggestions in mind. If you wish to provide the information the writer asks for, do so graciously. If the questions are numbered, number your responses correspondingly. If you cannot answer the questions, either because you don't know the answers or because you cannot divulge proprietary information, explain the reasons and offer to assist with other requests. Figure 14.7 shows a response to the inquiry letter in Figure 14.6.

The Sales Letter

A large, sophisticated sales campaign costs millions of dollars — for marketing surveys and consulting fees, printing, postage, and promotions. This kind of campaign is beyond the scope of this book. However, you may well have to draft a sales letter for a product or service.

The "you attitude" is crucial in sales letter. Your readers don't care why you want to sell your product or service. They want to know why they should buy it. You are asking them to spend valuable time studying the letter. You must therefore provide clear, specific information to help them understand what you are selling and how it will help them. Be upbeat and positive, but never forget that readers want facts.

Most writers of sales letters use a four-part strategy:

1. *Gain the reader's attention.* Unless the opening sentence seems either interesting or important, the reader will put the letter aside. To attract

the reader, use interesting facts, quotations, or questions. In particular, try to identify a problem that will interest your reader. A few examples of effective openings follow.

> How much have construction costs risen since your plant was built? Do you know how much it would cost to rebuild at today's prices?

> The Datafix copier is better than the Xerox—and it costs less, too. We'll repeat: It's better and it costs less!

> If you're like most training directors, we bet you've seen your share of empty promises. We've heard all the stories, too. And that's why we think you'll be interested in what Fortune said about us last month.

2. *Describe the product or service you are trying to sell.* What does it do? How does it work? What problems does it solve?

> The Datafix copier features automatic loading, so your people don't waste time watching the copies come out. Datafix copies from a two-sided original—automatically! And Datafix can turn out 80 copies a minute—which is 25 percent faster than our fastest competitor. . . .

3. *Convince your reader that your claims are accurate.* Refer to users' experience, testimonials, or evaluations performed by reputable experts or testing laboratories.

> In a recent evaluation conducted by Office Management Today, more than 85 percent of our customers said they would buy another Datafix. The next best competitor? 71 percent. And Datafix earned a "Highly Reliable" rating, the highest recommendation in the reliability category. All in all, Datafix scored higher than any other copier in the desk-top class. . . .

4. *Tell your reader how to find out more about your product or service.* If possible, provide a postcard that the reader can use to request more information or arrange for a visit from one of your sales representatives. Make it easy to proceed to the next step in the sales process.

Figure 14.8 provides an example of a sales letter.

The Claim Letter

A claim letter is a polite and reasonable complaint. If as a private individual or a representative of an organization you purchase a defective or falsely advertised product or receive inadequate service, your first recourse is a claim letter.

Davis Tree Care

1300 Lancaster Avenue
Berwyn, PA 19092

May 13, 19XX

Dear Homeowner:

Do you know how much your trees are worth? That's right—
your trees. As a recent purchaser of a home, you know how
much of an investment your house is. Your property is a big
part of your total investment.

Most people don't know that even the heartiest trees need
periodic care. Like shrubs, trees should be fertilized and
pruned. And they should be protected against the many kinds
of diseases and pests that are common in this area.

At Davis Tree Care, we have the skills and experience to keep
your trees healthy and beautiful. Our diagnostic staff is made
up of graduates of major agricultural and forestry universities,
and all of our crews attend special workshops to keep current
with the latest information in tree maintenance. Add this to
our proven record of 43 years of continuous service in the
Berwyn area, and you have a company you can trust.

May we stop by to give you an analysis of your trees—
absolutely without cost or obligation? A few minutes with one
of our diagnosticians could prove to be one of the wisest
moves you've ever made. Just give us a call at 865-9187 and
we'll be happy to arrange an appointment at your
convenience.

Sincerely yours,

Daniel Davis III

Daniel Davis III
President

FIGURE 14.8   Sales Letter

The purpose of the claim letter is to convince your reader that you are
a fair and honest customer who is justifiably dissatisfied. If it does, your
chances of receiving an equitable settlement are good. Most organizations
today pay attention to reasonable claims, because they realize that unhappy customers are bad business. In addition, claim letters indicate the
weak points in their product or service.

Writing a claim letter calls for a four-part strategy:

ROBBINS CONSTRUCTION, INC.

255 Robbins Place   Centerville, MO 65101   (417) 934-1850

August 19, 19XX

Mr. David Larsen
Larsen Supply Company
311 Elmerine Avenue
Anderson, MO 63501

Dear Mr. Larsen:

As steady customers of yours for over 15 years, we came to
you first when we needed a quiet pile driver for a job near a
residential area. On your recommendation, we bought your
Vista 500 Quiet Driver, at $14,900. We have since found, much
to our embarrassment, that it is not substantially quieter than
a regular pile driver.

We received the contract to do the bridge repair here in
Centerville after promising to keep the noise to under 90 db
during the day. The Vista 500 (see enclosed copy of bill of sale
for particulars) is rated at 85 db, maximum. We began our
work and, although one of our workers said the driver didn't
seem sufficiently quiet to him, assured the people living near
the job site that we were well within the agreed sound limit.
One of them, an acoustical engineer, marched out the next
day and demonstrated that we were putting out 104 db.
Obviously, something is wrong with the pile driver.

I think you will agree that we have a problem. We were able to
secure other equipment, at considerable inconvenience, to
finish the job on schedule. When I telephoned your company

FIGURE 14.9   Claim Letter

1. *Identify the product or service.* List the model numbers, serial numbers, sizes, and any other pertinent data.
2. *Explain the problem.* State explicitly the symptoms. What function does not work? What exactly is wrong with the service?
3. *Propose an adjustment.* Define what you want the reader to do: for example, refund the purchase price, replace or repair the item, improve the service.

Mr. David Larsen
Page 2
August 19, 19XX

that humiliating day, however, a Mr. Meredith informed me
that I should have done an acoustical reading on the driver
before I accepted delivery.

I would like you to send out a technician—as soon as
possible— either to repair the driver so that it performs
according to specifications or to take it back for a full refund.

Yours truly,

Jack Robbins

Jack Robbins, President

JR/lr

Enclosure

FIGURE 14.9   *Continued*

4. *Conclude courteously.* Say that you trust the reader, in the interest of
   fairness, to abide by your proposed adjustment.

Tone, the "you attitude," is just as important as content in a claim letter.
You must project a calm and rational tone. A complaint such as "I'm sick
and tired of being ripped off by companies like yours" will hurt your
chances of an easy settlement. If, however, you write, "I am very disap-
pointed in the performance of my new Eversharp Electric Razor," you

sound like a responsible adult. There is no reason to show anger in a claim letter, even if the other party has made an unsatisfactory response to an earlier one. Calmly explain what you plan to do, and why. Your reader will then more likely see the situation from your perspective. Figure 14.9 provides an example of a claim letter.

The Adjustment Letter

In an adjustment letter, you respond to a claim letter and tell the customer how you plan to handle the situation. Whether you are granting the customer everything the claim letter proposed, part of it, or none of it, your purpose remains the same: to show that your organization is fair and reasonable, and that you value the customer's business.

If you can grant the request, the letter will be simple to write. Express your regret about the situation, state the adjustment you are going to make, and end on a positive note by encouraging the customer to continue doing business with you.

If you cannot grant the request, try to salvage as much goodwill as you can. Obviously, your reader will not be happy. If your letter is carefully written, however, it can show that you have acted reasonably. In denying a request, you are attempting to explain your side of the matter, thus educating your reader about how the problem occurred and how to prevent it.

This more difficult adjustment letter generally has a four-part structure:

1. *Attempt to meet the customer on some neutral ground.* Consider an expression of regret—never an apology! You might even thank the customer for bringing the matter to the attention of the company. Never admit that the customer is right in this kind of adjustment letter. If you write, "We are sorry that the engine you purchased from us is defective," the customer would have a good case against you if the dispute ended up in court.

2. *Explain why your company is not at fault.* Most often, you explain to the customer the steps that led to the failure of the product or service. Do not say, "You caused this." Instead, use the less blunt passive voice: "The air pressure apparently was not monitored. . . ."

3. *Clearly state that your company, for the above-mentioned reasons, is denying the request.* This statement must come late in the letter. If you begin with it, most readers will not finish the letter, and you will not achieve your twin goals of education and goodwill.

4. *Try to create goodwill.* You might, for instance, offer a special discount on another, similar product. A company's profit margin on any one item is almost always large enough to permit attractive discounts as an inducement to continue doing business.

Figures 14.10 and 14.11 show examples of "good news" and "bad news" adjustment letters.

A reply to the claim letter shown in Figure 14.9.

Larsen Supply Company

311 Elmerine Avenue
Anderson, MO 63501

August 21, 19XX

Mr. Jack Robbins, President
Robbins Construction, Inc.
255 Robbins Place
Centerville, MO 65101

Dear Mr. Robbins:

I was very unhappy to read your letter of August 19 telling me about the failure of the Vista 500. I regretted most the treatment you received from one of my employees when you called us.

Harry Rivers, our best technician, has already been in touch with you to arrange a convenient time to come out to Centerville to talk with you about the driver. We will of course repair it, replace it, or refund the price. Just let us know your wish.

I realize that I cannot undo the damage that was done on the day that a piece of our equipment failed. To make up for some of the extra trouble and expense you incurred, let me offer you a 10 percent discount on your next purchase or service order with us, up to $1,000 total discount.

You have indeed been a good customer for many years, and I would hate to have this unfortunate incident spoil that relationship. Won't you give us another chance? Just bring in this letter when you visit us next, and we'll give you that 10 percent discount.

Sincerely,

Dave Larsen

Dave Larsen, President

FIGURE 14.10   "Good News" Adjustment Letter

Quality Video Products

February 3, 19XX

Mr. Dale Devlin
1903 Highland Avenue
Glenn Mills, NE 69032

Dear Mr. Devlin:

Thank you for writing us about the videotape you purchased on January 11, 19XX.

You used the videotape to record your daughter's wedding. While you were playing it back last week, the tape jammed and broke as you were trying to remove it from your VCR. You are asking us to reimburse you $500 because of the sentimental value of that recording.

As you know, our videotapes carry a lifetime guarantee covering parts and workmanship. We will gladly replace the broken videotape. However, the guarantee states that the manufacturer will not assume any incidental liability. Thus we are responsible only for the retail value of the blank tape.

However, your wedding tape can probably be fixed. A reputable dealer can splice tape so skillfully that you will hardly notice the break. It's a good idea to make backup copies of your valuable tapes.

Attached to this letter is a list of our authorized dealers in your area, who would be glad to do the repairs for you. We have already sent out your new videotape. It should arrive within the next two days.

Please contact us if we can be of any further assistance.

Sincerely yours,

Paul R. Blackwood

Paul R. Blackwood, Manager
Customer Relations

FIGURE 14.11 "Bad News" Adjustment Letter

WRITER'S CHECKLIST

The following checklist covers the basic letter format and the six types of letters discussed in the chapter.

Letter Format

1. Is the first page of the letter typed on letterhead stationery?

2. Is the date included?

3. Is the inside address complete and correct? Is the appropriate courtesy title used?

4. If appropriate, is an attention line included?

5. If appropriate, is a subject line included?

6. Is the salutation appropriate?

7. Is the complimentary close typed with only the first word capitalized? Is the complimentary close followed by a comma?

8. Is the signature legible and the writer's name typed beneath the signature?

9. If appropriate, are the reference initials included?

10. If appropriate, is an enclosure line included?

11. If appropriate, is a copy line included?

12. Is the letter typed in one of the standard formats?

Types of Letters

1. Does the order letter
   a. include the necessary identifying information, such as quantities and model numbers?
   b. specify, if appropriate, the terms of payment?
   c. specify the method of delivery?

2. Does the inquiry letter
   a. explain why you chose the reader to receive the inquiry?
   b. explain why you are requesting the information and to what use you will put it?
   c. specify by what date you need the information?
   d. list the questions clearly and, if appropriate, provide room for the reader's response?
   e. offer, if appropriate, the product of your research?

3. Does the response to an inquiry letter answer the reader's questions or explain why they cannot be answered?

4. Does the sales letter
    a. gain the reader's attention?
    b. describe the product or service?
    c. convince the reader that the claims are accurate?
    d. encourage the reader to find out more about the product or service?

5. Does the claim letter
    a. identify specifically the unsatisfactory product or service?
    b. explain the problem(s) clearly?
    c. propose an adjustment?
    d. conclude courteously?

6. Does the "good news" adjustment letter
    a. express your regret?
    b. explain the adjustment you will make?
    c. conclude on a positive note?

7. Does the "bad news" adjustment letter
    a. meet the reader on neutral ground, expressing regret but not apologizing?
    b. explain why the company is not at fault?
    c. clearly deny the reader's request?
    d. attempt to create goodwill?

EXERCISES

1. Write an order letter to John Saville, general manager of White's Electrical Supply House (13 Avondale Circle, Los Angeles, CA 90014). These are the items you want: one SB11 40-ampere battery backup kit, at $73.50; twelve SW402 red wire kits, at $2.50 each; ten SW400 white wire kits, at $2.00 each; and one SB201 mounting hardware kit, at $7.85. Invent any reasonable details about methods of payment and delivery.

2. Secure the graduate catalog of a university offering a graduate program in your field. Write an inquiry letter to the appropriate representative, asking at least three questions about the program the university offers.

3. You are the marketing director of the company that publishes this book. Draft a sales letter that might be sent to teachers of the course you are presently taking.

4. You are the marketing director of the company that makes your bicycle (calculator, stereo, running shoes, etc.). Write a sales letter that might be sent to retailers to encourage them to sell the product.

5. You are the recruiting officer for your college. Write a letter that might be sent to juniors in local high schools to encourage them to apply to the college when they are seniors.

6. You purchased four "D" size batteries for your cassette player, and they didn't work. The package they came in says that the manufacturer will refund the purchase price if you return the defective items. Inventing any reasonable details, write a claim letter asking for not only the purchase price, but other related expenses.

7. A thermos you just purchased for $8.95 has a serious leak. The grape drink you put in it ruined a $25.00 white tablecloth. Inventing any reasonable details, write a claim letter to the manufacturer of the thermos.

8. The gasoline you purchased from New Jersey Petroleum contained water and particulate matter; after using it, you had to have your automobile tank flushed at a cost of $50. You have a letter signed by the mechanic explaining what he found in the bottom of your tank. As a credit-card customer of New Jersey Petroleum, write a claim letter. Invent any reasonable details.

9. As the recipient of one of the claim letters described in Exercises 6 through 8, write an adjustment granting the customer's request.

10. You are the manager of a private swimming club. A member has written saying that she lost a contact lens (value $55) in your pool. She wants you to pay for a replacement. The contract that all members sign explicitly states that the management is not responsible for loss of personal possessions. Write an adjustment letter denying the request. Invent any reasonable details.

11. As manager of a stereo equipment retail store, you guarantee that you will not be undersold. If a customer who buys something from you can prove within one month that another retailer sells the same equipment for less money, you will refund the difference in price. A customer has written to you, enclosing an ad from another store showing that it is selling the equipment he purchased for $26.50 less than he paid at your store. The advertised price at the other store was a one-week sale that began five weeks after the date of his purchase from you. He wants his $26.50 back. Inventing any reasonable details, write an adjustment letter denying his request. You are willing, however, to offer him a blank cassette tape worth $4.95 for his equipment if he would like to come pick it up.

12. The following letters could be improved in both tone and substance. Revise them to increase their effectiveness, adding any reasonable details.

a.

Modern Laboratories, Inc.
DEAUVILLE, IN 43504

July 2, 19XX

Adams Supply Company
778 North Henson Street
Caspar, IN 43601

Gentlemen:

Would you please send us the following items:

one dozen petri dishes
one gross pyrex test tubes
three bunsen burners

Please bill us.

Sincerely,

Corey Dural

Corey Dural

CD/kw

b.

1967 Sunset Avenue
Rochester, N. Y. 06803

November 13, 19XX

Admissions Department
University of Pennsylvania Law School
Philadelphia, Pa. 19106

Gentlemen:

I am a senior considering going to law school. Would you
please answer the following questions about your law school:

1. How well do your graduates do?
2. Is the LSAT required?
3. Do you have any electives, or are all the courses required?
4. Are computer skills required for admission?

A swift reply would be appreciated. Thank you.

Sincerely yours,

Eileen Forster

Eileen Forster

c.

14 Wilson Avenue
Wilton, ME 04949

November 13, 19XX

Union Pacific Railroad
100 Columbia Street
Seattle, WA 98104

Attention: The President

Dear Sir:

I'm a college student doing a report on the future of mass transportation in the United States. Would you please tell me how many passenger miles your railroad traveled last year, the number of passengers, whether this was up or down from the year before, and the socioeconomic level of your riders?

My report is due in a week and a half. If you sent your answer in the next two days or so I'd be able to include your information in my report. Thank you in advance.

Very truly yours,

Jonathan Radley

Jonathan Radley

d.

Cutlass Vacuums, Inc.
Ridge Pike / Speonk, OR 97203

January 13, 19XX

Dear Service Station Owner:

I don't have to tell you that an indispensable tool for your garage is a good industrial-strength vacuum cleaner. It not only makes your garage better looking, but it makes it safer, too.

We at Cutlass are proud to introduce our brand-new Husky 450, which replaces our model 350. The 450 is bigger, so it has a greater suction power. It has a bigger receptacle, too, so you don't have to empty it as often.

I truly believe that our Husky 450 is the shop vacuum you've been looking for. If you would like further information about it, don't miss our ad in leading car magazines.

Yours truly,

Bob Wheeler

Bob Wheeler, President
Cutlass Vacuums, Inc.

e.

GUARDSMAN PROTECTIVE EQUIPMENT, INC.

3751 Porter Street
Newark, DE 19304

April 11, 19XX

Dear Smith Family:

A rose is a rose is a rose, the poet said. But not all home protection alarms are alike. In a time when burglaries are skyrocketing, can you afford the second-best alarm system?

Your home is your most valuable possession. It is worth far more than your car. And if you haven't checked your house insurance policy recently, you'll probably be shocked to see how inadequate your coverage really is.

The best kind of insurance you can buy is the Watchdog Alarm System. What makes the Watchdog unique is that it can detect intruders before they enter your home and scare them away. Scaring them away while they're still outside is certainly better than scaring them once they're inside, where your loved ones are.

At less than two hundred dollars, you can purchase real peace of mind. Isn't your family's safety worth that much?

If you answered yes to that question, just mail in the enclosed postage-paid card, and we'll send you a 12-page, fact-filled brochure that tells you why the Watchdog is the best on the market.

Very truly yours,

Jerry Wexler

Jerry Wexler, President

f.

19 Lowry's Lane
Morgan, TN 30610

April 13, 19XX

Sea-Tasty Tuna
Route 113
Lynchburg, TN 30563

Gentlemen:

I've been buying your tuna fish for years, and up to now it's been OK.

But this morning I opened a can to make myself a sandwich. What do you think was staring me in the face? A fly. That's right, a housefly. That's him you see taped to the bottom of this letter.

What are you going to do about this?

Yours very truly,

Seth Reeves

Seth Reeves

g.

Handyman Hardware, Inc.
Millersville, AL 61304

December 4, 19XX

Hefty Industries, Inc.
19 Central Avenue
Dover, TX 76104

Gentlemen:

I have a problem I'd like to discuss with you. I've been carrying your line of hand tools for many years.

Your 9″ pipe wrench has always been a big seller. But there seems to be something wrong with its design. I have had three complaints in the last few months about the handle snapping off when pressure is exerted on it. In two cases, the user cut his hand, one seriously enough to require nineteen stitches.

Frankly, I'm hesitant to sell any more of the 9″ pipe wrenches, but I still have over two dozen in inventory.

Have you had any other complaints about this product?

Sincerely yours,

Peter Arlen

Peter Arlen, Manager
Handyman Hardware

PA/sc

h.

Sea-Tasty Tuna

Route 113
Lynchburg, TN 30563

April 21, 19XX

Mr. Seth Reeves
19 Lowry's Lane
Morgan, TN 30610

Dear Mr. Reeves:

We were very sorry to learn that you found a fly in your tuna fish.

Here at Sea-Tasty we are very careful about the hygiene of our plant. The tuna are scrubbed thoroughly as soon as we receive them. After they are processed, they are inspected visually at three different points. Before we pack them, we rinse and sterilize the cans to ensure that no foreign material is sealed in them.

Because of these stringent controls, we really don't see how you could have found a fly in the can. Nevertheless, we are enclosing coupons good for two cans of Sea-Tasty tuna.

We hope this letter restores your confidence in us.

Truly yours,

Valarie Lumaris

Valarie Lumaris
Customer Service Representative

VL/ck

Enclosures

i.

Hefty Industries, Inc.

19 Central Avenue
Dover, TX 76104

December 11, 19XX

Mr. Peter Arlen, Manager
Handyman Hardware, Inc.
Millersville, AL 61304

Dear Mr. Arlen:

Thank you for bringing this matter to our attention.

In answer to your question—yes, we have had a few complaints about the handle snapping on our 9″ pipe wrench.

Our engineers brought the wrench back to the lab and discovered a design flaw that accounts for the problem. We have redesigned the wrench and have not had any complaints since.

We are not selling the old model any more because of the risk. Therefore we have no use for your two dozen. However, since you have been a good customer, we would be willing to exchange the old ones for the new design.

We trust this will be a satisfactory solution.

Sincerely,

Robert Panofsky

Robert Panofsky, President
Hefty Industries, Inc.

15

Job-Application Materials

For most of you, the first nonacademic test of your technical writing skills will come when you create your job-application materials. These materials will inform employers about your academic and employment experience, personal characteristics, and reasons for applying. But they will provide a lot more, too. Employers have learned that one of the most important skills an employee can bring to a job is the ability to communicate effectively. Therefore, potential employers look carefully for evidence of writing skills. Job-application materials pose a double hurdle: the tasks of showing employers both what you can do and how well you can communicate.

Some students think that once they get a satisfactory job, they will never again have to worry about résumés and application letters. Statistics show that they are wrong. The typical professional changes jobs more than five times. Although this chapter pays special attention to the student's first job hunt, the skills and materials discussed here also apply to established professionals who wish to change jobs.

Understanding the Five Ways to Get a Job

There are five traditional ways to get a professional-level position:

- *Through a college or university placement office.* Almost all colleges and universities have placement offices, which bring companies and students together. Generally, students submit a résumé—a brief listing of credentials—to the placement office. The résumés are then made available to representatives of business, government, and industry, who use the placement office to arrange on-campus interviews with selected students. Those who do best in the campus interviews are then invited by the representatives to visit the organization for a tour and another interview. Sometimes a third interview is scheduled; sometimes an offer is made immediately or shortly thereafter. The advantage of this system is twofold: first, it is free; second, it is easy. The student merely has to deliver a résumé to the placement office and wait to be contacted.

- *Through a professional placement bureau.* A professional placement bureau offers essentially the same service as a college placement office, but it charges either the employer (or, rarely, the new employee) a fee—often 10 percent of the first year's salary—when the client accepts a position. Placement bureaus cater primarily to more advanced professionals who are looking to change jobs.

- *Through a published job ad.* Published job ads generally offer the best opportunity for both students and professionals. Organizations advertise in three basic kinds of publications: public-relations catalogs (such as *College Placement Annual*), technical journals, and newspapers. If you are looking for a job, you should regularly check the major technical journals in your field and the large metropolitan newspapers—especially the Sunday editions. In responding to an ad, you must include with the résumé a job-application letter—one that highlights the crucial information on the résumé.

■ *Through an unsolicited letter to an organization.* You need not wait for a published ad. You can write unsolicited letters of application to organizations you would like to work for. The disadvantage of this technique is obvious: there might not be an opening. Yet many professionals favor this technique, because there are fewer competitors for those jobs that do exist, and sometimes organizations do not advertise all available positions. And sometimes an impressive unsolicited application can prompt an organization to create a position for you. Before you write an unsolicited application, make sure you learn as much as you can about the organization (see Chapter 4): current and anticipated major projects, hiring plans, and so forth. You should know as much as you can about any organization you are applying to, of course, but when you are submitting an unsolicited application you have no other source of information on which to base your résumé and letter. The business librarian at your college or university will be able to point out sources of information, such as the Dun and Bradstreet guides, the *F&S Index of Corporations*, and the indexed newspapers such as the *New York Times* and the *Wall Street Journal*.

■ *Through connections.* A relative or an acquaintance who can influence or at least point out a new position can help you get a job. Other good contacts include employers from your past summer jobs and faculty members in your field.

The college placement office, published ads, unsolicited letters, and connections are most useful for students. Too many students rely solely on the placement office, thereby limiting the range of possibilities to those organizations that choose to visit the college. The best system is to use the placement office and respond to published ads. If an attractive organization is not advertising, send an unsolicited application letter after researching them and finding out the name of the appropriate person and department.

Writing the Résumé

Whether it is submitted to a college placement office or sent along with a job-application letter, the résumé communicates in two ways: by its appearance and by its content.

Appearance of the Résumé

Because potential employers normally see the résumé before they see the person who wrote it, the résumé has to make a good first impression. Employers believe—often correctly—that the appearance of the résumé reflects the writer's professionalism. A sloppy résumé implies that the writer would do sloppy work. A neat résumé implies that the writer would do professional work. When employers look at a résumé, they see a sample of the documents they will be reading if they hire the writer.

Some colleges and universities advise students to have their résumé professionally printed. A printed résumé is attractive, and that's good—provided, of course, that the information on it is consistent with its professional appearance. Most employers agree, however, that a neatly typed or word-processed résumé photocopied on good-quality paper is just as effective. Using a laser printer can give you a professional-looking résumé at a very low cost.

People who photocopy a typed or word-processed résumé are more likely to tailor different versions to the needs of the organizations to which they apply—a good strategy. People who go to the trouble and expense of a professional printing job are far less likely to make up different résumés; they tend to submit the printed one to all their prospects. The résumé looks so good that they feel it is not worth tinkering with. This strategy is dangerous, for it encourages the writer to underestimate the importance of addressing the content of the résumé to a specific audience.

However they are reproduced, résumés should appear neat and professional. They should have

- ☐ *Generous margins.* Leave a one-inch margin on all four sides.
- ☐ *Clear type.* Use a typewriter or printer with clear, sharp, unbroken letters. Avoid strikeovers and obvious corrections. Avoid dot-matrix printers.
- ☐ *Symmetry.* Arrange the information so that the page has a balanced appearance.
- ☐ *Clear organization.* Use adequate white space. Make sure the spacing *between* items is greater than the spacing *within* an item. That is, the spacing between your education section and your employment section should be greater than the spacing within one of those sections. You should be able to see the different sections clearly if you stand and look at the résumé on the floor in front of your feet.

Use indentation clearly. When you arrange items in a vertical list, indent all second and subsequent lines (turnovers) of any item a few spaces. Notice, for example, that the following list from the computer-skills section of a résumé can be confusing.

Computer Experience

Systems: IBM AT/XT pc, Apple Macintosh, Apple IIe/II-Plus, Andover AC-256, Prime 360

```
Software: Lotus 1-2-3, DBase II and III, Multiplan, PFS
File/Report, Wordstar
Languages: Fortran, Pascal, BASIC
```

With the second lines indented, the arrangement is much easier to understand.

Computer Experience

```
Systems: IBM AT/XT pc, Apple Macintosh, Apple IIe/II-Plus,
    Andover AC-256,    Prime 360
Software: Lotus 1-2-3, DBase II and III, Multiplan, PFS File/Report,
    Wordstar
Languages: Fortran, Pascal, BASIC
```

Content of the Résumé

Although different experts advocate different approaches to résumé writing, everyone agrees on two things.

- ■ *The résumé must be completely free of errors.* Grammar, punctuation, usage, and spelling errors cast doubt on the accuracy of the information in the résumé. Ask for assistance after you have written the draft, and proofread the finished product at least twice.

- ■ *The résumé must provide clear and specific information, without generalizations or self-congratulation.* Your résumé is a sales document, but you are both the salesperson and the product. You cannot say, "I am a terrific job candidate," as if the product were a toaster or an automobile. Instead, you have to provide the specific details that will lead the reader to the conclusion that you are a terrific job candidate. You must *show* the reader. Telling the reader is graceless and, worse, unconvincing.

A résumé should be long enough to include all the pertinent information but not so long as to bore or irritate the reader. Generally, you should keep it to one page. If, however, you have special accomplishments — such as journal articles or patents — a two-page résumé is appropriate. If your information comes to just over one page, either eliminate or condense some material to make it fit onto one page, or modify the layout so that it fills a substantial part of the second page.

Elements of the Résumé

Almost every résumé has six basic sections:

- ☐ identifying information
- ☐ objective
- ☐ education
- ☐ employment

☐ personal information
☐ references

But your résumé should reflect one particular person: you. Some people use additional sections to convey special skills or background.

Identifying Information

Place your full name, address, and phone number at the top of the page. Generally, you should present your name in full capitals, centered at the top. Use your complete address, with the zip code. Use the two-letter state abbreviations used by the Post Office. (See Appendix C.) Also give your complete phone number, with the area code. For the telephone exchange, use numbers rather than letters.

If you have two addresses and phone numbers, list both and identify them clearly. An employer might call during an academic holiday to arrange an interview.

Objective

After the identifying information, add a statement of objective, in the form of a brief phrase or sentence—for example, "Objective: Entry-level position as a dietitian in a hospital." According to a recent study (Harcourt and Krizan 1989), over 90 percent of Fortune 500 personnel officers surveyed consider a statement of objective important because it gives the appearance that the writer has a clear sense of direction and goals. In drafting the statement, however, be careful that you state only goals or duties explicitly mentioned in the job advertisement, or at least clearly implied. If you unintentionally suggest that your goals are substantially different from the job responsibilities, the organization might infer that you would not be happy in the position and end your candidacy.

Education

If you are a student or a recent graduate, add the education section after identifying information and the statement of objective. If you have substantial professional experience, place the employment experience section before the education section.

Include the following information in the education section: the degree, the institution, its location, and the date of graduation. After the degree abbreviation (such as B.S., B.A., A.A., or M.S.), list the academic major (and, if you have one, the minor)—for example, "B.S. in Materials Engineering, minor in general business." Identify the institution by its full name: "The Pennsylvania State University," not "Penn State." Also include the city and state of the institution. If your degree has not yet been granted, write "Anticipated date of graduation" or "Degree expected in" before the month and year.

You should also list any other institutions you attended beyond the high-school level—even those at which you did not earn a degree. Students are sometimes uneasy about listing community colleges or junior colleges; they shouldn't be. Employers are generally impressed to learn that a student began at a smaller or less-advanced school and was able to transfer to a four-year college or university. The listing for other institutions attended should include the same information as the main listing. Arrange the entries in reverse chronological order: that is, list first the school you attended most recently.

In addition to this basic information, many students and recent graduates like to include more details about their educational experience. They feel, quite correctly, that the basic information alone implies that they merely endured an institution and received a degree. The following strategies can be useful in elaborating on your educational background:

- *List your grade-point average.* If your grade-point average is significantly above the median for the graduating class, list it. Or list your average in your major courses, if that is more impressive.

- *Include a list of courses.* Choose courses that will be of particular interest to your reader. Advanced courses in your major concentration might be appropriate, especially if the potential employer has mentioned those subjects in the job advertisement. Also useful would be a list of communications courses—technical writing, public speaking, organizational communications, and the like. A listing of business courses on an engineer's résumé—or the reverse—also shows special knowledge and skills. The only kind of course listing that is *not* particularly helpful is one that merely names the traditional required courses for your major. Make sure to list courses by title, not by number; employers won't know what Chemistry 250 is.

- *Describe a special accomplishment.* If you did a special senior design or research project, for example, a two- or three-line description would be helpful. Include in this description the title and objective of the project, any special or advanced techniques or equipment you used, and—if you know them—the major results. Such a description might be phrased as follows: "A Study of Composite Substitutes for Steel—a senior design project intended to formulate a composite material that can be used to replace steel in car axles." Even a traditional academic course in which you conducted a sophisticated project and wrote a sustained report can be described profitably. A project discussion makes you look less like a student—someone who takes courses—and more like a professional—someone who designs and carries out projects.

- *List any honors and awards you received.* Scholarships, internships, and academic awards all offer evidence of exceptional ability. If you have received a number of such honors, or some that were not exclusively academic, you might list them separately (in a section called "Honors" or "Awards") rather than in a subsection of the education section. Often, some information could logically be placed in two or even three different locations, and you must decide where it will make the best impression.

The education section is the easiest part of the résumé to adapt for different positions. For example, a student majoring in electrical engineering who is applying for a position that calls for strong communications skills can list communications courses in one version of the résumé. The same student can use a list of advanced electrical engineering courses in a résumé directed to another potential employer. As you compose the education section of your résumé, carefully emphasize what in your background meets the requirements for the particular job.

Employment

The employment section, like the education section, conveys at least the basic information about each job you've held: the dates of employment, the organization's name and location, and your position or title.

However, such a skeletal list would not be very informative or impressive. As in the education section, you should provide carefully selected details.

What readers want to know, after they have learned where and when you were employed, is what you actually did. Therefore, you should provide at least a two- to three-line description for each position. For particularly important or relevant jobs, write a longer description. Focus the description on one or more of the following factors:

- □ *Reports.* What kinds of documents did you write or assist in writing? List, especially, various governmental forms and any long reports, manuals, or proposals.
- □ *Clients.* What kinds of, and how many, clients did you transact business with as a representative of your organization?
- □ *Skills.* What kinds of technical skills did you learn or practice in doing the job?
- □ *Equipment.* What kinds of technical equipment did you operate or supervise? Mention, in particular, any computer skills you demonstrated, for they can be very useful in almost every kind of position.
- □ *Money.* How much money were you responsible for? Even if you considered your bookkeeping position fairly simple, the fact that the organization grossed, say, $2 million a year shows that the position involved real responsibility.
- □ *Personnel.* How many personnel did you supervise? Students sometimes supervise small groups of other students or technicians. Naturally, supervising shows maturity and responsibility.

Make sure you use the active voice—"supervised three workers"—rather than the passive voice—"three workers were supervised by me." The active voice emphasizes the verb, which describes the action. In thinking about

administered	discovered	operated
advised	edited	organized
analyzed	evaluated	oversaw
assembled	examined	performed
built	expanded	prepared
calibrated	hired	produced
collected	identified	provided
completed	implemented	purchased
conducted	improved	recorded
constructed	increased	reported
coordinated	installed	researched
corresponded	instituted	served
created	maintained	solved
delivered	managed	supervised
developed	monitored	trained
devised	obtained	wrote
directed		

FIGURE 15.1   **Strong Action Verbs for Use in Résumés**

your functions and responsibilities in your various positions, keep in mind the strong action verbs that clearly communicate your activities. Consider the strong verbs listed in Figure 15.1.

Here is a sample employment listing:

June–September 19XX: Millersville General Hospital, Millersville, TX. Student Dietician. Gathered dietary histories and assisted in preparing menus for a 300-bed hospital. Received "excellent" on all items in evaluation by head dietician.

In just a few lines, you can show that you sought and accepted responsibility and that you acted professionally. Do not write, "I accepted responsibility"; rather, present facts that lead the reader to that impression.

Naturally, not all jobs entail such professional skills and responsibilities. Many students find summer work as laborers, sales clerks, short-order cooks, and so forth. If you have not had a professional position, list the jobs you have had, even if they were completely unrelated to your career plans. If the job title is self-explanatory — such as waitress or service-station attendant — don't elaborate. Every job is valuable; you learn that you are expected to be someplace at a specific time, wear appropriate clothes, and perform some specific duties. Also, every job helps pay college expenses.

If you can say that you earned, say, 50 percent of your annual expenses through a job, employers will be impressed by your self-reliance. Most of them probably started out with nonprofessional positions. And don't forget that any job you have held can yield a valuable reference.

List your jobs in reverse chronological order on the résumé, to highlight those positions you have held most recently.

One further word: if you have held a number of nonprofessional positions as well as several professional positions, you can group the nonprofessional ones together in one listing:

Other Employment: Cashier (summer, 1989), salesperson
(part-time, 1989), clerk (summer, 1988).

This technique prevents the nonprofessional positions from drawing the reader's attention away from the more important positions.

Personal Information

This section of the résumé has changed considerably in recent years. Only a decade ago, most résumés listed the writer's height, weight, date of birth, and marital status. Most résumés today include none of these items. One explanation for this change is that federal legislation now prohibits organizations from requiring this information. Perhaps more important, most people now feel that such personal information is irrelevant to a person's ability.

The personal information section of the résumé *is* the appropriate place for a few items about your outside interests. Participation in community service organizations—such as Big Brothers/Big Sisters—or volunteer work in a hospital is, of course, extremely positive. Hobbies related to your career interests—for example, amateur electronics for an engineer—are useful, too. You also should list any sports, especially those that might be socially useful in your professional career, such as tennis, racquetball, and golf. Also point out any university-sanctioned activity—such as membership on a team, participation on the college newspaper, or election to a responsible position in an academic organization or a residence hall. In general, do not include activities that might create a negative impression: hunting, gambling, performing in a rock band. And always omit such activities as reading and meeting people—everybody reads and meets people.

References

You may list the names of three or four referees—people who have written letters of recommendation or who have agreed to speak on your behalf. Or you may simply say that you will furnish the names of the referees upon request. The length of your résumé sometimes dictates which approach to use. If the résumé is already long, the abbreviated form might be preferable. If it does not fill out the page, the longer form might be the one to use. However, each style has advantages and disadvantages that you should consider carefully.

Furnishing the referees' names appears open and forthright. It shows that you have already secured your referees and have nothing to hide. If one or several of the referees are prominent people in their fields, the reader is likely to be impressed. And, perhaps most important, the reader can easily phone the referees or write them a letter. Listing the referees makes it easy for the prospective employer to proceed with the hiring process. The only disadvantage is that it takes up space on the résumé that might be needed for other information.

Writing "References will be furnished upon request," on the other hand, takes up only one line. In addition, it leaves you in a flexible position. You can still secure referees after you have sent out the résumé. Also, you can send selected letters of reference to prospective employers according to your analysis of what they want. Using different references for different positions is sometimes just as valuable as sending different résumés. However, some readers will interpret the lack of names and addresses as evasive or secretive or assume that you have not yet asked prospective referees. A greater disadvantage is that if the readers are impressed by the résumé and want to learn more about you, they cannot do so quickly and directly.

What do personnel officers prefer regarding references? Data from the study referred to earlier (Harcourt and Krizan 1989) lead to the following conclusions: the most preferred strategy is to list the names of previous employers; the next most popular strategy is to indicate that the references are available upon request; and the least popular strategy is to list the names of professors. If you decide to include the listing, identify every referee by name, title, organization, mailing address, and phone number. For example:

Dr. Robert Ariel, Manager
Office Products Division
Northwest Supply, Inc.
1411 Seneca Avenue
Portland, OR 97202
(503) 666-6666

Remember that a careful choice of referees is as important as careful writing of the résumé. Solicit references only from persons who know your work best and for whom you have done the best work—for instance, a previous employer who you worked with closely, or a professor from whom you have received A's. It is unwise to ask prominent professors who do not know your work well, for the advantage of having a famous name on the résumé will be offset by the referee's brief and uninformative letter. Often, a young and less-well-known professor can write the most informative letter or provide the best recommendation.

After you have decided whom to ask, give the potential referee an opportunity to decline gracefully. Sometimes the person has not been as impressed with your work as you think. Also, if you simply ask, "Would you please write a reference letter for me?" the potential referee might accept and then write a lukewarm letter. It is better to follow up the first question with "Would you be able to write an enthusiastic letter for me?" or "Do you feel you know me well enough to write a strong recommendation?" If the potential referee shows any signs of hesitation or reluctance, you can withdraw the invitation at that point. The scene is a little embarrassing, but it is better than receiving a halfhearted recommendation.

Other Elements

The discussion so far has concentrated on the sections that appear on virtually everyone's résumé. Other sections are either discretionary or are appropriate only for certain writers.

If you are willing to relocate, state that explicitly. Many organizations today will find you a more attractive candidate if they know you are willing to move around as you learn the business.

If you are a veteran of the armed forces, include a military service section on the résumé. Define your military service as if it were any other job, citing the dates, locations, positions, ranks, and tasks. Often a serviceperson receives regular evaluations from a superior; these "marks" can work in your favor.

If you have a working knowledge of a foreign language, your résumé should include a *Language Skills* section. Language skills are particularly relevant if the potential employer has international interests and you could be useful in translation or foreign service.

Figures 15.2 and 15.3 provide examples of effective résumés. The job-application letters that accompany these résumés are shown in Figures 15.4 and Figure 15.5.

Figure 15.2 is a resume for a student seeking a full-time job. Carl Lockyer is what is known as a traditional student: he entered college right after high school and has proceeded on schedule. He has no dependents and is willing to relocate, so he makes that point clearly. Because he has some high-powered referees from his former employers, he chooses to list their names at the end of the resume.

Figure 15.3, on the other hand, was written by a nontraditional student: a single mother who has returned to school after having raised three children. She is applying for an internship position.

Notice that résumés often omit the *I* at the start of sentences. Rather than writing, "I prepared bids...," many people will write, "Prepared bids...." Whichever style you use, be consistent.

Carl Lockyer

3109 Vista Street   Philadelphia, PA 19136   (215) 895-2660

Objective: Entry-level position in signal processing.

Education
Bachelor of Science in Electrical Engineering
Drexel University, Philadelphia, PA
Anticipated June 1991

Grade-Point Average: 3.80 (of 4.0)

Senior-Design Project: "Enhanced Path-planning Software for Robotics"

Advanced Engineering Courses
Digital Signal Processing   Computer Hardware
Introduction to Operating Systems I, II System Design
Digital Filters   Computer Logic Circuits I, II

Employment
June 1990-   RCA Advanced Technology Laboratory, Moorestown, NJ
January 1991   Developed and tested new Fortran software to
               implement the automated production of VLSI
               standard cell family databooks. Assisted senior
               engineer in the design and testing of two Gate Array
               (2700 and 5000 gates) VLSI integrated chips.

June 1988-   RCA Advanced Technology Laboratory, Moorestown, NJ
January 1989   Verified and documented several integrated circuit
               designs. Used CAD software and hardware to
               simulate, check, and evaluate these designs. Gained
               experience on the VAX and Applicon.

Honors, Awards, and Organizations
Eta Kappa Nu (Electrical Engineering Honor Society)
Tau Beta Pi (General Engineering Honor Society)
Institute of Electrical and Electronic Engineers

Willing to relocate.

References
Mr. Albert Feller   Mr. Fred Borelli, Unit Manager Mr. Sam Shamir, Comptroller
Engineering Consultant   RCA Corporation   RCA Corporation
700 Church Road   Route 38   Route 38
Cherry Hill, NJ 08002   Moorestown, NJ 08057   Moorestown, NJ 08057
(609) 663-7256   (609) 866-6508   (609) 866-6588

FIGURE 15.2   Sample Résumé of a Traditional Student

ALICE P. LINDER
(215) 674-4006

1781 Weber Road
Warminster, PA 18974

Objective: A position in molecular research that uses my computer skills.

Education: Harmon College, West Yardley, PA
Major: Bioscience and Biotechnology
Expected graduation date: June 1991

Related Course Work:
General Chemistry I, II, III   Biology I, II, III
Organic Chemistry I, II, III   Statistical Methods for Research
Physics I, II, III   Technical Writing
Calculus I, II, II, IV

Employment Experience:

June 1990-present
(20 hours/week)

Smith Kline & French Laboratory, Upper Merion, PA
Analyze molecular data on E&S PS300, Macintosh,
and IBM PC computers. Write programs in Fortran,
and wrote an instructional computing manual.
Train and consult with scientists and deliver in-
house briefings.

June 1974-
January 1978

Anchor Products, Inc., Ambler, PA
Managed 12-person office in a $1.2 million
company. Also performed general bookkeeping
and payroll.

Volunteer Work:

October 1987-
Present

Children's Hospital of Philadelphia, Philadelphia, PA
Volunteer in the physical therapy unit. Assist
therapists and guide patients with their therapy.
Use play therapy to enhance strengthening
progress.

Honors:

Awarded three $4,000 tuition scholarships
(1988-90) from Gould Foundation.

**Additional
Information:**

Member, Harmon Biology Club, Yearbook Staff
Raised three school-age children.
Tuition 100% self-financed.

References:

Available upon request.

FIGURE 15.3   Sample Résumé of a Nontraditional Student

Writing the Job-Application Letter

The job-application letter is crucial because it is the first thing your reader sees. If the letter is ineffective, the reader probably will not bother to read the résumé.

If job candidates had infinite time and patience, they would make up a different résumé for each prospective employer, highlighting their appropriateness for that one job. But because most candidates don't have unlimited time and patience, they make up one or two different versions of their résumés. As a result, the typical résumé makes a candidate look only relatively close to the ideal the employer has in mind. Thus, candidates must use job-application letters to appeal directly and specifically to the needs and desires of particular employers.

Appearance of the Job-Application Letter

Like the résumé it introduces, the letter must be error-free and professional looking. A good job-application letter has all the virtues of any business letter: adequate margins, clear and uniform type, no strikeovers or broken letters. And, of course, it must conform to one of the basic letter formats (see Chapter 14).

Such advice is easy to give but hard to follow. The problem is that every job-application letter must be typed individually, because its content is unique. (Even if you use a word processor, the inside address and the salutation, at least, must be typed by hand.) In creating the job-application letter, therefore, you have a double burden: the one page that makes the greatest impression has to be composed and typed separately each time you apply for a position. Compared to job-application letters, résumés are simple: you have to make them perfect only a few times.

Content of the Job-Application Letter

Like the résumé, the job-application letter is a sales document. Its purpose is to convince the reader that you are an outstanding candidate who should be called in for an interview. Of course, you accomplish this purpose through hard evidence, not empty self-praise.

The job-application letter is *not* an expanded version of everything in the résumé. The key to a good application letter is selectivity. In the letter, you choose from the résumé two or three points of greatest interest to the potential employer and develop them into paragraphs. If one of your previous part-time positions called for specific skills that the employer is looking for, that position might be the subject of a substantial paragraph in the letter, even though the résumé devotes only a few lines to it. If you try to cover every point on your résumé, the letter will be fragmented. The reader then will have a hard time forming a clear impression of you, and the purpose of the letter will be thwarted.

In most cases, a job-application letter should fill up the better part of a page. Like all business correspondence, it should be single-spaced, with double spaces between paragraphs. For more-experienced candidates, the letter may be longer, but most students find that they can adequately describe their credentials in one page. A long letter is not necessarily a good letter. If you write at length on a minor point, you end up being boring. Worse still, you appear to have poor judgment. Employers are always seeking candidates who can say much in a short space.

Elements of the Job-Application Letter

Among the mechanical elements of the job-application letter, the inside address—the name, title, organization, and address of the recipient—is most important. If you know the correct form of this information from an ad, there is no problem. However, if you are uncertain about any of the information—the recipient's name, for example, might have an unusual spelling—you should verify it by phoning the organization. Many readers are very sensitive about such matters, so you should not risk beginning the letter with a misspelling or incorrect title. When you do not know who should receive the letter, do not address it to a department of the company—unless the job and specifically says to do so—because nobody in that department might feel responsible for dealing with it. Instead, phone the company to find out who manages the department. If you are unsure of the appropriate department or division to write to, address the letter to a high-level executive, such as the president of the organization. The letter will be directed to the right person. Also, because the application includes both a letter and a résumé, type the enclosure notation (see Chapter 14) in the lower left-hand corner of the letter.

The four-paragraph letter that will be discussed here is, naturally, only a very basic model. Because every substantial job-application letter has an introductory and a concluding paragraph, the minimum number of paragraphs for the job-application letter has to be four. But there is no reason that an application letter cannot have five or six paragraphs. The four basic paragraphs are as follows:

- ☐ introductory paragraph
- ☐ education paragraph
- ☐ employment paragraph
- ☐ concluding paragraph

Because this is such an important letter, you should plan it carefully. Choose information from your background that best responds to the needs of the potential employer. Draft the letter and then revise it. Let it sit for a while, then revise it again. Spend as much time on it as you can. Make each paragraph a unified, functional part of the whole letter. Supply clear transitions from one paragraph to the next.

The Introductory Paragraph

The introductory paragraph establishes the tone for the rest of the letter and captures the reader's attention. It also has four specific functions:

- *It identifies your source of information.* In an unsolicited application, all you can do is ask if a position is available. For most applications, however, your source of information is a published advertisement or an employee already working for the organization. If it is an ad, identify specifically the publication and its date of issue. If it is an employee, identify that person by name and title.

- *It identifies the position you are interested in.* Often, your reader will be unfamiliar with all the different ads the organization has placed; without the title, he or she will not know which position you are interested in.

- *It states that you wish to be considered for the position.* Although the context makes it obvious, you should mention it because the letter would be awkward without it.

- *It forecasts the rest of the letter.* Choose a few phrases that forecast the body of the letter, so that the letter flows smoothly.

These four aspects of the introductory paragraph need not appear in any particular order, nor need they each be covered in a single sentence. The following examples of introductory paragraphs demonstrate different ways to provide the necessary information.

Response to a job ad:

I am writing in response to your notice in the May 13 New York Times. I would like to be considered for the position in system programming. My studies at Eastern University in computer science, along with my programming experience at Airborne Instruments, would qualify me, I believe, for the position.

Unsolicited:

My academic training in hotel management and my experience with Sheraton International have given me a solid background in the hotel industry. Would you please consider me for any management trainee position that might be available?

Personal Contact:

Mr. Howard Alcott of your Research and Development Department has suggested that I write. He thinks that my organic chemistry degree and my practical experience with Brown Laboratories might be of value to you. Do you have an entry-level position in organic chemistry for which I might be considered?

As these sample paragraphs indicate, the important information can be conveyed in any number of ways. The difficult part of the introductory paragraph—and of the whole letter as well—is to achieve the proper tone: quiet self-confidence. Because the letter will be read by someone who is professionally superior to you, the tone must be modest, but it should not be self-effacing or negative. Never say, for example, "I do not have a very good background in finance but I'm willing to learn." The reader will take this kind of statement at face value and probably stop reading right there. You should show pride in your education and experience, while at the same time suggesting by your tone that you have much to learn.

The Education Paragraph

For most students the education paragraph should come before the employment paragraph because the content of the former will be stronger. If, however, your work experience is more pertinent than your education, discuss your work first.

In devising your education paragraph, take your cue from the job ad. What aspect of your education most directly fits the job requirements? If the ad stresses versatility, you might structure your paragraph around the range and diversity of your courses. Also, you might discuss course work in a field related to your major, such as business or communications skills. Extracurricular activities are often very valuable; if you were an officer in a student organization in your field, you could discuss the activities and programs that you coordinated. Perhaps the most popular strategy for developing the education paragraph is to discuss skills and knowledge gained from advanced course work in the major field.

Whatever information you provide, the key to the education paragraph is to develop one unified idea, rather than to toss a series of unrelated facts onto the page. Notice how each of the following education paragraphs develops a unified idea:

> At Eastern University, I have taken a wide range of courses in the sciences, but my most advanced work has been in chemistry. In one laboratory course, I developed a new aseptic brewing technique that lowered the risk of infection by over 40 percent. This new technique was the subject of an article in the Eastern Science Digest. Representatives from three national breweries have visited our laboratory to discuss the technique with me.

> To broaden my education at Southern, I took eight business courses in addition to my requirements for the Civil Engineering degree. Because your ad mentions that the position will require substantial client contact, I believe that my work in marketing, in particular, would be of special value. In an advanced marketing seminar, I used Pagemaker to produce a 20-page sales brochure describing one company's building structures for sale to industrial customers in our city.

> The most rewarding part of my education at Western University took place outside the classroom. My entry in a fashion-design

competition sponsored by the university won second place. More important, through the competition I met the chief psychologist at Western Regional Hospital, who invited me to design clothing for handicapped persons. I have since completed six different outfits. These designs are now being tested at the hospital. I hope to be able to pursue this interest once I start work.

Notice that each of these paragraphs begins with a topic sentence—a forecast of the rest of the paragraph—and uses considerable detail and elaboration to develop the main idea. Notice, too, a small point: if you haven't already specified your major and college or university in the introductory paragraph, be sure to do so in the education paragraph.

The Employment Paragraph

Like the education paragraph, the employment paragraph should begin with a topic sentence and then elaborate a single idea. That idea might be that you have had a broad background or that a single job has given you special skills that make you particularly suitable for the available one. The following examples show effective experience paragraphs.

For the past three summers and part-time during the academic year, I have worked for Redego, Inc., a firm that specializes in designing and planning industrial complexes. I began as an assistant in the drafting room. By the second summer, I was accompanying a civil engineer on field inspections. Most recently, I have used CAD to assist an engineer in designing and drafting the main structural supports for a 15-acre, $30 million chemical facility.

Although I have worked every summer since I was fifteen, my most recent position, as a technical editor, has been the most rewarding. I was chosen by Digital Systems, Inc., from among 30 candidates because of my dual background in computer science and writing. My job was to coordinate the editing of computer manuals. Our copy editors, who are generally not trained in computer science, need someone to help verify the technical accuracy of their revisions. When I was unable to answer their questions, I was responsible for getting the correct answer from our systems analysts and making sure the computer novice could follow it. This position gave me a good understanding of the process by which operating manuals are created.

I have worked in merchandising for three years as a part-time and summer salesperson in men's fashions and accessories. I have had experience running inventory-control software and helped one company switch from a manual to an on-line system. Most recently, I assisted in clearing $200,000 in out-of-date men's fashions: I coordinated a campaign to sell half of the merchandise at cost and was able to convince the manufacturer's representative to accept the other half for full credit. For this project, I received a certificate of appreciation from the company president.

Notice how these writers carefully define their duties to give their readers a clear idea of the nature and extent of their responsibilities.

Although you will discuss your education and experience in separate paragraphs, try to link these two halves of your background. If an academic course led to an interest that you were able to follow up in a job, make that point clear in the transition from one paragraph to the other. Similarly, if a job experience helped shape your academic career, tell the reader about it.

The Concluding Paragraph

The concluding paragraph of the job-application letter is like the ending of any sales letter: its function is to stimulate action. In this case, you want the reader to invite you for an interview. In the preceding paragraphs of the letter, you provided the information that should have convinced your reader to give you another look. In the last paragraph, you want to make it easy for him or her to do so. The concluding paragraph contains three main elements:

- ☐ a reference to your résumé
- ☐ a request for an interview
- ☐ your phone number

If you have not yet referred to the enclosed résumé, do so at this point. Then, politely but confidently request an interview—making sure to use the phrase "at your convenience." Don't make the request sound as if you're asking a personal favor. And be sure to include your phone number and the time of day you can be reached—even though the phone number is also on your résumé. Many employers will pick up a phone and call promising candidates personally, so make it as easy as possible for them to do so.

Following are examples of effective concluding paragraphs.

The enclosed résumé provides more information about my education and experience. Could we meet at your convenience to discuss further the skills and experience I could bring to Pentamax? A message can be left for me any time at (333) 444-5555.

More information about my education and experience is included on the enclosed résumé, but I would appreciate the opportunity to meet with you at your convenience to discuss my application. You can reach me after noon Tuesdays and Thursdays at (333) 444-5555.

Figures 15.4 and 15.5 show examples of effective job-application letters corresponding to the résumés presented in Figure 15.2 and 15.3.

The letter in Figure 15.5 needs some comment. The writer decided to discuss her nontraditional background—she is a single mother who has been out of school for more than 10 years—without overemphasizing it.

3109 Vista Street
Philadelphia, PA 19136

January 19, 19XX

Mr. Stephen Spencer, Director of Personnel
Department 411
Boeing Naval Systems
103 Industrial Drive
Wilmington, DE 20093

Dear Mr. Spencer:

I am writing in response to your advertisement in the January 16 Philadelphia Inquirer. Would you please consider me for the position in Signal Processing? I believe that my academic training at Drexel University in Electrical Engineering, along with my experience with RCA Advanced Technology Laboratory, would qualify me for the position.

My education at Drexel University has given me a strong background in computer hardware and system design. I have concentrated on digital and computer applications, developing and designing computer and signal-processing hardware in two graduate-level engineering courses I was permitted to take. For my senior-design project, I am working with four other undergraduates in enhancing the path-planning software for a infrared night-vision robotics application.

While working at RCA I was able to apply my computer experience to the field of the VLSI design. In one project I used my background in Fortran to develop and test new

FIGURE 15.4   Job-Application Letter

Accordingly, the first two paragraphs of the body discuss her qualifications for the internship position. She wants to make sure that the reader evaluates her as he would any candidate. Then she adds a paragraph that attempts to explain how her additional experience—as an office manager and a single mother—have enabled her to develop skills that would be of value to anyone in any field. She exploits her situation gracefully, always appealing to the reader's needs, not to his sympathy.

Mr. Stephen Spencer
Page 2
January 19, 19XX

Fortran software used in the automated production of VLSI standard cell family databooks. In another project, I used CAD software on a VAX to evaluate IC designs.

The enclosed résumé provides an overview of my education and experience. Could I meet with you at your convenience to discuss my qualifications for this position? Please write to me at the above address or leave a message any time at (215) 895-2660.

Yours truly,

Carl Lockyer

Carl Lockyer

Enclosure (1)

FIGURE 15.4 *Continued*

1781 Weber Road
Warminster, PA 18974

January 17, 19XX

Mr. Harry Gail
Fox Run Medical Center
399 N. Abbey Road
Warminster, PA 18974

Dear Mr. Gail:

Last April I contacted you regarding the possibility of an internship as a laboratory assistant at your center. Your assistant, Mary McGuire, told me then that you might consider such a position this year. With the experience I have gained since last year, I feel that I would be a valuable addition to your center in many ways.

At Harmon College, I have earned a 3.7 GPA in 36 credits in chemistry and biology; all but two of these courses have had laboratory components. One important skill stressed here is the ability to communicate effectively in writing and orally. Our sciences courses have extensive writing and speaking requirements; my portfolio includes seven research papers and lab reports of more than 20 pages each, and I have delivered four oral presentations, one of 45 minutes, to classes.

At Smith Kline & French, where I currently work part time, I analyze molecular data on an E&S PS300, a Macintosh, and an IBM PC. I have tried to remain current with the latest advances; my manager at Smith Kline has allowed me to attend two different two-day in-house seminars on computerized data analysis using SAS.

FIGURE 15.5   Job-Application Letter

Mr. Harry Gail
Page 2
January 17, 19XX

Having been out of school for more than a decade, I am well aware of how much the technology has changed. However, I feel that the interpersonal skills I developed as an office manager would be of benefit at Fox Run. In addition, as a single mother of three I know something about time management.

More information about my education and experience is included on the enclosed résumé, but I would appreciate the opportunity to meet with you at your convenience to discuss my application. If you would like any additional information about me or Harmon's internship program, please write to me at the above address or call me at (215) 674-4006.

Very truly yours,

Alice P. Linder

Alice P. Linder

Enclosure

FIGURE 15.5   *Continued*

Writing the Follow-up Letter

The final element in a job search is a follow-up letter. Your purpose in writing it is to thank the representative for taking the time to see you and to emphasize your particular qualifications. You can also take this opportunity to restate your interest in the position.

The follow-up letter can do more good with less effort than any other step in the job-application procedure. The reason is simple: so few candidates take the time to write the letter. Figure 15.6 provides an example of a follow-up letter.

1901 Chestnut Street
Phoenix, AZ 63014

July 13, 19XX

Mr. Daryl Weaver
Director of Operations
Cynergo, Inc.
Spokane, WA 92003

Dear Mr. Weaver:

Thank you for taking the time yesterday to show me your facilities and to introduce me to your colleagues.

Your advances in piping design were particularly impressive. As a person with hands-on experience in piping design, I can fully appreciate the advantages your design will have.

The vitality of your projects and the obvious good fellowship among your employees further confirm my initial belief that Cynergo would be a fine place to work. I would look forward to joining your staff.

Sincerely yours,

Albert Rossman

Albert Rossman

FIGURE 15.6 Follow-up Letter

WRITER'S CHECKLIST

The following checklist covers the résumé, the job-application letter, and the follow-up letter.

Résumé

1. Does the résumé have a professional appearance, with generous margins, clear type, a symmetrical layout, adequate white space, and effective indentation?

2. Is the résumé free of errors?

3. Does the identifying information section contain your name, address, and phone number?

4. Have you included a clear statement of your job objective?

5. Does the education section include your degree, your institution and its location, and your anticipated date of graduation, as well as any other information that will help your reader appreciate your qualifications?

6. Does the employment section include, for each job, the dates of employment, the organization's name and location, and your position or title, as well as a description of your duties and accomplishments?

7. Does the personal information section include relevant hobbies or activities, including extracurricular interests?

8. Does the references section include the names, job titles, organizations, mailing addresses, and phone numbers of three or four referees? If you are not listing this information, does the strength of the rest of the résumé offset the omission?

9. Does the résumé include any other appropriate sections, such as military service, language skills, or honors?

Job-Application Letter

1. Does the letter look professional?

2. Does the introductory paragraph identify your source of information and the position you are applying for, state that you wish to be considered, and forecast the rest of the letter?

3. Does the education paragraph respond to your reader's needs with a unified idea introduced by a topic sentence?

4. Does the employment paragraph respond to your reader's needs with a unified idea introduced by a topic sentence?

5. Does the concluding paragraph include a reference to your résumé, a request for an interview, and your phone number?

6. Does the letter include an enclosure notation?

Follow-up Letter

Does the letter thank the interviewer and briefly restate your qualifications?

EXERCISES

1. In a newspaper or journal, find a job advertisement for a position in your field for which you might be qualified. Write a résumé and a job-application letter in response to the ad.

2. Revise the following résumés to make them more effective. Make up any reasonable details.

   a.

   Bob Jenkins
   2319 Fifth Avenue
   Waverly, CT 01603
   Phone: 611-3356

Personal Data:	22 Years old Height 5′11″ Weight 176 lbs.
Education:	B.S. in Electrical Engineering University of Connecticut, June 1991

Experience:		
	6/91–9/91	Falcon Electronics Examined panels for good wiring. Also, I revised several schematics.
	6/90–9/90	MacDonalds Electrical Supply Co. Worked parts counter.
	6/89–9/89	Happy Burger Made hamburgers, fries, shakes, fish sandwiches, and fried chicken.
	6/88–9/88	Town of Waverly Outdoor maintenance. In charge of cleaning up McHenry Park and Municipal Pool picnic grounds. Did repairs on some electrical equipment.

Background: Born and raised in Waverly.
               Third baseman, Fisherman's Rest
               softball team.
               Hobbies: jogging, salvaging and
               repairing appliances, reading
               magazines, politics.

References: Will be furnished upon request.

b.

Harold Perkins

1415 Ninth Street   319 W. Irvin Avenue
Altoona, Pa. 16013   State College, Pa.
667-1415   304-9316

Education: Economics Major
               Penn State, 19XX
               Cum: 3.03

               Important courses: Econ. 412, 413, 501,
               516.
               Fin. 353, 614 (seminar), Mgt. 502-503.

Experience: Radio Shack, State College
               Bookkeeper. Maintained the books.

               Radio Shack, State College
               Clerk

               Holy Redeemer Hospital
               Volunteer

Organizations: Scuba Club
               Alpha Sigma Sigma—Pledge Master
               Treasurer, Penn State Economics
               Association

References:
George Williams   Arthur Lawson   Dr. John Tepper
Economics Dept.   Radio Shack   Holy Redeemer Hospital
McKee Hall   State College, Pa. Boalsburg, Pa.
Stage College, Pa.

3. Revise the following job-application letters to make them more effective. Make up any reasonable details.

a.

April 13, 19XX

Wayne Grissert
Best Department Store
113 Hawthorn
Atlanta, Georgia

Dear Mr. Grissert:

As I was reading the Sunday <u>Examiner</u>, I came upon your ad for a buyer. I have always been interested in learning about the South, so would you consider my application?

I will receive my degree in fashion design in one month. I have taken many courses in fashion design, so I feel I have a strong background in the field.

Also, I have had extensive experience in retail work. For two summers I sold women's accessories at a local clothing store. In addition, I was a temporary department head for two weeks.

I have enclosed a resume and would like to interview you at your convenience. I hope to see you in the near future. My phone number is 436-6103.

Sincerely,

Brenda Adamson
Brenda Adamson

---

b.

---

113 Holloway Drive
Nanuet, Oklahoma 61403

May 11, 19XX

Miss Betty Richard
Manager of Technical Employment
Scott Paper Co.
Philadelphia, Pa. 19113

Dear Miss Betty,

I am writing this letter in response to your ad in the <u>Philadelphia Inquirer</u>, March 23, regarding a job opening in research and development.

I am expecting a B.S. degree in chemical engineering this June. I am particularly interested in design and have a strong background in it. I am capable of working independently and am able to make certain important decisions pertaining to certain problems encountered in chemical processes. I am including my résumé with this letter.

If you think I can be of any help to you at your company, please contact me at your convenience. Thank you.

Sincerely,

Glenn Corwin

Glenn Corwin

4. Revise this follow-up letter to make it more effective. Make up any reasonable details.

914 Imperial Boulevard
Durham, NC 27708

November 13, 19XX

Mr. Ronald O'Shea
Division Engineering
Safeway Electronics, Inc.
Holland, MI 49423

Dear Mr. O'Shea:

Thanks very much for showing me around your plant. I hope I was able to convince you that I'm the best person for the job.

Sincerely yours,

Robert Wilcox

Robert Wilcox

REFERENCE

Harcourt, J., and Krizan, A. C. 1989. A comparison of résumé content preferences of Fortune 500 personnel administrators and business communication instructors. *Journal of Business Communication* 26, 2 (Spring), 159–176.

16

Instructions
and Manuals

A set of instructions is a process description written to help the reader perform a specific task. A process description discusses how a tire is changed. A set of instructions explains how to change the tire. Whereas a process description states that the car is jacked up so that the tire can be removed, a set of instructions goes into more detail: it explains how to assemble and use the jack effectively and safely.

A manual is a document consisting primarily of instructions. Often it is printed and bound, like a book. Manuals can be classified according to function. One common type is a procedures manual. Like a set of instructions, its function is to instruct. For example, it explains how to use a software program, maintain inventory for a particular store, or operate a piece of machinery. There are installation manuals, maintenance manuals, and repair manuals.

Effective instructions and manuals look easy to write, but they aren't. You have to make sure not only that your readers will be able to understand and follow your instructions easily, but also that in performing the tasks they won't damage any equipment or, more important, injure themselves or other people.

Understanding the Role of Instructions and Manuals

Instructions and manuals are central to technical writing, and you will probably be asked to write them or contribute to them often in your career. Many factors have tremendously increased the number and complexity of instructions and manuals produced every day.

Consumer products are much more complex than they used to be. A radio sold 30 years ago needed a simple sheet listing repair shops and perhaps a simple guarantee. Today a modular stereo system—complete with dual cassettes, programmable turntable, compact disc player, 5-band graphic equalizer, and ready-to-assemble walnut veneer rack—requires a long manual. The average personal computer requires about 4,000 pages of manuals; the average mainframe computer, about 50,000 (Mackin 1989).

Nonconsumer items, too, are much more complex. According to one expert (Muller 1990), the average military airplane in 1950 needed 2,000 pages of manuals; by 1976, the average aircraft came with 300,000 pages of manuals. Today that figure has probably doubled again.

At the workplace, procedures are also much more sophisticated. Buying a screwdriver at Sears requires that the clerk enter a string of numbers and letters so that the electronic cash register can help maintain an accurate inventory. Employee salaries and raises are determined by a complicated formula intended to make the process as fair as possible. Learning how to use the formula requires a manual. In manufacturing, workers routinely operate and monitor computers and robots.

Just a decade or two ago, little attention was paid to the quality of documentation — the instructions and manuals. The technical people who created the product or system were the experts. If the product did its job, no more was asked. The users had to cope as well as they could. Often companies did not provide any documentation at all, or they provided it six months or a year after the product was introduced. The main way to find out information about the product was to phone the service representative — the person who went out to the job site to help the user.

Today the goal is to make systems "user friendly." The competition for our dollars is so great that we can require that the product be easy to set up and use. When computers are compared these days, the quality of the manuals is one of the prime considerations. Major advertising campaigns for all kinds of products stress clear, simple, easy-to-use manuals. The failure of the Coleco Adam, a personal computer, was widely attributed to poor documentation. And the *New York Times Book Review* (Perrin 1989) even reviewed the owner's manuals for the 1990 Chevrolet S-10 and the Buick LeSabre. The reviewer criticized the Chevrolet manual for opaque sentences such as this: "The windshield defrosting and defogging system provides visibility through designated areas of the windshield during inclement weather conditions."

Documentation is no longer an afterthought but rather an integral part of the planning process. The documentation writers are part of the research-and-development team. And they take questions of audience and purpose seriously.

Analyzing the Writing Situation

The purpose of a set of instructions or of a manual is usually self-evident. Assembly instructions for a backyard swing helps the purchaser set up the swing. A maintenance manual for a computerized conveyor belt explains how and when to perform maintenance on it.

The question of audience, however, is difficult and subtle. In fact, when instructions and manuals are ineffective, chances are that the writer has inaccurately assessed the audience. Performing a function — such as assembling the backyard swing or maintaining the conveyor belt — is easy for the expert but not so easy for the person reading the documentation. A reader who doesn't know what a self-locking washer is or what the calibrations on a timing control mean might not be able to complete the process.

Before you start to write a set of instructions or a manual, think carefully about your audience's background and skill level. If you are writing to people who are experienced in the field, use technical vocabulary and make any reasonable assumptions about their knowledge. But if you are addressing laypersons, define technical terms and give your directions in more detail. Don't be content to write, "Make sure the tires are rotated properly." Define proper rotation and describe how to achieve it.

The best way to make sure you have assessed your audience effectively is to find someone whose background resembles that of your intended readers. Give this person a draft of your document and watch as he or she tries to perform the tasks. This process will give you valuable information about how clear your writing is. See Chapter 5 for a more detailed discussion of evaluating documents.

One other aspect of audience analysis should be mentioned. Inexperienced writers assume that people will read through the document carefully before beginning the first step. They won't; most people will merely glance at the whole set of instructions or the manual to see if anything catches their eye before they get started. For this reason, you should organize the document in a strict chronological sequence. This is particularly important when you want your reader to know of potential dangers. There are two generally accepted kinds of safety labels, described here from more urgent to less urgent:

- *Danger.* "Danger" warns the reader of an immediate hazard.

  DANGER: HIGH VOLTAGE. STAND BACK.

- *Warning.* "Warning" alerts the reader to a potential threat that could cause injury.

  WARNING: TO PREVENT SERIOUS EYE INJURY, WEAR SAFETY GOGGLES AT ALL TIMES!

In addition to these safety warnings is the *caution*, which alerts the reader of potential damage to equipment or failure of the process.

  CAUTION: Do not attempt to use nonrechargeable batteries in this charging unit; they could damage the charging unit.

Finally, there is the *note*, which is a tip to help the reader carry out the process easily.

  Note: Two different-size screws are provided. Be sure to use the 3/8″-long screws here.

If a danger, warning, or caution applies to the whole process, state it emphatically at the start of the instructions. If it applies only to a particular step, however, insert it before that step.

Safety messages should be seen and read easily. A government publication, *The Occupational Safety and Health Administration Guidelines* (Chapter XVII, Sections 1910.145 and 1926.155), describes proper standards for placing safety messages on products and manuals.

These standards address the following questions:

☐ Is the message prominently displayed so that the user sees it?
☐ Is the message large enough and clearly legible, under operating conditions?
☐ Are the graphics and the words of the message clear and informative?

Writing Effective Instructions

always #ed
use active voice

Instructions can be brief—a small sheet of paper—or extensive—twenty pages or more. Brief instructions could be produced by one or two people: a writer, or a writer and an artist. Sometimes a technical expert is added to the team. For more extensive instructions, other people—marketing and legal personnel, for example—could be added. The team could consist of ten or twenty professionals working with a budget of many thousands of dollars.

Like process descriptions, you will probably find it easiest to write instructions sequentially; that is, in the order in which they will appear. Many writers like to perform the task as they write up the instructions. In this way, they can detect any errors or omissions quickly and fix them easily.

The Structure of the Instructions

Regardless of the size of the project, most instructions are structured like process descriptions (see Chapter 9). The main difference is that the conclusion of a set of instructions is less a summary than an explanation of how to make sure the reader has followed the instructions correctly. Most sets of instructions contain the following three components:

☐ a general introduction that prepares the reader for performing the instructions
☐ the step-by-step instructions
☐ a conclusion

The General Introduction
The general introduction gives the readers the preliminary information they will need to follow the instructions easily and safely. In writing the introduction, answer these three questions about the process:

☐ Why should the reader carry out this task?
☐ What safety measures should the reader take, and what background information is necessary before he or she begins the process?
☐ What tools and materials will the reader need?

■ *Why should the reader carry out this task?* Often, the answer to this question is obvious and should not be stated. For example, the purchaser of a backyard barbecue grill does not need an explanation of why it should be assembled. Sometimes, however, readers need to be told why they should carry out the task. Many preventive maintenance chores—such as changing radiator antifreeze every two years—fall into this category.

If appropriate, answer two more questions:

When should the reader carry out the task? Some tasks, such as rotating tires or planting crops, need to be performed at particular times or particular intervals.

Who should carry out the task? Sometimes it is necessary to describe the person or persons who are to carry out a task. Some kinds of aircraft maintenance, for example, may be carried out only by persons certified to do that maintenance.

■ *What safety measures or other concerns should the reader understand?* In addition to the safety measures that apply to the whole task, state any tips that will make your reader's job easier. For example:

NOTE: For ease of assembly, leave all nuts loose. Give only 3 or 4 complete turns on bolt threads.

■ *What tools and materials will the reader need?* The list of necessary tools and equipment is usually included in the introduction so that the readers do not have to interrupt their work to hunt for another tool.

Following is a list of tools and materials from a set of instructions on replacing broken window glass:

You will need the following <u>tools</u> and <u>materials</u>:

Tools	Materials
glass cutter	putty
putty knife	glass of proper size
window scraper	paint
chisel	hand cleaner
electric soldering iron	work gloves
razor blade	linseed oil
pliers	glazier's points
paint brush	

The Step-by-Step Instructions

The step-by-step instructions are essentially like the body of the process description. There are, however, two differences.

First, always number the instructions. Each numbered step should define a single task that the reader can carry out easily, without having to refer back to the instructions. Don't write overloaded steps like this one:

1. Mix one part of the cement with one part water, using the trowel. When the mixture is a thick consistency without any lumps bigger than a marble, place a strip of about 1″ high and 1″ wide along the face of the brick.

On the other hand, if the step is too simple, the reader will be annoyed:

1. Pick up the trowel.

Second, always state the instructions in the imperative mood: "Attach the red wire. . . ." The imperative is more direct and economical than the indicative mood ("You should attach the red wire. . ." or "The operator should attach the red wire. . ."). Make sure your sentences are grammatically parallel. Avoid the passive voice ("The red wire is attached. . ."), because it can be ambiguous: Is the red wire already attached?

Keep the instructions simple and direct. However, do not omit the articles (*a, an, the*) to save space. Omitting the articles makes the instructions hard to read and, sometimes, unclear. In the sentence "Locate midpoint and draw line," for example, the reader cannot tell if "draw line" is a noun ("the draw line") or a verb and its object ("draw the line").

Be sure to include graphics in your step-by-step instructions. When appropriate, accompany each step with a photograph or diagram that shows what the reader is supposed to do. Some kinds of activities—such as adding two drops of a reagent to a mixture—do not need illustration. However, steps that require manipulating physical objects—such as adjusting a chain to a specified tension—can be clarified by graphics.

Figure 16.1 shows the extent to which a set of instructions can integrate words and graphics. This excerpt is from the operating-instructions booklet for a video cassette recorder. The booklet has already described the basic controls and functions of the device. At this point, numbered steps clearly indicate where to find the controls. See Chapter 11 for a discussion of graphics.

The Conclusion

Instructions sometimes do not require conclusions. Generally, however, the instructions conclude with maintenance tips. Another popular conclusion is a trouble-shooter's checklist, usually in the form of a table, that identifies and tells how to solve common problems.

Following is a portion of the trouble-shooter's guide included in the operating instructions of a lawn mower.

Problem	Cause	Correction
Mower does not start	1. Out of gas.	1. Fill the gas tank.
	2. "Stale" gas.	2. Drain the tank and refill it with fresh gas.
	3. Spark plug wire disconnected from spark plug.	3. Connect the wire to the plug.

Mower loses power.	1. Grass too high.	1. Set the mower in "higher-cut" position.
	2. Dirty air cleaner.	2. Replace the air cleaner.
	3. Buildup of grass, leaves, and trash.	3. Disconnect the spark plug wire, attach it to the retainer post, and clean the underside of the mower housing.

Some trouble-shooter's checklists refer the reader back to the page that discusses the action being described. For example, the Correction column in the lawn-mower trouble-shooting checklist might say "1. Fill the gas tank. See page 4."

Figure 16.2 shows a set of instructions on how to apply panels to stud walls. The instructions were written by the Home Center Institute of the National Retail Hardware Association.

Writing Effective Manuals

Most of what has been said about instructions applies to manuals. For example, you have to understand your audience and purpose, explain procedures clearly, and complement your descriptions with graphics.

However, because manuals are usually much more ambitious projects than instructions, they require much more careful planning. A bigger investment is at stake for the organization, not only in the costs of writing and producing the manual but in the potential effects of the published manual. A good manual will attract customers and reduce costs, for the organization will need fewer people to deal with questions in the field. A poor manual will be expensive to produce, will have to be revised more often, and will alienate customers.

The Process of Writing a Manual

Writing a manual, like writing a book, is so complex that the following discussion can provide only an introduction. If you are involved in creating a manual, consult the books listed in the bibliography (Appendix D) for detailed advice.

In some respects, writing a manual calls for the general writing process: prewriting, drafting, and revising. However, manuals require adjustments to the general process.

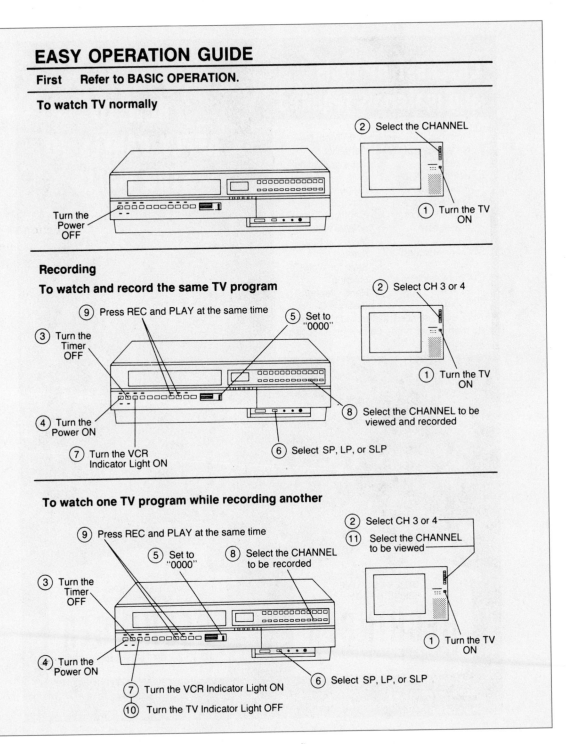

EASY OPERATION GUIDE

First   Refer to BASIC OPERATION.

To watch TV normally

② Select the CHANNEL

Turn the Power OFF

① Turn the TV ON

Recording

To watch and record the same TV program

② Select CH 3 or 4

⑨ Press REC and PLAY at the same time

⑤ Set to "0000"

③ Turn the Timer OFF

① Turn the TV ON

④ Turn the Power ON

⑧ Select the CHANNEL to be viewed and recorded

⑦ Turn the VCR Indicator Light ON

⑥ Select SP, LP, or SLP

To watch one TV program while recording another

② Select CH 3 or 4

⑪ Select the CHANNEL to be viewed

⑨ Press REC and PLAY at the same time

⑤ Set to "0000"

⑧ Select the CHANNEL to be recorded

③ Turn the Timer OFF

① Turn the TV ON

④ Turn the Power ON

⑦ Turn the VCR Indicator Light ON

⑩ Turn the TV Indicator Light OFF

⑥ Select SP, LP, or SLP

FIGURE 16.1   Instructions That Integrate Words and Graphics

Checklist of tools and materials required

- ☐ Hammer
- ☐ Nails (Several Sizes)
- ☐ Level
- ☐ Power Saw
- ☐ Moldings
- ☐ Plastic Film or Waterproof Paper
- ☐ Color-Matched Putty
- ☐ Countersinking Punch
- ☐ Square
- ☐ Fine-Tooth Saw
- ☐ Adhesive
- ☐ Tape or Folding Rule
- ☐ Proper Saw Blades
- ☐ Electrical Tape or Wire Nuts
- ☐ Magnetic Stud Finder
- ☐ Stain for Wood Molding
- ☐ Hand Cleaner
- ☐ Cleanup Cloths

ASK FOR detailed instructions on how to do many other home do-it-yourself projects at Hechinger.

HOW TO APPLY PANELS TO STUD WALLS

ANOTHER CUSTOMER SERVICE FROM

HECHINGER
The World's Most Unusual Lumber Yards

ANOTHER CUSTOMER SERVICE FROM

HECHINGER
The World's Most Unusual Lumber Yards

© 1976 by the Home Center Institute of National Retail Hardware Association. Reproduced by special permission.

FIGURE 16.2   Instructions with Graphics and Checklist

Here are some tips and instructions on installing paneling on stud walls. Take the time to read them thoroughly. Following these instructions can save time, money, and effort. It can also help you to end up with a neater, more satisfactory installation—with less waste.

1. STORING PANELS UNTIL THEY ARE INSTALLED

A. Store all panels in room temperature comparable to rooms where they are to be applied. This permits each panel to adjust to the temperature and humidity of the area.

B. Store flat with scraps of wood inserted between panels. Cover the stack. Storing flat reduces warping. Scraps of wood between the panels let each panel breathe and absorb room moisture. A cover over the stack keeps the panels clean and prevents marring.

2. MEASURING TO DETERMINE PANELS NEEDED

A. Measure each wall carefully to determine the number of panels required. Measure both height and width. Here's a good rule to follow:

B. Multiply total length of all walls by height. Subtract total area not to be paneled (windows, doors, etc.) from previous total. Add 5% for waste. If ceiling height is 8' or under, the table in Figure 1 can be used.

Figure 1

DETERMINING THE NUMBER OF 4 x 8 PANELS REQUIRED BASED ON PERIMETER OF ROOM

Room Perimeter	4' x 8' Panels Required	Room Perimeter	4' x 8' Panels Required
36 ft.	9	60 ft.	15
40 ft.	10	64 ft.	16
44 ft.	11	68 ft.	17
48 ft.	12	72 ft.	18
52 ft.	13	76 ft.	20
56 ft.	14	92 ft.	23

For example, if your room walls measured 16' x 16' x 18' x 18', the perimeter would be 68' requiring 17 4' x 8' panels based on the chart.

To allow for areas such as doors, windows, etc. use the deductions below and subtract from the total panels required.

For each door deduct ⅓ panel
For each window deduct ¼ panel
For each fireplace deduct ½ panel

Always use the next highest number of panels when the total perimeter is between the ranges shown in the table.

These figures are based on rooms with ceiling heights of 8' or less.

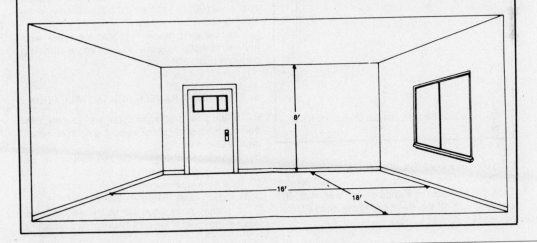

FIGURE 16.2 *Continued*

3. PREPARING FOR PANELING ON STUD WALLS

A.   Studs in walls are usually placed 16″ apart. To locate a stud, measure 16″ from wall and test by driving a 6 penny nail into the wall in an unobtrusive location, or use a magnetic stud finder to locate nails in studs (Figure 2).

B.   Locate all studs and mark the location with a chalk line. Use a level on a 2 x 4 to get all markings plumb and straight. Continue lines 2″ or 3″ out on the floor and ceiling (Figure 2).

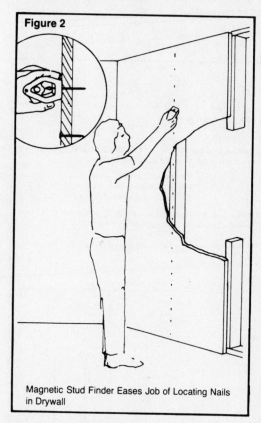

Figure 2

Magnetic Stud Finder Eases Job of Locating Nails in Drywall

C.   Remove all ceiling and floor moldings.

D.   Remove the light fixtures and all wall plates. Tape the exposed wire ends or use wire nuts. Be sure you turn off the electricity at the main control box panel before you begin.

4. PREPARING FOR PANELING ON NEW STUDS

E.   Be sure the studs are plumb and evenly spaced.

F.   Be sure there is a horizontal 2 x 4 at top and bottom (Figure 3).

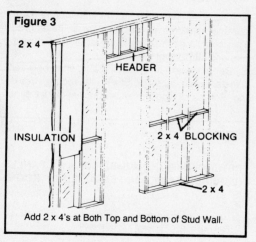

Figure 3

2 x 4

HEADER

INSULATION

2 x 4 BLOCKING

2 x 4

Add 2 x 4's at Both Top and Bottom of Stud Wall.

G.   Two-by-four blocking in the center helps prevent twisting and also gives added wall support (Figure 3).

H.   Building paper or #15 roofing felt is sometimes added as a protection against moisture. If added, use a staple gun and apply it horizontally, overlapping each course about 6″.

I.   On new walls, staple 1½″ or 2″ thick rock wool insulating bats between studs before applying paneling (Figure 3).

5. ARRANGING PANELS FOR INSTALLATION

A.   Stand panels up against the wall loosely, then back off for a study of the wood grain, color match, groove blend, etc.

B.   Rearrange to achieve the most attractive sequence of grain and tone.

C.   Use care in rearranging panels. Don't rub them against each other. This mars the finish.

D.   If panels have been stored in another area, leave in the room at least 24 hours before installing.

FIGURE 16.2   *Continued*

6. MEASURING AND CUTTING PANELS

A. Measure the floor-to-ceiling height at several points to detect any variations.

B. If height varies no more than ¼″, take the shortest height, subtract ¼″, and cut all panels to this measurement.

C. If height varies more than ¼″, cut each panel separately.

D. In either case, cut each panel ¼″ shorter than the actual ceiling height. Ceiling and floor moldings will cover height variations of up to 2″.

E. Start the panel installation in a corner (Figure 4). Be sure the panel is plumb. Use a level of sufficient length to get an accurate reading.

Figure 5

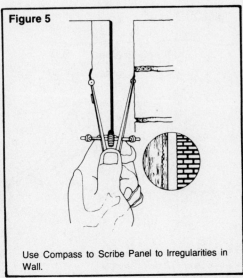

Use Compass to Scribe Panel to Irregularities in Wall.

Figure 4

Start Panels in a Corner. Use level to Plumb.

J. If cutting paneling with a hand saw, always cut from finished side (Figure 6). If using a power saw, always cut from back of panel (Figure 7).

Figure 6

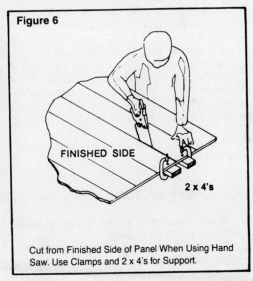

FINISHED SIDE

2 x 4's

Cut from Finished Side of Panel When Using Hand Saw. Use Clamps and 2 x 4's for Support.

F. The slightest irregularity in the wall can cause trouble. To allow for such irregularities, place the first panel in a plumb position in such a way that the panel and the wall can be spanned by your scribing compass (Figure 5). Draw a scribed line from top to bottom. Use a china marking pencil for scribing.

G. Cut to scribed line with coping saw where irregular cuts are necessary.

H. Double check all measurements before sawing.

I. Use a fine tooth, finish saw. Never cut paneling with a coarse saw.

K. After first corner panel is straight, proceed with installation by putting each panel firmly against the one already installed.

FIGURE 16.2 *Continued*

Figure 7

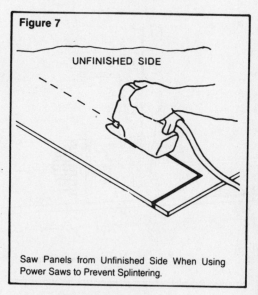

UNFINISHED SIDE

Saw Panels from Unfinished Side When Using Power Saws to Prevent Splintering.

7. APPLYING PANELS WITH NAILS

A. When nailing panels to furring strips, use 1″ nails.

B. If nailing directly to old wall, use 1½″ nails.

C. Start at one side and nail straight across to opposite wall.

D. Nail every 6″ along panel edges.

E. Space the nails 16″ apart across face of panels. This will place each nail on a stud.

F. Use a nailset to countersink all nails approximately ¹⁄₃₂″.

G. Fill countersink holes with matching-color putty stick.

H. If you use matching-color head nails, countersinking is not necessary.

8. APPLYING PANELS WITH ADHESIVES

A. If wall is papered, remove all paper before applying adhesive and paneling.

B. Do not use contact cement unless skilled in its application. Use the recommended cement—either rubber-base or plastic cement—and follow instructions carefully.

C. Apply a 3″ ribbon of adhesive along the furring strip lines and across bottom and top. Use serrated spreader or glue gun.

D. Nail loosely at top to hinge panel in proper location. Use scrap of lumber about 8″ long (Figure 8) to prop bottom of panel out from wall for 2 or 3 minutes, or until the adhesive gets "tacky."

E. Remove block after about 3 minutes, press the panel into position, and nail about every 6″ along base.

Figure 8

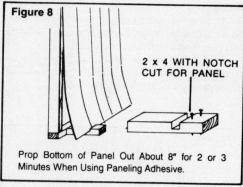

2 x 4 WITH NOTCH CUT FOR PANEL

Prop Bottom of Panel Out About 8″ for 2 or 3 Minutes When Using Paneling Adhesive.

9. HANDLING SPECIAL PANELING PROBLEMS

A. If a switch receptacle box requires a cutout in a panel, place the panel in its normal position over the box. Lay a soft wood block over the approximate location and tap the block soundly.

B. The outlet box will make its imprint on back of the panel. Drill a small pilot hole and cut the box outlet with a keyhole saw (Figure 9).

C. When a cut must be made for a door or window, measure the distance from the edge of the last panel installed to the door or window facing.

Figure 9

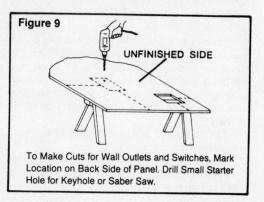

UNFINISHED SIDE

To Make Cuts for Wall Outlets and Switches, Mark Location on Back Side of Panel. Drill Small Starter Hole for Keyhole or Saber Saw.

FIGURE 16.2   *Continued*

D.   Next, place the panel in the correct position, plumb it, and mark that same measurement on the face of the panel with a china marking pencil. For example (Figure 10), suppose the distance from the edge of the last panel to the door facing was 24″. When positioned and plumbed, that measurement should be marked on the panel (A). Section C can then be cut.

E.   Be sure to measure from the floor to the top of the door (B), and transcribe this measurement to the panel to be cut. This should give you a perfect fit around the door.
F.   To cut around a fireplace, scribe around the fireplace as in the method shown in Figure 5, marking the irregular cut on the paneling. You can use quarter round to conceal small variations.

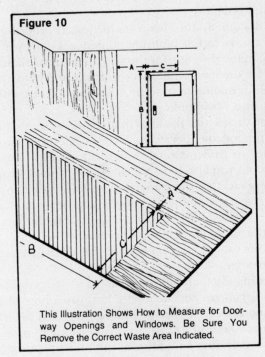

Figure 10

This Illustration Shows How to Measure for Doorway Openings and Windows. Be Sure You Remove the Correct Waste Area Indicated.

10. APPLYING THAT FINISHING TOUCH

A.   Apply proper moldings at top and bottom of panels.
B.   Cut the wood moldings with a fine-tooth saw and a miter box. Apply with small nails.
C.   Countersink and fill holes with matching-color putty stick. Clean with a dry rag if prefinished. Paint, stain, or varnish if unfinished.
D.   Metal moldings are available if desired.

11. IF TILE OR CARPETING IS TO BE LAID

A.   Install base molding first.
B.   Lay tile or carpet up to the base molding.
C.   Nail shoe molding over flooring tile.

FIGURE 16.2   *Continued*

The major reason is that manuals are almost always collaborative projects. A full-size manual is simply too big and requires too many skills for one person to write quickly: technical skills in the subject area, writing skills, artistic skills, as well as the knowledge of contracts and law that is required in an age of lawsuits. In addition, other perspectives are needed to prevent little problems from becoming big problems.

Prewriting

The most important stage of manual writing is prewriting. First, you must analyze your audience and purpose.

When you plan the documentation for a sophisticated procedure or system, you will almost always face a multiple audience. Often you will decide to produce a set of manuals—one for the user, one for the manager, one for the installer, one for the maintenance technicians, and so forth. Or you might decide on a main manual and some other, related documents, such as brochures, flyers, and workbooks.

Your primary purpose varies with the kind of manual. A user's manual helps a user carry out a process or operate a system. A reference manual enables a system designer to find detailed design information quickly. With most manuals, however, you also have a secondary purpose: to motivate your readers. The mere fact that the reader will be holding a 200-page manual means that you have to persuade that person to learn what has to be learned. In many cases the reader is uncomfortable with the new system. Your job is to make the task seem less overwhelming. A well-designed manual, with plenty of white space and graphics, will help. See Chapter 12 for a discussion of page design.

The next step is outlining the manual. As discussed in Chapter 3, outlining starts with brainstorming. For manuals, brainstorming almost always takes more than the fifteen or twenty minutes required for small reports. Often, you will need several brainstorming sessions to plan the dozens of discussions and graphics. In addition, the brainstorming sessions usually involve all or most of the people on the documentation team.

Drafting

Drafting a manual is much the same as drafting any other kind of technical document. Sometimes one person collects information from all the people on the documentation team and creates a draft from it. Sometimes different subject-area specialists write their own drafts, which are then revised or rewritten by one person or, for very large manuals, the writing team.

During the drafting stage the graphics are created or assembled.

Revising

Because manuals are bigger and more complex than most other kinds of technical writing, revision is a more complex process. In fact, revision is a series of different kinds of checks: from checks of technical information down to checks of structure, organization, emphasis, and style.

Revising manuals usually involves field testing. Because a manual is, above all, a practical teaching document, you and the other members of the documentation team want to make sure it works effectively. The general approach is to try out the manual on people with backgrounds similar to those of the eventual readers. See Chapter 5 for more information on document testing.

If you field test the manual carefully, you can learn which parts are effective and which aren't. Then you revise the ineffective portions and retest the manual, repeating the process until the manual is as effective as it can be or until time or money runs out.

The Structure of the Manual

There is no single way to structure a manual, of course. However, many organizations have found through experience that three components make the manual easier to use and therefore more effective:

- ☐ the front matter
- ☐ the body
- ☐ the back matter

The Front Matter

The front matter, which helps readers understand the content of the manual and how to use it, consists of everything before the body. All manuals have a cover or title page and a table of contents. In addition, most manuals contain a preface and a how-to-use-this-manual section. Sometimes, however, these last two items are simply combined in an introduction.

Whether to use a cover or just a title page is determined by the manual's intended use and size. Manuals that will receive some wear and tear—such as maintenance manuals for oil-rigging equipment—will be given a hard cover, usually made of a water-resistant material. Manuals used around an office usually don't need hard covers, unless they are so big that they need extra strength.

The title page, usually designed by a graphic artist, contains at least the title of the manual:

How to Fill Out Credit Application Forms

Or:

Filer: An Electronic Database Program

Or:

The Murphy 4330 Bottling Machine: Maintenance Manual

In addition, the title page contains any other identifying information that will help the readers see that they have received the right manual. For example:

Filer: An Electronic Database Program

For the IBM PC and All Compatible Models

Or:

The Murphy 4330 Bottling Machine: Maintenance Manual

For 1986–1990 Models

Sometimes a title page names the authors of the manual. For more information on titles and title pages, see Chapters 7, 12, and 19.

The table of contents is especially important in a manual because readers need to find particular information quickly. Readers consult manuals for specific information more than they read straight through; therefore, they need an extensive table of contents. The headings in an effective table of contents are easy to understand, and they focus on the task the reader wants to accomplish. For example, notice in the following excerpt how the writer has used *-ing* verb phrases to highlight the action:

DELETING, INSERTING, AND MOVING CELLS	31
Deleting Cells	31
Inserting Cells	32
Inserting Blank Columns and Rows	32
Sorting Rows	33

Another effective way to phrase headings is to use infinitive phrases preceded by *how*:

HOW TO DELETE, INSERT, AND MOVE CELLS	31
How to Delete Cells	31
How to Insert Cells	32
How to Insert Blank Columns and Rows	32
How to Sort Rows	33

For more information on headings, see Chapter 7.

The other introductory information, which sometimes precedes and sometimes follows the table of contents, takes a number of forms. The first page of text in the manual, for instance, can be called "Introduction" or have no heading at all. Or the first page of text can be called "Preface," which is a brief statement describing the manual. A manual with a preface often has a section called "How to Use This Manual" or some similar phrase.

Conventions

The following conventions are used throughout this manual:

- **Bold** usually indicates a term defined in the Glossary.
- *Italic* is used for emphasis.
- COMPUTER type indicates text that appears on the printer's control panel display.
- *UPPERCASE ITALIC* is used to indicate a *variable* choice in a menu item.
- (KEY) or (Key) indicates a control panel key.

Note 👆 Notes contain important information set off from the text.

Caution ✋ Caution messages indicate procedures which, if not observed, could result in damage to equipment.

Warning ✊ Warning messages indicate when a specific procedure or practice is not followed correctly, personal injury could occur.

FIGURE 16.3   Sample Conventions Defined

Regardless of what you call this front matter, it must answer several basic questions for your reader:

- ☐ Who should use this manual?
- ☐ What product, procedure, or system does it describe?
- ☐ What is its purpose?
- ☐ What are its major components?
- ☐ How should it be used?

These questions need not be answered in separate paragraphs. Sometimes several of the answers can be combined in a single sentence or paragraph.

One other question needs to be answered in some manuals: what does the typography signify? If you will be using type creatively in the manual, and you want your readers to understand what your conventions will be, define them in the front matter. Figure 16.3 (an excerpt from *LaserJet IIP User's Guide* 1989, p. iv) shows how the writers of one manual define their conventions.

Following is an example of an introduction that answers all the basic questions. The manual is entitled "Travel Procedures for DeJohn Nursing Service Employees":

Introduction

As a DeJohn Nursing Service representative, you will spend approximately 40 percent of each year on business-related overnight travel. Like any other organization, DeJohn has a specific set of procedures and regulations that its representatives are expected to follow when preparing for, and after returning from, business trips.

This manual has been designed to familiarize you with DeJohn travel procedures. It is divided into two parts: (1) a detailed discussion of travel regulations and procedures that apply to all business trips and (2) a detailed discussion of additional procedures that apply to trips that include training seminars.

In addition, a glossary and samples of filled-out travel forms are included in the appendixes. The manual concludes with an index.

Read the entire manual before your first trip for DeJohn. The manual contains valuable information that will help you make your trip pleasant and productive.

If you have any questions, please call Ms. Calkins in Personnel, x3229.

The following example, from the tutorial manual for a software program, divides the same kinds of information into a preface and a how-to-use-this-manual section.

PREFACE

Congratulations. You are now the owner of Data-Ease, a powerful data-base program for the Apple Macintosh® computer. With Data-Ease, you will soon be able to manipulate data in ways you never thought possible. You will be able to enter, list, categorize, summarize, calculate, and report data.

This is the Tutorial Manual. It will show you how to use those functions of Data-Ease that can be accessed to menu selections. Most of you will find that menu selections will fulfill almost all your data-base management needs.

The other manual, the Reference Manual, discusses some additional features of Data-Ease: the text editor and the E-ZEE programming language.

Before you turn to the next page, How to Use the Data-Ease Tutorial Manual, fill out the attached License Agreement. When we receive your

License Agreement, you are a member of the Data-Ease Users Group, which will entitle you to receive information and updates on this software package and our other new releases.

Good luck!

HOW TO USE THE DATA-EASE TUTORIAL MANUAL

The Data-Ease Tutorial Manual is structured so that you can read it like a book or refer to it like a reference source.

Everyone should read "Data-Ease in a Snap," the introductory chapter that provides an overview of the program. Even if you have used other data-bases in the past, you'll find powerful new features described here—and you'll find out how to read more about them.

The Tutorial Manual is divided into three major sections:

Section 1. Learning the Elementary Commands
Section 2. Learning the More Advanced Commands
Section 3. Using More than One Relation

If you want to skip from section to section, here is a brief summary of each chapter:

Chapter 1. Data-Ease in a Snap. Tells you how to start the program, use the basic commands, add a record, and print a report.
Chapter 2. Creating a Report.
Chapter 3. Printing a Report.

(NOTE: Chapters 9 and 10, on how to edit reports and on the advanced features of the screens, tell you more about the report generator.)

If at this point you want to create your own applications, read the following three chapters:

Chapter 8. Using the Structure of Your Relations
Chapter 9. Modifying the Shape of Your Relations
Chapter 10. Creating Your Own Relations

Chapter 4. Modifying the Content of Your Relation. Tells you how to manipulate records within a single relation.

For information on how to use the Query function, read the following three chapters:

Chapter 5. Using the Query
Chapter 6. Using Small Groups of Records
Chapter 7. Entering Data. Tells you how to enter data into more than one relation at the same time.

To learn how to create simple and complex screens, read the following three chapters:

Chapter 12. Formatting a Data Entry Screen
Chapter 13. Creating a Data Entry Screen
Chapter 14. Advanced Features of Screens

A Few Notes about Punctuation

We use single quotation marks to set off commands. When you see double quotation marks in this manual, type them.

One other point: Standard punctuation puts commas and periods within quotation marks. For clarity, we put them outside. So if you see a comma or a period as part of a command, type it.

The Body

The body of a manual might look like the body of a traditional report, or it might look radically different. Its structure, style, and use of graphics will depend on its purpose and audience. The body of a manual might have summaries and diagnostic tests to help the readers determine whether they have understood the discussion.

The body of a manual is structured according to the way the reader will use it. For example, if the manual describes a process that the reader is supposed to carry out, the manual will probably be structured chronologically. If the manual describes a system that the reader is supposed to understand, the manual might be structured from more important elements to less important elements. The various organizational patterns discussed in Chapters 5 and 10 are appropriate, but, as always, the writing situation might call for another pattern.

A long manual might have more than one "body"; that is, each chapter might be a self-contained unit with its own introduction, body, and conclusion.

The writing style in a manual must be clear. Simple, short sentences work best. Use the imperative when giving instructions. For some kinds of manuals, especially those used by readers untrained in the subject, common vocabulary and a rather informal style are appropriate. Contractions, casual vocabulary, and a sense of humor are common. A software manual, for example, might have a section titled "The Couch Potato's Guide to Spreadsheets." However, the reference manual for a computer system probably would be more formal.

Graphics are as important in manuals as they are in instructions. They break up the text and, in many situations, are easier to understand than words. Whenever you want your readers to perform some action with their hands, include a diagram or photograph showing the action being performed. Whenever possible, use appropriate tables and figures. Deere & Co., the farm-equipment manufacturer, uses the term *illustration* to describe the writing style of its owner/operator manuals: each step is an illustration with a brief explanatory comment.

Following are two samples from the bodies of manuals. The first sample is from the manual describing travel procedures used at DeJohn Nursing Services. The excerpt is from the section on how to make hotel arrangements.

2.1.1 How to Make Hotel Reservations

As with ground transportation, it is your responsibility to make hotel reservations for yourself for each stop on your trip. If you are not familiar with your destination, check with your co-workers or consult the Hotel Guide. If one of your co-workers has been to the area, he or she might be able to suggest an acceptable hotel close to the exhibit or seminar site. The Hotel Guide lists most of the hotels in the United States, Canada, Mexico, and the Caribbean. The Guide is divided alphabetically by country, then further divided by state or province, and then city or town.

DeJohn prefers that, whenever possible, you stay in a moderately priced hotel that charges between $60 and $100 per night for a single room. This is not, however, an unbreakable rule. In some areas, an average room will cost less; in other areas, it will cost more. DeJohn does not expect you to stay in an unacceptable room, or in one located far from the exhibit site just to save money.

This sample from a manual is similar to any other kind of technical writing. There are no graphics because the subject does not lend itself to them. The writer is trying to explain an idea that cannot be visualized.

The first portion of a chapter from the user's guide for the Hewlett-Packard LaserJet IIP printer is shown in Figure 16.4. It is an example of effective organization and development.

Notice that the chapter begins with an introduction that explains how the optional lower cassette relates to the standard paper tray, the Multi-Purpose tray. The second paragraph is like a preface to a manual in that it helps the reader understand whether—and how—to read the chapter. Finally, the introduction concludes with a brief overview of the topics included in the chapter.

The next section, "Questions answered in this chapter," is a more detailed preview of the chapter, complete with page numbers so that a reader interested in only a particular point can skip ahead. The next section, "About the Optional Lower Cassette," provides the background needed to understand how the unit works. This section concludes with a table indicating how to use the unit effectively.

7

Using the Optional Lower Cassette

Introduction

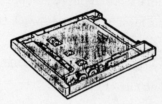

The basic LaserJet IIP printer comes with a Multi-Purpose (MP) tray that allows you to load up to 50 sheets of media. The Multi-Purpose tray can handle different types and sizes of media, giving you significant versatility in paper handling. In addition, you can purchase an optional Lower Cassette (LC) and tray that increases the paper-handling capacity of the printer and provides dual-bin capabilities.

This chapter is specifically about the optional Lower Cassette. The rest of this manual includes relevant information about the optional Lower Cassette when appropriate. *If your printer does not have an optional Lower Cassette, you can skip this chapter.*

This chapter includes information about:

- Configuring the control panel for the optional Lower Cassette.
- Loading paper into the optional Lower Cassette.
- Using different trays with the optional Lower Cassette.

Questions answered in this chapter

- How do I configure my printer so I can close the Multi-Purpose tray door and print just from the optional Lower Cassette? *See pages 7-3 and 7-6.*

- What types of media does the optional Lower Cassette support? *See page 7-16.*

- What additional trays are available for my optional Lower Cassette? *See page 7-17.*

- How many envelopes can I feed from the envelope tray? *See page 7-23.*

- Can I print on the back side of a piece of paper that has already been printed on by a LaserJet IIP printer? *Two sided printing (duplex) is NOT recommended.*

Using the Optional Lower Cassette 7-1

FIGURE 16.4   Using the Optional Lower Cassette

About the Optional Lower Cassette

The optional Lower Cassette comes with a booklet called *Optional Lower Cassette Installation Guide.* If you purchase your optional Lower Cassette with your printer, use this guide, along with the *Getting Started Guide* that comes with the printer, to set up your printer system. If you purchase your optional Lower Cassette after you have installed your printer, use the *Optional Lower Cassette Installation Guide* to set up your optional Lower Cassette, referring to the *Getting Started Guide* if necessary.

The optional Lower Cassette is not merely attached; rather it becomes an integral part of the printer. Because of this, the optional Lower Cassette enhances all the functionality available to the base printer. Installing this option adds a new control panel menu item TRAYS and new display messages. The printer commands on the *Technical Quick Reference Card* apply to the optional Lower Cassette as well as the base printer.

Table 7-1 lists the different ways you can use the optional Lower Cassette.

Table 7-1. Possible Uses for the Optional Lower Cassette

Use for jobs that:	The optional Lower Cassette:	Set TRAYS to:
Are large	Holds 250 sheets, increasing the paper capacity of the printer from 50 sheets to 300 sheets.	BOTH*
Use paper with different characteristics	Provides one kind of paper (such as plain) and the Multi-Purpose tray provides another kind (such as letterhead).	LC TRAY or MP TRAY*
Use two sizes of media	Provides one size (such as letter) and Multi-Purpose tray provides another size (such as envelopes).	BOTH*
Need only one paper source	Provides paper, allowing you to close up the Multi-Purpose tray and make the LaserJet IIP printer more compact.	LC ONLY*
Need an occasional manual feed	Feeds paper, with media occasionally being fed manually from the Multi-Purpose tray.	LC ONLY*

7-2   Using the Optional Lower Cassette

FIGURE 16.4   *Continued*

The Back Matter

The audience and purpose of the manual will determine what makes up the back matter. However, two items are common: a glossary and an index.

A glossary, discussed in Chapter 19, is an alphabetized list of definitions of crucial terms used in the document. An index is common for most manuals of twenty to thirty pages or more. The index, along with the table of contents, is a principal accessing tool. If a person wants to read about a particular item—such as the query function in a software package—the index will be the easiest way to find the references to it.

Creating an index is difficult because it involves deciding whether an item is important enough to be listed as a main entry or should be a subset of another entry. And making the entries consistent requires sustained concentration. Much of the labor involved in creating an effective index is being performed today by computers. Rather than having to read through the whole manual to find every reference to a term (such as "*Query function*") the indexer can simply use the search function to find the references. When the writers have used vocabulary consistently, this technique is very effective.

Appendixes to manuals can contain many different kinds of information. Procedures manuals often have flow charts or other graphics that picture the processes described in the body of the manual. These graphics can sometimes be removed from the manual and taped to the office wall for frequent reference. Users' guides often have diagnostic tests and reference materials—error messages and sample data lists for computer system, trouble-shooting guides, and the like. For a further discussion of appendixes, see Chapter 19.

The following excerpt (Plum 1987) is from a procedures manual for student aides working at the information desk in a university student center. The manual was written by a student. Although the table of contents covers the entire manual, only the first section of the manual is included here. Also included here is the letter of transmittal that accompanies the manual. For a discussion of letters of transmittal, see Chapter 19.

Brief marginal comments have been added.

403 Franklin Street
Philadelphia, PA 19104

December 3, 19XX

Mr. Thomas Williams
Director, Hamilton Student Center
Eastern University
Philadelphia, PA 19104

Dear Mr. Williams:

    I am enclosing a copy of the first draft of the <u>Hamilton Student Center Student Aide Manual</u> that I have written.

The writer explains the purpose of the manual.

    I wrote this manual because student aides who work at the Information Desk need a training and reference guide. Until now, we have had no such manual, and as a result staff members spend an inordinate amount of time training new student aides each quarter. This manual is designed to help shorten the training period.

This paragraph provides an overview of the manual.

    The manual is divided into three sections: Procedures, Resources, and Policies. Section 1 deals with the daily routine of the Information Desk, explaining the procedures that aides will be asked to perform. Section 2 discusses the resources available to student aides to help them work more efficiently. Section 3 sets forth the policies of the Hamilton Student Center. I have also included appendixes listing frequently called telephone numbers and answering frequently asked questions, along with floor plans of HSC, a map of campus, and a sample timesheet.

Here the writer explains why she has decided *not* to discuss a particular subject.

    I have deliberately omitted a discussion of the duties specific to the Game Room, because Game Room aides do little besides make change and provide a physical presence to

Mr. Thomas Williams
Page 2
December 3, 19XX

deter vandalism. However, because they are sometimes called upon to work at the Information Desk in the absence of the regular desk worker, I hope that this manual will be given to them as well as to the Information Desk aides.

    I have enjoyed writing this manual and welcome your suggestions for its revision. If you would like my help with other projects, such as a codification of the responsibilities of the Evening/Weekend Manager, please do not hesitate to contact me.

        Sincerely,

        Sandra D. Plum

        Sandra D. Plum

Enclosure

The writer concludes politely and offers her assistance with future projects.

HAMILTON STUDENT CENTER

STUDENT AIDE MANUAL

Prepared for: Thomas Williams, Director
Hamilton Student Center

by: Sandra D. Plum

December 3, 19XX

Notice the writer's use of capitalization, underlining, and indentation to clarify the hierarchical relationships among the various headings.

Notice too her use of leader dots to direct the reader's eye to the page number.

CONTENTS

An overview of the organization of the manual.

PREFACE

This manual is a training and reference guide to help you in your work at the Hamilton Student Center Information Desk.

This manual is divided into three sections.

- "Procedures" discusses the daily routine of the Information Desk, those duties that you will be called upon to perform on a regular basis.

- "Resources" lists some of the publications, phone numbers, and people available to help you do your job. The list is not complete but may help you discover some different places to look for answers to your own and other people's questions.

- "Policies" explains some general guidelines established by the Hamilton Student Center.

The appendixes should be quite useful to you and I hope that you will take time to look them over. Appendix A lists some frequently called phone numbers and Appendix B answers some frequently asked questions. Appendixes C and D are plans of Hamilton Student Center and the Eastern campus, respectively. Finally, Appendix E shows a sample timesheet used for payroll.

I hope this manual proves helpful to you in your job here at the Hamilton Student Center.

i

(Each first-level heading should start on a new page.)

The writer introduces the topics she will be describing in detail. Notice how the writer uses a wider left margin to emphasize that this is a subsection.

The writer clearly describes the reasons behind the procedures.

1.0 PROCEDURES

Your job as a student aide at the Information Desk involves many different tasks, some routine and some not at all routine. The more predictable things you are asked to do daily include signing in each time you work, answering the telephone, operating the cash register, signing out equipment, and doing photocopying.

1.1 Signing In

You <u>must</u> sign in each time you work. The sign-in sheets, located on the center post inside the Information Desk, are used by the secretary for calculating the payroll. If you do not sign in, you will not be paid.

There are two sign-in sheets. If you usually work in the Game Room, sign in on the sheet marked "Game Room"; if you usually work at the Information Desk, sign in on the "Main Desk" sheet. Even if you are covering a different area than you normally do, sign in on the sheet for the location to which you are normally assigned. This system facilitates bookkeeping because Game Room and Information Desk aides are paid from different payroll accounts.

Be sure to sign in on the correct line and for the correct days and times. If you are late, sign in for the time you do come in, and if you work overtime, sign out for the time you actually leave. Times must be recorded by the quarter-hour, however, so you may need to estimate slightly. (See Appendix E for a sample timesheet.)

1.2 Answering the Telephone

One of your primary duties is answering the telephone. Because many callers receive their first impression of Hamilton Student Center, and often Eastern itself, from the manner in which the Information Desk presents itself, it is very important to be courteous and friendly when answering the telephone.

Greet the caller by saying either, "Good morning (afternoon, evening), Hamilton Student Center," or "Hamilton Student Center. May I help you?" Do not merely state, "Hamilton Student Center." Some callers may be put off by the abruptness of this approach.

Respond to the person warmly; he or she is calling an information number and may need encouragement to state the question(s). Always strive to leave the caller with the feeling that HSC cares, and never hang up without either getting the information that the caller needs or referring him or her to a source that can help out (see Section 2.0, Resources).

2

1.2.1 Taking Messages

When a caller wishes to speak with a staff member who is unable to take the call, fill out a pink "WHILE YOU WERE OUT" phone message form. Be certain to record a telephone number where the caller can be reached, even if the person has left previous messages. (This prevents the problem of the recipient's needing to locate another message.) Put the message on the staff member's desk. If the call is for the Director or Assistant Director, and that person is busy, give the message to the secretary, and he or she will forward it when the recipient is available. Do not hold messages until the end of your shift.

1.2.2 Transferring Calls

There are two ways to transfer calls—one method for transferring calls within the office, and another for transferring calls to other offices within the University.

1.2.2.1 Inside the Office

Transferring calls to the secretary is simple because of the proximity of the telephones. Merely inform the secretary that a call is waiting and identify the incoming line (3515, 3516, or 3517).

3

However, transferring calls to the Director or Assistant Director is not so easy. First, try to ascertain if he or she is already on another call. If more than one button (line) on the telephone is lit, and you are not sure who is on the other line, check with the secretary before transferring the call. If neither of the other lines is lit, and you are reasonably sure that the staff member is both in the office and free to take a call, you may then transfer it by "buzzing" the office.

There are two buttons that buzz the offices. The button for the Assistant Director is on the telephone at the extreme right and is marked with her name. The button for the Director is mounted on the wall to the right of the telephone and is also marked.

To transfer a call, do the following:

1. Place the caller on hold by pressing the red "Hold" button firmly. The light on the extension line will begin to blink.
2. Press the appropriate buzzer briefly, using the following code:
   (a) one buzz for 3515
   (b) two buzzes for 3516
   (c) three buzzes for 3517
3. Check that the recipient picks up the phone. (The extension light will stop

4

For step-by-step instructions, the writer uses a numbered list in single-spaced format.

blinking.) If he or she does not, press in the extension button again, express your regret to the caller that you are unable to put through the call, and take a message so that the call may be returned later.

1.2.2.2 Outside the Office

If the caller wants to be connected to another office inside the University, first find out if he or she is telephoning from inside Eastern. If so, tell him or her the number of the extension desired—calls cannot be transferred from within the university itself. If the caller is from outside the university and would like to be transferred, do the following:

1. Tell the caller the extension number (complete with the Eastern prefix 695) to which you are transferring the call. Doing this will enable the caller to dial directly in case you accidentally disconnect the call.
2. Press the receiver button down firmly but briefly, and listen for a sound similar to a dial tone. This step must be repeated if you do not hear the sound.
3. Dial the extension number. If it rings, you may hang up. If the extension is busy, however, press the receiver again (which will reconnect you to the caller), ask the person to try the call again later, and repeat the telephone number if necessary.

5

1.3 Operating the Cash Register

A major service of the Information Desk is handling money transactions: providing change for visitors, staff, and students, selling parking tokens, and collecting fees for parking tickets and bowling. The cash register is simple to operate. You need only remember a few guidelines.

The cash register can be used as a storage device that does not monitor the flow of money through it, or it can be used to record (ring up) those transactions in which money is actually received over the amount that the drawer normally contains.

To use the register as a simple holding device, open it by pushing the "total" key only. To ring up a sum, do the following:

1. Enter the amount of the transaction on the tens, ones, tens-of-cents, and cents keys.
2. Push the appropriate numerical key for the account to which the sum is being deposited.
3. Hit the "total" key.

For example, to ring up $56.82 on key #1, you press the keys for $50 and $6 and 80¢ and 2¢, then key #1, and finally "total." The register drawer will open and the transaction will be recorded on the cash register tape inside the machine. The receipt button is kept depressed unless a receipt is needed. If the customer needs a receipt

6

of the transaction, flip the lever at the top right of the register before ringing up the transaction. Doing this will make the receipt button at the lower left of the register pop out, and the transaction will be recorded on a tape that will emerge from the receipt slot when you press the "total" key. If nothing has been rung up on a key, the receipt will list the amount as $0.00.

1.3.1 Making Change

When you make change, do not ring up the amount. Simply hit the "total" key, which will open the register, and substitute the denominations. If you need more change, ask the secretary or the desk clerk to get more from the safe. Always verify the amount you have received from the customer and count the change in front of the person.

1.3.2 Selling Tokens

The HSC Information Desk sells tokens for use in the Eastern Parking Garage. Each token costs $1.00, and two tokens are required for each stay at the facility. Do not ring up parking tokens; just open the register by hitting "total." The tokens are kept in a compartment at the far left side of the register drawer.

7

Money for tokens is also put in that compartment—not with the other cash.

1.3.3 Validating Parking Tickets

Parking tickets issued by Eastern University may be paid at the Information Desk. Each ticket must be validated by placing it in a slot at the left side of the register while the transaction is being rung up. Parking tickets are rung up on key #10.

To accept payment for a ticket, do the following:

1. Check the ticket for the total amount. This amount is usually $3.00 but may be much higher, depending upon the number and type of the outstanding tickets.
2. Flip the lever at the top right of the cash register in order to obtain a receipt for the ticket amount.
3. Place the ticket in the slot under the cash register at the left side. Be sure to push the ticket firmly against the register so the ticket will be properly validated.
4. Enter the amount of the ticket. Next hit key #10, and finally hit the "total" key.

The register drawer will open and a receipt will pop up from its slot. Put the cash or check and the validated ticket in the left-hand compartment where the tokens are stored. Give the cash-register receipt to the person as proof of payment.

8

WRITER'S CHECKLIST

The following questions cover instructions:

1. Does the introduction to the set of instructions
   a. state the purpose of the task?
   b. describe safety measures or other concerns that the readers should understand?
   c. list necessary tools and materials?

2. Are the step-by-step instructions
   a. numbered?
   b. expressed in the imperative mood?
   c. simple and direct?

3. Are appropriate graphics included?

4. Does the conclusion
   a. include any necessary follow-up advice?
   b. include, if appropriate, a trouble-shooter's guide?

The following questions cover manuals:

1. Does the manual include, if appropriate, a cover?

2. Does the title page provide all the necessary information to help readers understand whether they are reading the appropriate manual?

3. Is the table of contents clear and explicit? Are the items phrased to indicate clearly the task the reader is to carry out?

4. Does the other front matter clearly indicate
   a. the product, procedure, or system the manual describes?
   b. the purpose of the manual?
   c. the major components of the manual?
   d. the best way to use the manual?

5. Is the body of the manual organized clearly?

6. Are appropriate graphics included?

7. Is a glossary included, if appropriate?

8. Is an index included, if appropriate?

9. Are all other appropriate appendix items included?

10. Is the writing style clear and simple throughout the manual?

EXERCISES	**1.** Write a set of instructions for one of the following activities or for a process used in your field. Include appropriate graphics. In a brief note preceding the instructions, indicate your audience and purpose.

a. How to load film into a 35-mm camera
b. How to change a bicycle tire
c. How to parallel-park a car
d. How to study a chapter in a text
e. How to light a fire in a fireplace
f. How to make a cassette-tape copy of a CD
g. How to tune up a car
h. How to read a corporate annual report
i. How to tune a guitar
j. How to take notes in a lecture class

2. You work in the documentation department of a company that makes outdoor shutters. The engineering department head has just handed you the draft of a set of installation instructions that you see below. He has asked you to comment on their effectiveness. Write a memo to him, evaluating the instructions and suggesting improvements.

OUTDOOR SHUTTER INSTALLATION

1. Draw a 4″ to 6″ light vertical line from the top and bottom, 1″ from the sides of the shutter. Locate screw locations by positioning shutter next to window and marking screw holes. If six or eight screws are to be used, make sure they are equidistant.
2. Drill 3/16″ holes at marked locations. DO NOT DRILL INTO HOUSE. Drilling should be done on the ground or a workbench.
3. Mount the shutter screws.
   Notes: 1. For drilling into cement, use a cement bit.
   2. Always wear eye protection when drilling.
4. Make sure shutters are mounted with the top side up.
5. If you are going to paint shutters, do so before mounting them.

---

3. You work in the customer-relations department of the company that makes plumbing supplies. The products-development head has just handed you the draft of the installation instructions for a sliding tub door that you see on the next page. He has asked you to comment on their effectiveness. Write a memo to him, evaluating the instructions and suggesting improvements.

4. Write a brief manual for some process or system you are familiar with. In looking for topics, consider any activities you carry out at school. If, for instance, you are the business manager of the newspaper staff, you could write a manual describing how to perform the duties of that posi-

INSTALLATION INSTRUCTIONS

<u>CAUTION:</u> SEE BOX NO. 1 BEFORE CUTTING ALUMINUM HEADER OR SILL

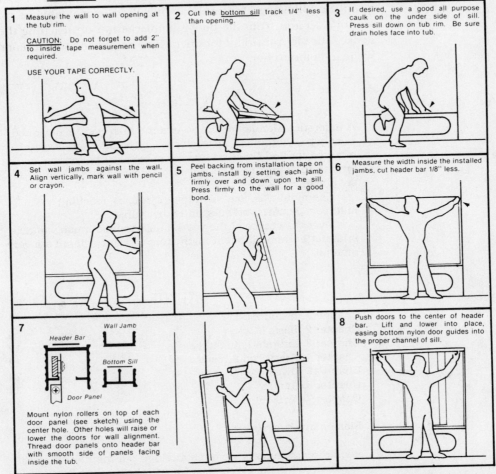

1 Measure the wall to wall opening at the tub rim.

<u>CAUTION:</u> Do not forget to add 2" to inside tape measurement when required.

USE YOUR TAPE CORRECTLY.

2 Cut the <u>bottom sill</u> track 1/4" less than opening.

3 If desired, use a good all purpose caulk on the under side of sill. Press sill down on tub rim. Be sure drain holes face into tub.

4 Set wall jambs against the wall. Align vertically, mark wall with pencil or crayon.

5 Peel backing from installation tape on jambs, install by setting each jamb firmly over and down upon the sill. Press firmly to the wall for a good bond.

6 Measure the width inside the installed jambs, cut header bar 1/8" less.

7

Header Bar

Wall Jamb

Bottom Sill

Door Panel

Mount nylon rollers on top of each door panel (see sketch) using the center hole. Other holes will raise or lower the doors for wall alignment. Thread door panels onto header bar with smooth side of panels facing inside the tub.

8 Push doors to the center of header bar. Lift and lower into place, easing bottom nylon door guides into the proper channel of sill.

TRIDOR MODEL ONLY:

To reverse direction of panels, raise panels out of bottom track and slide catches past each other thereby reversing direction so that shower head does not throw water between the panels.

HARDWARE KIT CONTENTS
TUDOR MODEL
4 nylon bearings
4 ball bearing screws # 8-32 × 3/8"
TRIDOR MODEL
6 nylon bearings
6 ball bearing screws # 8-32 × 3/8"

tion. Off-campus activities such as participation in civic groups or part-time jobs are other sources of ideas.

5. You work in the education department of a company that offers swimming and diving instruction. The department is currently revising the manual used by the company's swimming instructors. You have been given the following excerpt, including the preface, table of contents, and one section from the body: the discussion on teaching the backstroke. The vice president for operations has asked for your comments. Write a memo to her.

Preface

A Reference Manual for the Swimming Instructor is designed to help the swimming instructor teach the different strokes required for Red Cross certification. Because the manual is written in plain language, it is also a useful tool for beginning swimmers.

The manual covers the different strokes, teaching methods, games, workouts, and other subjects.

Over the course of the years, I have gained much valuable information from different instructors, to all of whom I am very grateful.

CONTENTS

Elementary Backstroke

Beginning swimmers are often afraid to put their heads in the water. Because the elementary backstroke involves keeping the swimmer's head out of the water at all times, it is a good stroke to teach first.

Initial Body Position

The swimmer starts in a horizontal, supine position. The back is kept almost straight, with the face always clear of the water. The arms are placed alongside the body, with the palms touching the thighs. The legs are together with the toes pointed forward. Keep the hips near the surface at all times.

Arm Stroke

The arm stroke for this stroke is continuous. At the start of the recovery phase, the palms are drawn upward alongside the body until they reach the armpits. The wrists are then rotated outward so that the fingers point away from the shoulders and the palms face the feet. A breath is taken at this point. Then the arms are put out straight toward the sides. The swimmer's body will look like a T. The student should not extend the arms or hands above the shoulder level. Then the palms and arms press back and down toward the feet. The student then recovers to the glide position.

Kick
    There are two kicks that can be used with the elementary backstroke. They are the frog kick and the inverted wedge.
    In the wedge kick, which is the favorite for all students except those with limited rotational ability in the hips or knees, the legs recover from the glide position by bending and separating at the knees. The heels move toward the bottom of the pool and toward the rear. The feet are relaxed with the heels almost touching. At the end of the recovery, the knees should be a little wider than the hips. The ankles should be dorsiflexed with the feet rotated outward.
    The next step is propulsion. The feet slide sideways. When they are under the knees they rotate slightly inwards. Then with a rounded, outward, backward, and then inward movement the feet and legs are brought back together. This movement starts slowly and then accelerates. The glide is then resumed. It is imperative that the emphasis be placed on the backward movement and not on the sideways movement.
    The frog kick must be performed in a plane horizontal to the surface. Recovery is begun by bending the legs and separating the knees laterally as the heels are drawn toward the rear. The heels are kept close to each other. Then the ankles are dorsally flexed and rotated outward so that the toes point away from and to the side of the body. The toes lead the legs to a fully extended position outside the hips. Without stopping, the legs press backward and inward with force. The glide position is then resumed.

REFERENCES

LaserJet IIP User's Guide. 1989. Boise, Id.: Hewlett-Packard Company.

Mackin, J. 1989. Surmounting the barrier between Japanese and English Technical Documents. *Technical Communication* 36, 4 (4Q), 346–351.

Muller, W. G. 1990. Usability research in the Navy. Quoted in Schriver, K. A., Document design from 1980 to 1989: Challenges that remain. *Technical Communication* 36, 4 (4Q), 316–331.

National Retail Hardware Association. 1976. *How to apply panels to stud walls.*

Perrin, N. 1989. Getting to know your synchronized input shaft. *New York Times,* December 24, 1989, Section 7, p. 18.

Plum, S. L. 1987. Hamilton Student Center student aide manual. Unpublished manual.

Technical Articles

Many popular magazines, from *Reader's Digest* to *Popular Mechanics* and *Personal Computing*, print journalistic articles about technical subjects. This chapter discusses technical articles, which in contrast are essays that are written by specialists—for other specialists—and that appear in professional journals. Technical articles are written principally to inform their readers, not to entertain them.

Types of Technical Articles

There are three basic kinds of technical articles:

☐ research articles
☐ review articles
☐ conference papers

A research article reports on a research project carried out by the writer in the laboratory or the field. A typical research article might describe the effect of a new chemical on a strain of bacterium, or the rate of contamination of a freshwater aquifer.

A review article, on the other hand, is an analysis of the published research on a particular topic. It might be, for example, a study of the advances in chemotherapy research over the last two years. A review article does not merely provide a bibliography of the important research; it also classifies and evaluates the work and perhaps suggests fruitful directions for future research.

A conference paper is the text of an oral presentation that the author gave at a professional conference. Because conferences provide a useful forum for communicating tentative or partially completed research findings, a paper printed in the "proceedings"— the published collection of the papers given at the meeting—is generally less authoritative and prestigious than an article published in a scholarly journal. However, proceedings do give the reader a good idea of the current research in the field.

Who Writes Technical Articles, and Why?

Technical articles are written by the people who actually perform the research or, in the case of review articles, by specialists who can evaluate the research of the others. Often, the only characteristic these authors share is that they themselves generated or gathered the information they communicate: a physicist at a research organization writes up her experiment in laser technology, just as a hospital administrator writes up his research on the impact of health maintenance organizations on the private hospital industry.

Technical articles are the basic means by which professionals communicate with each other. Most professional organizations hold national conferences (and some hold regional and international meetings as well) at which members present papers and discuss their collective interests. But meetings occur infrequently, and not all interested parties can attend. Only technical journals can provide a convenient and inexpensive way to transmit recent, authoritative information to all who need it. A technical article can be in a reader's hand less than two months after the writer sent it off to the journal; most books, which deal with broader subjects and contain more information than articles can, require at least two years and therefore are not useful for communicating new findings. And because it has been accepted for publication by a group of referees—an editorial board of specialists who evaluate its quality—a technical article is likely to be authoritative.

Because technical articles are a vital means of communication, their authors are highly valued and frequently rewarded by employers and professional organizations. In some professions—notably university teaching and research—regular and substantial publication is virtually a job requirement. In many other professions, occasional articles enhance the reputations and hasten the promotion and advancement of their authors.

Technical Articles and the Student

As an undergraduate you are unlikely to write an important technical article, although you might well write a journalistic article for a local publication about some subject you're studying in an advanced course or on a project one of your professors is working on. However, within a few years— probably far fewer than you think—you may find yourself applying for a job that requires publication in professional journals. Or you may realize while working on a project that you have perfected a new technique for carrying out a particular task and that your technique warrants publication.

In the meantime most of your contact with technical articles will be as a reader. Nonetheless, when you read technical articles, don't limit your attention to the information that they convey. Notice the conventions and strategies that they employ, keeping in mind that at some point your résumé might benefit from listing an article or two.

When the occasion for writing a technical article arises, you will want to know how to approach your task. Even a reader of technical articles might find some familiarity with publication procedures helpful: recognizing the editorial positions of the journals in your field and understanding how an author's article has come to be published may shed some light on apparent mysteries; moreover, knowing the standard structure of technical articles will help you skim possible sources for research.

Choosing a Journal

Having determined the topic of a possible article and perhaps having written a rough outline, a writer must think of a journal to which to submit the article. Why should anyone worry about placing an article that hasn't been written yet—wouldn't it be more logical to write it first and then find an appropriate journal? The answer would be yes if all journals operated in the same way. But they don't.

Most professional journals are run by volunteers. Although two or three persons at a journal might be salaried employees, the bulk of the work usually is done by eminent researchers and scholars whose principal goal is to strengthen the work in their fields. Consequently, these editors and their assistants are extremely serious about their work; they draw up and publish careful and comprehensive statements of purpose and editorial policies. They see themselves as addressing a particular audience, in a particular way, for a particular reason.

Editors expect a prospective author to take the journals they publish seriously and to follow their editorial policies. An article that does not follow stated editorial policies suggests that its author is careless, indifferent, or a bit smug. An article that violates the specified style of documentation may well be rejected without comment or explanation.

Sheer numbers contribute to this situation. Publication is so highly valued these days that few editors of reputable journals ever have a shortage of good articles from which to choose. The acceptance rate might be as low as 3 or 4 percent; rarely is it above 40 percent. Understandably, editors are unwilling to spend the time revising an article when they already have a dozen equally good ones that don't need stylistic revisions.

For these reasons, then, an author has more chance of success by choosing a journal *before* starting to write and then tailoring the article to its first audience—the editorial board of a particular journal.

What to Look For

If you plan to write a professional article, start in the library. You already know the leading journals in your field. Study a few recent issues of each.

First check the masthead, generally located on one of the first few pages, which includes the statement of ownership and editorial policy. The editorial board is often listed on the masthead as well. The editor is likely to have a column—also near the front of the issue—that sometimes discusses the kinds of articles the journal is (or isn't) looking for. Study the editor's comments and the journal's editorial policy carefully for crucial information that will save you considerable time, trouble, and expense. For instance, every once in a while a journal stops accepting submissions for a specified period (sometimes up to a year or two) because of a backlog of good articles. Of course, you would not write your article for that journal. Or you might learn that a particular journal does not accept unsolicited manuscripts: that is, it commissions all its articles from well-known authorities, or it publishes only staff-written articles. Again, this is not the journal for you. Or maybe a journal is planning a special issue devoted entirely to one subject. Some journals print *only* special issues, and they announce the subjects in advance. If you choose one of these journals, make sure your idea fits the subject before you write the article.

Not all the information in the editorial policy or the editor's column is negative, of course. Most journals *do* accept unsolicited articles and explain their requirements. Often, they will provide guidelines for sending in manuscripts (how to mail them, how many copies to send, etc.). Perhaps the most pertinent piece of information journals convey is the name of their preferred style sheet: the *Council of Biology Editors Style Manual, The Chicago Manual of Style*, the United States Government Printing Office *Style Manual*, or some other guide. If the journal specifies a style manual, find it and follow it. Editors notice these things immediately. Figure 17.1 provides an example of an editorial policy statement.

After you have studied a journal's editorial policies, read a number of articles to determine as much as you can about the following matters:

- *Article length.* Sometimes, journals print only articles that fall within a certain range, such as 4,000 to 6,000 words. Even if a range is not stated explicitly in the journal, the editors might have one in mind. If your article is going to be either much shorter or much longer than the average, you might want to write it for another journal.

- *Level of technicality.* Are the articles moderately technical or extremely technical? Do they include formulas, equations, figures?

- *Prose style.* How long are the paragraphs? Do the authors use the passive voice ("The mixture was added. . .") or active voice ("I added the mixture. . .")? Is the writing formal or somewhat informal?

- *Formal requirements.* Are the titles purely informative or are they "catchy"? Are subtitles included? Abstracts? Biographical sketches? Are the articles written with American or British spelling and punctuation?

The Query Letter

Once you are fairly certain of the journal you want to shape your article for, it's a good idea to find out what the editor thinks. Some editors will talk to you on the phone, but most prefer that you write a query letter.

The Sport Psychologist Editorial Statement

The Sport Psychologist (*TSP*) is published for educational sport psychologists (those who teach psychological skills to coaches and athletes) and for clinical sport psychologists (those who provide clinical services to athletes and coaches with psychological dysfunctions). The journal is also intended for those who teach sport psychology in academic institutions, and for coaches who have training in sport psychology. *TSP* focuses on the professional interests of sport psychologists as these pertain to the delivery of psychological services to coaches and athletes. It is international in scope, receptive to nonscientific methodologies, and refereed. In particular, *TSP* has the following sections:

Applied Research: *The Sport Psychologist* publishes research reports that focus directly upon the application of psychological services in sport. Case studies, field studies, and reports of evaluation research examining the effectiveness of educational and clinical psychological services in sport are highly appropriate. Articles about methodological and measurement issues are welcomed, as are reports that utilize qualitative and other emerging research methods.

Professional Practice: Published in this section are reviews that synthesize research findings into implications for practice. Articles are solicited that describe the methods used to teach the subject of sport psychology, including the teaching of psychological skills. Articles that report experiential learning, introspective observations, and clinical observations are also appropriate.

Profiles: This section publishes interviews with coaches, athletes, and other sport psychologists. It also contains editorials that address issues in the field, as well as feature stories on sport psychology in various countries.

Bulletin Board: This section reports international news in sport psychology, featuring news of the International Society of Sport Psychology, as well as news from national sport psychology societies. It also reports events that transpired at past conferences, announces forthcoming meetings, and lists new resources available to sport psychologists.

Books and Videos: Reviewed in this section are books and videos that are of interest to sport psychologists.

FIGURE 17.1   Editorial Policy Statement from *The Sport Psychologist*

Century Plastics

ALBION, MISSISSIPPI 34213

March 21, 19XX

Dr. William Framingham
Editor
Plastics Technology
1913 Vine Street
Fowler, MS 34013

Dear Dr. Framingham:

I would like to know if you would be interested in considering an article for Plastic Technology.

My work at Century Plastics over the last two years has concentrated on laminate foamboard premixes. As you know, one of the major problems with these premixes is the loss of the blowing agent. In my research, I devised a new compound that effectively retards these losses.

We are currently in the process of patenting the compound, and I would now like to share my findings. I think this compound will reduce fabrication costs by 15 percent, as well as improve the quality of the premix.

The article, which will be a standard report of a laboratory procedure, will run about 2,500 words. Currently I am a Senior Chemist at Century Plastics and the author of 14 articles in the leading plastics journals (including Plastics Technology, Winter 1982, Fall 1986, and Summer 1989).

If you think this article would interest your readers, please contact me at the above address or at (316) 437-1902.

Sincerely yours,

Anson Hawkins

Anson Hawkins, Ph.D.
Senior Chemist

FIGURE 17.2   Query Letter for an Article

A query letter is brief—usually less than a page. Your purpose is to answer concisely the following questions:

☐ What is the subject of the article?
☐ Why is the subject important?
☐ What is your approach to the subject (that is, will the article be the report of a laboratory procedure, a rebuttal of another article, etc.)?
☐ How long will the article be?
☐ What are your credentials?
☐ What is your phone number?

Figure 17.2 shows an example of a query letter.

Writing the Article

Most of this text has explained multiple-audience documents—those addressed both to people who know your subject and want to read all the details and to people who want to know only the basics. Although some readers of a technical article will skip one or several of the sections, for the most part you will be writing to fellow specialists who will read the whole article carefully. Because technical articles are intended for a national or international audience, a uniform structure has been developed to assist both writers and readers.

This structure is essentially the same as that of the *body* of the average report, except that an abstract generally precedes the article itself, with no other material intervening. The components of this structure are as follows:

☐ title of the article and name of the author
☐ abstract
☐ introduction
☐ materials and methods
☐ results
☐ discussion
☐ references

The following discussion describes each of these components and includes representative sections of a brief technical article from the journal *The Sport Psychologist* as illustration.

The sequence of composition is not the same as the sequence of presentation in a technical article. Many writers like to start with the materials-and-methods section, because that is relatively simple to write, and then proceed to the end of the article. Then they go back and write the introduction, for that section has to lead smoothly into the following sections. Finally, writers draft the abstract and the title.

The Development of Response Selection Accuracy in a Football Linebacker Using Video Training

Robert W. Christina, Jamie V. Barresi, and Paul Shaffner
The Pennsylvania State University

FIGURE 17.3   Title and Author Listing of a Technical Article

Title and Author

Technical articles tend to have long (and clumsy) titles, because of the need for specificity. Abstracting and indexing journals classify articles according to the key words in the title; therefore, be sure to include those terms that will enable the readers to find your article.

Following the title, put your name (and the name of any coauthors) and your institutional affiliation, as shown in Figure 17.3.

Notice that the title indicates what the article is about: the use of video training to help a football linebacker improve the accuracy of his response selection. Notice too in this title that the important nouns are modified by adjectives: "response selection accuracy," "football linebacker," and "video training." The title is extremely precise, which helps potential readers determine whether to study the article.

Abstract

Technical articles contain abstracts for the same reasons that reports do: to enable readers to decide whether to read the whole discussion. Most journals today prefer informative abstracts to descriptive ones (see Chapter 19 for a discussion of these two basic types). The following paragraph is the abstract for the article on football.

Abstract of a technical article

This study was undertaken to determine if response selection accuracy could be improved without sacrificing a football linebacker's response selection speed by practicing his response selection skills in relation to various offensive plays that were seen via a videotape from a viewing angle similar to what he would see in a game. The task required the linebacker to respond to the cues of the tight end and backfield play by manipulating a joystick as accurately and quickly as possible. The data revealed that there was an improvement in response selection accuracy without sacrificing response selection speed. This finding was interpreted as evidence that training using a videotape that displays a view of plays that is similar to what is seen in a game situation can be an effective method for improving the perceptual skills needed for response selection accuracy by a linebacker in a laboratory setting.

This abstract provides a clear overview of the article. The first sentence explains the purpose of the study: to determine whether viewing videotaped game situations can help a linebacker learn to make better choices without slowing down his reaction time. The second sentence describes the methods of the experiment; the third sentence, the results; the fourth, the conclusions the researchers drew from the results.

Introduction

An introduction to a technical article defines the problem that led to the investigation. To do so, the introduction generally must (1) fill in the background and (2) isolate the particular current deficiencies. Often, the introduction requires a brief review of the literature, that is, a summary of the major research, either on the background of the problem or on the problem itself. This literature review provides a context for the discussion that is to follow. Perhaps more important, the literature review gives credence to the article by demonstrating the author's thorough familiarity with the pertinent research.

Some writers like to end the introduction with a brief summary—usually just one sentence—of their main conclusion. This summary serves two functions: to enable some readers to skip parts of the discussion they don't need and to help other readers understand the discussion that follows.

The introduction to the article on football is reproduced here, with the paragraphs numbered. Paragraph 1 defines the critical terms *response selection* and *response execution* and explains the shortcomings of using traditional "game films." Paragraph 2 cites the literature supporting the idea that learning improves as the training comes to resemble the task to be performed. Paragraphs 3 and 4 cite the literature on video training in other sports; there apparently are no relevant studies about football. Paragraph 5 summarizes the researchers' hypothesis and forecasts the methods section that follows.

Introduction to a technical article

1 Many sports such as football require an athlete to select the correct response as quickly as possible and then execute it as accurately as possible. In these situations it is possible for the athlete to make an error in response selection and/or response execution (Schmidt, 1976, 1988). An error in response selection, which is the main focus of this study, occurs when the athlete chooses an incorrect response to execute, and an error in response execution occurs when the movements are not executed as planned. Response selection for a football linebacker is dependent on the player's ability to perceive the relevant cues generated by an opponent and to use them to select

the appropriate defensive response. There appears to be agreement among football coaches that improvement in response selection can be obtained not only through the traditional physical practice (i.e., football practice and drills) but also by cognitive study of opponents' actions on film. Currently, however, film study of opponents' actions involves a camera angle that provides a view from a press box located near the top of a stadium.

2   Although that viewing angle can be helpful in preparing a defensive player, it is inadequate because it does not allow the player to see the actions of an opponent from the same viewing angle that he is likely to experience in a game. Based on what is known about task similarity and the transfer of learning (for a review, see Fischman, Christina, & Vercruyssen, 1982) and what is recommended in terms of teaching for positive transfer (Christina & Corcos, 1988), it seems reasonable to argue that a player would be better served and more completely prepared if he could see the actions of an opponent from the same angle that he is likely to experience in a game. However, no evidence was found to support this argument for the sport of football.

3   Although the use of videotape or film that provides a player's view of the game in order to develop the perceptual skills necessary for accurate response selection in football was not found in the literature, it has been used in other sports such as tennis (Haskins, 1965), baseball (Burroughs, 1984), hockey (Salmela & Fiorito, 1979), and badminton (Abernethy, 1988; Abernethy & Russell, 1987; Jones & Miles, 1978). Haskins (1965) developed a training film to shorten the time necessary to perceive the direction of a tennis return. The training film showed a tennis player making a series of return shots to the opposite court. The view of the player was similar to what an opponent would see in a tennis match. Pre- and posttraining response times were determined from the analysis of film footage of actual match situations. The results revealed a significant improvement in response time after the use of the training film. Burroughs (1984) investigated the effectiveness of visual simulation training to enhance the visual pitch recognition and pitch location skills of collegiate batters. The film simulation showed a pitcher's delivery and pitch from the perspective of a right-handed batter. The results indicated that players who participated in the film simulation training session were significantly more accurate in perceiving the correct location of a pitch, but there were no significant improvements in the players' ability to distinguish between a fast straight ball and a breaking ball. Burroughs suggested that the absence of a training effect was possibly due to a ceiling effect, that is, the high pretest ability of the players, leaving little room for improvement.

4   Salmela and Fiorito (1979) used videotape to simulate the view of a goaltender in hockey. The videotape displayed a hockey player shooting a puck at the net (camera). Different conditions were set up by occluding different parts of the player's backswing prior to impact. In using this technique, Salmela and Fiorito found that goaltenders could accurately predict the location of various shots based only on the perceptual cues derived prior to impact. Abernethy (1988), Abernethy and Russell (1987), and Jones and Miles (1978) have also combined these simulation and occlusion techniques and determined differences in the use of perceptual cues between novice and expert badminton players, respectively.

5   Taken together, these studies suggest that training videos or films that provide a player's view of the game can be useful in identifying the relevant

cues used by inexperienced and experienced players and can have a positive training effect. Whether training videos or films have a similar effect on development of the perceptual skills needed for accurate response selection in the sport of football has yet to be determined. Thus the general intent of this study was to extend the previous research to the sport of football and, hence, increase the amount of motor learning research at applied levels as recommended by Christina (1987, 1989). Specifically, the purpose was to determine if training on a videotape simulation that (a) depicts a view of an offensive play executed against a defense and (b) is similar to what an outside linebacker sees in a game affects the response selection performance of an experienced outside linebacker in a laboratory setting. Essentially, the player viewed each play on the videotape and mentally practiced specific reactions to offensive formations and movements. It was reasoned that mental practice would reduce the attentional focus by emphasizing the cognitive-symbolic rather than the motor elements of the task (Feltz & Landers, 1983).

Materials and Methods

"Materials and methods" is a standard phrase used to describe the actual procedure the research followed. Basically, this section of the article has the same structure as a common recipe: these are the things you needed, and this is what you did. Often, the materials-and-methods section is clearly divided into its two halves, and the steps of the procedure are numbered and expressed in the past tense, without the subjects of the sentences (for example, "Mixed the. . .").

The materials-and-methods section of the *Sport Psychologist* article is presented here. Notice that the section is titled "Methods," not "Materials and methods." Even though a lot of equipment used in this study could be called "materials," that term would not be appropriate because the study uses a human subject, the football player.

The methods section of this article consists of three subsections: the subject and task, the apparatus, and the procedure itself. Subheadings such as these are common in technical articles because they clarify the organization of the section.

Materials-and-methods sections of a technical article

Method

Subject and Task

The male subject (22 years of age) was a senior outside linebacker on The Pennsylvania State University football team. He had 4 years of football experience in high school at various positions and 3 years of experience at the outside linebacker position at The Pennsylvania State University. We were informed by the football coaching staff that this linebacker appeared to be responding quickly enough in game situations but all too often was selecting the incorrect response to execute. Since the subject could adequately execute the physical aspect of his skill, it was thought that he could be helped in learning his assignments by reducing his attentional focus. Specifically, by isolating the response selection component of his task, a task and training program was designed in an attempt to enhance his response selection accuracy. The task required the subject to view a series of offensive players that were displayed on a video monitor. The plays were similar to those an outside linebacker would ordinarily experience in a game situation and they were seen from about the same visual angle. The subject was instructed to view the videotape simulation of each play and to respond as quickly and accurately as possible to the cues of the play by moving a joystick in the direction in which he would ordinarily move in a game situation.

Apparatus

The video display was recorded on an Eastman T-30 VHS tape cassette. Recording of the display was done with a Panasonic WV-555 camera. The visual angle subtended by this camera was approximately 54°. Recording of the display took place indoors at The Pennsylvania State University football practice facility. Seven members of the Penn State football team volunteered to participate in the video taping of 20 different offensive plays. The plays were scripted and the assignments of the tight end, fullback, tailback, quarterback, and offensive linemen were specified for each play.

The videotaping of a play was done in three steps. The first step involved placing the camera 1 yard opposite the tight end. The camera's position was at a height equivalent to eye level of the outside linebacker when he was in his defensive stance. Thus, from this position the subject had the viewing angle or perspective that was similar to what he would have from his outside linebacking position in a game situation. The tight end was videotaped in his stance for 3 seconds prior to his initial movement. When the 3 seconds had passed, a sound on the tape served as a marker indicating the beginning of the tight end's movement. The videotape footage then showed the tight end's release from his stance and movement from the line of scrimmage as specified by the play. This movement was in focus until the tight end passed the camera either to the right or left. The time from the tight end's first movement until he left the screen was recorded. This time of movement was a prerequisite for combining this sequence of videotape footage with the subsequent videotape footage of the backfield action.

The second part of the process involved videotaping the corresponding backfield and/or offensive line movement for each play. The camera was placed at a position that corresponded to the position of the outside linebacker had he made his first movement response correctly in relation to the tight end's movement. For instance, in most plays when the tight end steps in the direction of the tackle beside him (i.e., to the inside), the outside linebacker should step in the same direction. In the case of the tight end stepping toward the adjacent teammate, the camera was placed 1 yard (0.91 m) laterally in the direction of the tight end's movement and focused on the backfield.

The third part of the process involved combining the backfield videotape with the corresponding tight end videotape. Ordinarily with the tight end present, the outside linebacker cannot see all of the backfield action. Consequently, the time of the tight end's movement was noted and was deleted from the beginning of each backfield movement. This procedure provided the subject with a more realistic simulation because it meant that he would attend to the tight end and then be able to focus on the cues from the offensive backfield or the offensive line as they were executing their assignments. This was done to avoid the unrealistic situation of the outside linebacker attending to the initiation of both the tight end's movement and backfield's movement. In plays that did not involve a tight end, the camera was aligned as if a tight end were present but it was focused on the movements of the backfield. In these situations the 3-second premovement viewing was of the backfield. A sound marker on the tape corresponded with the first movement of the backfield. A constant foreperiod of 3 seconds was used for all plays. Although a constant foreperiod is not typical of what would occur in an actual game, it was used in this study to simplify the task and emphasize the response selection component of the task.

Following the videotaping session and splicing of videotape, each play was duplicated; this yielded a total of 40 plays, 20 different plays videotaped two times each. The 40 plays were assembled in a random order and a 20-second interval was placed between each one. The subject was seated at a desk of standard height in front of a television monitor (25-in. Zenith color receiver) and a VHS videocassette recorder (JVC, Model No. BR-3100U). A joystick (Joystick IIe, Model No. A2M2002) and the videocassette recorder were both connected to a digital oscilloscope (Nicolet Explorer III, Model No. 2090) that was placed on the right side of the television monitor. The sound signal corresponding to the beginning of each play was monitored by the videocassette recorder. This signal was relayed to the oscilloscope that initiated a timing device and recorded time data in milliseconds. The joystick was used to transmit the subject's movement signal to the oscilloscope and to indicate the response time of each movement. Each movement was displayed on the oscilloscope and was recorded and stored on a floppy disc.

Procedure

The subject trained 4 days a week for 4 weeks, which resulted in a total of 16 training sessions. The 16 sessions consisted of 8 practice and 8 test days that were alternated such that the first day was practice, the second day was test, and so forth. On the first practice day the subject was given instructions followed by a question and review period. Feedback was provided after each play on practice days but not on test days. The feedback after each play consisted of the subject's response time, response error, and response correction. The response times and the number of correct responses over the 40 plays were determined for each test day.

Each testing session consisted of 40 plays and was about 25 minutes in duration. The experimenter sat beside the subject and in front of the oscilloscope in order to observe the subject, record the data, and reset the oscilloscope after each play was shown. The subject was seated at a desk of standard height in front of the television monitor. He sat in a dimly lit room approximately 3 ft (0.91 m) from the monitor and his line of view was directly at the monitor. A joystick that could be moved in any direction was placed in front of him and was controlled with his right hand. The subject was instructed to hold the joystick and watch the monitor as each play was shown. His responses were made by moving the joystick in a direction that was compatible with the direction of the movements he would make in a game. For instance, if the tight end and backfield movement cues indicated that the subject should move to the right and then upfield, he was to move the joystick to the right and then forward away from his body. Similarly, plays that required movements to the left would be indicated by moving the joystick to the left. The movement of the joystick as well as the response time was displayed on the oscilloscope. Following every test session, the experimenter evaluated the discs and recorded response time and errors.

Results

The results of the procedure are what happened. Keep in mind that the word *results* refers only to the observable or measurable effects of the methods; it does not attempt to explain or interpret those effects.

Discussion

The discussion section answers two basic questions: (1) Why did the results happen? (2) What are the implications of the results? The two halves of the typical discussion section of an article are thus similar to the conclusion and recommendation sections of a report.

The football article uses a combined results-and-discussion section. Such a combination is common in brief technical articles.

This results-and-discussion section explains that the study confirmed the hypothesis: that the video training helped the linebacker learn to make better choices without hurting his response time.

Results-and-discussion
section of a technical
article

Results and Discussion

Correct responses were dependent on the cues of the tight end and the backfield. For example, if the tight end's initial movement was made in a lateral direction, toward the flank and along the line of scrimmage, the linebacker's correct response would be to move the joystick in a corresponding horizontal motion. Response errors were recorded when the subject's movements of the joystick in response to the cues of the tight end and backfield were not those movements specified by the cues. For instance, response error would occur if the subject moved the joystick vertically or in a direction opposite the movement of the tight end. The percentage of correct responses as a function of training days is shown in Figure 1. With the exception of Day 6, the subject made progressively more correct responses with each test day of training. This indicates that the accuracy of response selection improved with training. This is in agreement with what was reported by Burroughs (1984), who found significant improvements in a baseball batter's ability to correctly perceive the location of a pitched ball as a result of training with film that provided a view of a pitcher's delivery and pitch that was similar to what he would see in a game. What is not known, however, is whether the improvement in response selection accuracy in the present study was made at the expense of response selection speed. The answer to this question can be found by examining the response time means.

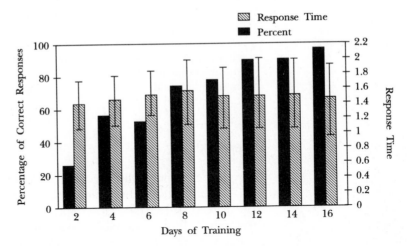

Figure 1. Percentage correct responses and mean response time (with standard deviations) as a function of day of training.

The response time for each play was the time from the onset of the first offensive cue until the initiation of the movement of the joystick for the second response. Thus it included both the reaction and movement time for the first cue and his reaction time to the second cue. In plays in which the tight end was present, reaction time for the first cue was the time from the onset of the tight end's movement until the subject's initiation of movement of the joystick. The reaction time for the second cue was the time from the onset of backfield movement to the initiation of movement of the joystick. In plays in which the tight end was not present, the reaction time for the first cue was the time from the onset of the backfield movement to the initiation movement of the joystick. Movement time for the first response was the duration from the onset of movement of the joystick to the completion of the movement. Figure 1 reveals that there was very little difference in the subject's response time as a result of 16 days of training.

Taken together, these results reveal that the subject's speed of responding was not appreciably affected by training. This finding was not too surprising since the subject had several years of experience at responding as quickly as possible to cues similar to those presented in the video. Thus it is very likely that he was already near or at his upper limit for the speed of responding to these video cues before training. The important point to note here is that although the subject's speed of responding stayed about the same with training, his accuracy of responding improved, as evidenced by the increase in percentage of correct responses with training. This clearly indicates that the subject was not trading speed (i.e., slowing down) to gain accuracy, but instead was maintaining his response speed while improving his response accuracy with training.

Thus, it was concluded that response selection accuracy of an experienced football linebacker could be improved significantly in a laboratory setting as a result of training by viewing videotape simulations of offensive plays that provided a viewing angle similar to what is seen in a game. Whether or not his improvement transferred from the laboratory to actual game conditions could not be determined because of an injury the linebacker sustained shortly after training and only two games into the season. However, two coaches were interviewed and, based on the linebacker's abbreviated season, they felt that his response selection accuracy had improved under game conditions following the 16 days of video training. However, further research is needed before the validity of their opinion regarding this transfer effect can be ascertained.

References

There are different styles for listing references, but each method has the same objective: to enable readers to locate the cited source, should they wish to read it. Journals either specify that writers use a particular style sheet (which in turn defines a style for references) or provide their own guidelines in the discussion of editorial policies near the front of the issue.

The format used in the football article highlights the names of the authors and the dates of publication of their works. The boldface number that follows each title of a journal article is the volume number; after the commas are the page numbers.

References section of a technical article

References

Abernethy, B. (1988). The effects of age and expertise upon perceptual skill development in a racquet sport. *Research Quarterly for Exercise and Sport*, **59**, 210–221.

Abernethy, B., & Russell, D.G. (1987). The relationship between expertise and visual search strategy in a racquet sport. *Human Movement Science*, **6**, 283–319.

Burroughs, W.A. (1984). Visual simulation training of baseball batters. *International Journal of Sport Psychology*, **15**, 117–126.

Christina, R.W. (1987). Motor learning: Future lines of research. In M. Safrit & H. Eckert (Eds.), *The cutting edge in physical education and exercise science research. American Academy of Physical Education Papers* (No. 20, pp. 26–41). Champaign, IL: Human Kinetics.

Christina, R.W. (1989). Whatever happened to applied research in motor learning? In J. Skinner, C. Corbin, D. Landers, P. Martin, & C. Wells (Eds.), *Future directions in exercise and sport science research* (pp. 411–422). Champaign, IL: Human Kinetics.

Christina, R.W., & Corcos, D.M. (1988). *Coaches guide to teaching sport skills*. Champaign, IL: Human Kinetics.

Feltz, D.L., & Landers, D.M. (1983). The effects of mental practice on motor skill learning and performance: A meta-analysis. *Journal of Sport Psychology*, **5**, 25–57.

Fischman, M.G., Christian, R.W., & Vercruyssen, M.J. (1982). Retention and transfer of motor skills. *Quest*, **33**, 181–194.

Haskins, M.J. (1965). Development of a response-recognition training film in tennis. *Perceptual and Motor Skills*, **21**, 207–211.

Jones, C.M., & Miles, T.R. (1978). Use of advance cues in predicting the flight of a lawn tennis ball. *Journal of Human Movement Studies*, **4**, 231–235.

Salmela, J.H., & Fiorito, P. (1979). Visual cues in ice hockey goaltending. *Canadian Journal of Applied Sport Sciences*, **4**(1), 51–59.

Schmidt, R.A. (1976). Control processes in motor skills. *Exercise and Sport Science Reviews*, **4**, 229–261.

Schmidt, R.A. (1988). *Motor control and learning: A behavioral emphasis* (2nd ed.). Champaign, IL: Human Kinetics.

Acknowledgments

The article on video training includes another section that has not been mentioned yet: an acknowledgment section expressing appreciation to persons or institutions that have assisted in the research.

Acknowledgment section
of a technical article

Acknowledgment

We wish to thank Jerry Sandusky, defensive coordinator of The Pennsylvania State University Football Team, for working with us on this project. Appreciation is also extended to John Palmgren for his technical assistance in developing the instrumentation.

Preparing and Submitting the Manuscript

Preparing and mailing the manuscript professionally will not guarantee that an article is accepted for publication, of course. However, an unprofessional approach will almost certainly ensure that it is *not* accepted.

Typing the Article

The article must be typed double-spaced. And remember, editors hate sloppy manuscripts, particularly ones that violate their stylistic guidelines. An unprofessional manuscript costs money to revise and shows bad manners.

One new area of concern is word processing. Many journals will not accept dot-matrix print, so use a letter-quality or laser printer.

Many journals prefer that writers send their articles electronically, using a modem. This saves money for the journal: no retyping and proofreading expenses.

If the journal has stated its preferences, follow them to the letter. Most journals have clear guidelines about margins, pagination, spacing, number of copies, and so on. Figure 17.4, for example, shows the instructions section from *The Sport Psychologist.*

If the journal has provided no typing guidelines, double-space everything and type on only one side of 8½-by-11-inch nonerasable bond paper. Use a good electric typewriter or a word processor with a letter-quality or laser printer. Also use a fresh ribbon. Send a photocopy with the original, and *make sure you keep a photocopy.* The percentage of manuscripts lost in the mail or in the journal's office is probably quite small, but when it happens to people who didn't keep a copy, they start to ask fundamental questions about life.

Instructions to Contributors

In preparing manuscripts for publication in *The Sport Psychologist*, authors should adhere to the guidelines in the *Publications Manual of the American Psychological Association* (3rd ed., 1983). Copies are found in most university libraries or may be obtained through the Order Department, American Psychological Association, 1200 17th St. N.W., Washington, D.C. 20036. All articles must be preceded by an abstract, not to exceed 150 words, typed on a separate page. Special attention should be given to the preparation and accuracy of references. The manuscript must be double-spaced including the abstract, references, and any block quotations. Manuscripts are subject to editing for sexist language. All figures must be professionally prepared and camera-ready; freehand and typewritten lettering will not be accepted. If photos are used, they should be black and white, clear, and glossy.

Three copies of the manuscript including the original should be submitted to either editor: Daniel Gould, Department of Exercise and Sport Science, University of North Carolina at Greensboro, Greensboro, NC 27412; or Glyn Roberts, University of Illinois, Department of Kinesiology, 906 S. Goodwin Ave., Urbana, IL 61801. All copies should be clear, readable, and on paper of good quality. A dot matrix or unusual typeface is acceptable only if it is clear and legible. Dittoed and mimeographed copies will not be considered. Manuscripts must *not* be submitted to another journal at the same time. Authors are advised to check carefully the typing of the final copy and to retain a copy of the manuscript to guard against loss. Manuscripts will not be returned to authors. Manuscripts are read by two reviewers, with the review process taking 8 to 12 weeks. There are no page charges to contributors. Authors of manuscripts accepted for publication must transfer copyright to Human Kinetics Publishers, Inc.

A blind review process is used to evaluate manuscripts. With each copy of the manuscript, authors are requested to submit a separate cover sheet including the title, name of author(s), institutional affiliation(s), running head, date of submission, and full mailing address and telephone number of the author who is to receive the galley proofs. The first page of the manuscript should omit the author's name and affiliation but include the title and the date of submission. Footnotes that identify the author should be typed on a separate page. Every effort should be made to see that the manuscript itself contains no clues to the author's identity.

FIGURE 17.4   Instructions to Contributors Statement from *The Sport Psychologist*

Brief reports are limited to 7 pages. Articles in the Applied Research and Professional Practice categories generally should not exceed 25 pages. They will be judged according to their applied focus, contribution to knowledge, presentation of information, appropriateness of the discussion, interpretation of ideas, and clarity of writing. In addition, Applied Research articles will be judged on their methodology/design and data analysis. Authors are expected to have their raw data and descriptive statistics available throughout the editorial review process and are responsible for providing elaboration upon request.

FIGURE 17.4   *Continued*

Writing a Cover Letter

Publishing conventions require a cover letter as a courtesy. This letter should be brief, for your purpose is not to "sell" the article; any boasting would probably work against you. The abstract will tell your readers everything they want to know.

If the editor has responded to a query letter, work this fact into the first paragraph. If you didn't inquire first, simply state that you are enclosing an article that you would like the editor to consider. In the next paragraph, you might wish to define briefly the subject or approach of your article and give its title if you haven't done so already. Then conclude the letter politely. Figure 17.5 shows a sample cover letter.

Submitting the Article

If you are sending the article, rather than submitting it electronically, follow the journal's mailing guidelines—about the number of copies to include, for example. Some journals request that you enclose a self-addressed, stamped envelope (SASE); others ask for loose postage. You might wish to state, in your cover letter, that you are enclosing an SASE or postage.

To keep the article from getting mangled, it's a good idea to enclose it between pieces of cardboard, especially if you are enclosing photographs or other artwork. If you are sending a total of about one hundred pages or more, use a padded envelope, which you can buy at the post office.

Some journals state that they acknowledge received manuscripts. If the journal you are writing to doesn't, enclose a self-addressed postcard and ask the editor to mail it after receiving your article.

Century Plastics

ALBION, MISSISSIPPI 34213

April 3, 19XX

Dr. William Framingham
Editor
Plastics Technology
1913 Vine Street
Fowler, MS 34013

Dear Dr. Framingham:

Thank you for replying promptly to my inquiry about the article on laminate foamboard premixes. As you requested, I am submitting the article for your consideration.

The article provides a step-by-step procedure for using Century Plastics' new compound in reducing blowing agent losses.

I look forward to learning your decision.

Sincerely yours,

Anson Hawkins, Ph.D.
Senior Chemist
(316) 437-1902

FIGURE 17.5 Cover Letter for a Technical Article

Receiving a Response from the Journal

When your article arrives at a journal, an editor reads it. If the subject matter is within the scope of the journal and the article appears authoritative and reasonably consistent with the specified stylistic guidelines, the editor will send it out to two or three specialists in your subject.

What the Reviewers Look For

There is little mystery about what reviewers look for when they read an article. They want to determine, basically, if the subject of the article is appropriate for publication in the journal, if the article is technically sound, and if it is reasonably well written. These three questions might be expanded as follows.

First, does the article fall within the scope of the journal? If it doesn't, the article will be rejected, regardless of its quality. If the article does fall within the journal's scope, the reviewers will want to see a convincing case that the subject is important enough to warrant publication.

Second, is the article sound? Is the problem defined clearly, and is the literature review comprehensive? Would the readers be able to reproduce your work on the basis of the methods-and-materials section? Are the results clearly expressed, and do they flow logically from the methods? Is the discussion clear and coherent? Have the proper conclusions been drawn from the results, and have the implications of your research been clearly suggested?

Third, is the article well written? Does it conform to the journal's stylistic requirements? Is it free of errors in grammar, punctuation, and spelling?

The Editor's Response

While the reviewers are studying your article, you wait—anywhere from three weeks to six months or a year. When you receive an acknowledgment, the editor will usually tell you how long the review process should take. If you haven't received an acknowledgment within three or four weeks, call or write the editor to find out if your manuscript arrived.

Eventually, you will receive a response from the editor: an acceptance, a request to revise the article and resubmit it, or a rejection.

- *Acceptance.* Very few articles—usually fewer than 5 percent—are accepted outright. If you are one of the select few, congratulations. The acceptance letter might tell you the issue the article will appear in. Some journals edit the author's manuscripts; you might be asked to review the edited manuscript and to proofread the galleys—the first stage of typeset proof, which is later corrected and cut into pages. Other journals don't edit the manuscripts or ask the authors to proofread.

- *Request to revise the article.* Before being published, most articles are revised according to the reviewers' recommendations. Sometimes reviewers call for a total rethinking of the article; sometimes, minor repairs. Some journals call this response a provisional acceptance: they promise to accept your article if you make the specified revisions.

Other journals merely invite you to send it in again. If the reviewers' comments seem reasonable to you, revise the article and send it in. If they don't, you might want to consider trying another journal. Keep in mind, however, that another journal might require other revisions.

■ *Rejection.* The majority of submitted articles are simply rejected. The journal will not print it and does not want to see another version of it. All rejections hurt. The worst offer no explanation: "We do not feel your article is appropriate for our journal." These curt rejections are frustrating, because you wonder if anyone read the article that you worked so hard on. Some journals include the reviewers' comments or excerpt them. If the criticism is valid, you don't feel too bad. If it isn't, you feel miserable.

Most experienced writers would advise that you develop a thick skin—but not too thick. Many important articles are rejected by several journals before they are accepted. Still, you should try to learn from the rejections. Study the reviewers' comments (if you received any) and try to react objectively. Consider it free advice, not a personal insult. If you haven't received any comments, analyze the article to see if you can discover any flaws. If you do, correct them and submit the article to another journal.

The first article is the hardest to write, and the one most likely to be rejected. After you publish two or three, you will develop a system that works well for you and learn to accept rejection as an inevitable part of publication. Knowing that most articles never get printed, you can take understandable pride in your successes.

WRITER'S CHECKLIST

1. Does the article fall within the scope of the journal?

2. Is the subject of the article sufficiently important to justify publication in the journal?

3. Does the abstract effectively summarize the problem, methods, results, and conclusions?

4. Is the article technically sound?
   a. Is the problem clearly defined?
   b. Is the literature review comprehensive?
   c. Is the materials-and-methods section complete and logically arranged? Would it be sufficient to enable researchers to reproduce the technique?
   d. Are the results clear? Do they seem to follow logically from the methods?
   e. Is the discussion clear and coherent? Has the writer drawn the proper conclusions from the results and clearly suggested the implications of the research?

5. Is the article reasonably well written?
   a. Does it conform (at least basically) to the journal's stylistic requirements?
   b. Is the writing clear and unambiguous? Is it free of basic errors of grammar, punctuation, and spelling? In short, can it be understood easily?

EXERCISES

1. You are an intern working for a major journal in your field. The editor has asked you to analyze the degree to which its contents correspond to its stated editorial policy. Write the editor a memo in response to this request. First, using the editorial policy statement, identify what the editor says the journal publishes. Consider such questions as range of subjects covered, type of article, writing style (paragraph and sentence length, technicality, etc.), and format (type of abstract, documentation style, etc.). Then, survey two or three articles in the issue. How closely do the articles reflect the editorial policy statement?

2. Write a memo to your instructor analyzing the quality of an article in your field, according to the criteria listed in the Writer's Checklist. Include a photocopy of the article.

REFERENCE

Christina, R. W., J. V. Barresi, and P. Schaffner. 1990. "The development of response selection accuracy in a football linebacker using video training." *Sport Psychologist* 4, 11–17.

18

Oral Presentations

Understanding the Types of Oral Presentations

Although you will occasionally be called on for impromptu oral reports on some aspect of your work, most often you will have the opportunity to prepare what you have to say. This chapter covers two basic kinds of oral presentations that are prepared in advance: the extemporaneous presentation and the scripted presentation. In an extemporaneous presentation, you might refer to notes or an outline, but you actually make up the sentences as you go along. Regardless of how much you have planned and rehearsed the presentation, you create it as you speak. A scripted presentation is completely written in advance; you simply read your text.

This chapter will not discuss the memorized oral presentation. Memorized presentations are not appropriate for most technical subjects because of the difficulty involved in trying to remember technical data. In addition, few people other than trained actors can memorize oral presentations of more than a few minutes.

The extemporaneous presentation is preferable for all but the most formal occasions. At its best, it combines the virtues of clarity and spontaneity. If you have planned and rehearsed your presentation sufficiently, the information will be accurate, complete, logically arranged, and easy to follow. And if you can think well on your feet without grasping for words, the presentation will have a naturalness that will help your audience concentrate on what you are saying, just as if you were speaking to each person individually.

The scripted presentation sacrifices naturalness for increased clarity. Most people sound stilted when they read from a text. But, obviously, a complete text increases the chances of saying precisely what you intend.

Understanding the Role of Oral Presentations

In certain respects, oral presentations are inefficient. For the speaker, preparing and rehearsing the presentation generally take more time than writing a document would. For the audience, physical conditions during the presentation — such as noises, poor lighting or acoustics, or an uncomfortable temperature — can interfere with the reception of information.

Yet oral presentations have one big advantage over written presentations: they permit a dialogue between the speaker and the audience. The listeners can offer alternative explanations and viewpoints, or simply ask questions that help the speaker clarify the information.

Oral presentations are therefore an increasingly popular means of technical communication. You can expect to give oral presentations to three different types of listeners: clients and customers, colleagues in your organization, and fellow professionals in your field.

Oral presentations are a valuable sales technique. Whether you are trying to interest clients in a silicon chip or a bulldozer, you will present its features and its advantages over the competition's. Then, after the sale, you will likely provide detailed oral operating instructions and maintenance procedures to the users. In both cases, extemporaneous presentations would probably be the more effective.

You might give similar information orally within your organization, too. If you are the resident expert on a mechanism or a procedure, you may have to instruct your fellow workers, both technical and nontechnical. After you return from an important conference or from an out-of-town project, your supervisors will, whether or not you submit a written trip report, want an opportunity to ask questions: you must anticipate them and prepare answers. If you have an idea for improving operations at your organization, you probably will first write a brief informal proposal, then present the idea orally to a small group of managers. In an hour or less, they can determine whether it is prudent to devote resources to studying the idea.

Oral presentations to fellow professionals generally are given at technical conferences and are often read from manuscript. You might speak on your own research project or on a team project carried out at your organization. Or you may be invited to speak to professionals in other fields. If you are an economist, for example, you might be invited to speak to realtors about interest rates.

The opportunities for oral presentations increase as you assume greater responsibility within an organization. For this reason, many managers and executives seek professional instruction. You might not have had much experience in public speaking, and perhaps your few attempts have been difficult. A few writers can produce effective reports without outlines or rough drafts, and a few natural speakers can talk "off the cuff" effortlessly. For most of us, however, an oral presentation, like a document, requires deliberate and careful preparation.

Preparing to Give an Oral Presentation

Preparing to give an oral presentation requires four steps:

- ☐ assessing the speaking situation
- ☐ preparing an outline or note cards
- ☐ preparing the graphics
- ☐ rehearsing the presentation

Assessing the Speaking Situation

The first step in preparing an oral presentation is to assess the speaking situation: audience and purpose are as important to oral presentations as they are to written reports.

Analyzing the audience requires careful thought. Who are the people who make up the audience? How much do they know about your subject? You must answer these questions in order to determine the level of technical vocabulary and concepts appropriate to the audience. Speaking over an audience's head puzzles them; oversimplifying and thereby appearing condescending insults them.

If the audience is relatively homogeneous—for example, a group of landscape architects or physical therapists—you can easily choose the correct level of technicality. If, however, the group contains individuals of widely differing backgrounds, you should parenthetically define technical words and concepts that some of your listeners likely don't know. Ask yourself the same kinds of questions you would ask about a group of readers: Why is the audience there? What do they want to accomplish as a result of having heard your presentation? Are they likely to be hostile, enthusiastic, or neutral? A presentation on the virtues of free trade, for instance, will be received one way by conservative economists and another way by American steelworkers.

Analyzing your purpose requires equally careful thought. Are you attempting to inform your audience, or to both inform and persuade them? If you are explaining how windmills can be used to generate power, you have one type of presentation. If you are explaining how *your* windmills are an economical means of generating power, you have another type.

Finally, the audience and purpose will affect the strategy—the content and the form—of your presentation. You might have to emphasize some aspects of your subject or ignore some altogether. You might have to arrange topics to accommodate the audience's needs.

As you are planning, don't forget the time allowed for your speech. At most professional conferences, the organizers clearly state a maximum time, such as 20 or 30 minutes, for each speaker. If the question-and-answer period is part of your allotted time, plan accordingly. Even at an informal presentation, you probably will have to work within an unstated time limit that you must figure out from the speaking situation. Claiming more than your share of an audience's time is rude and egotistical, and eventually they will simply stop paying attention to you.

How much material can you communicate in a given period of time? Most speakers need a minute or more to deliver a double-spaced page of text effectively. Regardless of whether you are making an extemporaneous or scripted presentation, keep this guideline in mind.

Preparing an Outline or Note Cards

After assessing your audience, purpose, and strategy, prepare an outline or a set of note cards. Keep in mind that an oral presentation should, in general, be simpler than a written version of the same material, for the listeners cannot stop and reread a paragraph they do not understand. Keep statistics and equations, for example, to a minimum.

You must structure the oral presentation logically. The organizational patterns presented in Chapter 10 are as useful for structuring an oral presentation as they are for structuring a written document.

Your introduction must gain and keep the audience's attention. An effective introduction might define the problem that led to the project, offer an interesting fact, or present a brief quotation from an authoritative figure in the field or a famous person not generally associated with the field (for example, Abraham Lincoln on the value of American coal mines). All these techniques should lead into a clear statement of your purpose, scope, and organization. If none of these techniques is appropriate, you can begin directly by defining the purpose, scope, and organization. A forecast of the major points is useful for long or complicated presentations. Don't be fancy. Use the words *scope* and *purpose*. And don't try to enliven the presentation by adding a joke. Humor is usually inappropriate in technical presentations.

The conclusion, too, is crucial, for it emphasizes the major points of the talk and clarifies their relationships with one another. Without a conclusion many otherwise effective oral presentations would sound like a jumble of unrelated facts and theories.

With these points in mind, write your outline or note cards just as you would for a written report. Your own command of the facts will determine the degree of detail necessary. Many people prefer a sentence outline (see Chapter 5) because of its specificity; others feel more comfortable with a topic outline, especially for note cards.

Figure 18.1 shows a combined sentence/topic outline, and Figure 18.2 shows a topic outline. (The outline could of course be written on note cards.) The speaker is a specialist in waste-treatment facilities. The audience is a group of civil engineers interested in gaining a general understanding of new developments in industrial waste disposal. The speaker's purpose is to provide this information and to suggest that his company is a leader in the field.

Notice the differences in both content and form between these two versions of the same outline. Whereas some speakers prefer to have full sentences before them, others find that full sentences interfere with spontaneity. (For presentations read from manuscript, of course, the outline is only preliminary to writing the text.)

OUTLINE

Describing, to a group of civil engineers, a new method of treatment and disposal of industrial waste.

I. Introduction
  A. The recent Resource Conservation Recovery Act places stringent restrictions on plant engineers.
  B. With neutralization, precipitation, and filtration no longer available, plant engineers will have to turn to more sophisticated treatment and disposal techniques.

II. The Principle Behind the New Techniques
  A. Waste has to be converted into a cementitious load-supporting material with a low permeability coefficient.
  B. Conversion Dynamics, Inc., has devised a new technique to accomplish this.
  C. The technique is to combine pozzolan stabilization technology with traditional treatment and disposal techniques.

III. The Applications of the New Technique
  A. For new low-volume generators, there are two options.
    1. Discussion of the San Diego plant.
    2. Discussion of the Boston plant.
  B. For existing low-volume generators, Conversion Dynamics offers a range of portable disposal facilities.
    1. Discussion of Montreal plant.
    2. Discussion of Albany plant.
  C. For new high-volume generators, Conversion Dynamics designs, constructs, and operates complete waste-disposal management facilities.
    1. The Chicago plant now processes up to 1.5 million tons per year.
    2. The Atlanta plant now processes up to 1.75 million tons per year.
  D. For existing high-volume generators, Conversion Dynamics offers add-on facilities.
    1. The Roanoke plant already complies with the new RCRA requirements.
    2. The Houston plant will be in compliance within six months.

IV. Conclusion
  The Resource Conservation Recovery Act will require substantial capital expenditures over the next decade.

FIGURE 18.1   Sentence/Topic Outline for an Oral Presentation

---

OUTLINE

Describing, to a group of civil engineers, a new method of treatment and disposal of industrial waste.

Introduction:
- Implications of the RCRA

Principle Behind New Technique
- reduce permeability of waste
- pozzolan stabilization technology and traditional techniques

Applications of New Technique
- for new low-volume generators (San Diego, Boston)
- for existing low-volume generators (Montreal, Albany)
- for new high-volume generators (Chicago, Atlanta)
- for existing high-volume generators (Roanoke, Houston)

Conclusion

FIGURE 18.2   Topic Outline for an Oral Presentation

Preparing the Graphics

Graphics fulfill the same purpose in an oral presentation that they do in a written one: they clarify or highlight important ideas or facts. Statistical data, in particular, lend themselves to graphical presentation, as do representations of equipment or processes. As always, graphics should be immediately clear and self-explanatory. In addition, they should present a single idea, not be overloaded with more information than the audience can absorb. Remember that your listeners have not seen the graphic before and won't get to linger over it.

In choosing a medium for the graphic, consider the room in which you will give the presentation. The people in the back and near the sides of the room must be able to see each graphic clearly and easily (a flip chart, for instance, would be ineffective in an auditorium). If you make a transparency from a page of text, enlarge the picture or words; what is legible on a printed page is usually too small to see on a screen. In general, 18- or 24-point type is best for transparencies. See Chapter 12 for more information on type size.

A good rule of thumb is to have a different graphic for every 30 seconds of the presentation. Changing from one to another helps you keep the presentation visually interesting, and it helps you signal transitions to your audience. It is far better to have a series of simple graphics than to have one complicated one that stays on the screen for 10 minutes.

After you have created your graphics, double-check them for accuracy and correctness. Spelling errors are particularly embarrassing when the letters are six inches tall.

Following is a list of the basic media for graphics, with the major features cited.

- *Computer imaging:* images appear on a computer screen or on a projector attached to the computer, or can be transferred to slides, videotape, or film.

  Advantages

  It has a very professional appearance.
  The speaker can program the graphics to produce any combination of static or dynamic images, from simple graphs to sophisticated, three-dimensional images.
  The speaker can modify the image live.

  Disadvantages

  The equipment is not commonly available.
  Preparing the graphics is very time consuming.

- *Slide projector:* projects previously prepared slides onto a screen.

  Advantages

  It has a very professional appearance.
  It is versatile — can handle photographs or artwork, color or black-and-white.
  With a second projector, the pause between slides can be eliminated.
  During the presentation, the speaker can easily advance and reverse the slides.

  Disadvantages

  Slides are expensive.
  The room has to be kept relatively dark during the slide presentation.

- *Overhead projector:* projects transparencies onto a screen.

  Advantages

  Transparencies are inexpensive and easy to draw.
  Speakers can draw transparencies "live."
  Overlays can be created by placing one transparency over another.
  Lights can remain on during the presentation.
  The speaker can face the audience.

  Disadvantages

  They are not as professional-looking as slides.
  Each transparency must be loaded separately by hand.

■ *Opaque projector:* projects images on paper onto a screen.

Advantages

It can project single sheets or pages in a bound volume.
It requires no expense or advance preparation.

Disadvantages

The room has to be kept dark during the presentation.
It cannot magnify sufficiently for a large auditorium.
Each page must be loaded separately by hand.
The projector is noisy.

■ *Poster:* a graphic drawn on oak tag or other paper product.

Advantages

It is inexpensive.
It requires no equipment.
Posters can be drawn or modified "live."

Disadvantage

It is ineffective in large rooms.

■ *Flip chart:* a series of posters, bound together at the top like a loose-leaf binder; generally placed on an easel.

Advantages

It is relatively inexpensive.
It requires no equipment.
The speaker can easily flip back and forth.
Posters can be drawn or modified "live."

Disadvantages

It is ineffective in large rooms.

■ *Felt board:* a hard, flat surface covered with felt, onto which paper can be attached using doubled-over adhesive tape.

Advantages

It is relatively inexpensive.
It is particularly effective if speaker wishes to rearrange the items on the board during the presentation.
It is versatile — can handle paper, photographs, cutouts.

Disadvantage

It has an informal appearance.

- *Chalkboard*

  Advantages

  It is almost universally available.
  The speaker has complete control—he or she can add, delete, or modify the graphic easily.

  Disadvantages

  Complicated or extensive graphics are difficult to create.
  It is ineffective in large rooms.
  It has a very informal appearance.

- *Objects:* such as models or samples of material that can be held up or passed around through the audience.

  Advantages

  They are very interesting for the audience.
  They provide a very good look at the object.

  Disadvantages

  Audience members might not be listening while they are looking at the object.
  It can take a long while to pass an object around a large room.
  The object might not survive intact.

- *Handouts:* photocopies of written material given to each audience member.

  Advantages

  Much material can be fit on paper.
  Audience members can write on their copies and keep them.

  Disadvantage

  Audience members might read the handout rather than listen to the speaker.

A word of advice: before you design and create any graphics, make sure the room in which you will be giving the presentation has the equipment you need. Don't walk into the room carrying a stack of transparencies only to learn that there is no overhead projector. Even if you have arranged beforehand to have the necessary equipment delivered, check to make sure it is there; if possible, bring it with you.

Rehearsing the Presentation

Even the most gifted speakers have to rehearse. It is a good idea to set aside enough time to rehearse your speech several times. For the first rehearsal of an extemporaneous presentation, don't worry about posture or voice projection. Just try to compose your presentation out loud with your outline before you. In this first rehearsal, your goal is to see if the speech makes sense—if you can explain all the points you have listed and can forge effective transitions from point to point. If you have any trouble, stop and try to figure out the problem. If you need more information, get it. If you need a better transition, create one. You might well have to revise your outline or notes. This is very common and no cause for alarm. Pick up where you left off and continue through the presentation, stopping again where necessary to revise the outline. When you have finished your first rehearsal, put the outline away and do something else.

Come back again when you are rested. Try the presentation once more. This time, it should flow more easily. Make any necessary revisions in the outline or notes. Once you have complete control over the structure and flow, check to see if you are within the time limits.

After a satisfactory rehearsal, try the presentation again, under more realistic circumstances—if possible, in front of people. The listeners might offer constructive advice about parts they didn't understand or about your speaking style. If people aren't available, use a tape recorder and then evaluate your own delivery. If you can visit the site of the presentation to get the feel of the room and rehearse there, you will find giving the actual speech a little easier.

Once you have rehearsed your presentation a few times and are satisfied with it, stop. Don't attempt to memorize it—if you do, you will surely panic the first time you forget the next phrase. During the presentation, you must be thinking of your subject, not about the words you used during the rehearsals.

Rehearse a written-out presentation in front of people, too, if possible, or use a tape recorder.

Giving the Oral Presentation

Most professional actors freely admit to being nervous before a performance, so it is no wonder that most technical speakers are nervous. You might well fear that you will forget everything or that nobody will be able to hear you. These fears are common. But signs of nervousness are much less apparent to the audience than to the speaker. And after a few minutes most speakers relax and concentrate effectively on the subject.

All this sage advice, however, is unlikely to make you feel much better as you wait to give your presentation. Take several deep breaths as you sit there; this will help relieve some of the physical nervousness. When the moment arrives, don't jump up to the lectern and start speaking quickly. Walk up slowly and arrange your text, outline, or note cards before you. If water is available, take a sip. Look out at the audience for a few seconds before you begin. It is polite to begin formal presentations with "Good morning" (or "Good afternoon," "Good evening") and to refer to the officers and dignitaries present. If your name has not been mentioned by the introductory speaker, identify yourself. In less formal contexts, just begin your presentation.

So the audience will listen to you and have confidence in what you are saying, project the same attitude that you would in a job interview: restrained self-confidence. Show interest in your topic and knowledge about your subject. You can convey this sense of control through your voice and your body.

Your Voice

Inexperienced speakers encounter problems with five aspects of vocalizing: volume, speed, pitch, articulation, and nonfluencies.

- *Volume.* The acoustics of rooms vary greatly, so it is impossible to be sure how well your voice will carry in a room until you have heard someone speaking there. In some well-constructed auditoriums, speakers can use a conversational volume. Other rooms require a real effort at voice projection. An annoying echo plagues some rooms. These special circumstances aside, in general more people speak too softly than too loudly. After your first few sentences, ask if the people in the back of the room can hear you. When people speak into microphones, they tend to bend down toward the microphone and end up speaking too loudly. Glance at your audience to see if you are having volume problems.

- *Speed.* Nervousness makes people speak more quickly. Even if you think you're speaking at the right rate, you might be going a little too fast for some of your audience. Remember, you know where you're going. Your listeners, however, are trying to understand new information. For particularly difficult points, slow down for emphasis. After finishing a major point, pause before beginning the next point.

- *Pitch.* In an effort to control their voices, many speakers end up flattening their pitch. The resulting monotone is boring—and, for some listeners, actually distracting. Try to let the pitch of your voice go up or down as it would in a normal conversation. In fact, experienced speakers often exaggerate pitch variations slightly.

- *Articulation.* The nervousness that goes along with an oral presentation tends to accentuate sloppy pronunciation. If you want to say the word "environment," don't say "envirament." Say "nuclear," not "nucular." Don't drop final g's. Say "trying," not "tryin'." A related pronunciation problem concerns technical words and phrases—especially the important ones. When a speaker uses a phrase over and over, it tends to get clipped and becomes difficult to understand. Unless you articulate carefully, "Scanlon Plan" will end up as "Scanluhplah," or "total dissolved solids" will be heard as "toe-dizahved sahlds."

■ *Nonfluencies.* Avoid such meaningless fillers as "you know," "okay," "right," "uh," and "um." These phrases do not disguise the fact that you aren't saying anything; they call attention to it. A thoughtful pause is better than an annoying verbal tic.

Your Body

Besides listening to what you say, the audience will be looking at you. Effective speakers know how to use their bodies to help the listeners follow the presentation.

Your eyes are perhaps most important. People will be looking at you as you speak; it is only polite to look back. This is called "eye contact." For small groups, look at each listener randomly; for larger groups, be sure to look at each segment of the audience frequently during your speech. Do not stare at your notes, at the floor, at the ceiling, or out the window. Even if you are reading your presentation, you should be well enough rehearsed to allow the frequent eye contact that indicates how the audience is receiving your presentation. You will be able to tell, for instance, if the listeners in the back are having trouble hearing you.

Your arms and hands also are important. Use them to signal pauses and to emphasize important points. When referring to graphics, point to direct the audience's attention.

Avoid mannerisms—those physical gestures that serve no useful purpose. Don't play with your jewelry or the coins in your pocket. Don't tug at your beard or fix your hair. These nervous gestures can quickly distract an audience from what you are saying. Like verbal mannerisms, physical mannerisms are often unconscious. Constructive criticism from friends can help you pinpoint them.

Answering Questions After the Presentation

On all but the most formal occasions, an oral presentation is followed by a question-and-answer period.

When you invite questions, don't abruptly say, "Any questions?" This phrasing suggests that you don't really want any questions. Instead, say something like this: "If you have any questions, I'd be happy to try to answer them now." If asked politely, people will be much more likely to ask; therefore, you will more likely communicate your information effectively.

In fielding a question, first make sure that everyone in the audience has heard it. If there is no moderator to do this job, you should ask if people have heard the question. If they haven't, repeat or paraphrase it yourself, perhaps as an introduction to your response: "Your question about the relative efficiency of these three techniques. . . ."

If you hear the question but don't understand it, ask for a clarification. After responding, ask if you have answered the question adequately.

If you understand the question but don't know the answer, tell the truth. Only novices believe that they ought to know all the answers. If you have some ideas about how to find out the answer — by checking a certain reference text, for example — share them. If the question is obviously important to the person who asked it, you might offer to meet with him or her after the question-and-answer period to discuss ways to give a more complete response, such as through the mail.

If a belligerent member of the audience rejects your response and insists on restating his or her original point, offer politely to discuss the matter further after the session. This strategy will prevent the person from boring or annoying the rest of the audience.

If it is appropriate to stay after the session to talk individually with members of the audience, offer to do so. Don't forget to thank them for their courtesy in listening to you.

SPEAKER'S CHECKLIST

This checklist covers the steps involved in preparing to give an oral presentation.

1. Have you assessed the speaking situation — the audience and purpose of the presentation?

2. Have you determined the content of your presentation?

3. Have you shaped the content into a form appropriate to your audience and purpose?

4. Have you prepared an outline or note cards?

5. Have you prepared graphics that are
   a. clear and easy to understand?
   b. easy to see?

6. Have you made sure that the presentation room will have the necessary equipment for the graphics?

7. Have you rehearsed the presentation so that it flows smoothly?

8. Have you checked that the presentation will be the right length?

EXERCISES

1. Prepare a five-minute presentation, including graphics, for one of the following contexts. For each presentation, your audience is the other students in your class, and your purpose is to introduce them to an aspect of your academic field.
   a. Define a key term or concept in your field.
   b. Describe how a particular piece of equipment is used in your field.
   c. Describe how to carry out a procedure common in your field.

2. Prepare a five-minute oral version of your proposal for a final report topic.

3. After the report is written, prepare a five-minute oral presentation based on your findings and conclusions.

4. Write a memo to your instructor in which you analyze a recent oral presentation of a guest speaker at your college or a politician on television.

PART

FOUR

Technical Reports

Formal Elements
of a Report

Because most reports today reach people with different backgrounds and needs, the structure and organization must make information accessible. A comprehensive table of contents and a well-written executive summary are two ways to do so.

This chapter discusses the traditional formal elements of a report in business, government, and industry. Few reports will have all the elements in the order in which they are covered here; most organizations have their own format preferences. You should therefore study the style guide used in your organization. If there is no style guide, study a few of the reports in your organization's files. Successful reports are the best teaching guides; ask a colleague to suggest some samples.

The following elements will be discussed here:

☐ letter of transmittal
☐ title page
☐ abstract
☐ table of contents
☐ list of illustrations
☐ executive summary
☐ glossary and list of symbols
☐ appendix

Two crucial elements of a report are not discussed in this chapter: the body, which varies according to the type of report, and the documentation, which can also vary. For discussions of the body, refer to the individual chapters on the various types of reports: proposals (Chapter 20), progress reports (Chapter 21), and completion reports (Chapter 22). For documentation—the system of citing sources of information used in a report—see Appendix B.

Writing the Formal Elements

Chapter 5 discusses the writing process: the activities that begin with choosing a topic and then proceed to brainstorming, outlining, drafting, and finally revising. That chapter argues that most technical writers do not write a document straight through from the first sentence to the last.

The same concept applies to the formal elements of a report. The components are not written in the same order in which they appear. The letter of transmittal, for instance, is the first thing the principal reader sees, but it was probably the last to be created. The reason is simple: the transmittal letter cannot be written until the document to which it is attached has been written. Many writers like to include the title of the document in the first paragraph of the transmittal letter; consequently, even the title has to be decided upon before the letter can be written. In the same way, the title cannot be chosen until the body of the report is complete, because the title must reflect accurately the contents of the report.

Usually, the body of the report is written before any of the other formal elements. After that, the sequence makes little difference. Many writers create the two summaries—the executive summary and the abstract—then the appendixes, the glossary and list of symbols, and finally the table of contents, title page, and letter of transmittal.

As you become a more experienced writer, you will develop strategies that work best for you.

Using the Word Processor

A word processor simplifies assembling the formal elements of a report.

For one thing, with a word processor you have a much better sense of the length of the various elements because your draft is typed, not handwritten. Some formal elements have length restrictions. A word processor lets you see exactly how long each element is so that you can expand or contract it to meet your requirements.

In addition, the copy function on a word processor helps you create the different summaries and the transmittal letter. You can make a copy of the body of the report and then eliminate the details that do not belong in the particular formal element you are creating. For instance, if the problem statement in the body is two paragraphs long, you can reduce it to a few sentences for the executive summary by eliminating most of the technical details. If the methods section will not be treated in the transmittal letter, you can simply erase it.

Creating the formal elements by cutting material from the body is not only faster than writing them from scratch, it is also more accurate because you don't introduce any technical errors. If in the body of the report you say that the Library of Congress catalogs 180,000 books each year, the number will remain 180,000—not be changed to 18,000 or 1,800,000—in the executive summary.

Writing the Letter of Transmittal

The letter of transmittal introduces the purpose and content of the report to the principal reader. The letter is attached to the report or simply placed on top of it. Even though the letter might contain no information that is not included elsewhere in the report, it is important, because it is the first thing the reader sees. It establishes a courteous and graceful tone for the report. Letters of transmittal are customary even when the writer and the reader both work for the same organization and ordinarily communicate by memo.

The letter of transmittal lets you emphasize whatever your reader will find particularly important in the report. It also enables you to point out any errors or omissions. For example, you might want to include some information that was gathered after the report was typed or printed.

Transmittal letters generally contain most of the following elements:

- ☐ a statement of the title and the purpose of the report
- ☐ a statement of who authorized or commissioned the project and when
- ☐ a statement of the methods used in the project (if they are noteworthy) or of the principal results, conclusions, and recommendations
- ☐ an acknowledgment of any assistance you received in preparing the materials
- ☐ a gracious offer to assist in interpreting the materials or in carrying out further projects

Figure 19.1, on page 516, provides an example of a transmittal letter. (For a discussion of letter format, see Chapter 14.)

Writing the Title Page

The only difficulty in creating the title page is to think of a good title. The other usual elements—the date of submission and the names and positions of the writer and the principal reader—are simply identifying information.

A good title informs without being unwieldy. It answers two basic questions: What is the subject of the report? and What type of report is it? Examples of effective titles follow:

Choosing a Microcomputer: A Recommendation

An Analysis of the Kelly 1013 Packager

Open Sea Pollution-Control Devices: A Summary

Note that a convenient way to define the type of report is to use a generic term—such as *analysis, recommendation, summary, review, guide,* or *instructions*—in a phrase following a colon. For more information on titles, see Chapter 7.

If you are creating a simple title page, center the title about a third of the way down the page. Then add the reader's and the writer's names and positions, the organization's name, and the date. Tags such as "Prepared for" are common. Figure 19.2 provides a sample of a simple title page.

ALTERNATIVE ENERGY, INC.

Bar Harbor, ME 00314

April 3, 19XX

Rivers Power Company
15740 Green Tree Road
Gaithersburg, MD 20760

Attention: Mr. J. R. Hanson
             Project Engineering Manager

Subject: Project # 619-103-823

Gentlemen:

We are pleased to submit "A Proposal for the Riverfront
Energy Project" in response to your request of February 6,
19XX.

The windmill design described in the attached proposal uses
the most advanced design and materials. Of particular note is
the state-of-the-art storage facility described on pp. 14–17. As
you know, storage limitations are a crucial factor in the
performance of a generator such as this.

In preparing this proposal, we inadvertently omitted one
paragraph on p. 26 of the bound proposal. That paragraph is
now on the page labeled 26A. We regret this inconvenience.
If you have any questions, please do not hesitate to call us.

Yours very truly,

Ruth Jeffries

Ruth Jeffries
Project Manager

RJ/fj
Enclosure

FIGURE 19.1   Letter of Transmittal

PETROLEUM PRICES AT THE END OF THE CENTURY:
A FORECAST

Prepared for: Harold Breen, President

Reliance Trucking Co.

by: Adelle Byner, Manager

Purchasing Department

Reliance Trucking Co.

April 19, 19XX

RELIANCE TRUCKING CO.
moving with the times since 1942

FIGURE 19.2   Simple Title Page

Most organizations have their own formats for title pages. Often, an organization will have different formats for different kinds of reports. Figure 19.3, on the next two pages, shows the complex title-page format used for research reports at one company.

This company wants so much information that it requires two pages. Notice that, on the first page, the title of the report would be followed by a statement of whether the report is a progress report or final report. On the second page, the word *distribution* would be followed by the list of those who are to receive the report. Like many companies, this one in fact sends the complete report to relatively few people (approximately four or five) and to the files. The summary report (the executive summary) goes to many more readers (approximately twenty or thirty).

Writing the Abstract

address readers familiar w/ subject

An abstract is a brief technical summary of the report—usually no more than 200 words. Like the abstract that accompanies published articles (see Chapter 17), the abstract of a report addresses readers familiar with the technical subject and needing to know whether to read the full report. Therefore, in writing an abstract you can use technical terminology freely and refer to advanced concepts in your field. (For managers who want a summary focusing on the managerial implications of a project, many reports that contain abstracts also contain executive summaries. See the discussion on page 528.)

Because abstracts can be useful before and after a report is read—and can even be read in place of the report—they are duplicated and kept on file in several locations within an organization: the division in which the report originated and many or all of the higher-level units of the organization. And, of course, a copy of the abstract is attached to or placed within the report. It is not unusual to find six or eight copies of an abstract somewhere in an organization. To facilitate this wide distribution, some organizations type abstracts on special forms.

The two basic types of abstracts are usually called descriptive and informative abstracts. The descriptive abstract is seen most often where space is at a premium. Some government proposals, for example, call for descriptive abstracts to be placed at the bottom of the title page. The informative abstract, on the other hand, is seen most often when more comprehensive information is called for. If you are ever called upon to write "an abstract," write an informative one, not a descriptive one.

RESEARCH REPORT COVER PAGE

(Company Confidential)

Standard Technical Report No. _____

Date Issued _____

Security (Check One) _____ RC _____ C

Originating R&D Department _____

Location (Facility, City, State) _____

Group or Division _____

WILSON CHEMICALS, INC.

_____

_____

(Title)

_____

(Indicate Progress or Final Report)

Work Done By:

Report Written By:

Supervisor:

R&D Director:

Previous Related Reports:

Department Overhead Number:

Project Number:

Period Covered:

Notebook Number(s)

PROPRIETARY INFORMATION

FOR AUTHORIZED COMPANY USE ONLY

FIGURE 19.3   Complex Title Page (page 1)

AUTHOR

_____

SUPERVISOR

_____

DIRECTOR

_____

DISTRIBUTION

Complete Report

Summary Report

FIGURE 19.3   Complex Title Page (page 2)

ABSTRACT

"Design of a Radio-based System for Distribution Automation"

by Brian D. Crowe

At this time, power utilities' major techniques of monitoring their distribution systems are after-the-fact indicators such as interruption reports, meter readings, and trouble alarms. These techniques are inadequate in two ways. One, the information fails to provide the utility with an accurate picture of the dynamics of the distribution system. Two, after-the-fact indicators are expensive. Real-time load monitoring and load management would offer the utility both system reliability and long-range cost savings. This report describes the design criteria we used to design the radio-based system for a pilot program of distribution automation. It then describes the hardware and software of the system.

FIGURE 19.4   Descriptive Abstract

The Descriptive Abstract

The descriptive abstract, sometimes called the topical, indicative, or table-of-contents abstract, does only what its name implies: it describes what the report is about. It does not provide the important results, conclusions, or recommendations. It simply lists the topics covered, giving equal coverage to each. Thus, the descriptive abstract essentially duplicates the major headings in the table of contents. Figure 19.4 provides an example of a descriptive abstract.

The Informative Abstract

The informative abstract presents the major information that the report conveys. Rather than merely listing topics, it states the problem, the scope and methods (if appropriate), and the major results, conclusions, and recommendations.

The informative abstract includes three elements:

■ *The identifying information.* The name of the report, the writer, and perhaps the writer's department.

- *The problem statement.* One or two sentences that define the problem or need that led to the project. Many writers mistakenly omit the problem statement, assuming that the reader knows what the problem is. The *writer* knows, being intimately involved with the project, but the readers likely don't. Without an adequate problem statement to guide them, many readers will be unable to understand the abstract.

- *The important findings.* The final three or four sentences — the biggest portion of the abstract — state the crucial information contained in the report. Generally, this means some combination of results, conclusions, recommendations, and implications for further projects. Sometimes, however, the abstract presents other information. For instance, many technical projects focus on new or unusual methods for achieving results that have already been obtained through other means. In such a case, the abstract will focus on the methods, not the results.

Following are the introduction, conclusion, and recommendation from the report (Crowe 1985) whose descriptive abstract appears in Figure 19.4. The excerpts are annotated to show how the writer used these excerpts to create his descriptive abstract.

The writer describes the problem, first in general terms and then in particular terms.

Introduction

At this time, power utilities' major techniques of monitoring their distribution systems are after-the-fact indicators such as interruption reports, meter readings, and trouble alarms. This system is inadequate in two ways.

One, the information fails to provide the utility with an accurate picture of the dynamics of the distribution system. To ensure enough energy for our customers, we have to overproduce. Last year we overproduced by 7 percent. This worked out to a loss of $273,000.

Two, after-the-fact indicators are expensive. Meter readings for our "easy-to-access" customers cost $20/year. Currently, 12,000 of our customers are classified as "difficult-to-access." Meter readings for each of these customers average $80/year, for an annual cost of $960,000. If we could reduce costs by installing radio-based monitoring systems on these 12,000 residences as a trial project, we could realize substantial savings while we perfect the system.

The writer describes the purpose of the report and the project.

This report describes a project to design a radio-based system for a pilot project. If the radio-based units are technically and economically feasible, they will be installed on the 12,000 residences for a one-year study.

System Description

[Here the writer describes the hardware and software in detail. The basic system, which uses packet-switching technology, consists of a base unit (built around a personal computer), a radio link, and a remote unit.]

Conclusions

The writer presents the four major conclusions.

The radio-based distribution monitoring system described in this report meets our four criteria: reliability, size, cost, and ease of use. It is more accurate than the after-the-fact indicators currently used, it is small enough to replace the existing meters, it would pay for itself in 3.9 years (see Appendix C, p. 14), and it is simple to use.

Recommendation

The writer presents the major recommendation and links it to the long-range corporate goals.

We recommend that the pilot program begin with the purchase and installation of the 12,000 meters. This pilot program is the first step toward our long-range goal: total automation of the distribution monitoring system.

Total automation would increase the reliability of the readings while drastically reducing labor costs. In addition, total automation would enable us to save money by not overgenerating energy; we could base our generation on accurate, timely data, rather than on rough estimates derived from weather forecasts and precedent.

The informative abstract based on this excerpt appears in Figure 19.5. An informative abstract resembles a descriptive abstract in that it contains identifying information and a problem statement. But whereas the descriptive abstract gives equal emphasis to most of the topics listed in the table of contents, the informative abstract concentrates on the important findings.

ABSTRACT

"Design of a Radio-based System for Distribution Automation"

by Brian D. Crowe

At this time, power utilities' major techniques of monitoring their distribution systems are after-the-fact indicators such as interruption reports, meter readings, and trouble alarms. This system is inadequate in that it fails to provide the utility with an accurate picture of the dynamics of the distribution system, and it is expensive. This report describes a project to design a radio-based system for a pilot project. The basic system, which uses packet-switching technology, consists of a base unit (built around a personal computer), a radio link, and a remote unit. The radio-based distribution monitoring system described in this report is more accurate than the after-the-fact indicators currently used, it is small enough to replace the existing meters, it would pay for itself in 3.9 years, and it is simple to use. We recommend installing the basic system on a trial basis.

FIGURE 19.5   Informative Abstract

The distinction between descriptive and informative abstracts is not absolute. Sometimes you will have to combine elements of both in a single abstract. For instance, you are writing an informative abstract but the report includes 15 recommendations, far too many to list. You might decide to identify the major results and conclusions, as you would in any informative abstract, but add that the report contains numerous recommendations, as you would in a descriptive abstract.

Writing the Table of Contents

Far too often, good reports are marred because the writer fails to create a useful table of contents that enables different readers to turn to specific pages to find the information they want. No matter how well organized the

Table of Contents

Introduction	1
Materials	3
Methods	4
Results	19
Recommendations	23
References	26
Appendixes	28

FIGURE 19.6   Ineffective Table of Contents

report itself may be, a table of contents that does not make the structure clear will be ineffective.

Most people will have read the abstract when they turn to the table of contents to find one or two items in the body of the report. Because a report usually has no index, the table of contents will provide the only guide to the report's structure, coverage, and pagination. The table of contents uses the same headings that appear in the report itself. To create an effective table of contents, therefore, you first must make sure the report has effective headings — and that it has enough of them. If the table of contents shows no entry for five or six pages, the report could probably be divided into additional subunits. In fact, some tables of contents have a listing — or several listings — for every page in the report.

Insufficiently specific tables of contents generally result from the overuse of generic headings (those that describe entire classes of items) in the report. Figure 19.6 shows how inadequate a table of contents can become if it simply lists generic headings.

To make the headings more informative, combine generic and specific items, as in the following examples:

Recommendations: Five Ways to Improve Information Retrieval

Materials Used in the Calcification Study

Results of the Commuting-Time Analysis

Then build more subheadings into the report. For example, in the "Recommendations" example, make a separate subheading for each of the five recommendations.

Once you have created a clear system of headings within the report, transfer them to the contents page. Use the same format — capitalization,

CONTENTS

FIGURE 19.7   Effective Table of Contents

underlining, indentation, and style (outline or decimal)—that you use in the body. (See the discussion of headings in Chapters 5 and 7.)

If you are using a word processor, you have more format choices: bold-face, italics, different size type, and so forth. See Chapter 12 for more information on using the word processor in page design.

A word processor makes it simple to transfer headings from the body of the report to the table of contents. Some software programs will actually

FIGURE 19.7 *Continued*

do the job for you automatically. But even the simplest software helps you make the table of contents quickly and easily—without introducing errors. Simply make a copy of the report. Then scroll through the copy, erasing the text. What you have left are the headings. If they are inconsistent you will notice immediately and be able to fix the problems.

Figure 19.7 shows how you can combine generic and specific headings and use the resources of the typewriter to structure a report effectively.

The report is titled "Methods of Computing the Effects of Inflation in Corporate Financial Statements: A Recommendation."

The table of contents in Figure 19.7 works well for several reasons. First, managers can find the executive summary quickly and easily. Second, all the other readers can find the information they are looking for because each substantive section is listed separately. A specific table of contents also gives the readers a clear idea of the scope and structure of the report before they start to read.

A note about pagination is necessary. The abstract page of a report is generally not numbered, but other preliminary elements (for example, a preface or acknowledgments page) are numbered with lowercase roman numerals (i, ii, and so forth) centered at the bottom of the page. The report itself is generally numbered with Arabic numerals (1, 2, and so on) in the upper right-hand corner of the page. Some organizations include on each page the total number of pages (1 of 17, 2 of 17, and so forth) so the readers will be sure they have the whole document.

Writing the List of Illustrations

A list of illustrations is a table of contents for the figures and tables of a report. (See Chapter 11 for a discussion of figures and tables.) If the report contains figures but not tables, the list is called a *list of figures*. If the report contains tables but not figures, the list is called a *list of tables*. If the report contains both figures and tables, figures are listed separately, before the list of tables, and the two lists together are called a *list of illustrations*.

Some writers will begin the list of illustrations on the same page as the table of contents; others prefer a separate page for the list of illustrations. If the list of illustrations begins on a separate page, it is listed in the table of contents. Figure 19.8 provides an example of a list of illustrations.

Writing the Executive Summary

The executive summary (sometimes called the *epitome*, the *executive overview*, the *management summary*, or the *management overview*) is a one-page condensation of the report. It is addressed to managers, who have to cope with a tremendous amount of paper crossing their desks every day. Managers do not need or want a detailed and deep understanding of the various projects undertaken in their organizations; this kind of understanding would in fact be impossible for them because of limitations in time and knowledge. What managers *do* need is a broad understanding of the projects and how they fit

LIST OF ILLUSTRATIONS

Figures

Figure 1	Compilation-Execution of FORTRAN Program	3
Figure 2	Example of EXEC Statement to Invoke Compiler	6
Figure 3	Example of Compilation Feature	8
Figure 4	Text of FORTXCL Catalogued Procedure	9
Figure 5	Structure of Segment	14

Tables

Table 1	Job Control Statements	2
Table 2	Types of Compiler Options	4
Table 3	Contents of Object List	7
Table 4	Return Codes of Loader	10
Table 5	ANS Control Characters for Line Feed Control	11

FIGURE 19.8   List of Illustrations

together into a coherent whole. Consequently, a one-page (double-spaced) maximum for the executive summary has become a standard.

The special needs of managers dictate a two-part structure for the executive summary:

■ *Background.* Because managers are not necessarily technically competent in the writer's field, the background of the project is discussed clearly. The specific problem or opportunity is stated explicitly—what was not working or not working effectively or efficiently, or what potential modification of a procedure or product had to be analyzed.

■ *Major findings and implications.* Managers are not interested in the details of the project, so the methods—often the largest portion of the report—rarely receive more than one or two sentences. The conclusions and recommendations, however, are discussed in a full paragraph.

For instance, if the research and development division at an automobile manufacturer has created a composite material that can replace steel in engine components, the technical details of the report might deal with the following kinds of questions:

How was the composite devised?
What are its chemical and mechanical structures?
What are its properties?

The managerial implications, on the other hand, involve other kinds of questions:

Why is this composite better than steel?
How much do the raw materials cost? Are they readily available?
How difficult is it to make the composite?
Are there physical limitations to the amount we can make?
Is the composite sufficiently different from similar materials to prevent any legal problems?
Does the composite have other possible uses in cars?

The executives don't care about chemistry; they want to know how this project can help them make a better automobile for less money.

For several reasons, the executive summary poses a great challenge to writers. First, the brevity of the executive summary requires an almost unnatural restraint. Having spent some weeks or even months collecting data on a complex subject, writers find it difficult to reduce all that information to one page.

Second, the executive summary usually is written specifically for a non-technical audience. Most writers were trained to write to specialists in their field. Beginning with their training in high-school science courses, writers were asked to address an audience (their teachers) that knew at least as much about the subject as they did. Learning and using the technical vocabulary was an important part of this training. In communicating with nonspecialists, however, writers must avoid or downplay specialized vocabulary and explain the subject without referring to advanced concepts.

Third, and most important, writers are so used to thinking in terms of supporting their claims with hard evidence that the notion of simply making a claim—and then not substantiating it—seems almost sacrilegious. Although some managers would probably like to read the full report, they simply don't have the time. They must trust the writer's technical accuracy.

Unfortunately, it is easier to define why executive summaries are a challenge to write than it is to suggest ways to write them easily. Implicit in this discussion is an obvious point: you must try to ignore most of the technical details and think, instead, of the manager's needs. Keep in mind the expression *the bottom line* when you focus on the managerial implications of the project. And keep in mind the following questions as you draft the executive summary:

■ *What was the background of the study: the problem or the opportunity?* In describing problems you studied, focus on specific evidence. For most managers, the best evidence includes costs and savings. Instead of writing that the equipment we are now using to cut metal foil is ineffective, write that the equipment jams on the average of once every 72 hours, and that every time it jams we lose $400 in materials and $2,000 in productivity as the workers have to stop the production line. Then add up these figures for a monthly or annual total.

In describing opportunities you researched, use the same strategy. Your company uses thermostats to control the heating and air conditioning. Research suggests that if you had a computerized energy-management system you could cut your energy costs by 20–25 percent. If your energy costs last year were $300,000, you could save $60,000–$75,000. With these figures in mind, the readers have a good understanding of what motivated the study.

■ *What methods did you use to carry out the research?* In most cases, your principal reader does not care how you did what you did. He or she assumes you did it competently and professionally. However, if you think your reader is interested, include a brief description — no more than a sentence or two.

■ *What were the main findings?* The findings are the results, conclusions, and recommendations. Sometimes your readers understand your subject sufficiently and want to know your principal results — the most important data from your study. If so, provide them. Sometimes, however, your readers would not be able to understand the technical data or would not be interested. If so, go directly to the conclusions — the inferences you draw from the data. For example, the results of the feasibility study on a computerized energy-management system might contain a description of the different hardware and software that make up the system. If a brief description would be informative, include it. Otherwise, go directly to the conclusion. That is, answer the question, Would the system be cost effective? Finally, most managers want your recommendations — your suggestions for further action. Recommendations can range from the most conservative — do nothing — to the most ambitious — proceed with the project. Often, a recommendation will call for further study of an expensive alternative.

After you have drafted the executive summary, give it to someone who has had nothing to do with the project — preferably someone outside the field. That person should be able to read the page and understand what the project means to the organization.

Placement of the executive summary can be important. The current practice in business and industry is to place it before the detailed discussion. To highlight the executive summary further, writers commonly make it equal in importance to the entire detailed discussion. They signal this strategy in the table of contents, which shows the report divided into two units: the executive summary and the detailed discussion. In the traditional organization of a report, the executive summary would be unit one and the detailed discussion units two through six or seven.

Figure 19.9 shows an effective executive summary. Notice the differences between this executive summary and the informative abstract (Figure 19.5). The abstract focuses on the technical subject: whether the new radio-based system can effectively monitor the energy usage. The executive summary concentrates on whether the system can improve operations *at this one*

EXECUTIVE SUMMARY

Presently, we monitor our distribution system using after-the-fact indicators such as interruption reports, meter readings, and trouble alarms. This system is inadequate in two respects. First, it fails to give us an accurate picture of the dynamics of the distribution system. To ensure enough energy for our customers, we must overproduce. Last year we overproduced by 7 percent, or a loss of $273,000. Second, it is expensive. Escalating labor costs for meter readers and the increased number of "difficult-to-access" residences have led to higher costs. Last year we spent $960,000 reading the meters of 12,000 such residences. This report describes a project to design a radio-based system for a pilot project on these 12,000 homes.

The basic system, which uses packet-switching technology, consists of a base unit (built around a personal computer), a radio link, and a remote unit.

The radio-based distribution monitoring system described in this report is feasible because it is small enough to replace the existing meters and because it is simple to use. It would provide a more accurate picture of our distribution system, and it would pay for itself in 3.9 years. We recommend installing the system on a trial basis. If the trial program proves successful, radio-based distribution-monitoring techniques will provide the best long-term solution to the current problems of inaccurate and inexpensive data collection.

FIGURE 19.9   Executive Summary

company. The executive summary describes the symptoms of the problem at the writer's organization in financial terms.

After a one-sentence paragraph describing the system design — the results of the study — the writer describes the findings in a final paragraph. Notice how this last paragraph clarifies how the pilot program relates to the overall problem described in the first paragraph.

Writing the Glossary and List of Symbols

A glossary is an alphabetical list of definitions. It is particularly useful if your audience includes readers unfamiliar with the technical vocabulary used in your report.

Instead of slowing down the detailed discussion by defining technical terms as they appear, you can use an asterisk or some other notation to inform your readers that the term is defined in the glossary. A footnote at the bottom of the page on which the first asterisk appears serves to clarify this system for readers. For example, the first use of a term defined in the glossary might occur in the following sentence in the detailed discussion of the report: "Thus the positron* acts as the. . . ." At the bottom of the page, add

> *This and all subsequent terms marked by an asterisk are defined in the Glossary, page 26.

Although the glossary is generally placed near the end of the report, right before the appendixes, it can also be placed right after the table of contents. This placement is appropriate if the glossary is brief (less than a page) and defines terms that are essential for managers likely to read the body of the report. Figure 19.10 provides an example of a glossary.

A list of symbols is structured like a glossary, but rather than defining words and phrases, it defines the symbols and abbreviations used in the report. Don't hesitate to include a list of symbols if some of your readers might not know what your symbols and abbreviations mean or might misinterpret them. Like the glossary, the list of symbols may be placed before the appendixes or after the table of contents. Figure 19.11 provides an example of a list of symbols (in this case, abbreviations).

Writing the Appendix

An appendix is any section that follows the body of the report (and the list of references or bibliography, glossary, or list of symbols). Appendixes conveniently convey information that is too bulky to be presented in the

---

GLOSSARY

byte: a binary character operated upon as a unit and, generally, shorter than a computer word.

error message: an indication that an error has been detected by the system.

hard copy: in computer graphics, a permanent copy of a display image that can be separated from a display device. For example, a display image recorded on paper.

parameter: a variable that is given a constant value for a specified application and that may denote that application.

record length: the number of words or characters forming a record.

FIGURE 19.10 Glossary

LIST OF SYMBOLS

CRT	cathode-ray tube
H_z	hertz
rcvr	receiver
SNR	signal-to-noise ratio
uhf	ultra high frequency
vhf	very high frequency

FIGURE 19.11 List of Symbols

body or that will interest only a few readers. Appendixes might include maps, large technical diagrams or charts, computations, computer print-outs, test data, and texts of supporting documents.

Appendixes, which are usually lettered rather than numbered (Appendix A, Appendix B, etc.), are listed in the table of contents and are referred to at the appropriate points in the body of the report. Therefore, they are accessible to any reader who wants to consult them.

Remember that an appendix is titled "Appendix," not "Figure" or "Table," even if it would have been so designated had it appeared in the body.

WRITER'S CHECKLIST

1. Does the **transmittal letter**
   a. clearly state the title and, if necessary, the subject and purpose of the report?
   b. state who authorized or commissioned the report?
   c. briefly state the methods you used?
   d. summarize your major results, conclusions, or recommendations?
   e. acknowledge any assistance you received?
   f. courteously offer further assistance?

2. Does the **title page**
   a. include a title that both suggests the subject of the report and identifies the type of report it is?
   b. list the names and positions of both you and your principal reader?
   c. include the date of submission of the report and any other identifying information?

3. Does the **abstract**
   a. list your name, the report title, and any other identifying information?
   b. clearly define the problem or opportunity that led to the project?
   c. briefly describe (when appropriate) the research methods?
   d. summarize the major results, conclusions, or recommendations?

4. Does the **table of contents**
   a. clearly identify the executive summary?
   b. contain a sufficiently detailed breakdown of the major sections of the body of the report?
   c. reproduce the headings as they appear in your report?
   d. include page numbers?

5. Does the **list of illustrations** (or tables or figures) include all the graphics in the body of the report?

6. Does the **executive summary**
   a. clearly state the problem or opportunity that led to the project?
   b. explain the major results, conclusions, recommendations, and/or managerial implications of your report?
   c. avoid technical vocabulary and concepts that the managerial audience is not likely to know?

7. Does the **glossary** include definitions of all the technical terms your readers might not know?

8. Does your **list of symbols** include all the symbols and abbreviations your readers might not know?

9. Do your **appendixes** include the supporting materials that are too bulky to present in the report body or that will interest only a small number of your readers?

EXERCISES

1. For each of the following letters of transmittal, write a one-paragraph evaluation. Consider each letter in terms of clarity, comprehensiveness, and tone.

   a. From a report by an industrial engineer to her company president:

   > Dear Mr. Smith:
   >
   > The enclosed report, "Robot and Machine Tools," discusses the relationship between robots and machine tools.
   >
   > Although loading and unloading machine tools was one of the first uses for industrial robots, this task has only recently become commonly feasible. Discussed in this report are concepts that are crucial to remember in using robots.
   >
   > If at any time you need help understanding this report, please let me know.
   >
   > > Sincerely yours,

   b. From a report by a military engineer to his commanding officer:

   > Dear General Smith:
   >
   > Along with Milwaukee Diesel, we are pleased to submit our study on potential fuel savings and mission improvement capabilities that could result from retrofitting the Air Force's C-3 patrol aircraft with new, high-technology recuperative or intercooled-recuperative turboprop engines.
   >
   > Results show that significant benefits can be achieved, but because of weight and drag installation penalties, the new recuperative or intercooled-recuperative engines offer little additional savings relative to a new conventional engine.
   >
   > > Sincerely,

   c. From a report by a marketing consultant to his client, a company that makes food additives:

   > Dear Mr. Smith:
   >
   > Enclosed is the report you requested on the effects consumer preferences will have on the future use of natural flavors.
   >
   > The major object of this project was to assess the strength of the apparent consumer desire for naturally flavored food products stemming from the controversy surrounding artificial additives. The lesser objective was to examine the other factors affecting the development and production of natural flavors.
   >
   > I have found that food purchasers seem to be more interested in the taste of their foods than they are in the

19 — Formal Elements of a Report ——— 537

source of their ingredients. Moreover, the use of natural flavors is complicated by problems such as high cost, lack of availability, and inconsistent quality. Therefore, I have recommended that Smith & Co. not enter the natural flavor market at this time.

    If you require any additional information, please do not hesitate to contact me.

Sincerely yours,

2. For each of the following informative abstracts, write a one-paragraph evaluation. How well does each abstract define the problem, methods, and important results, conclusions, and recommendations involved?

a. From a report by a consulting audio engineer to his client, the owner of a discotheque:

"Proposal to Implement an Amplifier Cooling System"

    In professional sound-reinforcement applications, such as discotheques, the sustained use of high-powered amplifiers causes these units to operate at high temperatures. These high temperatures cause a loss in efficiency and ultimately audible distortion, which is undesirable at disco volume levels. A cooling system of forced air must be implemented to maintain stability. I propose the use of separate fans to cool the power amplifiers, and of the best components to eliminate any distorting signals.

---

b. From a report by an electrical engineer to his manager:

"Design of a New Computer Testing Device"

    The modular design of our new computer system warrants the development of a new type of testing device. The term modular design indicates that the overall computer system can be broken down into parts or modules, each of which performs a specific function. It would be both difficult and time consuming to test the complete system as a whole, for it consists of 16 different modules. A more effective testing method would check out each module individually for design or construction errors prior to its installation into the system. This individual testing process can be accomplished by the use of our newly designed testing device.

    The testing device can selectively call or "address" any of the logic modules. To test each module individually, the device can transmit data or command words to the module. Also, the device can display the status or condition of the module on a set of LED displays located on the front panel of the device. In addition, the device has been designed so that it can indicate when an error has been produced by the module being tested.

---

c. From a report by a computer consultant to her client, the chief executive of a hospital:

"A Recommendation for a New System of Monitoring Patients with Implanted Cardiac Pacemakers"

The monitoring system used today to test pacemaker patients is inadequate at the data distribution and storage end. Often doctors receive 20 to 50 reports a week from individual tests on patients who, because of clerical and administrative difficulties, cannot be evaluated. This report recommends a digital time-sharing system for data storage and instant recall of test results at hospital locations throughout the country. With the current system and the recommendations in this report, pacemaker monitoring will reduce data-storage procedures, extend the useful life of individual pacemakers, provide doctors with current and reliable information on individual heart conditions, and supply patients with personal support and professional care.

3. For each of the following tables of contents, write a one-paragraph evaluation. How effective is each table of contents in highlighting the executive summary, defining the overall structure of the report, and providing a detailed guide to the location of particular items?

a. From "An Analysis of Corporations v. Sole Proprietorships":

CONTENTS

b. From "Recommendation for a New Incentive Pay Plan: the Scanlon Plan":

c. From "Initial Design of a Microprocessor-Controlled FM Generator":

4. For each of the following executive summaries, write a one-paragraph evaluation. How well does each executive summary present concise and useful information to the managerial audience?

   a. From "Analysis of Large-Scale Municipal Sludge Composting as an Alternative to Ocean Sludge-Dumping":

   > Coastal municipalities currently involved with ocean sludge-dumping face a complex and growing sludge management problem. Estimates suggest that treatment plants will have to handle 65 percent more sludge in 1995 than in 1985, or approximately seven thousand additional tons of sludge per day. As the volume of sludge is increasing, traditional disposal methods are encountering severe economic and environmental restrictions. The EPA has banned all ocean sludge-dumping as of next January 1. For these reasons, we are considering sludge composting as a cost-effective sludge management alternative.
   >
   > Sludge composting is a 21-day biological process in which waste-water sludge is converted into organic fertilizer that is aesthetically acceptable, essentially pathogen-free, and easy to handle. Composted sludge can be used to improve soil structure, increase water retention, and provide nutrients for plant growth. At $150 per dry ton, composting is currently almost three times as expensive as ocean dumping, but effective marketing of the resulting fertilizer could dramatically reduce the difference.

b. From "Recommendation for a New Incentive Pay Plan: the Scanlon Plan":

Summary

The incentive pay plan, Payment by Results, in operation at Cargo Corporation of America's Trenton plant, is not working effectively. Since the implementation of this plan, union and management have experienced increasing conflicts. The employee turnover ratio and absenteeism both have risen markedly, and productivity has increased only slightly.

An extensive research project was conducted with the objective of providing a solution to the plant's problem. Originally, four possible solutions were considered: (1) a revised Payment by Results, (2) Fixed Hourly Wage (no incentive plan at all), (3) Measured Daywork, (4) Scanlon Plan. After careful consideration and investigation, the best alternative was found to be the Scanlon Plan.

The Scanlon Plan is a companywide bonus system to reward increases in productivity and efficiency. The plan has two major parts: (1) a base productivity norm and (2) a system of work committees. The base productivity norm is based on performance over a period compared to a base period. Any improvement over the base period is called a bonus pool to be shared by all employees. Many different formulas have been designed to calculate the norm, but the most common is total payroll divided by the sales value of production (sales-value of production method).

Most companies implement the Scanlon Plan with departmental production committees, all reporting to one central screening committee. The individual committees meet regularly and discuss suggestions for improvements. The production committee meets formally only once a month for an hour or so. At this meeting they review the suggestions that deal with operating improvements. Any issues that they do not have the authority to deal with—for example, union matters, wages, and bonuses—are forwarded to the central screening committee. This committee is composed of employees from a representative cross-section of the company, including union officials. The committee does not vote on suggestions but does thoroughly discuss all points of view when there is a disagreement. After careful consideration, management makes the final decision.

Managers who have implemented the plan have found that employees are much more compromising toward change, particularly technological change. Experience has shown that thorough discussion of planned changes enables employees to realize that the new technology is not going to bring with it all sorts of ills that cannot be solved satisfactorily.

Companies that have used the plan have not experienced any resentment from the presiding union, and actually relations have improved. In many union-management relationships, the most difficult problem for the union leader is to get top management to sit down and discuss their various problems. A basic prerequisite of the Scanlon approach is that management be "willing to listen."

The Scanlon Plan is not for companies that are seeking a gimmick that will solve their problems. It is not a substitute for good management. The message of the plan is simple: operations improvement is an area in which management, the union, and employees can work together without strife.

c. From "Applying Multigroup Processing to the Gangloff Billing Account":

Gangloff Accounting is divided into seven geographical areas. Previously, end-of-the-month billing was processed for each group separately. Multigroup processing allows for processing all the groups simultaneously.

Multigroup processing is beneficial to both the data processing department and the accounting department in many ways. Running all the groups together will cut the number of job executions from 108 to 14, an 87 percent difference. This difference accounts for a 60 percent saving in computer time and a 30 percent saving in paper. With multigroup processing, all sense switches would be eliminated. With the sense switches done away with and the tremendous decrease in computer operator responsibilities, the chance of human error in the execution will diminish significantly.

6. Beginning on the next page is the body of a report (Karody 1985) entitled "Determination of an Optimal Casein and Soy Protein Mixture for Wilson Labs Rats." The writer was a student nutritionist employed by Wilson Labs. The principal reader is Dr. George Breyer, Head of Nutritional Sciences at Wilson labs.

On the basis of this material, write

a. a transmittal letter to Dr. Breyer
b. a title page
c. an informative abstract (for the body sections)
d. a table of contents
e. an executive summary

II. DETERMINATION OF AN OPTIMAL CASEIN AND SOY PROTEIN MIXTURE

A. INTRODUCTION

The purpose of this research was to determine an optimal ration of casein and soy protein for our rat feed. The present formulation contains a 10 percent weight of a complete protein source, casein. This formula has been used since the feed was designed in 1966. By December of 19XX, a 25 percent price increase of casein is expected. This will directly affect the production cost of the product. In order to avoid a price increase, an alternative formulation was considered.

The nutritional quality of a protein depends on the quantity, availability, and proportions of the essential amino acids that make up the protein. A complete protein, such as casein, can provide growth and a positive nitrogen balance because it contains all the essential amino acids for a growing animal. The Amino Acid Score or Amino Acid Index method (Oser 1959) shows that casein provides a ratio of essential amino acids similar to the requirement for a growing rat. On the other hand, soy protein appears to limit the sulfur-containing amino acids: L-cisteine and methionine. Soy is an incomplete protein, unable to promote growth by itself, and of inferior protein quality when compared to casein. The Amino Acid Index also shows that adding casein to a soy-protein diet should complement the deficiency of soy's limiting amino acids, increasing its protein quality. But the Amino Acid Index method makes an unreliable prediction since it does not account for the absorption, availability, and metabolic interactions of the amino acids.

Bioassays, on the other hand, compensate for these factors by measuring the efficiency of the biological use of dietary proteins as sources of the essential amino acids under standard conditions. Many of these biological methods are based on the effects of the quality and quantity of protein on animal growth. Several biological assays have been proposed to measure protein quality. The Protein Efficiency Ratio (PER) was chosen for this research because it is widely used, is the official method in the United States and in Canada, and is simple and inexpensive.

B. METHODOLOGY

The PER method was first used in 1917 by Osborne and Mendel in their studies to establish protein quality (Osborne et al. 1919). This method requires that energy intake be adequate and the protein be fed at an adequate but not excessive level in order to promote growth. This is compatible with both the 10 percent protein level and the caloric content of our feed. PER is a measure of weight gain of a growing animal divided by its protein consumption. The official government procedure requires a number of factors to be modified in order to standardize the reference casein and test diets. These factors include the level of the protein intake; type and amount of dietary fat; fiber levels; strain, age, and sex of rats; assay life-span; room conditions; etc. All the biological factors were followed in our research, but instead of using the official casein diet we used our current feed as our standard.

1. Diet Preparation

Prior to the arrival of the rats, four kg of each diet were prepared with the following protein content:

Diet 1:   25% casein   75% soy protein
Diet 2:   50% casein   50% soy protein
Diet 3:   75% casein   25% soy protein
Diet 4:   100% casein   (our current feed)

With a fixed total-protein level of 10 percent, the diets were isocaloric. All other factors and ingredients of the diets were fixed at the current proportions of our current feed product. The following is the standard percentage weight formulation used for the reference product:

Ingredients	Percentage Weight
Protein (100% casein)	10%
Starch	72
Oil (corn-cottonseed)	9
Mineral Mixture*	5
Water	2
Cellulose	1
Vitamin Mixture*	1
	100%

*The composition of the salt and the vitamin mixtures is given in Appendix A.

2. Rat Bioassay

We used weanling male, Sprague-Dawly rats, weighing 48 to 60 grams. The animals were housed in the individual galvanized steel cages (7″ width; 7″ height; 15″ length) in Room 505 of the Animal Research Department. In Room 505 temperature (72–75°F) and lighting (12 hrs light/12 hrs dark) were controlled. Assay groups were assembled in lots of 5 rats per diet in a random manner so the weight difference was minimized. Throughout a period of 28 days, feed and water were provided ad libitum, after a fasting period of 36 hours. Food consumption and weight gains of individual rats were recorded weekly. Data were collected in an automatic electronic balance. Food disappearance, considered as food intake, was determined as the difference between the weights of the food cup before and after it was filled, with corrections made for food spilled. Protein intake was calculated by multiplying the percentage of protein in the diet (10%) times the total food intake, divided by 100. PER was calculated as the weight gain divided by the amount of protein consumed. The raw data for PER, weight gain, food intake, and protein intake are given in Appendix B.

C. RESULTS

The PER, weight gain, food intake, and protein intake values for the assay are shown in Table 1. Data in Table 1 were statistically manipulated using the MINITAB system. A two-sample t-test was used to check for significant differences of the means (α = 0.05). PER values range from 1.6 to 2.6. The PER values from the 50 percent casein, 50 percent soy diet and our standard feed proved by this statistical manipulation to have no significantly different means. The other two diets have different PER values than our standard. The food intake, and subsequently the protein intake, did not vary in the standard or in the test diets.

Table 1

Weight Gain, Food Intake, Protein Intake, and PER for Our Current Rat Feed and Test Diets

Diet	Weight Gain* (grs)	Food Intake (grs)	Protein Intake (grs)	PER
25% casein-75% soy prot.	52.96 ± 7.62	331 ± 70.6	33.1 ± 7.06	1.6 ± 0.22
50% casein-50% soy prot.	88.80 ± 5.56	370 ± 50.2	37.0 ± 5.25	2.4 ± 0.38
75% casein-25% soy prot.	101.79 ± 9.84	352 ± 43.4	35.1 ± 4.34	2.9 ± 0.21
100% casein (our current feed)	93.86 ± 5.94	362 ± 32.3	36.1 ± 3.2	2.6 ± 0.45

*Mean ± Standard Deviation

D. CONCLUSION

Our results show that a mixture of 50 percent casein and 50 percent soy protein has the same PER value (around 2.5) as our standard product. Therefore, it can be concluded that they are dietary protein sources with the same biological protein quality. Moreover, this optimal protein ratio could substitute for casein in the current feed formula without altering the expected nutritional performance of the product. At the same time—as described in the executive summary—modifying our current protein to 50 percent casein/50 percent soy will reduce by 27 percent, or $3,200 per year, our current rat feed costs.

E. RECOMMENDATION

If the Production and Quality Control Managers have no objection, the change in formulation of our current feed might be a feasible solution to avoid the increase in its production cost.

REFERENCES

Crowe, B. 1985. Design of a Radio-based System for Distribution Automation. Unpublished manuscript.

Korody, E. 1985. Determination of an Optimal Casein and Soy Mixture for Wilson Labs Rats. Unpublished manuscript.

20

Proposals

Most projects undertaken by organizations, as well as most changes made within organizations, begin with proposals. If Technical Documentation, Inc., for example, wants to write a set of procedures manuals for Mayer Contractors, it submits a bid in the form of an *external proposal*. Similarly, when a local police department replaces its fleet of patrol cars, it solicits bids; automobile manufacturers submit external proposals detailing the cost, specifications, and delivery schedules. In short, when one organization wants to sell goods or services to another, the seller must write a persuasive external proposal.

An employee who suggests to his or her supervisor that the organization purchase a new word processor or restructure a department writes a similar but generally less elaborate document—the *internal proposal*.

This chapter will discuss both types of proposals. Because the external proposal is the more formal and detailed of the two, it will be discussed at greater length and will serve as the model for the internal proposal, which borrows formal aspects of the external proposal according to the demands of the internal situation.

Writing the External Proposal

No organization produces all the products or provides all the services it needs. Paper clips and company cars have to be purchased. Offices have to be cleaned and maintained. Sometimes, projects that require special expertise—such as sophisticated market analyses or feasibility studies— have to be carried out. With few exceptions, any number of manufacturers would love to provide the paper clips or the cars, and a few dozen consulting organizations would happily conduct the studies.

For those seeking the product or service, it is almost always a buyer's market. In order to get the best deal, most organizations require that their potential suppliers compete for the business. In a proposal, the supplier attempts to make the case that it deserves the contract.

A vast network of contracts spans the working world. The United States government, the world's biggest customer, spent about $300 billion in 1989 on work farmed out to organizations that submitted proposals. The defense and aerospace industries, for example, are almost totally dependent on government contracts. But proposal writing is by no means limited to government contractors. One auto manufacturer buys engines from another, and a company that makes spark plugs buys its steel from another company. In fact, most products and services are purchased on a contractual basis.

External proposals are either solicited or unsolicited. A *solicited proposal* originates with a request from a potential customer. An *unsolicited proposal* originates with the potential supplier.

When an organization wants to purchase a product or service, it publishes one of two basic kinds of statements. An IFB—"information for bid"—is used for standard products. When the federal government needs pencils, for instance, it lets suppliers know that it wants to purchase, say, one million no. 2 pencils with attached erasers. The supplier that offers the lowest bid wins the contract. The other kind of published statement is an RFP—"request for proposal." An RFP is issued when the product or service is customized rather than standard. Police cars are likely to differ from the standard consumer model: they might have different engines, cooling systems, suspensions, and upholstery. The police department's RFP might be a long and detailed set of technical specifications. The supplier that can provide the automobile most closely resembling the specifications—at a reasonable price—will probably win the contract. Sometimes, the RFP is a more general statement of goals. The customer is in effect asking the suppliers to create their own designs or describe how they will achieve the specified goals. The supplier that offers the most persuasive proposal will probably win the contract.

Most organizations issue RFPs and IFBs in newspapers or send them in the mail. Government RFPs and IFBs are published in the journal *Commerce Business Daily* (see Figure 20.1). The vast majority of solicitations appear in such publications.

An unsolicited proposal looks essentially like a solicited proposal except, of course, that it does not refer to an RFP. Even though the potential customer never formally requested the proposal, in almost all cases the supplier was invited to submit the proposal after the two organizations met and discussed the project informally. Because proposals are expensive to write, suppliers are very reluctant to submit them without any assurances that the potential customer will study them carefully. Thus, the term *unsolicited* is only partially accurate.

External proposals—both solicited and unsolicited—can culminate in contracts of several different types: a flat fee for a product or a one-time service; a leasing agreement; or a "cost-plus" contract, under which the supplier is reimbursed for the actual cost plus a profit set at a fixed percentage of the costs.

The Element of Persuasion

An electronics company that wants a government contract to build a sophisticated radar device for a new jet aircraft might submit a three-volume, 2,000-page proposal. A stationery supplier offering an unusual bargain, on the other hand, might submit to a medium-sized company a simple statement indicating the price for which it would deliver a large quantity of 20-pound, 8½-by-11-inch, 25 percent cotton-fiber bond paper, along with an explanation of why the deal ought to be irresistible. One factor links these two very different kinds of proposals: both will be analyzed carefully and skeptically. The government officials will worry about whether the supplier will be able to live up to its promise: to build, on

Commander, U.S. Army Missile Command, Procurement Director-ate, Redstone Arsenal, AL 35898-5280
R – SPOUSE/CHILD ABUSE PREVENTION PLAN SOL AMSMIR-ACFO-0001 DUE 032591 POC (PCF)Dot Padley, Contract Specialist, (205)876-3368, AMSMI-PC-FEA Mike Thomason, Contracting Officer, (205)842-7441. Synopsis No.: 504-91. R499-Consultant Service to provide specific spouse/child abuse prevention plan which complies with regulatory guidance set forth in Army Regulation 608-18 for Redstone Arsenal. 1 Job, 1 Jul 91 thru 30 Sep 91, Purchase Request AMSMIR-ACFO-0001. In accordance with FAR 6.302-1 this is a sole source procurement. The intended source is Behavioral Sciences Associates, Inc. All responsible sources may submit a quotation which shall be considered by this agency. See Note(s): 22. (0077)

Houisng & Urban Development, 501 E. Polk St., Tampa, Fl 33602- 3945
R – REAL ESTATE PROPERTY SALES CLOSINGS SOL HUD - 38-91-067 POC Contact, Doris Nadeau, 813-228-2551. Contracting Officer, William J. Schroeder, 813-118-2551 Real Estate Property Sale Closings - Service - Prepare Forms, Conduct Closing & Wire Funds to U. S. Treasury, for properties in Polk, Hardee, Highlands and DeSoto Counties, Florida. Bonding Required. RFP's are available, max term is three years in periods of one year. Indefinite quty contr. Contract commencement approx 05-01-91. (0077)

S Utilities and Housekeeping Services

Contracting Officer, Pine Bluff Arsenal, Pine Bluff, AR 71602-9500
S – PROVIDE ELECTRIC SERVICE TO PINE BLUFF ARSENAL POC Ken Mc-Clain, 501/543-3021. This presolicitation notice is an attempt to locate interested suppliers of electric service requriements of Pine Bluff Arsenal, AR - Estimated annual demand is 8,000 KW with estimated annual consumption 36,000 MWH. Points of delivery are Pine Bluff Arsenal Sub-Station A, B and C. - Service description: Substation A: supply a three phase, four wire, 60 hertz, alternating current at 13,800 Y/7970 volts. Voltage of contractors high tension line is 13,800 volts. Normal capacity of sub-station is 5,000 KVA: overload capacity is 62,500 KVA for four hours; lightning arrestors; station type: switching apparatus; fused disconnects. Substation B: supply a three phase four wire, 60 hertz, alternating current at 13,800 Y/29,700 volts. Voltage of the contractor's high tension line is 13,800 volts. Normal capacity of the substation is 10,000 KVA: overload capacity is 12,000 KVA for four hours; lightning arrestors; station type; switching apparatus: fused disconnects. Substation C: supply a three phase, three wire, 60 hertz, alternating current at 115,000 volts. Voltage of the contractor's high tension line shall be 115KV. Substation Transformers: Normal capacity 10,000 KVA; delta connected, high side 115 KV, low side 13,200 GRD. WYE 7,620 volts. Lightning Arrestors: Station type class. Switching apparatus: high side interrupting capacity 20 KA-RMS symmetrical for technical questions contact Mr. Roger Johnson, A/C (501) 543-3257. Interested sources are requested to respond to this notice in writing to: Contracting, SMCPB-POC-P; Pine Bluff Arsenal; Plairview Complex, Bldg. 17-120; Pine Bluff, AR 71602-9500 – Response should cite CBD Presolicitation Notice. – Proposal issuance contingent upon synopsis publication after results of this market survey to identify potential sources. (0077)

Social Security Administration, Office of Acquisition and Grants, P.O. Box 7696, Baltimore, Maryland 21207-0696. Attn:/Cherylene Wright
S – JANITORIAL SERVICES SOL IFB-91-0831 DUE 050891 POC Cherylene Wright tel 301-965-9529 Provide all management supervision, labor, materials, supplies and equipment required to perform janitorial and related services at the Social Security Admi-

---

U.S. Forest Service, Redding Contracting Unit, 2400 Washington Avenue, Redding, CA 96001
S – TRACTOR PILING Sol. R5-14-91-51. Due 050691. Contact Nancy Ruffner, 916/246-5220/Contracting Officer, Janet Peterson, 916/246-5375. Project consists of approximately 321 acres of concentration tractor piling and 5 acres of firelane tractor piling on the McCloud Ranger District. Shasta-Trinity National Forest, Siskiyou County, California. Contract time varies between 20 and 110 calendar days and estimated start date varies between May 13 and July 29, 1991 depending on the item. See numbered notes 1 and 3. (074)

U.S. Army Corps of Engineers, Portland District, P.O. Box 2946, Portland, OR 97208-2946
S – LAWN MOWING SERVICES Sol DACW57-91-B-0032. Due 050791. Contact Point, Lisa Eddington, 503/326-5816. This action originally publicized in the CBD dated approximately during the week of March 11 will include two option periods in addition to the basic period. Performance period for basic will be May 1, 1991 or date of award whichever is later, thru September 30, 1991. The two additional option periods will run from March 1 to September 30 of each year. All other information publicized in the original synopsis is correct. SIC 0782. (077)

USDA - Forest Service, Modoc National Forest, 441 No. Main Street, Alturas CA 96101
'S – MECHANICAL PLANTATION RELEASE & THINNING Sol R5-09-91-05. Due 050691. Contact Loren W. Ambers, Contracting Officer at 916/233-5811. Contractor is to furnish the necessary labor, tools, equipment, transportation, supervision and incidentals to cut down and masticate brush and excess trees in plantations on the Doublehead Ranger District. The contract will total 122 acres. The contract time will be 180 calendar days with an estimated start date sometime the latter part of May. (074)

T Photographic, Mapping, Printing and Publication Services

U.S. Department of Commerce, Patent and Trademark Office, Office of Procurement - Box 6, Washington, D.C. 20231
T – FULL COLOR/PART COLOR PHOTOPRINTS SOL 51-PAPT-1-00018 DUE 051691 POC Naomi Sorrell, Contract Administrator, 703/557-0014, David Zelnick, Contracting Officer Fulfill all Patent and Trademark Office (PTO) requirements for high-quality color/part color photoprints of plant patent drawings issuing each week and reorder requirements, as necessary. Photoprints on reorders will require combining furnished text pages with photoprints and side stitching (2 staples) to complete the patent set. Color is a key element in a plant patent and must be accurately reproduced. A fixed-price-requirements contract is anticipated. The period of performance is expected to be from the date of the contract award through 09/30/92 with two (2) annual renewable options. On April 15, 1991, Solicitation No. 51-PAPT-1-00018 will be issued and available for pickup, at which time each prospective offeror may inspect examples of the original plant patent color and part color drawings and some of the finished photoprints. On May 03, 1991, upon request, each prospective offeror will be loaned six (6) representative original plant patent color and part color drawings (hereafter referred to as (GFP) Government/furnished property), for use in preparing pre-award color and part color photoprints samples to accompany the technical proposal. No later than May 16, 1991, offerors shall submit proposals which includes photoprint samples, and return the GFP. Prospective offerors must return the GFP on this date whether or not a proposal is submitted, or be liable for replacement costs up to $100.00 per original drawing. The contact person is Naomi Sorrell, (703)557-0014. The pickup and delivery point is Patent and Trademark Office. Office of Procurement. Crystal Park Bldg. 1 - Room 806, 2011 Crystal Drive, Arlington, VA 22202. This procurement is a 100% small business set-aside. (0077)

FIGURE 20.1   An Extract from *Commerce Business Daily*

schedule, the best radar device at the best price. With perhaps a dozen suppliers competing for the contract, the officials will know only that a number of companies want the work; they can never be sure—not even after the contract has been awarded—that they have made the best choice. The office manager would be reluctant to spend more than has been budgeted for the short term in order to save money over the long term and would no doubt be suspicious about the paper's quality.

The key to proposal writing, then, is persuasion. The writers must convince the readers that the *future benefits* will outweigh the *immediate and projected costs*. Basically, external proposal writers must clearly demonstrate that they

☐ understand the reader's needs
☐ are able to fulfill their own promises
☐ are committed to fulfilling their own promises

Understanding the Reader's Needs

The most crucial element of the proposal is the definition of the problem or opportunity the project is intended to respond to. This would seem to be mere common sense: how can you expect to write a successful proposal if you do not demonstrate that you understand the reader's needs? Yet the people who evaluate proposals—whether they be government readers, private foundation officials, or managers in small corporations—agree that an inadequate or inaccurate understanding of the problem or opportunity is the most common weakness of the proposals they see.

Sometimes, the writer of the RFP fails to convey the problem or opportunity. More often, however, the writers of the proposal are at fault. The suppliers might not read the RFP carefully and simply assume that they understand the client's needs. Or perhaps the suppliers, in response to a request, know that they cannot satisfy a client's needs but nonetheless prepare proposals detailing a project they can complete, hoping either that the readers won't notice or that no other supplier will come closer to responding to the real problem. It is easy for suppliers to concentrate on the details of what *they* want to do rather than on what the customers need. But most readers will toss the proposal aside as soon as they realize that it does not respond to their situation. If you are responding to an RFP, study it thoroughly. If there is something in it you don't understand, get in touch with the organization that issued it; they will be happy to clarify it, for a bad proposal wastes everybody's time. Your first job as a proposal writer is to demonstrate your grasp of the problem.

If you are writing an unsolicited proposal, analyze your audience carefully. How can you define the problem or opportunity so that your reader will understand it? Keep in mind the reader's needs (even if the reader is oblivious to them) and, if possible, the reader's background. Concentrate on how the problem has decreased productivity or quality or on how your ideas would create new opportunities. When you submit an unsolicited proposal, your task in many cases is to convince your readers that a need exists. Even when you have reached an understanding with some of your customer's representatives, your proposal will still have to persuade other officials.

Describing What You Plan to Do

Once you have shown that you understand why something needs to be done, describe what you plan to do. Convince your readers that you can respond to the situation you have just described. Discuss your approach to the subject: indicate the procedures and equipment you would use. Create a complete picture of how you would get from the first day of the project to the last. Many inexperienced writers believe that they need only convince the reader of their enthusiasm and good faith. Unfortunately, few readers are satisfied with assurances—no matter how well intentioned. Most look for a detailed, comprehensive plan that shows that the writer has actually started to do the work.

Writing a proposal is a gamble. You might spend days or months putting it together, only to have it rejected. What can you do with an unsuccessful proposal? If the rejection was accompanied by an explanation, you might be able to learn something from it. In most cases, however, all you can do is file it away and absorb the loss. In a sense, the writer takes the first risk in working on a proposal, which, statistically, is likely to be rejected. The reader who accepts a proposal also takes a risk in authorizing the work, for he or she does not know whether the work will be satisfactory.

No proposal can anticipate all of your readers' questions about what you plan to do, of course. But the more planning you have done before you submit the proposal, the greater the chances you will be able to do the work successfully if you get the go-ahead. A full discussion of your plan suggests to your readers that you are interested in the project itself, not just in winning the contract or in receiving authorization.

Demonstrating Your Professionalism

After showing that you understand the reader's needs and have a well-conceived plan of attack, demonstrate that you are the kind of person—or yours is the kind of organization—that is committed to delivering what is promised. Many other persons or organizations could probably carry out the project. You want to convince your readers that you have the pride, ingenuity, and perseverance to solve the problems that inevitably occur in any big undertaking. In short, you want to show that you are a professional.

You show your professionalism by describing your credentials and work history. Who are the people in your organization who have the qualifications and experience to carry out the project? What equipment and facilities do you have that will enable you to do the work? What management structure will you use to maintain coordination and keep all the different activities running smoothly? What similar projects have you completed successfully? In short, make the case that you know how to make this project work because you have made similar projects work.

Another major element of this professionalism is a work schedule, sometimes called a task schedule. This schedule—which usually takes the form of a graph or chart—shows when the various phases of the project will be carried out. In one sense, the work schedule is a straightforward piece of information that enables your readers to see how you would apportion your time. But in another sense, it reveals more about your attitudes toward your work than about what you will actually be doing on any given day. Anyone with even the slightest experience with projects knows that things rarely proceed according to plan: some tasks take more time than anticipated, some take less. A careful and detailed work schedule is actually another way of showing that you have done your homework, that you have attempted to foresee any problems that might jeopardize the success of the project.

Related to the task schedule is quality control. Your readers will want to see that you have established procedures to evaluate the effectiveness and efficiency of your work on the project. Sometimes, quality-control procedures consist of technical evaluations carried out periodically by the project staff. Sometimes, the writer will build into the proposal provisions for on-site evaluation by recognized authorities in the field or by representatives of the potential client.

Most proposals conclude with a budget—a formal statement of how much the project will cost.

The Structure of the Proposal

Most proposals follow a basic structural pattern. If the authorizing agency provides an IFB, an RFP, or a set of guidelines, follow it to the letter. If guidelines have not been supplied, or you are writing an unsolicited proposal, use the following conventional structure:

- ☐ summary
- ☐ introduction
- ☐ proposed program
- ☐ qualifications and experience
- ☐ budget
- ☐ appendixes

As with most technical documents, the sequence of composition is not the same as the sequence of presentation. The first section to write is the introduction, which includes the reader's needs. Until you know the reader's needs, you cannot write anything else. After you have drafted the introduction, do the proposed program and proceed to the end: the qualifications and experience, appendixes, and finally the budget. Once you have completed all these items, draft the summary.

Summary

For any proposal of more than a few pages, provide a brief summary. Many organizations impose a length limit—for example, 250 words—and ask

Summary

American Metal Foundry, Inc., proposes to modify St. Thomas College's pneumatic high-pressure air gun (PHPAG) to increase its usefulness in destructive testing research. The project entails redesigning the barrel of the gun and designing and manufacturing the receiver tank, which would hold the test specimens and catch the flying projectiles.

These modifications would enable St. Thomas College to carry out sophisticated destructive testing research in its materials laboratory.

This project, which would take 17 days, would cost $8,066. The unit would be guaranteed to meet specifications after it is delivered and installed.

FIGURE 20.2   Summary of a Proposal

the writer to type the summary, single-spaced, on the title page. The summary is crucial, because in many cases it will be the only item the readers study in their initial review of the proposal.

The summary covers the major elements of the proposal but devotes only a few sentences to each. To write an effective summary, first define the problem in a sentence or two. Next, describe the proposed program. Then provide a brief statement of your qualifications and experience. Some organizations wish to see the completion date and the final budget figure in the summary; others prefer that these data be displayed separately on the title page along with other identifying information about the supplier and the proposed project.

Figure 20.2 shows an effective summary taken from a proposal (Breen 1985) submitted by a metal foundry to a college to modify an air gun owned by the college. A sample internal proposal is included at the end of this chapter.

Introduction

The body begins with an introduction. Its function is to define the background and the problem or the opportunity.

In describing the background, you probably will not be telling your readers anything they don't already know (except, perhaps, if your proposal is unsolicited). Your goal here is to show them that *you* understand the problem or opportunity: the circumstances that led to the discovery, the relationships or events that will affect the problem and its solution, and so forth.

Introduction

American Metal Foundry, Inc., proposes to modify St. Thomas College's pneumatic high-pressure air gun (PHPAG) to increase its usefulness in destructive testing research. The project entails redesigning the barrel of the gun and designing and manufacturing the receiver tank, which would hold the test specimens and catch the flying projectiles.

St. Thomas has a 15 ft. PHPAG capable of firing projectiles at speeds in excess of 1,200 ft./sec. A specimen of a desired material is placed in front of the barrel. A projectile is then loaded and fired at the specimen. The effect of the impact is then measured. The results of this testing yield valuable information about the material.

The problems that this proposal addresses are as follows:

1. Although the barrel of the PHPAG has an inside diameter large enough to accommodate a 2.5 in. diameter projectile, the barrel has an insert that is used to fire .60 caliber shells. This insert prevents the use of larger shells.

2. The PHPAG does not have a receiving tank. Therefore, it cannot accommodate the larger projectiles necessary for more advanced destructive testing.

Modifying the barrel of the PHPAG and building a receiving tank will allow St. Thomas to carry out high-level destructive testing in its materials laboratory.

FIGURE 20.3 Introduction to a Proposal

In discussing the problem, be as specific as possible. Whenever you can, *quantify*. Describe it in monetary terms because the proposal itself will include a budget of some sort and you want to convince your readers that spending money on what you propose is wise. Don't say that a design problem is slowing down production; say that it is costing $4,500 a day in lost productivity.

Figure 20.3 is the introduction to the air-gun proposal.

Proposed Program

Once you have defined the problem or opportunity, you have to say what you are going to do about it. As noted earlier, the proposed program demonstrates clearly how much work you have already done. Be specific. You won't persuade by saying that you plan to "gather the data and analyze it." How will you gather it? What techniques will you use to analyze it? Every word you say—or don't say—will give your readers evidence on which to base their decision. If you know your business, the proposed program will show it. If you don't, you'll inevitably slip into meaningless generalities or include erroneous information that undermines the whole proposal.

If your project concerns a subject written about in the professional literature, show your familiarity with the scholarship by referring to the pertinent studies. Don't, however, just toss a bunch of references onto the page. For example, don't write, "Carruthers (1989), Harding (1990), and Vega (1990) have all researched the relationship between acid-rain levels and ground-water contamination." Rather, use the secondary research to sketch in the necessary background and provide the justification for your proposed program. For instance:

> Carruthers (1989), Harding (1990), and Vega (1990) have
>
> demonstrated the relationship between acid-rain levels and
>
> ground-water contamination. None of these studies, however,
>
> included an analysis of the long-term contamination of the aquifer.
>
> The current study will consist of. . . .

You might include just one reference to secondary research. However, if your topic has been researched thoroughly, you might devote several paragraphs or even several pages to secondary scholarship. Figure 20.4 shows the proposed program for the air-gun project. In this example, the writer generally describes the process his organization will perform (see Chapter 9 for a discussion of process descriptions). Notice that this proposal does not discuss any secondary research, for the project does not involve any.

Qualifications and Experience

After you have described how you would carry out the project, demonstrate your ability to undertake it. Unless you can convince your readers that you can turn an idea into action, your proposal will not be persuasive.

The more elaborate the proposal, the more substantial the discussion of qualifications and experience has to be. For a small project, a few paragraphs describing your technical credentials and those of your coworkers will usually suffice. For larger projects, the résumés of the project leader—often called the principal investigator—and the other important participants should be included.

External proposals should also include a discussion of the qualifications of the supplier's organization. Essentially similar to a discussion

Proposed Procedure

The project entails two major stages:

1. redesigning the barrel of the gun

2. designing and manufacturing the receiver tank, which would hold the test specimens and catch the flying projectiles.

1. Redesigning the Gun

The barrel of the gun has an inside diameter large enough to accommodate a 2.5 in. diameter projectile. However, the barrel has been fitted with an insert used to fire .60 caliber shells. This insert prevents the use of larger shells. We propose to remove this insert and remachine the barrel so that it can accommodate the larger projectiles.

2. Designing and Manufacturing the Receiver Tank

Adding a receiver tank involves a number of steps. We propose to design and manufacture a tank large enough to tolerate the enormous impact involved in destructive testing. We would install the gun in one side of the tank. At the insertion point we would install gaskets and flanges to keep the gun and tank airtight during firing. Finally, after testing the unit we would mount the tank to the floor in your laboratory. We would then repeat the testing to ensure that the unit performs according to specifications.

FIGURE 20.4   **Proposed Program of a Proposal**

of personnel, this section outlines the pertinent projects the supplier has completed successfully. For example, a company bidding for a contract to build a large suspension bridge should describe other suspension bridges it has built. The discussion also focuses on the necessary equipment and facilities the company already possesses, as well as the management structure that will ensure successful completion of the project. Everyone knows that young, inexperienced persons and new firms can do excellent work. But when it comes to proposals, experience wins out almost every time. Figure 20.5 shows the qualifications-and-experience section of the air-gun proposal.

Qualifications

American Metal Foundry, Inc., has been performing high-quality metal machining work since 1956. Our engineering staff consists of experts in stress analysis, mechanics, and machine design and fabrications. Our technicians combine state-of-the-art technical expertise and old-fashioned pride in their workmanship. In fact, we have won seven Certificates of Merit from the Association of Metal Foundries for our products.

Our team leader, Dr. Karen Mair, has over 14 years' experience in designing and fabricating metals and other materials. She is supported by a fully professional support staff of engineers, technicians, and clerical personnel. Dr. Mair has successfully completed projects for dozens of organizations, including Valley View Hospital, Parker State University, Hawkins Ford-Mercury, and Demling Nursery. For a complete listing of her credentials, see her résumé in Appendix B. Please see Appendix D for a full description of similar projects we have undertaken.

FIGURE 20.5   Qualifications-and-Experience Section of a Proposal

Budget

Good ideas aren't good unless they're affordable. The budget section of a proposal specifies how much the proposed program will cost.

Budgets vary greatly in scope and format. For simple internal proposals, the writer adds the budget request to the statement of the proposed program: "This study will take me about two days, at a cost—including secretarial time—of about $400" or "The variable-speed recorder currently costs $225, with a 10-percent discount on orders of five or more." For more complicated internal proposals and for all external proposals, a more explicit and complete budget is usually required.

Most budgets are divided into two parts: *direct costs* and *indirect costs.* Direct costs include such expenses as salaries and fringe benefits of program personnel, travel costs, and any necessary equipment, materials, and supplies. Indirect costs cover the intangible expenses that are sometimes called overhead. General secretarial and clerical expenses not devoted exclusively to the proposed program are part of the overhead, as are other operating expenses such as utilities and maintenance costs. Indirect costs

Budget Itemization

For period August 1, 19XX to August 17, 19XX.

Direct costs

1. Salaries and Wages

Personnel	Title	Time	Amount
Karen Mair	Design Engineer	2 weeks	$1,900
Ed Smith	Chief Machinist	2 weeks	1,200
	Typist	2 days	175
	Total Salaries and Wages		$3,275

2. Supplies and Materials

Tank	$2,300
Miscellaneous Materials	400
Total Supplies and Materials	$2,700
Subtotal	$5,975

Indirect Costs

35% of $5,975	$2,091
Total Cost	$8,066

FIGURE 20.6   Budget Section of a Proposal

are usually expressed as a percentage—ranging from less than 20 percent to more than 100 percent—of the direct expenses. In many external proposals, the client imposes a limit on the percentage of indirect costs.

Figure 20.6 shows an example of a budget statement.

Appendixes

Many different types of appendixes might accompany a proposal. The airgun proposal would include, among other items, descriptions of other projects the supplier has already completed. Another popular kind of appendix is the supporting letter—a testimonial to the supplier's skill and integrity, written by a reputable and well-known person in the field. Two other kinds of appendixes deserve special mention: the task schedule and the evaluation description.

The task schedule is almost always drawn in one of two graphical formats. The Gantt chart is a horizontal bar graph, with time displayed on the horizontal axis and tasks shown on the vertical axis. (See Chapter 11 for a discussion of bar graphs.) A milestone chart is a horizontal line that represents time; the tasks are written in under the line. Both Gantt and milestone charts can include prose explanations, if necessary. Figure 20.7 shows that the Gantt chart is more informative than the milestone chart in

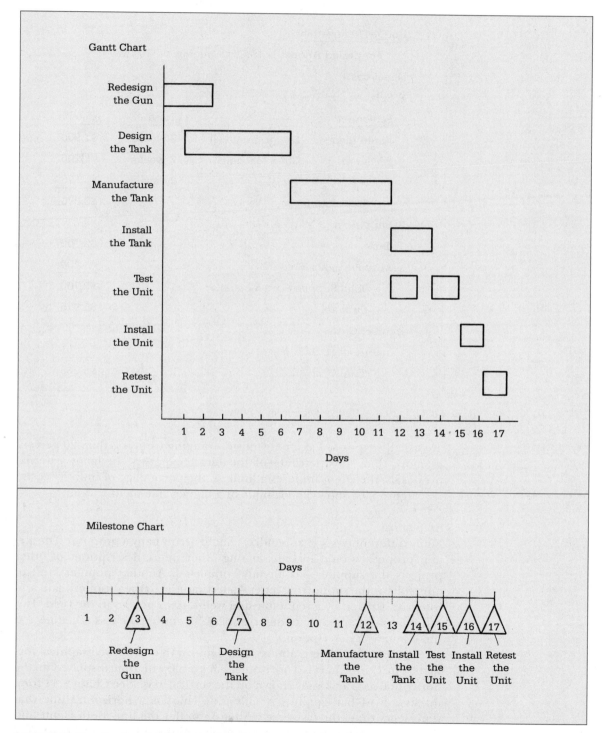

FIGURE 20.7   Gantt and Milestone Charts

Appendix A. Evaluation

We will submit to St. Thomas College two progress reports: one at the end of week one, and one at the end of week two. In addition, we would welcome your inspection of the progress on the project at any time. We will guarantee that the unit will work according to the agreed specifications once it is installed at your facility.

FIGURE 20.8   Evaluation-Techniques Section of a Proposal

that only the Gantt chart can indicate several tasks being performed simultaneously; the milestone chart can indicate only the dates on which the various tasks are to be completed.

Much less clear-cut than the task schedule is the description of evaluation techniques. In fact, the term *evaluation* means different things to different people, but in general an evaluation technique can be defined as any procedure for determining whether the proposed program is both effective and efficient. Evaluation techniques can range from simple progress reports to sophisticated statistical analyses. Some proposals provide for evaluation by an outside agent—a consultant, a testing laboratory, or a university. Other proposals describe evaluation techniques that the supplier itself will perform, such as cost/benefit analyses.

The subject of evaluation techniques is complicated by the fact that some people think in terms of *quantitative evaluations*—tests of measurable quantities, such as production increases—whereas others think in terms of *qualitative evaluations*—tests of whether a proposed program is improving, say, the workmanship of a product. And, of course, some people imply both qualitative and quantitative testing when they refer to evaluations. An additional complication is that projects can be tested both while they are being carried out (*formative evaluations*) and after they have been completed (*summative evaluations*).

When an RFP calls for "evaluation," experienced writers of proposals know it's a good idea to get in touch with the sponsoring agency's representatives to determine precisely what they mean. Figure 20.8 is a description of the evaluation techniques for the air-gun project.

The STOP Technique

The STOP technique for writing long proposals and other kinds of technical documents is becoming very popular in business and industry. STOP (Strategic Thematic Organization of Proposals) was devised by Hughes

Aircraft in 1962. The technique is based on brainstorming and a special structure for the document itself.

Chapter 5 discusses brainstorming, the process of quickly generating a list of ideas and topics that might go into the document. With long proposals—those of 100 pages or more—brainstorming usually involves several people, including subject-area specialists, writers, editors, illustrators, and contract specialists, all working together in the same room. In the STOP technique, and in related techniques, brainstorming sessions are held frequently, not just once at the start of the project. Between sessions, the proposal-writing team creates storyboards. Often attributed to film director Alfred Hitchcock, a storyboard is a sketch with a brief prose statement or notes. Hitchcock used storyboards to coordinate his shots with his script before he actually began shooting the film. His goal was to avoid wasting time and film once the cast and crew were assembled on location.

The goal is the same in the STOP technique. The participants bring their storyboards to the brainstorming sessions, where they test them out on the other participants. The goal is to make sure the content and emphasis will be right—before the actual writing process begins. The storyboards, once they are approved, become the draft of the proposal.

The structure of a STOP proposal calls for two-page units (called topics) of text and visuals. As Figure 20.9 shows, the left-hand page of each topic consists of a heading, a thesis statement to unify the discussion, and the discussion itself, of 200–700 words. The right-hand page consists of any remaining text from the left-hand page and a graphic: a sketch, photograph, diagram, flow chart, etc. Longer topics that cannot be reduced to a two-page topic are partitioned until they are concise enough. Topics are then gathered into sections: units of from one to seven topics.

The STOP technique forces the proposal team to use a systematic process in creating documents. The emphasis is on prewriting and on devising small, easy-to-read units with conveniently located graphics.

Writing the Internal Proposal

One day, while you're working on a project in the laboratory, you realize that if you had a new centrifuge you could do your job better and more quickly. The increased productivity would save your company the cost of the equipment in a few months. You call your supervisor and tell him about your idea. He tells you to send him a memo describing what you want, why you want it, what you're going to do with it, and what it costs; if your request seems reasonable, he'll try to get you the money.

Your memo is an internal proposal—a persuasive argument, submitted within an organization, for carrying out an activity that will benefit the organization, generally by saving it money. An internal proposal is simply

Section 1 - Technical Approach

2. ENCODING 2-NANOSECOND PULSE DATA USING COMMERCIAL MODULES
REP Ref: 2.3

Though optimum receiver performance would require development of 500-MHz A/D converter technology, the proposed pulse stretching approach obtains nearly optimum performance using commercially available components and proven techniques.

The result of applying incoming IF signals to a microscan receiver is a compressed output pulse exhibiting a wide bandwidth. In this case the 500-MHz IF input signal bandwidth results in a 250-MHz baseband output with minor variations to do the weighting scheme selected. Processing this wide bandwidth to obtain phase data and associating the phase and frequency digital words in real time, as is required in an Electronic Surveillance Measure (ESM) application, presents several key challenges. Chief among these is encoding the amplitude of 2 nanosecond pulses, which requires a Nyquist sampling rate of 500 MHz.

The amplitude encoding is required to perform the direction finding function. The output pulse from the receiver has a sin x/x response. The response from the bulk two quadrature phase detector outputs may approximate a d.c. level during the peak of the sin x/x response depending on the consistency of the phase relationships between mainlobe and sidelobes and the quality of the limiter. To assure best phase detection, the peak of the amplitude sin x/x response must be used to key sampling of the phase data during the slightly wider time interval of the main lobe.

If moderate cost, 8-bit 500 MHz A/D converter technology were available, direct conversion of R and I channel phase data would be an attractive implementation. Data words obtained at times other than amplitude peaks would be discarded while data captured at the peak would be passed on for further processing. This would provide a comprehensive parameter measurement implementation, and would result in performance sufficient to handle any real or potential threat environment now envisioned. Though Hughes is one of the leaders in the field of high-rate A/D conversion, including application of both silicon and GaAs technologies along with design of special-purpose sine-cosine encoders found in IFMs, it is apparent that developing new 500-MHz A/D technology is beyond the intended scope of this program. However, the potential performance for the proposed IF receiver may also be realized using an alternate design approach.

To facilitate a design employing mainly commercially available modules, the infrequent occurrence of the output data may be exploited. Though the output data bandwidth is large, the information of interest is relatively sparse. As an example, in a 200-nanosecond scan time only five events of interest may be encountered, implying a maximum A/D conversion rate of 25 MHz in the R and I channels. Many modular A/D converters are available to provide this performance level, including the TDC 1007J by TRW, the MATV-0820 by Computer Labs, and the ADC-TV8b by Datel. The key to encoding the 250-MHz pulse output is capturing of the 250-MHz pulse output and preserving it for low-rate A/D conversion.

The pulse stretching approach which offers full fidelity sampling for these special signals is shown in the figure. In this approach, the output voltage from the phase detector is sampled using a voltage switch. The resulting 2 ns pulse is then stretched using a delay line and summation approach to enable the lower speed conversion. Given a known and fixed pulse sample width, a coaxial delay line, and a wide bandwidth summing amplifier previously developed by Hughes, a 25 ns stretched version of the input pulses is obtained. This output is then smoothed by the inherent low pass of the commercially available sample-and-hold.

The salient difference between this sampling system and directly sampling and holding the pulse is that wider bandwidths are obtainable when the goal is to pulse sample instead of holding on a capacitor because resistive loading of the

HUGHES FULLERTON
Hughes Aircraft Company
Fullerton, California

sampling gate reduces the bridge time constants. In addition, the bandwidth of low impedance buffer amplifiers is more readily extended when compared to high impedance buffers required to hold voltage.

Further circuit optimization, including additional low-pass smoothing of the output pulse, is expected to result in the required 8-bit performance. The baseline subsystem incorporates one sample and stretch network for the R channel and one for I permitting capture events spaced approximately every 10 to 30 nanoseconds depending on the acquisition time of the sample and hold. Though the baseline configuration does not include a multiple stretching network operating in parallel, the design does not preclude addition of that capability in the future.

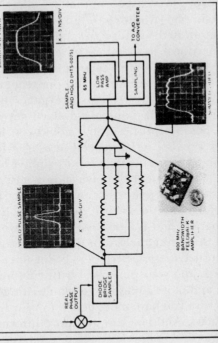

VIDEO PULSE SAMPLE

SAMPLE AND HOLD (HTS-0025)

REAL PHASE OUTPUT

DIODE BRIDGE SAMPLER

65 MHz LOW PASS AMP

SAMPLING

TO A/D CONVERTER

400 MHz BANDWIDTH FEEDBACK AMPL I=F R

X = 5 NS/DIV

Pulse Stretching Network. This approach has demonstrated the performance required to condition narrow pulses for sampling, holding, and A/D conversion by commercially available modules. The photographs document the critical waveforms, indicating a 7-bit settling performance.

1-3

1-2

FIGURE 20.9   STOP Topic © 1983 IEEE.

an argument made by someone within an organization, about how to improve some aspect of that organization's operations. The argument can be simple—to purchase an inexpensive piece of office equipment—or complicated—to hire an additional employee or even add an additional department to the organization. The nature of the argument determines the format. A simple request might be conveyed orally, either in person or on the phone. A more ambitious request might require a brief memo. The most ambitious requests are generally conveyed in formal proposals. Often, organizations use dollar figures to determine the format of the proposal. A proposal that would cost less than $1,000 to implement, for instance, is communicated in a brief form, whereas a proposal of more than $1,000 requires a report similar to an external proposal.

The element of persuasion is just as important in the internal proposal as it is in the external proposal. The writer must show that he or she understands the organization's needs, has worked out a rational proposed program, and is a professional who would see that the job gets done.

A careful analysis of the writing situation is the best way to start, for as usual every aspect of the document is determined by your reader's needs and your purpose. Writing an internal proposal is both more simple and more complicated than writing an external proposal. It is simpler because you have more access to your readers than you would to external readers. And you can get more information, more easily. However, you might find it more difficult to get a true sense of the situation in your own organization. Some of your coworkers might not be willing to tell you directly if your proposal is a long shot. Another danger is that when you identify the problem you want to solve, you are often criticizing—directly or indirectly—someone at your organization who instituted the system that needs to be revised or who failed to take necessary actions earlier.

An unsuccessful external proposal does not linger; it was a gamble and you lost. An unsuccessful internal proposal lingers because more of your colleagues know about it. Therefore, before you write an internal proposal, discuss your ideas thoroughly with as many potential readers as you can. In this way you will increase your chances of finding out what the organization really thinks of your idea before you commit it to paper.

Sample Internal Proposal

Following is an internal proposal (Banham 1990). The author was a student working as a co-op student for an engineering firm, Willis-Martin.

The progress report written after the project was under way is included in Chapter 21 (page 586); the completion report is in Chapter 22 (beginning on page 610). Marginal notes have been added.

This proposal was written as a memo. If it had been longer, it probably would have taken the form of a report.

To save space, this memo has not been divided into pages. For a discussion of memos, see Chapter 13.

DATE: January 4, 1990
TO: Don Leith, Manager, Audio/Video Systems Engineering, Willis-Martin, Inc.
FROM: Mark Banham, Audio Engineer Co-op, Willis-Martin, Inc.
SUBJECT: Proposal to design the first phase of the public-address system for the Sutherland International Airport in Sydney.

Purpose

This sentence describes the purpose of the memo and the purpose of the proposed investigation.

This proposal recommends the allocation of approximately three months of my time to coordinate and participate in the engineering design of the public-address system for the Sutherland International Airport in Sydney.

Summary

The background.

The proposal.

Overview of the proposal.

Cost of the project.

The theoretical background.

Willis-Martin was recently awarded a $1.2 million subcontract to design and install a public-address distributed loudspeaker system for the Sutherland International Airport. It is now necessary to assign personnel to perform the West Terminal Building stage of this design. At the present, I would be available to support this project full-time. With the assistance of one other engineer and the drafting department, Phase I of the design could be completed well in advance of the six months prescribed by National Electronics Company, Ltd. (NECL), the main electrical contractor and our managing agent on this job. The bonus would be $44,000, in addition to the contracted price of $80,000, if we are able to adhere to the design schedule. This proposal contains a detailed plan of personnel and design needs for the completion of Phase 1 in 90 days. The project would cost approximately $16,548.

Introduction

The specific challenges to be met.

As Martin (1982) has shown, any sound-system design requires three major tasks at the start of the project:

- A detailed acoustical analysis of the site must be completed using as-built drawings and a reliable liaison at the site. The construction details of the building are needed to determine equipment-mounting conditions and ambient-noise levels.
- Extensive architectural measurements must be made from the drawings. These measurements are used to determine the average square footage per room by ceiling height. The required number of loudspeakers can be estimated from these data.
- Calculations of speaker placement and cable runs must be made to determine specification sound-quality demands and equipment needs in terms of real dollars.

For large projects, such a design can take thousands of engineering hours because of the laborious drafting and calculations required.

The most important obstacle encountered in this design will be adherence to the task schedule. Communication with the contracting agents in Sydney presents the common problem of lag time encountered with great distance. Because Sydney is eight hours ahead of us, communication with NECL will be somewhat difficult. In the past, Willis-Martin has also lost valuable time on such projects by failure to coordinate the drafting tasks with the engineering design. If we can improve on the present method of prioritizing drafting jobs, we can easily meet the six-month time frame and thereby win the $44,000 bonus.

Proposed Procedure

I would begin by conducting a detailed analysis of the reflected ceiling plans of the West Terminal Building. Then I would analyze the electrical and mechanical drawings to produce a rough estimate of the amount of coverage and quantity of equipment that needs to be ordered. NECL has arranged to fax all necessary information for the design, and a Willis-Martin employee visit to the site will fortunately not be necessary. As the analysis progresses, NECL would provide on-site inspection of all areas that are not clear on the as-built drawings.

Measurements of the West Terminal Building would take approximately two weeks. The Terminal is 100 meters long and six stories high, a total of about 1.2 million square meters of coverage. Such a project would demand time-consuming acoustical analysis of each section of the building to determine ambient-noise conditions, area of coverage per grade of service, floor and ceiling finishes for reverberation calculations, and architectural ceiling details for speaker-mounting considerations.

The specifications call for two grades of service. These grades pertain to constancy of sound pressure levels (spl) and articulation loss values. In order to know where to place the speakers to meet these specs, detailed calculations must be made. This can be done using existing Willis-Martin software, SPL, written in BASIC. Cable estimates would be made by calculating line-loss values depending on the power tap of each speaker and the conduit runs already installed by the General Contractor on the job. Lotus 1-2-3 should work effectively for these power and cable estimations. A detailed conduit riser diagram would be developed during completion of this task.

As the systems on the six levels of the Terminal are

Notice that the writer describes the opportunity in monetary terms.

Specific statements of what the writer would do.

The writer shows that he has thought the problem through carefully.

designed, we could effectively coordinate the drafting department's role in this project. Each homerun of speakers must be plotted on the existing reflected ceiling plans, and all audio functional details (i.e., power amplifiers and control systems) must be drafted on our own title block. To reduce the present waiting time, at least one full-time draftsperson could be used on this job. The draftsperson's familiarity with the project would increase productivity as well.

Willis-Martin has completed designs of this nature before, but never of this scale. The contracted fee we would receive for this initial stage is $80,000, after equipment costs, a potentially lucrative job if handled properly. If we could earn the $44,000 bonus as well, a substantial profit will be assured. This design presents Willis-Martin with an opportunity to establish itself as a major supplier in the international market.

Costs

The costs incurred by the design team would be:

Mark Banham	504 hours at $12.00/hr. =	$ 6,048.
Additional Engineering	250 hours at $15.00/hr. =	3,750.
Drafting	1000 hours at $6.75/hr. =	6,750.
	Total	$16,548.

Credentials

I have been an audio/video systems engineer for Willis-Martin for one year and have completed several smaller distributed loudspeaker designs in that time period, including the Tropicana Hotel and Casino sound system, and Remmington Park Racetrack System. I also specialize in communications design at Drexel University.

Task Schedule

Notice that the writer lists the tasks specifically and that he uses a parallel structure—the imperative mood—for each task.

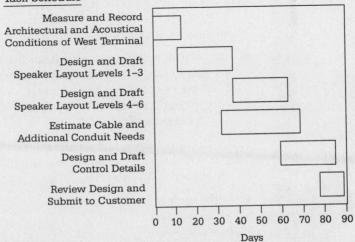

Measure and Record Architectural and Acoustical Conditions of West Terminal

Design and Draft Speaker Layout Levels 1–3

Design and Draft Speaker Layout Levels 4–6

Estimate Cable and Additional Conduit Needs

Design and Draft Control Details

Review Design and Submit to Customer

0 10 20 30 40 50 60 70 80 90

Days

References

Martin, Daniel W. 1982. Sound Reproduction and Recording
    Systems. In Electronics Engineers' Handbook, ed. D. Fink
    and D. Christiansen, New York: McGraw-Hill, Inc.
Kissinger, Corey, Sound System Engineer, Willis-Martin, Inc.
    Personal interview. Philadelphia, 2 December 1989.
Ritz, James, Chief Audio Engineer, Willis-Martin, Inc. Personal
    interview. Philadelphia, 3 December 1989.

WRITER'S CHECKLIST

The following checklist covers the basic elements of a proposal. Any guidelines established by the recipient of the proposal should of course take precedence over these general suggestions.

1. Does the summary provide an overview of
   a. the problem or the opportunity?
   b. the proposed program?
   c. your qualifications and experience?
   d. the expected completion date?
   e. the budget?

2. Does the introduction define
   a. the background leading up to the problem or the opportunity?
   b. the problem or the opportunity itself?

3. Does the description of the proposed program
   a. cite the relevant professional literature?
   b. provide a clear and specific plan of action?

4. Does the description of qualifications and experience clearly outline
   a. your relevant skills and past work?
   b. those of the other participants?
   c. your department's (or organization's) relevant equipment, facilities, and experience?

5. Do the appendixes include the relevant supporting materials, such as a task schedule, a description of evaluation techniques, and evidence of other successful projects?

6. Is the budget
   a. complete?
   b. accurate?

EXERCISES

1. Write a proposal for a research project that will constitute the major assignment in this course. Start by defining a technical subject that interests you. (See Chapter 4 for a discussion of choosing a topic.) Using abstract journals and other bibliographic tools, create a bibliography of articles and books on the subject. Then make up a reasonable real-world context: for example, you could pretend to be a young civil engineer whose company is considering buying of a new kind of earth-moving equipment. Address the proposal to your supervisor, requesting authorization to investigate the advantages and disadvantages of this new piece of equipment.

2. The following draft of a proposal (Dennis 1990) was written by a student who was working part time as a lab assistant for a company that uses large turbines in its manufacturing facility. You are J. Smith, the reader. Write a memo back to John Dennis, explaining to him the strengths and weaknesses of the proposal and recommending appropriate revisions.

Tyler Manufacturing

To: J. Smith, Research and Development
From: J. Dennis, NDT Lab
Subject: Non Destructive Evaluation of Plasma Sprayed Coatings
Date: 8 May 19XX

Purpose

This memo describes a proposal to investigate a method for evaluating the corrosion resistant properties of plasma coated turbines and shaft before damage caused by corrosion hinders operation.

Summary

Turbines and shafts are periodically coated with plasma sprays as protection against corrosion. After being coated the turbines are in virtually continuous operation until time to recoat the shaft and turbine. At times it is necessary to shutdown turbine operation due to damage caused by corrosion. The shutdown, dismantling, and repair of turbines is very costly and could be avoided if the quality of the corrosion resistant plasma coating could be determined prior to operation. Ultrasonic evaluation of the plasma coating is suggested as the most promising technique.

It will require 80 hours to develop a method to adequately determine the quality of plasma coatings. Eastern University's Materials Lab will be hired to determine the actual concentrations of sample coatings by destructive means. Our NDT lab will then tabulate ultrasonic data from the samples and be able to predict the quality of plasma coated turbine and shafts in the future.

Introduction

We are at times forced to shutdown operation of our turbines because of damage caused by corrosion. It is often found that a weakness in the corrosion resistant plasma coating is the cause of the corrosion. The shutdown and repair of the turbines, of course, is very costly and could be avoided if the quality of the plasma coating was satisfactory prior to turbine operation.

A plasma coating is a mixture of aluminum and polyester sprayed onto turbine shafts and turbines in order to resist corrosion. Because the plasma is sprayed onto a metallic surface there is no guarantee of evenly distributed polyester throughout the coating. A thin distribution of polyester in the coating does not adequately protect the metal from corrosion. Currently there is no non-destructive way of determining the percent of polyester in the plasma sprayed coatings. However, research has been documented by D. P. Almond, in NDT International (Dec. 1980), on the behavior of ultra sound through plasma coatings. By measuring ultrasonic characteristics through known concentrations of plasma coatings it will be possible to predict unknown concentrations of polyester and make the necessary adjustments.

Proposed Procedure

According to Almond, ultrasonic wave characteristics, specifically, velocity and attenuation, will be different for different concentrations of polyester in the plasma. Sound waves travel faster through the metal particles of the plasma and attenuation is much greater through high concentrations of polyester.

I propose that we fabricate samples of plasma coatings and hire Eastern University's Materials Lab to determine accurately by destructive methods the concentration of polyester for a range of concentrations. In our Non Destructive Laboratory we can measure velocity and attenuation characteristics through known concentrations of polyester and accumulate data. From the data we can reliably predict the polyester concentrations on our plasma coated turbines and make adjustments.

Turbines and shafts can easily be inspected on location by technicians from the NDT lab.

Costs

This project would require 8 hours to fabricate the samples, 32 hours to destructively evaluate the polyester concentrations and 40 hours to accumulate ultrasonic data, for a total of 80 hours.

Fabrication of Samples:	$100
Destructive Evaluation (Polyester Concentration):	200
Ultrasonic Tabulation:	200
Total:	$500

Credentials

I have worked with ultrasonics in the NDT lab for five months and I am quite familiar with the techniques required to formulate the necessary data.

Eastern University's Material Lab has excellent facilities for determining the polyester concentrations very accurately.

Reference

Almond, D. P., Dec. 1980, NDT International, "Ultrasonic Testing of Plasma Sprayed Coatings," p. 291–295.

3. The following proposal (Roberts 1990) was written by a student who was working part time as a lab assistant for a company that designs and manufactures security devices and systems for the banking industry. You are A. Partridge, the reader. Write a memo back to Linda Roberts, explaining to her the strengths and weaknesses of the proposal and recommending appropriate revisions.

DATE:	October 18, 1990
TO:	A. Partridge, Director of Development, Securities Inc.
FROM:	L. Roberts, Development Design Engineer, Securities Inc.
SUBJECT:	Proposal to Design a Digital Flexible SafePac™, FlexPac, that looks more realistic than the existing SafePac™.

Purpose

This memo describes a proposal a digital flexible SafePac™ that will provide a more realistic pack and update the technology of the current SafePac™ system.

Summary

There has been a recent request from the Marketing Department to make the existing SafePac™ look more realistic. It seems robbers know what to look for and can pick out the larger and more rigid pack from a standard pack of billnotes. This leaves our customers to face an angry robber or a live pack or both. In this memo, I propose a new digital receiver design along with a packaging, battery, and smoke and tear gas generator redesign. The result will be a new flexible pack that will closely model a standard pack of billnotes. This new FlexPac will cost not more than $30.00 per pack to manufacture. The whole project can be done at a cost $17,000.00 to the company in eight months.

Problem Definition

The bank security business is becoming more and more competitive everyday evidenced by the increasing number of companies coming out with systems similar to SafePac™. Although SafePac™ is the market's leader in sales, there is a growing need for a more realistic receiver module. It seems that in recent months we have been getting reports that the robbers know what to look for when they rob a bank and they can pick out the SafePac™ from a standard pack of billnotes. When this occurs not only does the SafePac™ fail as a deterrent, but it also poses a dangerous situation for the customer who might come to harm from the angered robber or the pack itself or both. So far Securities Inc. has only had to clean up a few banks. However, if something worse were to happen the name of Securities Inc. could be dealt a heavy blow both legally as well as for our reputation for safety.

The existing module is made up of a comformally coated electronics module, a battery, and a smoke and tear gas generator placed in a cut out stack of 200 billnotes. The entire package is rigid and twice as large as a standard pack of bills. The analog circuitry used in the existing pack has already been laid out so that space efficiency is maximized, leaving the physical dimensions of the components as the limiting factor. The only way to reduce the size of the current electronics would be to put the existing design onto a piece of ceramic in the form of a hybrid. This process is expensive. One can count on at least a $20,000 tooling and layout charge. Then each hybrid will cost approximately $30 as compared to the $22.50 it costs for the current electronics and electronics assembly (a 33% increase).

Another alternative to reduce the size of the receiver is to design a new digital receiver that will work with the existing transmitter and the rest of the system. This new design would be implemented in the form of an ASIC, Application Specific

Integrated Circuit, which brings the size of the majority of the electronics to roughly the size of a postage stamp and the height of a nickel. Due to the fast paced advancements in this technology it is now possible to produce a fully tested ASIC layout from a paper design for $10,000 and actual units for $20 per in quantities of 100,000.

Reducing the size of the smoke and tear gas generator is trivial. A design was proposed two years ago that fits this module in a 1.5″ × 2.0″ × 0.4″ area while still producing at least two thirds the smoke and tear gas of the existing canister. Bill Evans, Head of the Pyrotechnics Group, has assured me of this design and could have prototypes within a month with some notice.

This proposed new receiver design will fit in half the area of the present design while maintaining the high operating standards. Packaging studies will also be done to take full advantage of the reduced size and make it as realistic as possible. Since the problem of a realistic appearance of the receiver pack is one shared between all of our competitors, I would not be surprised if they too were looking into some kind of reduction scheme. As it is, our packs look the most realistic of what is available in the industry. It seems the first realistic pack to come could earn itself a large portion of the skeptic consumer (ie, California, Arizona where currency security systems have not been able to sell well in the past).

Proposed Procedure

The new FlexPac design must at least meet or exceed the SafePac™ system on all specifications including performance, reliability and cost. The following outlines what will be expected of the FlexPac design.

Electrical Specifications:

1. Circuit must not consume more than 30mA of current in the active state.
2. Circuit must be able to provide 900J of energy to the firing squib.
3. Acceptable bandwidth for the ASIC must be within ±2 hertz.
4. ASIC must operate between 3.5 to 7VDC.
5. ASIC must be programmable to change time delays in the firing circuit and the timeout reset.
6. All components outside the ASIC must be surface mount, pick and place components.
7. Battery life must be no less than 90mAH.
8. Internal impedance of the battery must not exceed 500Ω.
9. Equivalent series resistance, ESR, of the firing capacitor must not exceed 100Ω.

Mechanical Specifications:

1. All electronics, the smoke and tear gas generator, and the battery must fit into a 4.0″ × 2.0″ × 0.4″ cutout made within a stack of 100 bills.
2. The smoke and tear gas generator must produce no less than two thirds the amount of smoke, dye, and tear gas than that of the existing canister.
3. Pack must have two points of flexture symmetrical to each other 1.25″ from the center of the pack that are able to bend to 45° in either direction.
4. Pack must have a reed switch in such a place that will allow the new pack to be used with the current base plate.

Financial Specifications for Development Group:

1. A non-recurring engineering cost not to exceed $12,000.00 for the guaranteed layout and 100 prototypes of the paper design ASIC.
2. Cost of each additional ASIC not to exceed $25.00 in production runs of 10,000 and less.
3. A non-recurring engineering cost not to exceed $2000.00 for smoke and tear gas canister development and at least 50 prototypes.
4. A non-recurring engineering cost not to exceed $2000.00 for battery repackaging study and 100 prototypes of the battery and the battery holder.
5. A non-recurring engineering cost not to exceed $1000.00 for packaging study for the pack itself and materials for 20 packs.
6. All other costs (ie, Technician time, lab equipment time) to be taken care of in the allotted development budget.

Financial Specifications for Production:

1. Total cost of Flexpac (product ready for shipping) not to exceed the cost of the existing pack by more than 15%.

Credentials

I have been working with SafePac™ for two years now and I am very familiar with every aspect of the design. I have also designed digital circuits of this magnitude with complete success in the past. The ASIC will be produced outside by a custom chip manufacturer (probably Sony or Texas Instruments). As I mentioned earlier, the smoke and tear gas generator has already been designed for a previous project. The rest of the project is just a matter of repackaging all of these components.

Task Schedule

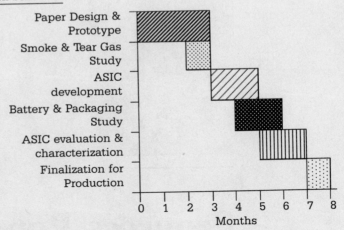

References

Grabel, Arvin. Design Your Own ASIC Using Your Own PC. BYTE, June 1987, p. 29–34.

Millman, Jacob. ASICs Are Cutting VSLI Down to Size. Electronics World, February 1988.

Troy, E. ASICs, Ten Pounds of Circuit in a Five Pound Bag. EDN July 13, 1988.

Rex July, Midwest Sales Representative "Incident Report" to L. Caporoni, March 2, 1988.

Cailer, James, Northeast Sales Representative "Incident Report" to L. Caporoni, July 23, 1988.

Kaahn, Chris, Northeast Sales Representative "Incident Report" to L. Caporoni, May 7, 1988.

Oxenfell, Tim, Head Marketing Executive for SafePac™, Interview with author, Valley Forge, May 13, 1988.

Oxenfell, Tim, Head Marketing Executive for SafePac™, Interview with author, Valley Forge, July 30, 1988.

Oxenfell, Time, Head Marketing Executive for SafePac™, Interview with author, Valley Forge, August 1, 1988.

REFERENCES

Banham, M. 1990. Proposal to design the first phase of the public-address system for the Sutherland International Airport in Sydney. Unpublished document.

Breen, D. 1985. Proposal to modify St. Thomas College's airgun. Unpublished document.

Dennis, J. 1990. Non destructive evaluation of plasma sprayed coatings. Unpublished document.

Roberts, L. 1990. Proposal to design a digital flexible SafePac™, FlexPac, that looks more realistic than the existing SafePac™. Unpublished document.

21

Progress Reports

A progress report communicates to a supervisor or sponsor the current status of a project that has been begun but is not yet completed. As its name suggests, a progress report is an intermediate communication—between the proposal (the argument that the project be undertaken) and the completion report (the comprehensive record of the completed project).

Although progress reports sometimes describe the entire range of operations of a department or division of an organization—such as the microcellular research unit of a pharmaceutical manufacturer—they usually describe a single, discrete project, such as the construction of a bridge or an investigation of excessive pollution levels in a factory's effluents.

Progress reports allow the persons working on a project to check in with their supervisors or sponsors. Supervisors are vitally interested in the progress of their projects, because they have to integrate them with other present and future commitments. Sponsors (or customers) have the same interest, plus an additional one: they want the projects to be done right—and on time—because they are paying for them.

The schedule for submitting progress reports is, of course, established by the supervisor or the sponsor. A relatively short-term project—one expected to take a month to complete, for example—might require a progress report after two weeks. A more extensive project usually requires a series of progress reports submitted on a fixed schedule, such as every month.

The format of progress reports varies widely. A small internal project might require only brief memos, or even phone calls. A small external project might be handled with letters. For a larger, more formal project—either internal or external—a formal report (see Chapter 19) generally is appropriate. Sometimes a combination of formats is used: for example, reports at the end of quarters and memos at the end of each of the other eight months. (See Chapter 13, "Memos," and Chapter 14, "Letters," for discussion of these formats.) This chapter discusses the strategy that applies to progress reports of every length and format.

The following example might help to clarify the role of progress reports. Suppose your car is old and not worth much. When you bought it for $600 three years ago, you knew you would be its third—and last—owner. Recently you've noticed some problems in shifting. You bring the car to your local mechanic, who calls within an hour to report that a small gasket has to be replaced, at a total cost of $25. Pleased with the mechanic's progress report, you tell him to do the work; you'll pick up the car later in the afternoon.

If all projects in business and industry went this smoothly—with no major technical problems and no unanticipated costs—every progress report would be simple to write and a pleasure to read. The writer would merely photocopy the task schedule from the original proposal, check off the tasks completed, and send the page on to the supervisor or sponsor. The progress report would be a happy confirmation that the project personnel had indeed accomplished what they promised they would do by a certain date. The reader's task would be simple and pleasant: to congratulate the project personnel and tell them to continue.

Unfortunately, most projects don't go so smoothly. Suppose that you had received a different phone call from the mechanic. Your transmission is ruined, he tells you; a rebuilt transmission will cost $300; a new one, $400. "What do you want me to do?" he asks. You tell him you'll call back in a few minutes. You sort through your options:

☐ bring your car to another mechanic for another estimate
☐ have the rebuilt transmission installed
☐ have the new transmission installed
☐ take the bus to the garage, retrieve the license plates, and sell the car for scrap

This version of the story illustrates why progress reports are crucial. Managers—you, in the case of the car—need to know the progress of their projects so that they can make informed decisions when things go wrong. The problem might not be what it was thought to be when the proposal was written, or unexpected difficulties might have hampered the work. Perhaps equipment failed; perhaps personnel changed; perhaps prices went up. An experienced manager could list a hundred typical difficulties.

The progress report is, to a large extent, the original proposal updated in the light of recent experience. If the project is proceeding smoothly, you simply report the team's accomplishments and future tasks. If the project has encountered difficulties—if the anticipated result, the cost, or the schedule has to be revised—you need to explain clearly and fully what happened and how it will affect the overall project. Your tone should be objective, neither defensive nor casual. Unless ineptitude or negligence caused the problem, you're not to blame. Regardless of what kind of news you are delivering—good, bad, or mixed—your job is the same: to provide a clear, honest, and complete account of your activities and to forecast the next stage of the project.

Writing the Progress Report

Progress reports vary considerably in structure because of differences in format and length. Written as a one-page letter, a progress report is likely to be a series of traditional paragraphs. As a brief memo, it might also contain section headings. As a report of more than a few pages, it might contain the elements of a formal report.

Regardless of these differences, most progress reports share a basic structural progression. The writer

☐ introduces the readers to the progress report by explaining the objectives of the project and providing an overview of the whole project
☐ summarizes (if appropriate) the discussion

SUMMARY

In November, Phase I of the project—to define the desired capabilities of the testing device—was completed.

The device should have two basic characteristics: versatility and simplicity of operation.

1. Versatility. The device should be able to test different kinds of logic modules. To do this, it must be able to "address" each module and ask it to perform the desired task.
2. Simplicity of Operation. The device should be able to display the response from the module being tested so that the operator can tell easily whether the module is operating properly.

Phase II—to design the device—is under way. We would like to meet with Research and Development to discuss a problem we are having in designing the device so that it can "address" each module. Please see "Discussion" and "Conclusion," below.

FIGURE 21.1   Summary of a Progress Report

□ discusses the work already accomplished and the work that remains to be done, and speculates on the future promise and problems of the project
□ provides a conclusion that evaluates the progress of the project

If appropriate, appendixes are attached to the report.

In composing the progress report, leave the summary for last. As with all technical documents, you cannot accurately summarize material that you have not yet written. However, you can write the other elements of the standard progress report in the sequence you will present them.

The best way to draft the progress report is to start with a copy of your proposal; many of the elements of the progress report are taken directly from it. If you are writing manually, photocopy the proposal and then cut and paste your changes. If you are working on a word processor, copy your proposal. The statement of the problem or opportunity that led to the project, for instance, can often be presented intact. The discussion of the past work is often a restatement of a portion of the proposed program, with the future tense changed to past. You might want to reproduce your task schedule—modified to show your current status—and your bibliography.

INTRODUCTION

This is the first monthly progress report on the project to develop a new testing device for our computer system. The goal of the project is to create a device that can debug our current and future systems quickly and accurately.

The project is divided into four phases:

    I. define the desired capabilities of the device

    II. design the device

    III. manufacture the device

    IV. test the device

FIGURE 21.2   Introduction to a Progress Report

Summary

Progress reports of more than a few pages likely contain a summary section, which, like all other summaries, provides a brief overview of the discussion for those readers who do not need all the technical details. Often, the summary enumerates the accomplishments achieved during the reporting period and then comments on the current work. Notice how the summary shown in Figure 21.1 calls the reader's attention to a problem described in greater detail later on in the progress report.

Introduction

The introduction provides background information. First, of course, it identifies the document as a progress report and identifies the period of time the report covers. If more than one progress report has been (or will be) submitted, the introduction places the report in the proper sequence—for example, as the third quarterly progress report. Second, the introduction states the objectives of the project. These objectives have already been defined in the proposal, yet they are generally repeated in each progress report for the benefit of readers who are likely following the progress of a number of projects simultaneously. And third, the introduction briefly states the phases of the project. Figure 21.2 provides an example of an introduction to a progress report.

Discussion

The discussion elaborates points listed in the summary. The discussion serves readers who want a complete picture of the team's activities during the period covered; many readers, however, will not bother to read the discussion unless the summary highlights an unusual or unexpected development during the reporting period.

Of the several different methods of structuring the discussion section, perhaps the simplest is the past work/future work scheme. After describing the problem, the writer describes all the work that has been completed in the present reporting period and then sketches in the work that remains. This scheme is easy to follow and easy to write.

Often the discussion is structured according to the tasks involved in the project. If the project requires that the researchers work on several of these tasks simultaneously, this structure is particularly effective, for it enables the writer to describe, in order, what has been accomplished on each task. Often, the task-oriented structure incorporates the past work/future work structure:

III. Discussion
    A. The Problem
    B. Task I
        1. past work
        2. future work
    C. Task II
        1. past work
        2. future work

Figure 21.3 exemplifies the standard chronological progression—from the problem to past, present, and future work. Notice, however, that the writer combines generic and specific phrases in the headings.

Conclusion

A progress report is, by definition, a description of the present status of a project. The reader will receive at least one additional communication—the completion report—on the same subject. The conclusion of a progress report, therefore, is more transitional than final.

In the conclusion of a progress report, you evaluate how the project is proceeding. In the broadest sense, you have one of two messages:

☐ Things are going well.
☐ Things are not going as well as anticipated.

If the news is good, convey your optimism, but avoid overstatement.

DISCUSSION

The Problem

Isolating and eliminating design and production errors has always been a top priority of our company. Recently, the sophistication of new computer systems we are producing has outpaced our testing capabilities. We have been asked to develop a testing device that can quickly and accurately debug our current and projected systems.

Currently, we have no technique for testing the individual logic modules of our new systems. Consequently, we cannot test the system until all of the modules are in place. When we do discover a problem, such as a timing error in one of the signals, we have to disassemble the system and analyze each of the modules through which the signal flows. Even after we have discovered a problem, we cannot know if that is the only problem until we reassemble the system and test it again. Although this testing method is effective—we are well within acceptable quality standards—it is very inefficient.

The solution to this problem is to develop a device for testing the logic modules individually before they are installed in the system. That is the overall objective of the project. Phase I of the project involved our determining the desired capabilities of this device.

Work Completed: Phase I—Determining the Desired Characteristics of the Testing Device

The first characteristic of the testing device must be versatility. Over the next two years, we will be introducing three new systems. This rate of introduction is expected to continue at least through 19XX. Although we cannot foresee the specific components of these new systems, all are expected to incorporate the latest large-scale integration techniques, in conjunction with microprocessor control units.

FIGURE 21.3   Discussion Section of a Progress Report

This basic structure will enable us to employ separate logic modules, each of which performs a specific function. The modules will be built on separate logic cards that can be tested and replaced easily. To achieve this versatility, the testing device should be able to "address" each module automatically and ask it to perform the desired task. Because the testing device will be asked to handle many logic modules, it should be able to distinguish between the different modules in order to ask them to perform their appropriate functions.

The second characteristic of the testing device should be simplification of operation: the ability to display for the operator the response from the module being tested. After the testing device transmits different command and data lines to the module, it should be able to receive a status or response word and communicate it to the operator.

Future Work

We are now at work on Phase II—designing the device to reflect these desired characteristics.

We are analyzing ways to enable the device to "address" automatically the various logic modules it will have to test. The most promising approach appears to be to equip each logic module with a uniform integrated circuit—such as the 45K58, a four-bit magnitude comparitor—that can be wired to produce a unique word that indicates the board address of that module.

The display capability appears to be a simpler problem. Once the module being tested has executed a command, it will generate a status word. The testing device will receive this word by sending an enable signal to the status enable pin on the unit holding the module. Standard LCD indicators on the front panel of the testing device will display the status word to the operator.

FIGURE 21.3   *Continued*

CONCLUSION

Phase I of the project has been completed successfully and on schedule.

We are now at work on Phase II and hope to complete the basic schematic of the testing device within two weeks. The one aspect of Phase II that is giving us trouble is the question of versatility. Although we can equip our future systems with a uniform integrated circuit that can be wired to produce unique identifiers, this procedure will create future headaches for R&D. We have arranged to meet with R&D next week to discuss this problem.

FIGURE 21.4   Conclusion of a Progress Report

Overstated

We are sure the device will do all that we ask of it, and more.

Realistic

We expect that the device will perform well and that, in addition, it might offer some unanticipated advantages.

Beware too, of promising early completion. Such optimistic forecasts are rarely accurate, and of course it is always embarrassing to have to report a failure after you have promised success.

On the other hand, don't panic if the preliminary results are not as promising as had been anticipated, or if the project is behind schedule. Readers know that the most sober and conservative proposal writers cannot anticipate all problems. As long as the original proposal contained no wildly inaccurate computations or failed to consider crucial factors, don't feel personally responsible. Just do your best to explain what happened and the current status of the work. If you suspect that the results will not match earlier predictions—or that the project will require more time, personnel, or equipment—say so, clearly. Don't lie in the hope that you eventually will work out the problems or make up the lost time. If your news is bad, at least give your reader as much time as possible to deal with it effectively.

Figure 21.4 shows the conclusion for the computer-testing device progress report.

Because the design of the testing device will affect the future design of the new computer systems, the writer has wisely decided to ask for technical assistance—from the Research and Development Department.

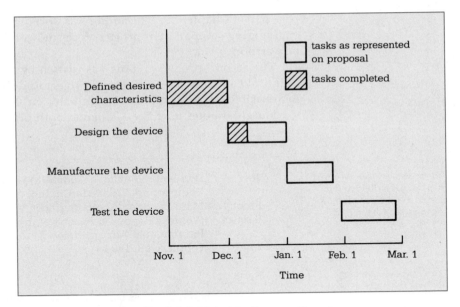

FIGURE 21.5   Updated Task Schedule of a Progress Report

Appendixes

In the appendixes to the report, include any supporting materials that you feel your reader might wish to consult: computations, printouts, schematics, diagrams, charts, tables, or a revised task schedule. Be sure to cross-reference these appendixes in the body of the report, so that the reader can consult them at the appropriate stage of the discussion.

Figure 21.5 shows an updated task schedule. The writer has taken the original task schedule from the proposal and added crosshatching to show the tasks that have already been completed.

Sample Progress Reports

Two sample progress reports are included here:

1. "Progress Report on Phase I of the Sutherland International Airport Project" (Banham 1990).
2. "Progress Report: Removing Miscible Organic Pollutants from Water Effluents" (Charles 1983).

The first progress report was written as a follow-up to the proposal on page 565 of Chapter 20. (The completion report on this study begins on page 610 of Chapter 22.) The author was working as a co-op student for an engineering company. His project was a study of the acoustical needs of a new airport.

The second progress report was written by a scientist employed by a chemical company. The subject is a project to devise a method of removing miscible organic pollutants from water effluents.

Marginal notes have been added to both progress reports.

To save space, this memo has not been divided into pages. For a discussion of memos, see Chapter 13.

The purpose statement defines the purpose of the memo, identifies the period the memo covers, and states the general purpose of the investigation.

The summary clearly identifies the tasks that have been completed and the work that is still in progress.

The introduction fills in the background, indicating the relationship between Phase I and the other phases.

The writer adds "Work in Progress" to his heading because he doesn't want to mislead his reader. The section begins with an overview.

The writer explains his principal accomplishments.

To: Don Leith, Manager, Audio/Video Systems Engineering, Willis-Martin, Inc.

From: Mark Banham, Audio Engineer, Willis-Martin, Inc.

Subject: Progress Report on Phase I of the Sutherland International Airport Project

Date: February 18, 1990

Purpose

This is the progress report on the first month's work on Phase I of the project to design the public address and background music system for the Sutherland International Airport in Sydney.

Summary

In the first month's work on Phase I, we have completed measuring the West Terminal floor surface area and analyzed the floor and ceiling finishes. We are currently modifying the Willis-Martin BASIC program SPL, to do the necessary calculations for the loudspeaker layout design. We anticipate that Phase I will be completed successfully within the 90-day target date.

Introduction

In June, 1989, Willis-Martin was awarded a $1.2 million contract by the National Electronics Company, Ltd. (NECL) to design and install the public address and background music system for the Sutherland International Airport in Sydney. Phase I of the project, which began in December, 1989, called for the design; Phase II calls for the installation; Phase III calls for comprehensively testing and modifying the system.

Work Completed and Work in Progress

Following are my analysis of the West Terminal Building acoustics and a discussion of my progress on the loudspeaker requirement calculations.

West Terminal Building Acoustics

The West Terminal Building, which houses the major public areas of this airport, such as arrival and departure lobbies and a parking garage, is approximately 5 million square feet in area. The diversity of the types of spaces included in the sound system coverage required detailed categorizing. On each level, we measured the floor space from the plans and indicated all surface finish materials. The reflected ceiling plans served as a guide to the elevations above the finished floor, which determined the speaker power needed to service each area. The specifications of this job called for two different grades of quality of sound. These grades are directly related to the uniformity of sound pressure within any room. Deviations in the sound pressure level, which represent changes in volume, decrease audibility.

a. Grade of Service 1

In those areas designated as Grade 1, the spl deviation was limited to ± 3dB. Some areas included in this group are the public areas where flight information must be heard, and special official sections. The deviation in spl is directly a function of the loudspeaker spacing. About 67% of the space in the West Terminal was specified at Grade of Service 1.

b. Grade of Service 2

Areas requiring this grade of service included employee areas and parking garages. The spl is permitted to vary by ± 4.5dB in these sections. Such a requirement means a greater spacing between speakers and therefore less equipment and lower costs. Approximately 33% of the West Terminal Building called for Grade of Service 2.

Loudspeaker Requirement Calculations

We are using three major assumptions in modifying the SPL to calculate the number of speakers required for a job like this:

- the spl in any room must be maintained within some tolerance
- the articulation losses of the grid of speakers at any spacing must fall within acceptable limits
- a variable reverberation-adjustment factor must be factored into the program

We have experienced some minor problems rewriting the SPL program to handle the calculations, but we have asked for assistance from an expert programmer and expect to have the program modified and running within the next three days.

Future Work

Once the loudspeaker layout design is completed, we will turn to the following other tasks:

The writer describes the task he is currently working on.

The writer clearly explains the difficulty, what he is doing about it, and how it will affect the progress of the project.

The writer lists the future tasks.

- estimating the cable requirements
- determining the optimum cable routing
- estimating homeruns and line-power needs
- determining the control and power system needs

Conclusion

The writer indicates when the completion report will be submitted.

Phase I of the project is proceeding according to schedule and without any significant delays. If we can get the modified SPL program up and running in the next few days, as we anticipate, we should have no problem completing Phase I within the 90-day target period. You should receive the Phase I completion report by March 15, 1990.

Updated Task Schedule

The writer updates his task schedule.

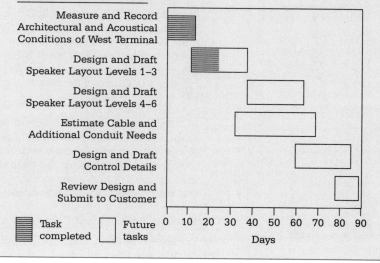

The type of report.

The subject of the report.

PROGRESS REPORT:

REMOVING MISCIBLE ORGANIC POLLUTANTS

FROM WATER EFFLUENTS

Submitted by: Martin Charles
Biologist Class I
Marshall Chemicals, Inc.

to: Dr. Helen Jenners
Chief, Biological
Division
Marshall Chemicals, Inc.

May 11, 19XX

The writer clearly defines the nature of this report and places this progress report within the sequence of reports.

The writer defines the purpose of the study.

The writer fills in the background of the problem.

The writer repeats the goal of the project.

INTRODUCTION

This progress report describes my findings after the first week of the investigation. The second progress report will follow in one week. The project is expected to be completed within two weeks, at which time a completion report will be submitted.

The purpose of this project is to devise a method for removing miscible organic pollutants from water effluents.

DISCUSSION

The Problem

Currently, there are two methods of removing miscible organic pollutants from water effluents: column filtration and absorbent screens.

Column filtration, in which the effluent is pumped through an absorbent-packed column, is ineffective for rivers and streams, because of their great flowrates. The result is that a large percentage of the effluent remains unfiltered.

Absorbent screens, in which the effluent passes through an absorbent screen, are ineffective because the screens create resistance, causing the effluent to flow around, rather than through, them.

The goal of this project is to devise a more effective method than column filtration or absorbent screens for removing miscible organic pollutants from water effluents.

This introductory
paragraph forecasts the
discussion that follows.

Notice how the writer
uses chronology to
emphasize the coherence
of the discussion.

Work Completed

To date, we have devised the general principle for the new method—the use of floatable absorbent particles—and selected an absorbent.

The general principle behind the new method is to introduce floatable absorbent particles into the bottom of the stream or river. As the particles float to the surface, they come in contact with and absorb pollutants. Particle floatability is limited, to allow for resubmergence in turbulent water conditions. At a calm section of water downstream, the particles are surface-skimmed.

We then tested three absorbents—Absorbtex 115, Filtrasorb 553, and activated carbon—to determine cost-effectiveness. These data are listed below.

Absorbent	Cost/Lb	Absorption/ Lb Absorbent	Lbs Absorbed/$1
Absorbtex 115	$9.50	20.9 lb CCl_4/lb	2.2
Filtrasorb 553	$7.65	18.6 lb CCl_4/lb	2.4
Activated Carbon	$2.89	10.9 lb CCl_4/lb	3.7

These data show that although activated carbon is the least-effective absorbent, it is—by far—the most cost-effective.

Future Work

In our preliminary tests, each of the three absorbents presented different technical problems. Because of the

Here the writer explains
how the nature of the
problem affected his
research plan.

2

substantial cost advantage of activated carbon, we decided to address the technical problems associated with that absorbent before investigating the other two absorbents.

In our experiments with activated carbon, we discovered that it loses buoyancy after short periods of contact with water. The necessary flotation must be induced by attaching the carbon to a buoyant material. Currently we are testing two materials—polyethylene and paraffin—and one other alternative: affixing the carbon to the outside of glass spheres.

In this paragraph, as well as in the next, the writer uses chronology effectively.

The next phase of study will involve determining the optimum size for the carbon particles. The smaller the size, the greater the surface area, of course, per gram of carbon. On the other hand, smallness increases the risk of water saturation, which causes the particles to sink and makes recovery impossible.

Finally, a method of introducing the carbon into the stream or river must be devised. If coated carbon is the absorbent, weighted milk jugs with water-soluble caps will probably be effective.

Conclusion

The writer attempts to forecast the future of the project.

We foresee no special problems in completing the next phases of the project. The second progress report should answer the questions of coatings and particle size. We expect to conclude the project successfully on time.

3

WRITER'S CHECKLIST

Even though progress reports vary considerably in format and appearance, the basic strategy behind them remains the same. The purpose of this checklist is to help you make sure you have included the major elements of a progress report.

1. Does the summary
   a. present the major accomplishments of the period covered by the report?
   b. present any necessary comments on the current work?
   c. direct the reader to crucial portions of the discussion section of the progress report?

2. Does the introduction
   a. identify the document as a progress report?
   b. indicate the period the progress report covers?
   c. place the progress report within the sequence of any other progress reports?
   d. state the objectives of the project?
   e. outline the major phases of the project?

3. Does the discussion
   a. describe the problem that motivated the project?
   b. describe all the work completed during the period covered by the report?
   c. describe any problems that arose, and how they were confronted?
   d. describe the work remaining to be done?

4. Does the conclusion
   a. accurately evaluate the progress on the project to date?
   b. forecast the problems and possibilities of the future work?

5. Do the appendixes include the supporting materials that substantiate the discussion?

EXERCISES

1. Write a progress report describing the work you are doing for the major assignment you proposed in Chapter 20.

2. The following progress report, titled "The Future of Municipal Sludge Composting," was written by Walter Prentice, an engineer working for the waste resources department of a medium-sized city. Assume the role of his reader, the head of his department. Write him a memo, evaluating the report from the points of view of clarity, comprehensiveness, and style.

Introduction

Sludge composting is a 21-day process by which wastewater sludge is converted into organic fertilizer that is aesthetically acceptable, pathogen-free, and easy to handle. Composted sludge can be used to improve soil structure, increase the soil's water retention, and provide nutrients for plant growth.

Discussion

Sludge composting is essentially a two-step process:

1. Aerated-Pile Composting
   Dump trucks deliver the dewatered raw sludge to the compost site. Approximately 10 tons of sludge is dumped on a 25 yd³ bed of bulking agent (usually woodchips). A front-end loader mixes the sludge into the bulking agent. The mixture is then placed on a compost pad and covered with a blanket of unscreened compost 1 ft thick. This layer is applied to insulate the sludge-bulking agent for ambient temperatures and for preventing the escape of odors from the pile. The air and odors are sucked out of the bulking agent base by an aeration system of pipes under the compost pad. After three weeks, the sludge in the aerated compost pile is essentially free of pathogens and stabilized.
2. Drying, Screening, and Curing the Composted Sludge
   After the aerated pile composting is completed, the pile if spread out and harrowed periodically until it is dry enough to screen. Screening is desirable, because it recovers 80 percent of the costly bulking agent for reuse with new sludge. The screened compost is stored for at least 30 days before being distributed for use. During the curing period, the compost continues to decompose, ensuring an odor- and pathogen-free product.

Future Work

The next step is to determine the cost of sludge composting. Sludge composting using the aerated-compost-pile method is estimated to cost between $35 and $50 per dry ton ($35 for a 50-dry-ton per day operation, $50 for a 10-dry-ton per day operation). These estimates include all facilities, equipment, and labor necessary to compost at a site separate from the treatment plant. Not included are the costs of sludge dewatering, transportation to and from the site, and runoff treatment. These additional factors can raise the cost of composting to $160 per dry ton.

The breakdown of the capital costs for composting follows:

1. Site development—one acre of land is required for every three dry tons per day capacity. Half of the site should be surfaced. Asphalt paving costs about $60,000 per acre. In addition, the site requires electricity for the aeration blowers.
2. Equipment—front-end loader, trucks, tractors, screens, blowers, and pipes are required.
3. Labor—labor represents between a third and a half of the operating costs. Labor is estimated to cost $6 per hour, with five weeks of paid sick or vacation time.

However, a potential market exists for compost. Although the high levels of certain heavy metals in the compost restrict its use in some cases, compost can be used by businesses such as nurseries, golf courses, landscaping, and surface mining. Transportation costs can be high, however.

Conclusion

This information comes from published reports by federal agencies and journal articles. Since the government banned ocean sludge dumping, composting has become a viable method of waste treatment, and much has been written about it.

Before we can reach a final decision about whether composting would be economically justifiable for the city of Corinth, we must add our numbers to the costs above. The process should take about two more weeks, provided that we can get all the information.

3. The following progress report, titled "Hiring Accounting Help or Buying an Electronic Register for *The Crow's Nest*: A Progress Report," was written by a college student who worked part-time in a tavern/restaurant in a major city. In an essay, evaluate the report from the points of view of clarity, completeness, and writing style.

Background

The arrival of the Saratoga for a 30-month renovation program has increased our business about 60 percent. As a result, the preparation of daily reports and inventory calculations has become a lengthy and cumbersome task. What used to take two hours a day now takes almost four. We have two alternatives: hire a part-time accountant (such as a local university student), or invest in an electronic cash register system. Following is a report on my first week's findings in the investigation of these two alternatives.

Work Completed

The going rate for an accounting student is about $4.50/hour. At two hours per night, seven nights per week, our annual costs would be about $3,500, including the applicable taxes. If the student were to take over all your bookkeeping tasks—about four hours per night—the cost would be about $7,000 annually.

Both local colleges have told me on the phone that we would have no trouble locating one or more students who would be interested in such work. Break-in time would probably be short; they could learn our system in a few hours.

The analysis of the electronic cash registers is more complicated. So far, I have figured out a way to compare the various systems and begun to gather my information. Five criteria are important for our situation:

1. overall quality
2. cost
3. adaptability to our needs
4. dealer servicing
5. availability of buy-back option

To determine which machines are the most reputable, several magazine articles were checked. Five brands—NCR, TEC, Federal, CASIO, and TOWA—were on everyone's list of best machines.

I am now in the process of visiting the four local dealers in business machines. I am asking each of them the same questions—about quality, cost, versatility, frequency-of-repair records, buy-back options, etc.

Work Remaining

Although I haven't completed my survey of the four local dealers, one thing seems certain: a machine will be cheaper than hiring a part-time accountant. The five machines range in price from $1,000 to $2,000. Yearly maintenance contracts are available for at least some of them. Also, buy-back options are available, so we won't be stuck with a machine that is too big or obsolete when the Saratoga repairs are complete.

I expect to have the final report ready by next Tuesday.

REFERENCES

Banham, M. 1990. Progress report on Phase I of the Sutherland International Airport Project. Unpublished document.

Charles, M. 1983. Progress report: Removing miscible organic pollutants from water effluents. Unpublished document.

■

22

■

Completion Reports

A completion report is generally the culmination of a substantial research project. Two other reports often precede it. A proposal (see Chapter 20) argues that the writer or writers be allowed to begin and carry out a project. A progress report (see Chapter 21) describes the status of a project that is not yet completed; its purpose is to inform the sponsors of the project how the work is proceeding. The completion report, written when the work is finished, provides a permanent record of the entire project, including the circumstances that led to its beginning.

Understanding the Functions of a Completion Report

A completion report has two basic functions. The first is immediate documentation. For the sponsors of the project, the report provides the necessary facts and figures, linked by narrative discussion, which enable them to understand how the project was carried out, what it found, and, most important, what those findings mean. All completion reports lead, at least, to a discussion of results. For example, a limited project might call for the writer to determine the operating characteristics of three competing models of a piece of lab equipment. The heart of that completion report will be the presentation of those results. Many completion reports call for the writer to analyze the results and present conclusions. The writer of the report on the lab equipment might inform the readers which of the three machines appears to be the most appropriate for his organization's needs. And finally, many completion reports go one step further and present recommendations: suggestions about how to proceed in light of the conclusions. The writer of the lab equipment report might have been asked to recommend which of the three machines—if any—should be purchased.

The completion report also serves as a future reference in three common situations.

First, personnel change: a new employee is likely to consult filed reports to determine the kinds of projects the organization has completed recently. Reports are not only the best source of this information—they are often the only source, because the employees who participated in the project might have left the organization.

Second, an organization contemplating a major new project usually wants to determine how the new project would affect existing procedures or operations. Old completion reports are the best source of this information, too. For example, if the owner of an office complex wants to computerize the temperature control of his buildings, he will bring in an expert to consult the reports on the electrical wiring and the heating, ventilating, and air-conditioning systems to determine whether computerization is feasible both technically and economically.

Third, and perhaps most important, if a problem develops after the project has been completed, employees will turn first to the project's completion report to figure out what went wrong. An analysis of a breakdown in a production line requires the technical description of the production line—the completion report written when the line was implemented.

In these three situations, completion reports are valuable long after the projects they describe have been completed.

Understanding the Types of Completion Reports

Terminology varies from organization to organization. Some organizations consider procedures manuals and policy descriptions to be completion reports because they report on the completion of projects: to establish procedures or policies. Some organizations consider a proposal to be a completion report because it completes a recognizable phase of a project. And because most external proposals—and many internal proposals—are never funded, they are really unintentional completion reports.

In general, however, completion reports fall into two broad categories: physical research reports and feasibility reports.

Physical Research Reports

A physical research report concerns a project involving substantial primary research carried out in a lab or in the field. Primary research lies at the heart of the scientific method: you begin with a hypothesis, conduct experiments to test it, record your results, and determine whether your hypothesis was correct as it stands or needs to be modified or rejected. Primary research means conducting experiments and generating original data (see Chapter 4).

For instance, you work for the Navy. Your supervisors are considering building the hulls of minesweeper ships from glass-reinforced plastic—GRP—instead of wood. GRP would seem to offer several advantages, but first its properties have to be established. How strong is it? How flexible? How well does it withstand temperature variations? The list of questions is long. Each question will be answered by physical research on GRP in the laboratory, on mock-ups of hulls, and even on computer simulations of the material.

Feasibility Reports

A feasibility report documents a study that evaluates at least two alternative courses of action. For example, should our company hire a programmer to write a program we need, or should we have an outside company write it for us? Should we expand our product line to include a new item, or should we make changes in an existing product?

A feasibility report might answer questions of possibility. We would like to build a new rail line to link our warehouse and our retail outlet, but if we cannot raise the cash, the project is not possible at this time. Even if we have the money, do we have the necessary permission from government authorities? If we do, are the soil conditions adequate for the rail line?

A feasibility report might consider questions of economic wisdom. Even if we can raise the money to build the rail line, is it wise? If we use up all our credit on this project, what other projects will have to be postponed or canceled? Is there a less expensive—or a less risky—way to achieve the same goals as a rail link?

Finally, a feasibility report might consider how an action will be received by interested parties. For instance, if your company's workers have recently accepted a temporary wage freeze, they might view the rail link as an unnecessary expenditure. The truckers' union might see it as a threat to their job security. Some members of the general public might be interested parties. Any sort of large-scale construction might affect the environment. Even though your plan might be perfectly acceptable according to its environmental-impact statement—the study required by the government—some citizens might disagree with the statement or might still oppose the project on aesthetic grounds. Whether or not you agree with the objections, going ahead with the project might create adverse publicity.

Feasibility studies often involve primary research. The primary research on GRP might be the first phase in a feasibility study comparing the existing material, wood, with GRP. The primary research answers the technical question of whether GRP would be an effective substitute for wood in the hulls of minesweepers. The feasibility study answers the question of whether it is possible and wise to use GRP.

Structuring the Completion Report

Like proposals and progress reports, completion reports have to make sense without the authors there to explain them. The difficulty is that you can never be sure when your report will be read—or by whom. All you can be sure of is that some of your readers will be managers who are *not* technically competent in your field and who need only an overview of the project, and that others will be technical personnel who *are* competent in your field and who need detailed information. There will be yet other readers, of course—such as technical personnel in related fields—but in most cases the divergent needs of managers and of technical personnel are all you need to consider.

To accommodate these two basic types of readers, completion reports today generally contain an executive summary that precedes the body (the full discussion). These two elements overlap in their coverage but remain independent; each has its own beginning, middle, and end. Most readers will be interested in one of the two, but probably not in both. As a formal report, the typical completion report will contain other standard elements:

title page
abstract
table of contents
list of illustrations
executive summary
glossary
list of symbols
body
appendix

This chapter will concentrate on the body of a completion report; the other elements common to most formal reports are discussed in Chapter 19.

Writing the Body of the Completion Report

The body of a typical completion report contains the following five elements:

1. introduction
2. methods
3. results
4. conclusions
5. recommendations

Some writers like to draft these elements in the order in which they will be presented. These writers like to compose the introduction first because they want to be sure that they have a clear sense of direction before they draft the discussion and the findings. Other writers prefer to put off the introduction until they have completed the other elements of the body. Their reasoning is that in writing the discussion and the findings they will inevitably have to make some substantive changes; therefore, they would have to revise the introduction, if they wrote it first. In either case, brainstorming and careful outlining are necessary before you begin to draft.

The Introduction

The first section of the detailed discussion is the introduction, which enables the readers to understand the technical discussion that follows. Usually, the introduction contains most or all of the following elements:

☐ A statement of the problem or opportunity that led to the project: what was not working, or not working well, in the organization, or what improvements in the operation of the organization could be considered if more information were known.

☐ A statement of the purpose of the report, what exactly it is intended to accomplish.

☐ A description of the background of the project, the facts readers need before they can understand the report.

☐ A statement of the scope of the document, those aspects of the problem or opportunity included in the project and those excluded.

☐ An explanation of the organization of the report, so readers will understand where you are going and why.

☐ A review of the relevant literature, either internal reports and memos or external published articles or even books that help your readers understand the context of your work.

☐ Definitions of key terms that your readers will need before they can follow the discussion.

For more information on introductions, see Chapter 7.

Figure 22.1 shows an introduction to the body of a completion report. The subject of the report (MacBride 1986) is an investigation to determine whether a retail store can increase its sales and reduce its heating bills by instituting an energy-management system. The writer was a student working for the store in its Engineering and Planning Department. The report is titled "Recommendation for Minimizing the Steam Consumption and Regulating the Ambient Temperature at Bridgeport Department Store." Marginal notes have been added.

The Methods

In the methods section of the report, you describe the technical tasks or procedures you performed. If you are reporting on a physical research project carried out in the lab or the field, this section will closely resemble the discussion section of a traditional lab report. If several research methods were available to you, begin by describing why you chose the method(s) you did. Either list the equipment and materials before you describe the research or mention them within the description itself. The preliminary listing is more common when some of the readers are going to duplicate the research. If they simply want to understand what you did, the listing is probably not necessary.

For a feasibility study, you should also begin by justifying your methods. Describe what you did: experiments, observations, theoretical studies, interviews, library research, and so forth. You want to show that you have conducted your research professionally. This will increase not only your readers' ability to understand the findings that follow but also your credibility.

INTRODUCTION

The background.

Bridgeport Department Store, built in 1912, is an expensive building to heat because of its high ceilings, large windows, and lack of wall insulation. Every increase in heating costs hits us harder than it hits our competitors in more modern facilities. Therefore, we must make sure we are using the most modern and cost-effective methods for regulating our store temperature.

Currently, we have an energy-management system, but it regulates only electricity for lights and air-conditioning, not our steam used for heating.

The problem, discussed in monetary terms.

In 1986, Bridgeport experienced a 6 percent increase in steam consumption over 1985 figures. This rise, combined with a 7 percent utility rate increase, resulted in a 15 percent ($67,000) rise in related operating expenses.

In addition, the number of customer complaints relating to the uneven temperatures throughout the building rose by 135 percent over the 1985 figures.

The purpose of the investigation.

The purpose of this study is to determine whether it would be cost effective to expand the capabilities of our current energy-management system to include the monitoring and regulating of steam consumption.

The scope of the investigation.

This study was restricted to changes in software. We do not have the funds to finance hardware changes to our main computer system, which was installed earlier this year.

The relevant literature.

Recent articles by Cottrell and by Gorham on energy management indicate that an effective energy management system can reduce steam usage by 20–25 percent. Donovan and Rossiter, writing on the relationship between store climate and consumer buying behavior, suggest that the more comfortable the shoppers, the longer they stay and the more money they spend.

FIGURE 22.1   Introduction to a Completion Report

METHODS

Before studying the possible enhancement to our current energy-management system, we devised a set of technical, management/maintenance, and financial criteria by which to evaluate the enhancement.

Then we studied the literature on energy-management systems (see References, page 11). We phoned Cottrell and Gorham to follow up on points they raised in their articles. We visited two sites (Jarvis Electronics and the Foundation Society)—local facilities that are mentioned in the literature.

Next we studied our internal records on energy usage and the blueprints, plans, and installation and operating manuals for our existing system. Then we surveyed our site to determine exactly how difficult it would be to convert to an effective energy-management system.

FIGURE 22.2   Methods Section of a Completion Report

Figure 22.2 shows the brief methods section from the report on the energy-management system at the department store.

The Results

The results are the data you discovered or created. You should present the results objectively, without commenting on them; save the interpretation of the results—the conclusion—for later. If you intermix results and conclusions, your readers might be unable to follow your reasoning process. Consequently, they will not be able to tell whether your conclusions are justified by the evidence—the results.

Just as the methods section answers the question "What did you do?" the results section answers the question "What did you see?"

The nature of the project will help you decide how to structure the results. For physical-research reports, you can often present the results as a brief series of paragraphs and graphics. If in the methods section you described a series of tests, in the results section you simply report the data in the same sequence you used for the methods. For feasibility studies, a comparison-and-contrast structure (see Chapter 10) is generally the most accessible. When you are evaluating a number of different alternatives, the whole-by-whole pattern, with the best alternative first, might be most appropriate. When you are evaluating only a few alternatives, the part-by-part structure might work best. Of course, you must always try to consider the needs of your readers. How much they know about the subject, what they plan to do with the report, what they anticipate your recommendation will be—these and many other factors will affect your decision on how to structure the results.

For instance, suppose that your company is considering installing a word-processing system. In the introduction you have already discussed the company's current system and its disadvantages. In the methods section you have described how you established the criteria to apply to the available systems, as well as the research procedures you carried out. In the results section, you provide the details of each system.

Figure 22.3 is an excerpt from the results section of the report on the energy-management system for the department store.

The Conclusions

The conclusions are the implications—the "meaning" of the results. Drawing valid conclusions from results requires great care. Suppose, for example, that you work for a company that manufactures and sells clock radios. Your records tell you that in 1990, 2.3 percent of the clock radios your company produced were returned as defective. An analysis of company records over the previous five years yields these results:

Year	Percentage Returned as Defective
1989	1.3
1988	1.6
1987	1.2
1986	1.4
1985	1.3

One obvious conclusion can be drawn: a 2.3 percent defective rate is a lot higher than the rate for any of the last five years. And that conclusion is certainly a cause for concern.

But do those results indicate that your company's clock radios are less well made than they used to be? Perhaps—but in order to reach a reasonable conclusion from these results, you must consider two other factors. First, you must account for consumer behavior trends. Perhaps consumers

Major Technical Components of an Efficient Energy-Management System

There are two main technical components to an efficient energy-management system: a network of sensors and customized software.

Sensors

Because climate is a function of humidity, entropy, and temperature levels, a network of sensors would have to be installed at strategic points throughout the building. These sensors would form the communication link to and from a particular area. Our energy-management system would continuously measure humidity, entropy, temperature, and steam-demand levels during a standard time interval. These actual climate measurements would be compared with predetermined target levels every few minutes. On the basis of these comparisons, the system would begin the appropriate action (open or close the steam valves) to maintain the target temperature.

Software

A computer program is needed to process the signals from the sensors, determine whether adjustments are required, and transmit instructions. This program would compare the parameter readings to the predetermined target levels, make the necessary adjustments (open or close the steam valves) for the target climate to be maintained, and record these measurements as data. These data could then be used to analyze current steam efficiency and potential energy-management opportunities.

Tasks Involved in Converting Our System

As discussed in the system description, a sophisticated energy-management system is based on a set of sensors and

FIGURE 22.3   Excerpt from the Results Section in the Body of a Completion Report

customized software. This section discusses the results of our analysis of the tasks required to convert our system to a sophisticated, computerized energy-management system.

Sensors

One complication is that the current system employs steam valves that are operated by a pneumatic signal system. The system works on the principle of air pressure against a diaphragm; the expansion and contraction of the diaphragm opens and closes the valve. However, a sophisticated energy-management system can recognize only digital signals. Therefore, pneumatic-to-digital converters would have to be purchased and installed so that our system could regulate steam flow throughout the building. These converters could be installed by our electricians. The cost of the units, installed, would be less than $3,500. See Appendix C, page 13, for descriptions and cost figures for the converters.

Software

The program required to monitor and regulate the steam would be quite similar to the current program used to monitor and regulate the air-conditioning units throughout the building. For this reason, developing the steam-regulating program would require only about four person-weeks of work by our programming staff. This would cost less than $2,000. In addition, several of the subprograms used by the air-conditioning system could also be used directly by the steam-heating program. This would reduce the amount of computer memory space needed. Therefore, the memory space currently available would be sufficient. See Appendix D, page 16, for the details of the programming needs.

FIGURE 22.3 *Continued*

> **CONCLUSION**
>
> Incorporating our current steam-delivery system into an efficient energy-management system would cost less than $6,000 and could be accomplished in less than two weeks without our having to employ any additional personnel. This action would be the most cost-effective way to contain rising energy costs at the Bridgeport Department Store.

FIGURE 22.4   Conclusion Section from the Body of a Completion Report

were more sensitive to quality in 1990 than they had been in previous years. A general increase in awareness — or a widely reported news item about clock radios — might account in part or in whole for the increase in consumer complaints. (Presumably, other manufacturers of similar products have experienced similar patterns of returns if general consumer trends are at work.) Second, you must examine your company's policy on defective clock radios. If a new, broader policy was instituted in 1990, the increase in the number of returns might imply nothing about the quality of the product. In fact, the clock radios sold in 1990 might even be better than the older models. In other words, beware of drawing hasty conclusions. Examine all the relevant information.

Just as the results section answers the question "What did you see?" the conclusions section answers the question "What does it mean?"

For more information on conclusions, see Chapter 7.

The conclusion of the energy-management system report is shown in Figure 22.4.

The Recommendations

Recommendations are statements of action. Just as the conclusions section answers the question "What does it mean?" the recommendations section answers the question "What should we do now?"

The recommendations section is always placed at the end of the body; because of its importance, however, the recommendations section is often also summarized — or inserted verbatim — after the executive summary.

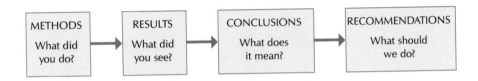

If the conclusion of the report leads to more than one recommendation, use a numbered list. If the report leads to only one recommendation, use traditional paragraphs.

Of more importance than the form of the recommendations section are its content and tone. Remember that when you tell your readers what you think they ought to do next, you must be clear, comprehensive, and polite. If the project you are describing has been unsuccessful, don't simply recommend that your readers "try some other alternatives." Be specific: What other alternatives do you recommend, and why?

And keep in mind that when you recommend a new course of action, you run the risk of offending whoever formulated the earlier course. Do not write that your new direction will "correct the mistakes" that have been made recently. Instead, write that your new action "offers great promise for success." If the previous actions were not proving successful, your readers will probably already know that. A restrained, understated tone is not only more polite but also more persuasive: you appear to be interested only in the good of your company, not in personal rivalries.

Figure 22.5 shows the recommendations section of the energy-management system report.

This report delivers good news: a relatively small expenditure will save the company a lot of money. Often, however, feasibility reports such as this one yield mixed news or bad news; that is, none of the available options would be an unqualified success, or none of the options would work at all. Don't feel that a negative recommendation reflects negatively on you. If the problem being studied were easy to solve, it probably would have been solved before you came along. Give the best advice you can, even if that advice is to do nothing. The last thing you want to do is to recommend a course of action that will not live up to the organization's expectations.

> **RECOMMENDATION**
>
> We recommend that Bridgeport Department Store incorporate the steam-delivery system into its existing energy-management system as soon as practicable. The expenditure would pay for itself in less than one month. The Engineering and Planning Department stands ready to undertake this task.

FIGURE 22.5 Recommendation Section from the Body of a Completion Report

Willis-Martin, Inc.   AUDIO/VIDEO SYSTEMS DIVISION

March 15, 1990

Mr. Don Leith, Manager
Audio/Video Systems Engineering
Willis-Martin, Inc.
11 Harshaw Boulevard
Philadelphia, PA 19131

Dear Mr. Leith:

Overview of the report.

Attached is the completion report for Phase I of the Sutherland International Airport sound system design. The report describes our adherence to the contract with the National Electronics Co., Ltd., and discusses our potential profit in future phases.

Summary of Phase I of the project.

Phase I of the design has been completed within the limit of 90 days, for which we have earned a $44,000 bonus, and the engineers working on the project are now in a good position to quickly and effectively complete the remaining three phases.

Through the use of existing Willis-Martin software, we are able to lay out the 20,000 speakers needed for Phase I without serious problem. The layout was carefully designed to meet and exceed the standards required in the contract. This was done at minimal sacrifice to cost, and the project is currently running within the prescribed budget.

Sample Completion Report

The following completion report was written by the student working for an engineering company. His project was to determine the acoustical needs of a new airport. The proposal for this project is included in Chapter 20 (page 565); the progress report is in Chapter 21 (page 610). Marginal comments have been added.

Mr. Don Leith
Page 2
March 15, 1990

The recommendation.

The design phase of the project is proceeding on course, and we anticipate that the job will be completed by January of 1991. I recommend that we continue with the present staff of engineers and draftspersons to meet our time goals.

A polite conclusion to the letter.

If you have any questions regarding this report, I will be glad to discuss them at your convenience.

Sincerely,

Mark R. Banham

Mark R. Banham
Audio Engineer

Title of the report.

SUTHERLAND INTERNATIONAL AIRPORT

SOUND SYSTEM DESIGN:

PHASE I COMPLETION REPORT

Submitted to: Mr. Don Leith, Manager

                  Audio/Video Systems Engineering

                  Willis-Martin, Inc.

Name and position of
the writer.

Submitted by: Mark Banham, Audio Engineer

                  Audio/Video Systems Division

                  Willis-Martin, Inc.

March 15, 1990

ABSTRACT

"Sutherland International Airport Sound System Design: Phase I Completion Report," Mark Banham, Audio Engineer, Willis-Martin, Inc., Audio/Video Systems Division.

The design of Phase I of this project was completed within the proposed 90 days, and we are now in a position to effectively complete the entire design well within the specified time constraints. Phase I involved the layout of 20,000 speakers in the West Terminal Building, the largest in the airport complex. An analysis of the existing acoustical conditions of the building showed substantial ambient noise problems. In addition to airplane noise, the architecture of the building included detailed metal panel ceilings and large, open public areas with fountains. The design team was able to overcome these problems and meet the specifications of constancy of sound-pressure levels. We accomplished this by using Willis-Martin's software to calculate proper spacing of cone diaphragm and horn loudspeakers. Proper cabling was designed to service all areas, and the conduit system can now be laid out. We suggest maintaining the current design team for the remainder of the project phases. The continuity offered would save time and permit us to produce a consistent, high-quality design integrated throughout the whole airport.

Overview of the project.

Methods and results.

Recommendations.

Prefatory pages are numbered with lower-case Roman numerals, centered horizontally.

i

The table of contents shows an effective use of boldface, capitalization, underlining, and indentation.

CONTENTS

iii

Overview of the project.

I. EXECUTIVE SUMMARY

The design of the public address and background music system for the Sutherland International Airport in Sydney, presents Willis-Martin with the opportunity to become a leader in the international audio systems market. At this date, Phase I of the design has been completed successfully and on time. The design marks a substantial effort by staff engineers to produce a professional sound system within the parameters prescribed by our contract with the National Electronics Company, Ltd.

The methods and results of the project, written to address the manager's needs.

The first phase of the job covered the most laborious task in the entire project: the layout of 20,000 speakers in the West Terminal Building. We spent two weeks examining the acoustics of the building. Then, using existing Willis-Martin software and newly developed spreadsheets for this job, we completed the loudspeaker layout within the 90 days proposed at the onset of this project. For the 5 million square feet of coverage, we used 4- and 8-inch cone speakers, and constant-directivity horn arrays. While the specifications were quite stringent regarding deviations on sound-pressure levels, we are able to meet and exceed all design parameters.

The recommendation, again from the manager's perspective.

We look forward to Phases II and III with great optimism. Having already earned the $44,000 bonus for early completion of Phase I, we feel we are in an excellent position to earn further bonuses on the next phases of the total project. By maintaining the present team of design engineers and draftspersons working the job, Willis-Martin will save a tremendous amount of time. Because the gear-up time and learning curve for this job were substantial, we should seek to eliminate repeating those costly factors. Additionally, we would like to maintain the integrity and continuity of the entire airport design so that our final product will exemplify our skills at designing large sound systems such as this.

2

II. INTRODUCTION

Background of the project.

In June, 1989, Willis-Martin was awarded a $1.2 million contract by the National Electronics Company, Ltd. (NECL) to design and install the public address and background music system for the Sutherland International Airport in Sydney. Full-scale design on Phase I of the project began in December, 1989, and was completed within the proposed 90 days. The sound system design involved analysis of the largest building in the airport complex, the West Terminal, for acoustic conditions. Loudspeaker design was completed first, followed by cable and power-amplifier estimates. This design marks the first of its scale for Willis-Martin in terms of quantity and quality of equipment design required. Therefore, completion of the first phase warrants a careful review in order to assess our performance and chances of completing this project within the bounds of profit and contract adherence.

Methods, results, and conclusions.

Overview of the organization of the report.

This report describes the methods used to design the speaker layout and to estimate the cable needs. The report then describes the design itself, beginning with a detailed look at the West Terminal design and proceeding to a financial analysis of Phase I and problems encountered during Phase I. Finally, the report makes recommendations regarding a carryover of the design team to the next phases of the project.

3

III. METHODS

Overview of the section.

In completing Phase I, we performed two major tasks:

- designing the loudspeaker layout
- estimating the amount of cable needed

A. Loudspeaker Layout Design Techniques

Detailed discussion of methods and their rationale.

The design of any loudspeaker layout must consider a large number of variables. However, the size of the space being serviced, along with the ambient noise and floor and ceiling finishes, are of primary concern. The most laborious task of this project was to measure the entire West Terminal floor surface area in divisions of ceiling height. These divisions marked changes in speaker design because the room volume changed with an increase in ceiling height.

Along with these measurements, the engineers noted the floor and ceiling finishes (i.e., carpet under metal, stone under acoustic tile). The importance of this information can be seen when considering the reverberation of sound on carpet versus that on stone. In an airport, flight information must be heard, and, therefore, articulation loss due to reverberation is extremely important.

To calculate the number of speakers necessary to service an area, the specification of sound-pressure level (spl) must be computed. To prevent substantial variation in spl as the

4

listener moves through a room, the speakers must be positioned so that the contribution of each, at any one point, does not vary the spl at that point by more than the specified decibel deviation. The speaker spacing is usually designed to fall into some type of grid pattern that may be repeated easily. The BASIC program SPL, designed by Willis-Martin, allows the necessary spacing calculations to be made.

Cross-reference to the appendix.

We chose to use two kinds of speakers: cones and constant-directivity horns. See the Appendix, page 21, for a schematic of the horns.

1. Cone Speakers

Cone-diaphragm speakers generally have a very good polar response. This response is a measure of the deviation in spl in dB as the listener moves from directly under the speaker to various angles of incidence of the sound wave. The majority of surface area in the airport will be serviced by cone speakers, which contain a transformer in a backbox to tap the line at a desired power level. The frequency response of all cone speakers used for this airport is sufficient to handle the paging announcements and background music clearly.

Cone speakers are used for most areas of moderate ceiling height and low ambient noise. Two types of cones were used in this design: the PRO-8A, an 8-inch, 8-watt backbox speaker;

5

and the Soundolier FC104, a 4-inch, 4-watt speaker. The method of choosing the proper speaker was carefully determined. The exhaustive data analysis of the as-built construction conditions of the building led to subsets of West Terminal areas by ceiling height. Within each area, a minimum power and clarity of sound were determined according to the contract specifications. The speaker best suited to each area was chosen, and the program gave suggested spacing parameters. Depending on the ceiling architecture, a square grid was followed in the layout.

2. Constant-Directivity Horns

For high ceiling areas, above 14 meters, some type of horn configuration must be used. The horn loudspeaker, unlike the cone, offers high efficiency at high power. While the polar response drops off considerably as one moves away from center, the horn provides little articulation loss and good frequency response. This airport contains several sections of very high ambient noise and ceiling height. The two public sections of the Arrival and Departure Levels have 40-meter ceilings and fountains. These factors call for large horn arrays to overcome the ambient conditions. All stairwells and parking garages must also have horns installed. The major disadvantage of using horns is that they are very expensive compared to cone speakers (Kissinger 1989).

6

B. Cable Estimating

To estimate the cable needed to supply power to the speakers, we studied the cable-routing diagram supplied by NECL with our contract. The existing cable tray allowed for easy routing of the 12 and 14 AWG wire used.

One of the most important aspects of cable estimating is accuracy. To increase accuracy, we paid particular attention to the homeruns (lines of connected speakers running to one power amplifier).

7

Overview of the section.

IV. RESULTS: THE DESIGN

We have successfully completed the sound system design for approximately 5 million sq. ft. of floor space, within the time limit of 90 days. The following discussion will cover four factors:

- technical analysis of Phase I
- financial analysis of Phase I
- problems encountered during Phase I
- carryover to future phases

Overview of the section.

A. Technical Analysis of Phase I

The following discussion will cover four topics:

- the West Terminal Building as-built acoustics
- the loudspeaker requirement calculations
- the cable and conduit estimates
- the control and power systems

1. West Terminal Building Acoustics

The West Terminal Building houses the major public areas of this airport, such as arrival and departure lobbies, and a parking garage. The diversity of the types of spaces included in the sound system coverage required detailed categorizing. On each level, we measured the floor space from the plans and indicated all surface-finish materials. The reflected ceiling

8

plans served as a guide to the elevations above the finished floor, which determined the speaker power needed to service each area. The specifications of this job called for two different grades of quality of sound. These grades are directly related to the uniformity of sound pressure within any room. Deviations in the sound pressure level, which represent changes in volume, decrease audibility.

a. Grade of Service 1

In those areas designated Grade 1, the spl deviation was limited to ± 3dB. Some areas included in this group are the public areas where flight information must be heard, and special official sections. The deviation in spl is directly a function of the loudspeaker spacing, discussed below. About 67% of the space in the West Terminal was specified at Grade of Service 1.

b. Grade of Service 2

Areas requiring this grade of service included employee areas and parking garages. The spl is permitted to vary by ± 4.5dB in these sections. Such a requirement means a greater spacing between speakers and therefore less equipment and lower costs. Approximately 33% of the West Terminal Building called for Grade of Service 2.

9

Overview of the section.

2. Loudspeaker Requirement Calculations

There are three major considerations in calculating the number of speakers required for a job like this (Martin 1982):

- sound-pressure levels
- articulation losses
- reverberation losses

Each of these factors was examined carefully to help produce a cost effective yet professional design.

a. Constancy of Sound-Pressure Levels

As discussed previously, the spl in any room must be maintained within some tolerance. The calculations which produced suggested speaker spacing in a room considered the polar response of the proposed speaker. For areas of ceiling height less than 7.5 meters, we used the 4-inch speaker; for areas between 7.5 and 14 meters, we used the 8-inch speaker. Therefore, in determining the speaker spacing for a room of ceiling height 3 meters (most hallways have this height), the polar-response data for the 4-inch speaker would be input to the SPL program. By making the decision of which speakers would best serve certain areas, based largely on the criteria of power dissipation capabilities, we reduced the spacing design task to a simple repetitious run of the SPL program.

10

b. Articulation-Loss Factors

The SPL program also considered the articulation losses of the grid of speakers at any spacing. The specifications of the job indicated acceptable levels of articulation loss for each grade of service, and thus a determination was made after each run of the program as to the acceptability of the articulation losses. When the deviation in spl was within acceptable limits, but the articulation losses were not, we would change the suggested speaker spacing for that area until acceptable data were recorded.

c. Reverberation Factors

Finally, we accounted for the floor and ceiling finishes, which were rigorously recorded for every room in the building. The SPL program contains a variable reverberation-adjustment factor as part of the input. For areas of minimal reverberation, such as a chapel with carpet on the floor and an acoustic tile ceiling, the adjustment for reverberation effect on audibility was negligible. In other words, there would be little destructive interference of echoing waves in a room of this type. This would correspond to a factor of 1 in the program input, not significantly altering the output.

11

However, we encountered many areas of high metal panel ceilings on stone floors, and a maximum factor of 5 would properly alter the output to show how articulation loss increased with the destructive reverberation that accompanies such rooms (Ritz 1989).

3. Cable and Conduit Estimates

The contract called for estimated cable requirements and suggested routing of that cable. However, the NECL-supplied electrical drawings included the existing cable trays in the building and suggested cable riser routes. This essentially reduced our job to estimating homeruns and line-power needs. The line losses were calculated by means of a complicated spreadsheet that we designed to handle this problem. The spreadsheet calculated the type of cable needed to supply a certain number of speakers being tapped at a designated power level. The gage of wire required for each homerun line varied with the number of speakers on the line and the power tap.

4. Control and Power Systems

With the cable requirements and conduit routing known, we needed to determine which speakers would be powered at variable levels, and how many speakers could be powered by each amplifier. The central power room, located on the service

12

level, was slated to hold up to 200 two-channel, 200-watt power amplifiers and control equipment.

a. Homerun Calculations

To calculate the required number of homeruns, we used the spreadsheet data. This spreadsheet, in addition to determining line losses, considered the contribution of all speakers in a grid pattern in any room and calculated the power tap needed. This calculation was done through a series of other calculations considering the ambient noise and number of speakers in the room. If 200 speakers were designated to go in one area, this spreadsheet would determine the number of separate homeruns needed and the maximum number of speakers on each run. This number was limited to the power per channel of the respective amplifier in the main control room. We designed the homeruns in an interleaving pattern to prevent total loss of service if a power amplifier should blow out.

b. Amplifiers

The amplifiers used for this job were mostly Altec and Ivie 200-watt amplifiers. The specifications called for approximately 200 amplifiers, and by the end of the design of this phase we have used all 200. The 20

13

equipment racks in the main control room were spaced to properly allow for the substantial heat dissipation associated with a high-power system.

B. Financial Analysis of Phase I

The writer explains the rationale for the discussion of finances.

While the engineering aspects of this job were not limited to strict contractor budget, we did make substantial efforts to reduce equipment costs and installation costs to the benefit of both Willis-Martin and our client. The open-ended design cost attitude of the contractor left substantial room for innovation in the design, but we were interested also in meeting the client's deadline and thereby earning the bonus.

1. Equipment Costs

It is the policy of this company to utilize top-of-the-line, current equipment in all designs, and this job was no exception. We made every effort to minimize the use of constant-directivity horn arrays, however, as these are the most expensive devices included in the job. We were successful in limiting their use to two 40-meter ceiling public areas on the Arrival and Departure Levels. This limiting did not adversely affect the design because we found a low-cost EV-102 paging horn for use in all stairways and the parking garage. Total equipment costs for this phase of the project were estimated at $150,000.

14

2. Adherence to Task Schedule

Engineering time on the job was well within the estimated figures, and the company should be in a position to earn substantial bonuses if the remaining phases can also be completed ahead of schedule. The excellent coordination by the Drafting Department is largely the reason we were able to maintain the proposed task schedule. Having one full-time draftsperson on the job at all times led to more efficiently completed drawings and less time wasted on bringing new draftspersons up to speed.

C. Problems Encountered During Phase I

The writer discusses the problems he encountered.

The problems encountered on the job, with the exception of normal technical design obstacles, were limited to construction-drawing inconsistencies. We encountered a number of areas on the reflected ceiling plans that differed substantially from the architectural boundaries on the zoning diagrams. The zoning diagrams were not as-built drawings; rather, they were new drawings drafted by the General Contractor to designate areas of sound-system coverage. To clear up these inconsistencies, it was necessary to communicate with NECL in Sydney more than had originally been anticipated. Because of the eight-hour time difference, phone conversations were sometimes difficult. In addition,

15

NECL was very slow, sometimes two weeks, in responding to the many fax messages sent on this subject.

D. Carryover to Future Phases

An important function of this report is to analyze what we have learned from Phase I that can be applied to future phases of the project. With only one building in the airport complex completed, it is beneficial to understand how we can eliminate problems in upcoming stages and make use of some of the groundwork that was laid so far.

The two most important factors to consider are the learning-curve effects and the developed software.

1. Learning-Curve Effects

There was a substantial learning curve accompanying this job. First, we learned how to effectively acquire data from the construction drawings. Second, we developed a method for effectively organizing this data in spreadsheet form. Third, we produced a substantial number of different speaker-layout grids that may be applied to similar areas in future phases of the design. And fourth, we gained the expertise to quickly and accurately design large areas of coverage by strict methodology. All of these factors will come into play in future loudspeaker designs.

16

The writer analyzes what he learned in Phase I that could be carried over to future phases.

2. Developed Software

Some rather complicated software was developed and used for this design. Listed below are the various programs developed or modified to make this and future designs easier:

- SPL—sound pressure level constancy calculations for loudspeaker spacing, written in Microsoft™ BASIC
- As-Built Acoustics Lotus 1-2-3™ spreadsheet, which allows construction data to be sorted into various subsets including room volume and surface finishes.
- Line Losses Lotus 1-2-3™ spreadsheet, which allows the user to calculate the power losses of speaker cables and the number of homeruns needed to service an area.

17

V. CONCLUSION

The writer provides concluding comments.

The design team chosen to perform Phase I of this design completed the task within the bonus period. Most importantly, the team did its work with a high degree of professionalism. The Sutherland Airport design required substantial effort because of the substantial size and ambient noise factors. In addition to airplane noise, the architects have placed fountains and other public-area distractions throughout the West Terminal, posing great design barriers. By careful calculations, we overcame these factors and produced a high-quality system.

The team of engineers has submitted the final drawings for this phase and is awaiting approval by NECL. With the expected approvals, we should be capable of using the knowledge we have gained from this phase to effectively complete the remaining phases.

VI. RECOMMENDATIONS

The writer provides his recommendation and its rationale.

There are seven smaller buildings remaining in this airport complex to be serviced with sound, including the Concourses and the Control Tower. We recommend that the design team currently working on this job be maintained for these designs. This will minimize learning time required at the beginning of any design project and provide the continuity necessary to maintain the integrity and quality of the entire airport sound system.

VII. REFERENCES

Martin, Daniel W. 1982. Sound Reproduction and Recording Systems. In <u>Electronics Engineers' Handbook</u>, ed. D. Fink and D. Christiansen, 19-1-87. New York: McGraw-Hill, Inc.

Kissinger, Corey, Sound System Engineer, Willis-Martin, Inc. Personal interview. Philadelphia, 2 December 1989.

Ritz, James, Chief Audio Engineer, Willis-Martin, Inc. Personal interview. Philadelphia, 3 December 1989.

Mark Banham's note: "This appendix would give the technical reader more information on the equipment chosen for the design. By including this section, the quality of the design can be better understood. Generally, manufacturer's cut-sheets on the equipment would be sufficient for this purpose. Here are samples of important characteristics of cone and horn loudspeakers that would appear on these cut-sheets."

APPENDIX:

SAMPLE LOUDSPEAKER AND POWER AMPLIFIER SPECIFICATIONS

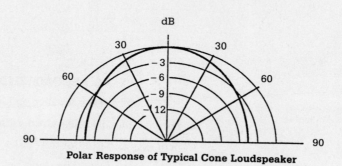

Polar Response of Typical Cone Loudspeaker

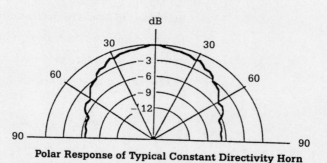

Polar Response of Typical Constant Directivity Horn

WRITER'S CHECKLIST

The following checklist applies only to the body of the completion report. The checklist pertaining to the other report elements is included in Chapter 19.

1. Does the introduction
   a. identify the problem that led to the project?
   b. identify the purpose of the project?
   c. identify the scope of the project?
   d. explain the organization of the report?

2. Does the methods section
   a. provide a complete description of your methods?
   b. list or mention the equipment or materials?

3. Are the results presented clearly, and objectively, and without interpretation?

4. Are the conclusions
   a. presented clearly?
   b. drawn logically from the results?

5. Are the recommendations stated directly, diplomatically, and objectively?

EXERCISES

1. Write the completion report for the major assignment you proposed in Chapter 20.

2. Following is a report titled "A Procedure for the Aseptic Propagation of Yeast" written by William J. DeMedio, a chemist working at a brewery (DeMedio 1980). It was submitted to the brewmaster, William Miller. Assume the role of William Miller and write a memo to Demedio analyzing the effectiveness of the report.

American Brewing Company

May 28, 19XX

Mr. William Miller
Brewmaster
American Brewing Company
Center and Elm Streets
Philadelphia, PA 19123

Dear Mr. Miller:

This report covers the observations and results of the near-beer yeast propagation procedure, originally proposed April 19, 19XX. The procedure was found to be highly effective in preventing yeast infection.

The procedure uses the logarithmic growth pattern of the yeast, as well as certain sterilization procedures fitted to the brewery, to grow yeast aseptically from test tube slants to the eight barrel pitching amount.

The propagation is very inexpensive: $840.00 for labor and equipment. It is also a rapid procedure and can be started in a day's notice. The procedure should prevent any troubles associated with infected near-beer brews.

If you have any questions concerning this report, please feel free to contact me. I shall be happy to discuss them with you, at your convenience.

Sincerely,

William DeMedio

William DeMedio
Chemist
Brewing Laboratory

ABSTRACT

The brewing laboratory was given the task of developing a method for the aseptic propagation of near-beer yeast from test-tube slants. Vague details of the propagation were sent from Switzerland, but it was the task of the brewing laboratory to develop a method that would fit the needs of the brewery. A highly successful and inexpensive method was developed, which involved specialized sterilization procedures and the utilization of the logarithmic growth pattern of the yeast. The total cost of the method (material and labor) was $840.00. If this is compared to the cost of an infected brew, which would be a direct consequence of an improper yeast propagation, one sees a considerable savings. In my opinion, the method of yeast propagation developed in the brewing laboratory is the safest and least expensive way to grow yeast.

William DeMedio
Chemist
Brewing Laboratory

CONTENTS

I. EXECUTIVE SUMMARY

If a near-beer yeast culture becomes infected, serious problems can result. First, an entire new brew would be ruined, resulting in a loss of the range $700–$1,000 for materials alone. Second, the company would have to go through government clearance procedures in order to dispose of the brew. Third, draining the beer would increase the sewage costs somewhat. Finally, if infected near beer were canned and sold, the company would be liable to severe fines.

An infected brew can be prevented by pitching it with infection-free yeast.

The brewing laboratory was given the task of developing a procedure for the aseptic propagation of near-beer yeast from test-tube slants to an amount feasible for pitching a commercial brew. A successful and inexpensive method was developed, utilizing the growth patterns of yeast. The procedure was shown to be extremely effective in preventing costly infection.

As well as being safe, the method was also very inexpensive. Total labor costs amounted to $640.00, while equipment costs were $200.00. (Please refer to Appendix 1, on page 12.) Also, the equipment that was purchased is a permanent investment, since it can be used in repeated propagations.

The procedure takes eight days to produce an amount of yeast feasible for pitching a 160-barrel brew. This is enough beer to can 5,000 cases. The cost of propagation is a small fraction (6%) of the total cost of production of near beer.

The propagation is based on the vague details sent from Switzerland. (Please see Appendix 2, on page 13.) It was the task of the brewing laboratory to develop a method that both meets the requirements and fits the needs of the brewery.

II. THE PROPAGATION PROCEDURE

1. Background

The method of propagation contained four major steps. The first step was the transfer of a loopful of each of the two yeast strains into individual 100-ml bottles of wort, each containing 5 percent sucrose (saccharose). After two days of fermentation time, each 100-ml sample of yeast culture was transferred into individual 1,000-ml wort samples, containing 5 percent sucrose. Again, these were allowed to ferment for two days. In the third step of the propagation, each 1,000-ml sample of yeast culture was transferred into individual 10,000-ml wort samples, containing 5 percent sucrose. Finally, in the fourth step of the propagation, the two strains of yeast were mixed in eight barrels of plain wort. This was allowed to ferment for two days to obtain the objective: eight barrels of aseptic pitching yeast. The procedure obviously involved taking advantage of the logarithmic growth pattern of yeast.

In order to prevent infection, standard aseptic technique was used in all yeast transfers. (Please see H. J. Benson, Microbiological Applications, W. C. Brown and Co., 1979, p. 78, for an example of standard aseptic technique.)

Three germicides were used in order to sterilize utensils: heat, iodophor, and anhydrous isopropyl alcohol. First, each utensil was rinsed in a 50 PPM iodophor solution. Then, each was rinsed clean with anhydrous isopropyl alcohol in sparing amounts. Finally, the utensils were placed in an oven at 100°C to maintain sterility. Presterilized absorbent cotton was used to plug all flasks in the propagation.

In order to maintain the pH at the bacteria-inhibiting 4.0–4.5 level, water corrective was employed (85° phosphoric acid) in the final steps of the propagation. A negligible amount was used.

The wort used was ordinary lager kettle wort. In order to differentiate between strains, one was called strain "A" and the other was called strain "B."

2. Step One

a) Preparation: First, we sterilized three 250-ml centrifuge bottles. Then, we cleared sufficient bench space and lit two bunsen burners and a propane torch for a hood effect. We then got all materials ready: two 7-g packets of sucrose, 2 yeast slants in a test-tube stand, inoculating loops, towels. Towels were used to handle all hot material.

b) Wort Obtaining: Two technicians were needed for this procedure. First, one technician took a 2½-gallon bucket, with a 12-foot rope attached, to the kettle. With another technician assisting, he then lowered the bucket into the hot (214°F) wort. The bucket was allowed to fill with wort, and then the technicians carefully raised it out of the kettle. This was the wort to be used as a medium for growing the yeast. The technicians then took this hot wort up to the laboratory, where they placed it on the stand. Since the wort was boiling, it was already sterile.

c) Preparation of 100-ml Wort Media: Two technicians were needed for this procedure. One technician carefully placed 7 g of sucrose into each of two centrifuge bottles, marked "A" and "B." A third bottle, marked "C," was used as a control. After placing the sucrose into the sterile bottles, the technician placed a sterile ½" × 5" stainless steel funnel into the mouths of bottles "A" and "B." Another technician dipped hot wort fresh from the kettle and filled first the "A" bottle, and then the "B" bottle up to the 100-ml mark. The "B" bottle funnel was then placed in the mouth of the "C" bottle and this was filled to the 100-ml mark. A sterile thermometer was placed in the "C" bottle, and all three bottles were placed in the refrigerator. When the "C" bottle registered 70°F, all three were taken from the refrigerator.

d) Inoculation: One slat was marked "A" and the other "B" to differentiate them. A loopful of the yeast from the "A" slant was transferred into the bottle marked "A." The same was done with the slant marked "B." The bench area had been cleaned in advance with a 50 PPM iodophor solution. After inoculation, all of the bottles were placed in a dark place at 70°F. The uninoculated "C" bottle was used to determine whether any infection could have invaded the "A" or "B" bottles, with the idea that if the "C" bottle remained uninfected, so would the "A" and "B" bottles.

e) Fermentation: The bottles were allowed to ferment for two days. Bubbles appeared in the "A" bottle after 24 hours and in the "B" bottle after 28 hours. The bottles were periodically swirled to disperse the yeast. Each bottle had a very clean smell. After two days, no growth was seen in the "C" bottle, nearly confirming the asepticness of bottles

"A" and "B." A simple taste test showed no infection present in the "C" bottle.

3. Step two

a) Preparation: We cleared and prepared a bench area as in "Step One—Preparation," on page 4, except for obtaining the slants, test-tube stand, and inoculating loops. Also, we used three 2,000-ml Erlenmeyer flasks and two 1" × 7" stainless steel funnels in this step.

b) Wort Obtaining: The wort was obtained in exactly the same way as in "Step One—Wort Obtaining," on page 5.

c) Preparation of 1,000-ml Wort Media: The method of preparation was the same as in "Step One—Preparation of 100-ml Wort Media," on page 5, except that 2,000-ml Erlenmeyer flasks, 1,000-ml wort samples, 70-g packets of sucrose, and 1" × 7" funnels were used.

d) Inoculation: Each 100-ml yeast culture was transferred into its respective flask, marked "A" and "B." This was done aseptically by simply pouring the contents of each bottle into the other flask.

e) Microbial Yeast Examination: A wet mount slide was prepared from the residue in each bottle marked "A" and "B." The slide was observed under high power. A good number (¾) of the yeast cells showed signs of budding. There was no sign of any infection microbes.

f) Fermentation: The observations on maintenance of the flasks were exactly the same as in "Step One—Fermentation," on page 6. The only difference was size.

4. Step Three

a) Preparation: The preparation for Step Three was the same as that mentioned in "Step Two—Preparation," on page 7, except that 12,000-ml Florence flasks, 10,000-ml wort samples, and 700-g packets of sucrose were obtained. Also, 1½" × 12" stainless steel funnels were used.

b) Wort Obtaining: The wort was obtained in the same way as mentioned in "Step One—Wort Obtaining," on page 5. Several trips to the kettle were necessary in order to get enough. It was found that one 2½ gallon bucketful of wort is approximately 10,000-ml of wort.

c) Preparation of 10,000-ml Wort Media: The preparation of media for step three is the same as in "Step One—Preparation of 100 ml Wort Media," on page 5, except 12,000-ml Florence flasks, 700-g sucrose packets, and 1½" × 12" funnels were used.

d) Inoculation: We transferred the 1,000-ml yeast cultures into their respective flasks as in the method described in "Step Two—Inoculation," on page 7.

e) Microbial Yeast Examination: The same wet mount procedure was performed here as in "Step Two—Microbial Yeast Examination," on page 8. Again, we found much yeast activity and little infection.

f) Fermentation: The same procedure and observations were made here as in "Step Two—Fermentation," on page 8.

5. Step Four

a) Preparation: An eight-barrel yeast tank was filled with wort by pumping it directly from the kettle. The lines, pump, and yeast tank were sterilized with 50 PPM iodophor in advance. The wort was adjusted to pH 4.3 by adding 4 ounces of 85 percent H_3PO_4 in a slurry. (Please refer to Appendix 3, on page 13, to see the experiment used to determine this figure.)

b) Inoculation: The two 10,000-ml yeast cultures were taken to the yeast tank and poured directly into it together. The tank was then covered. No sucrose was added to the wort.

c) Microbial Yeast Examination: We periodically observed wet mounts of the wort-yeast mixture in the tank. No signs of infection were observed, and most yeast cells showed activity.

d) Fermentation: The inoculated wort in the yeast tank was allowed to ferment for two days at 70°F. At the end of two days, we had reached our objective: eight barrels of aseptic near-beer yeast. This was then given to the fermenting-room men, who pitched a 160-barrel brew with it.

III. CONCLUSION

According to our results, we can conclude that the procedure we have developed is very effective in preventing the infection of near-beer yeast. This procedure is also very inexpensive, involving little cost in labor and materials. The propagation takes only eight days to perform and can be started with a day's notice. Overall, we conclude that the procedure is safe, effective, inexpensive, and fits the needs of the brewery well.

APPENDIX 1. Costs of Equipment

Item	Price
Case of four 12,000-ml Florence Flasks	$ 50.00
6 Stainless-Steel Funnels	20.00
1 Carton Wax Pencils	3.00
Case of four 2,000-ml Erlenmeyer Flasks	25.00
Platinum Inoculating Loops	102.00
TOTAL	$200.00

APPENDIX 2. Original Propagation Instructions

BIRELL—Yeast Propagation

- The two yeast strains A and B must be propagated in separate flasks up to the 10-liter stage under strict sterile conditions.
- The following procedure is recommended:
- Inoculate from the agar slant tube into 100 ml of hopped wort, containing 5 percent sucrose.
- After two days at 20 °C, pour the 100-ml culture into flasks containing 1 liter of wort with 5 percent sucrose.
- Pour the 1-liter culture, after two days at 20 °C, into flasks containing 10 liters of wort with 5 percent sucrose.
- Pitch 5 ml of wort (as sterile as possible) with the 10-liter cultures of strains A and B.

Dr. K. Hammer

APPENDIX 3. Data from Experiment to Change the pH of Lager Wort from 5.10 to 4.30

Water corrective [85 % H_3PO_4 (aq)] was added in varying amounts to one gallon of lager wort, and the pH was measured. The purpose was to find out how much corrective is required to lower the pH of lager wort from 5.10 to 4.3–4.5 range.

Trial	Amt. 85% H_3PO_4 Added	pH of Wort	Comments
1	0.00	5.10	Start
2	0.50	4.47	Considerable change
3	1.0	4.30	OK

CONCLUSION

We must add 1.00 ml corrective to one gallon lager wort to change the pH from 5.10 to 4.3–4.5, or we must add four ounces to eight barrels of lager wort.

REFERENCES

Banham, M. 1990. Sutherland International Airport sound system design: Phase I completion report. Unpublished document.

DeMedio, W. 1980. A procedure for the aseptic propagation of yeast. Unpublished document.

MacBride, M. 1986. Recommendation for minimizing the steam consumption and regulating the ambient temperature at Bridgeport Department Store. Unpublished document.

APPENDIX

A

Handbook

This handbook concentrates on style, punctuation, and mechanics. Where appropriate, it defines common errors directly after discussing the correct usage.

Many of the usage recommendations made here are only suggestions. If your organization or professional field has a style guide that makes different recommendations, you should of course follow it.

Also, note that this is a selective handbook. It cannot replace full-length treatments, such as the handbooks often used in composition courses.

Sentence Style

Avoid Sentence Fragments

A sentence fragment is an incomplete sentence. Most sentence fragments are caused by one of two problems:

- a missing verb

Fragment	The pressure loss caused by a worn gasket.
Complete Sentence	The pressure loss was caused by a worn gasket.
Complete Sentence	The pressure loss caused by a worn gasket was identified and fixed.

Fragment	A 486-series computer equipped with a VGA monitor.
Complete Sentence	It is a 486-series computer equipped with a VGA monitor.
Complete Sentence	A 486-series computer equipped with a VGA monitor will be delivered today.

- a dependent element used without an independent clause

Fragment	Because the data could not be verified.
Complete Sentence	Because the data could not be verified, the article was not accepted for publication.
Complete Sentence	The article was not accepted for publication because the data could not be verified.

Fragment	Delivering over 150 horsepower.
Complete Sentence	Delivering over 150 horsepower, the two-passenger coupe will cost over $22,000.
Complete Sentence	The two-passenger coupe will deliver over 150 horsepower and cost over $22,000.

Avoid Comma Splices

A comma splice is the error in which two independent clauses are joined, or spliced together, by a comma. Independent clauses can be linked correctly in three different ways:

▪ by a comma and a coordinating conjunction

Comma Splice   The 909 printer is our most popular model, it offers an unequalled blend of power and versatility.

Correct   The 909 printer is our most popular model, for it offers an unequalled blend of power and versatility.

In this case, a comma and one of the coordinating conjunctions (*and, or, not, but, for, so, yet*) link the two independent clauses. The coordinating conjunction explicitly states the relationship between the two clauses.

▪ by a semicolon

Comma Splice   The 909 printer is our most popular model, it offers an unequalled blend of power and versatility.

Correct   The 909 printer is our most popular model; it offers an unequalled blend of power and versatility.

In this case, a semicolon is used to link the two independent clauses. The semicolon creates a somewhat more distant relationship between the two clauses than the comma-and-coordinating conjunction link; the link remains implicit.

▪ by a period or other terminal punctuation

Comma Splice   The 909 printer is our most popular model, it offers an unequalled blend of power and versatility.

Correct   The 909 printer is our most popular model. It offers an unequalled blend of power and versatility.

In this case, the two independent clauses are separate sentences. Of the three ways to punctuate the two clauses correctly, this punctuation suggests the most distant relationship between them.

Avoid Run-on Sentences

A run-on sentence (sometimes called a fused sentence) is a comma splice without the comma. In other words, two independent clauses appear without any punctuation between them. Any of the three strategies for fixing comma splices fixes run-on sentences.

Run-on sentence   The 909 printer is our most popular model it offers an unequalled blend of power and versatility.

Correct   The 909 printer is our most popular model, for it offers an unequalled blend of power and versatility.

Correct   The 909 printer is our most popular model; it offers an unequalled blend of power and versatility.

Correct   The 909 printer is our most popular model. It offers an unequalled blend of power and versatility.

Avoid Ambiguous Pronoun Reference

Pronouns must refer clearly to the words or phrases they replace. Ambiguous pronoun references can lurk in even the most innocent-looking sentences:

Unclear   Remove the cell cluster from the medium and analyze it.

Analyze what, the cell cluster or the medium?

Clear   Analyze the cell cluster after removing it from the medium.
Clear   Analyze the medium after removing the cell cluster from it.
Clear   Remove the cell cluster from the medium. Then analyze the cell cluster.
Clear   Remove the cell cluster from the medium. Then analyze the medium.

Ambiguous references can also occur when a relative pronoun such as *which*, or a subordinating conjunction such as *where*, is used to introduce a dependent clause:

Unclear   She decided to evaluate the program, which would take five months.

What would take five months, the program or the evaluation?

Clear   She decided to evaluate the program, a process that would take five months.

By replacing "which" with "a process that," the writer clearly indicates that the evaluation will take five months.

Clear   She decided to evaluate the five-month program.

By using the adjective "five-month," the writer clearly indicates that the program will take five months.

Unclear   This procedure will increase the handling of toxic materials outside the plant, where adequate safety measures can be taken.

Where can adequate safety measures be taken, inside the plant or outside?

Clear   This procedure will increase the handling of toxic materials outside the plant. Because adequate safety measures can be taken only in the plant, the procedure poses risks.
Clear   This procedure will increase the handling of toxic materials outside the plant. Because adequate safety measures can be taken only outside the plant, the procedure will decrease safety risks.

As the last example shows, sometimes the best way to clarify an unclear pronoun is to split the sentence in two, eliminate the problem, and add clarifying information. Clarity is always the primary characteristic of good technical writing. If more words will make your writing clearer, use them.

Ambiguity can also occur at the beginning of a sentence:

Unclear   Allophanate linkages are among the most important structural components of polyurethane elastomers. They act as cross-linking sites.

What act as cross-linking sites, allophanate linkages or polyurethane elastomers?

Clear   Allophanate linkages, which act as cross-linking sites, are among the most important structural components of polyurethane elastomers.

The writer has changed the second sentence into a clear nonrestrictive modifier.

Your job is to use whichever means — restructuring the sentence or dividing it in two — will best ensure that the reader will know exactly which word or phrase the pronoun is replacing.

If you use a pronoun to begin a sentence, be sure to follow it immediately with a noun that clarifies the reference. Otherwise, the reader might be confused.

Unclear   The new parking regulations require that all employees pay for parking permits. These are scheduled to be discussed at the next senate meeting.

What are scheduled to be discussed, the regulations or the permits?

Clear   The new parking regulations require that all employees pay for parking permits. These regulations are scheduled to be discussed at the next senate meeting.

Compare Items Clearly

When comparing or contrasting items, make sure your sentence clearly communicates the relationship. A simple comparison between two items often causes no problems: "The X3000 has more storage than the X2500." However, don't let your reader confuse a comparison and a simple statement of fact. For example, in the sentence "Trout eat more than minnows," does the writer mean that trout don't restrict their diet to minnows or that trout eat more than minnows eat? If a comparison is intended, a second verb should be used: "Trout eat more than minnows do." And if three items are introduced, make sure that the reader can tell which two are being compared:

Ambiguous   Trout eat more algae than minnows.
Clear   Trout eat more algae than they do minnows.
Clear   Trout eat more algae than minnows do.

Beware of comparisons in which different aspects of the two items are compared:

Illogical	The resistance of the copper wiring is lower than the tin wiring.
Logical	The resistance of the copper wiring is lower than that of the tin wiring.

In the illogical construction, the writer contrasts "resistance" with "tin wiring" rather than the resistance of copper with the resistance of tin. In the revision, the pronoun *that* is used to substitute for the repetition of "resistance."

Use Adjectives Clearly

In general, adjectives are placed before the nouns that they modify: "the plastic washer." Technical writing, however, often requires clusters of adjectives. To prevent confusion, use commas to separate coordinate adjectives, and use hyphens to link compound adjectives.

Adjectives that describe different aspects of the same noun are known as coordinate adjectives:

portable, programmable CD player
adjustable, step-in bindings

In this case, the comma replaces the word *and.*

Note that sometimes an adjective is considered part of the noun it describes: "electric drill." When one adjective is added to "electric drill," no comma is required: "a reversible electric drill." The addition of two or more adjectives, however, creates the traditional coordinate construction: "a two-speed, reversible electric drill."

The phrase "two-speed" is an example of a compound adjective—one made up of two or more words. Use hyphens to link the elements in compound adjectives that precede nouns:

a variable-angle accessory
increased cost-of-living raises

The hyphens in the second example prevent the reader from momentarily misinterpreting "increased" as an adjective modifying "cost" and "living" as a participle modifying "raises."

A long string of compound adjectives can be confusing even if hyphens are used appropriately. To ensure clarity in such a case, put the adjectives into a clause or phrase following the noun:

Unclear	an operator-initiated, default-prevention technique
Clear	a technique initiated by the operator for preventing default

In turning a string of adjectives into a phrase or clause, make sure the adjectives cannot be misread as verbs. Use the pronouns *that* and *which* to prevent confusion:

> *Confusing*   The good experience provides is often hard to measure.
> *Clear*   The good that experience provides is often hard to measure.

Maintain Number Agreement

Number disagreement commonly takes one of two forms in technical writing: (1) the verb disagrees in number with the subject when a prepositional phrase intervenes; (2) the pronoun disagrees in number with its referent when the latter is a collective noun.

Subject-Verb Disagreement

A prepositional phrase does not affect the number of the subject and the verb. The following examples show that the object of the preposition can be plural in a singular sentence, or singular in a plural sentence. (The subjects and verbs are italicized.)

> *Incorrect*   The *result* of the tests *are* promising.
> *Correct*   The *result* of the tests *is* promising.

> *Incorrect*   The *results* of the test *is* promising.
> *Correct*   The *results* of the test *are* promising.

Don't be misled by the fact that the object of the preposition and the verb don't sound natural together, as in *tests is* or *test are*. Grammatical agreement of subject and verb is the primary consideration.

Pronoun-Antecedent Disagreement

The problem of pronoun-antecedent disagreement crops up most often when the antecedent, or referent, is a collective noun—one that can be interpreted as either singular or plural, depending on its usage:

> *Incorrect*   The *company* is proud to announce a new stock option plan for *their* employees.
> *Correct*   The *company* is proud to announce a new stock option plan for *its* employees.

In this example, "the company" acts as a single unit; therefore, the singular verb, followed by a singular pronoun, is appropriate. When the individual members of a collective noun are stressed, however, plural pronouns and verbs are appropriate: "The inspection team have prepared their reports." Or, "The members of the inspection team have prepared their reports."

Punctuation

The Period

Periods are used in the following instances.

1. At the end of sentences that do not ask questions or express strong emotion:

   The lateral stress still needs to be calculated.

2. After some abbreviations:

   M.D.
   U.S.A.
   etc.

   (For a further discussion of abbreviations, see p. 667.)

3. With decimal fractions:

   4.056
   $6.75
   75.6%

The Exclamation Point

The exclamation point is used at the end of a sentence that expresses strong emotion, such as surprise or doubt:

The nuclear plant, which was originally expected to cost $1.6 billion, eventually cost more than $8 billion!

Because technical writing requires objectivity and a calm, understated tone, technical writers rarely use exclamation points.

The Question Mark

The question mark is used at the end of a sentence that asks a direct question:

What did the commission say about effluents?

Do not use a question mark at the end of a sentence that asks an indirect question:

He wanted to know whether the procedure had been approved for use.

When a question mark is used within quotation marks, the quoted material needs no other end punctuation:

"What did the commission say about effluents?" she asked.

The Comma

The comma is the most frequently used punctuation mark, as well as the one about whose usage many writers most often disagree. Following are the basic uses of the comma.

- To separate the clauses of a compound sentence (one composed of two or more independent clauses) linked by a coordinating conjunction (*and, or, nor, but, so, for, yet*):

  > Both methods are acceptable, but we have found that the Simpson procedure gives better results.

  In many compound sentences, the comma is needed to prevent the reader from mistaking the subject of the second clause for an object of the verb in the first clause:

  > The RESET command affects the field access, and the SEARCH command affects the filing arrangement.

  Without the comma, the reader is likely to interpret the coordinating conjunction "and" as a simple conjunction linking "field access" and "SEARCH command."

- To separate items in a series composed of three or more elements:

  > The manager of spare parts is responsible for ordering, stocking, and disbursing all spare parts for the entire plant.

  The comma following the second-to-last item is required by most technical-writing style manuals, despite the presence of the conjunction *and*. The comma clarifies the separation and prevents misreading. For example, sometimes in technical writing the second-to-last item will be a compound noun containing an *and*.

  > The report will be distributed to Operations, Research and Development, and Accounting.

- To separate introductory words, phrases, and clauses from the main clause of the sentence:

  > However, we will have to calculate the effect of the wind.

  > To facilitate trade, the government holds a yearly international conference.

  > Whether the workers like it or not, the managers have decided not to try the flextime plan.

  In each of these three examples, the comma helps the reader follow the sentence. Notice in the following example how the comma actually prevents misreading:

  > Just as we finished eating, the rats discovered the treadmill.

The comma is optional if the introductory text is brief and cannot be misread.

> *Correct*   First, let's take care of the introductions.
> *Correct*   First let's take care of the introductions.

- To separate the main clause from a dependent clause:

  > The advertising campaign was canceled, although most of the executive council saw nothing wrong with it.

  > Most accountants wear suits, whereas few engineers do.

- To separate nonrestrictive modifiers (parenthetical clarifications) from the rest of the sentence:

  > Jones, the temporary chairman, called the meeting to order.

- To separate interjections and transitional elements from the rest of the sentence:

  > Yes, I admit your findings are correct.

  > Their plans, however, have great potential.

- To separate coordinate adjectives:

  > The finished product was a sleek, comfortable cruiser.

  > The heavy, awkward trains are still being used.

  The comma here takes the place of the conjunction *and*. If the adjectives are not coordinate — that is, if one of the adjectives modifies the combination of the adjective and the noun — do not use a comma:

  > They decided to go to the first general meeting.

- To signal that a word or phrase has been omitted from an elliptical expression:

  > Smithers is in charge of the accounting; Harlen, the data management; Demarest, the publicity.

  In this example, the commas after "Harlen" and "Demarest" show that the phrase "is in charge of" has been omitted.

- To separate a proper noun from the rest of the sentence in direct address:

  > John, have you seen the purchase order from United?

  > What I'd like to know, Betty, is why we didn't see this problem coming.

- To introduce most quotations:

  > He asked, "What time were they expected?"

- To separate towns, states, and countries:

  Bethlehem, Pennsylvania, is the home of Lehigh University.

  He attended Lehigh University in Bethlehem, Pennsylvania, and the University of California at Berkeley.

  Note the use of the comma after "Pennsylvania."

- To set off the year in dates:

  August 1, 1993, is the anticipated completion date.

  Note the use of the comma after "1993." If the month separates the date from the year, the commas are not used, because the numbers are not next to each other:

  The anticipated completion date is 1 August 1993.

- To clarify numbers:

  12,013,104

  (European practice is to reverse the use of commas and periods in writing numbers: periods are used to signify thousands, and commas are used to signify decimals.)

- To separate names from professional or academic titles:

  Harold Clayton, Ph.D.
  Marion Fewick, CLU
  Joyce Carnone, P.E.

  Note that the comma also follows the title in a sentence:

  Harold Clayton, Ph.D., is the featured speaker.

Common Errors

- No comma between the clauses of a compound sentence:

Incorrect	The mixture was prepared from the two premixes and the remaining ingredients were then combined.
Correct	The mixture was prepared from the two premixes, and the remaining ingredients were then combined.

- No comma (or just one comma) to set off a nonrestrictive modifier:

Incorrect	The phone line, which was installed two weeks ago had to be disconnected.
Correct	The phone line, which was installed two weeks ago, had to be disconnected.

- No comma separating introductory words, phrases, or clauses from the main clause, when misreading can occur:

Incorrect	As President Canfield has been a great success.
Correct	As President, Canfield has been a great success.

■ No comma (or just one comma) to set off an interjection or a transitional element:

Incorrect	Our new statistician, however used to work for Konaire, Inc.
Correct	Our new statistician, however, used to work for Konaire, Inc.

■ Comma splice (a comma used to "splice together" independent clauses not linked by a coordinating conjunction):

Incorrect	All the motors were cleaned and dried after the water had entered, had they not been, additional damage would have occurred.
Correct	All the motors were cleaned and dried after the water had entered; had they not been, additional damage would have occurred.
Correct	All the motors were cleaned and dried after the water had entered. Had they not been, additional damage would have occurred.

For more information on comma splices, see p. 647.

■ Superfluous commas:

Incorrect	Another of the many possibilities, is to use a "First in, first out" sequence.

In this sentence, the comma separates the subject, "Another," from the verb, "is."

Correct	Another of the many possibilities is to use a "first in, first out" sequence.
Incorrect	The schedules that have to be updated every month are, 14, 16, 21, 22, 27, and 31.

In this sentence, the comma separates the verb from its complement.

Correct	The schedules that have to be updated every month are 14, 16, 21, 22, 27, and 31.
Incorrect	The company has grown so big, that an informal evaluation procedure is no longer effective.

In this sentence, the comma separates the predicate adjective "big" from the clause that modifies it.

Correct	The company has grown so big that an informal evaluation procedure is no longer effective.
Incorrect	Recent studies, and reports by other firms confirm our findings.

In this sentence, the comma separates the two elements in the compound subject.

Correct	Recent studies and reports by other firms confirm our findings.
Incorrect	New and old employees who use the processed order form, do not completely understand the basis of the system.

In this sentence, a comma separates the subject and its restrictive modifier from the verb.

Correct   New and old employees who use the processed order form do not completely understand the basis of the system.

The Semicolon

Semicolons are used in the following instances.

- To separate independent clauses not linked by a coordinating conjunction:

  The second edition of the handbook is more up-to-date; however, it is more expensive.

- To separate items in a series that already contains commas:

  The members elected three officers: Jack Resnick, president; Carol Wayshum, vice-president; Ahmed Jamoogian, recording secretary.

In this example, the semicolon acts as a "supercomma," keeping the names and titles clear.

Common Error

Use of a semicolon when a colon is called for:

Incorrect   We still need one ingredient; luck.
Correct   We still need one ingredient: luck.

The Colon

Colons are used in the following instances.

- To introduce a word, phrase, or clause that amplifies, illustrates, or explains a general statement:

  The project team lacked one crucial member: a project leader.

  Here is the client's request: we are to provide the preliminary proposal by November 13.

  We found three substances in excessive quantities: potassium, cyanide, and asbestos.

  The week had been productive: fourteen projects had been completed and another dozen had been initiated.

Note that the text preceding a colon should be able to stand on its own as a main clause:

Incorrect   We found: potassium, cyanide, and asbestos.
Correct   We found potassium, cyanide, and asbestos.

- To introduce items in a vertical list, if the sense of the introductory text would be incomplete without the list:

   We found the following:

   > potassium
   > cyanide
   > asbestos

- To introduce long or formal quotations:

   The president began: "In the last year. . ."

Common Error

Use of a colon to separate a verb from its complement:

Incorrect	The tools we need are: a plane, a level, and a T-square.
Correct	The tools we need are a plane, a level, and a T-square.
Correct	We need three tools: a plane, a level, and a T-square.

The Dash

Dashes are used in the following instances.

- To set off a sudden change in thought or tone:

   The committee found — can you believe this? — that the company bore *full* responsibility for the accident.

   That's what she said — if I remember correctly.

- To emphasize a parenthetical element:

   The managers' reports — all ten of them — recommend production cutbacks for the coming year.

   Arlene Kregman — the first woman elected to the board of directors — is the next scheduled speaker.

- To set off an introductory series from its explanation:

   Wetsuits, weight belts, tanks — everything will have to be shipped in.

   When a series *follows* the general statement, a colon replaces the dash:

   Everything will have to be shipped in: wetsuits, weight belts, and tanks.

Note that typewriters and many word processors do not have a key for the dash. In typewritten or word-processed text, a dash is represented by two uninterrupted hyphens. No space precedes or follows the dash.

Common Error

Use of a dash as a "lazy" substitute for other punctuation marks:

Incorrect	The regulations—which were issued yesterday—had been anticipated for months.
Correct	The regulations, which were issued yesterday, had been anticipated for months.

Incorrect	Many candidates applied—however, only one was chosen.
Correct	Many candidates applied; however, only one was chosen.

Parentheses

Parentheses are used in the following instances.

- To set off incidental information:

  Please call me (x3104) when you get the information.

  Galileo (1546–1642) is often considered the father of modern astronomy.

  H. W. Fowler's *Modern English Usage* (New York: Oxford University Press, 2nd ed., 1987) is the final arbiter.

- To enclose numbers and letters that label items listed in a sentence:

  To transfer a call within the office, (1) place the party on HOLD, (2) press TRANSFER, (3) press the extension number, and (4) hang up.

  Use both a left and a right parenthesis—not just a right parenthesis—in this situation.

Common Error

Use of parentheses instead of brackets to enclose the writer's interruption of a quotation (see the discussion of brackets):

Incorrect	He said, "The new manager (Farnham) is due in next week."
Correct	He said, "The new manager [Farnham] is due in next week."

The Hyphen

Hyphens are used in the following instances.

- In general, to form compound adjectives that precede nouns:

  general-purpose register
  meat-eating dinosaur
  chain-driven saw

  Note that hyphens are not used after words that end in *ly*:

  newly acquired terminal

  Also note that hyphens are not used when the compound adjective follows the noun:

The Woodchuck saw is chain driven.

Many organizations have their own preferences about hyphenating compound adjectives. Check to see if your organization has a preference.

- To form some compound nouns:

  vice-president
  editor-in-chief

- To form fractions and compound numbers:

  one-half
  fifty-six

- To attach some prefixes and suffixes:

  post-1945
  president-elect

- To divide a word at the end of a line:

  We will meet in the pavil-
  ion in one hour.

Whenever possible, avoid such breaks; they annoy some readers. When you do use them, check the dictionary to make sure you have divided the word *between* syllables.

The Apostrophe

Apostrophes are used in the following instances.

- To indicate the possessive case:

  the manager's goals
  the foremen's lounge
  the employees' credit union
  Charles's T-square

For joint possession, add the apostrophe and the *s* to only the last noun or proper noun:

  Watson and Crick's discovery

For separate possession, add an apostrophe and an *s* to each of the nouns or proper nouns:

  Newton's and Galileo's ideas

Make sure you do not add an apostrophe or an *s* to possessive pronouns: *his, hers, its, ours, your, theirs.*

▪ To form contractions:

> I've
> can't
> shouldn't
> it's

The apostrophe usually indicates an omitted letter or letters. For example, *can't* is *can(no)t*, *it's* is *it (i)s*.

Some organizations discourage the use of contractions; others have no preference. Find out the policy your organization follows.

▪ To indicate special plurals:

> three 9's
> two different JCL's
> the why's and how's of the problem

As in the case of contractions, it is a good idea to learn the stylistic preferences of your organization. Usage varies considerably.

Common Error

Use of the contraction *it's* in place of the possessive pronoun *its*.

Incorrect	The company does not feel that the problem is it's responsibility.
Correct	The company does not feel that the problem is its responsibility.

Quotation Marks

Quotation marks are used in the following instances.

▪ To indicate titles of short works, such as articles, essays, or chapters:

> Smith's essay "Solar Heating Alternatives"

▪ To call attention to a word or phrase that is being used in an unusual way or in an unusual context:

> A proposal is "wired" if the sponsoring agency has already decided who will be granted the contract.

Don't use quotation marks as a means of excusing poor word choice:

> The new director has been a real "pain."

▪ To indicate direct quotation, that is, the words a person has said or written:

> "In the future," he said, "check with me before authorizing any large purchases."
> As Breyer wrote, "Morale *is* productivity."

Do not use quotation marks to indicate indirect quotation:

Incorrect   He said that "third-quarter profits would be up."
Correct   He said that third-quarter profits would be up.
Correct   He said, "Third-quarter profits will be up."

Related Punctuation

Note that if the sentence contains a "tag"—a phrase identifying the speaker or writer—a comma is used to separate it from the quotation:

John replied, "I'll try to fly out there tomorrow."
"I'll try to fly out there tomorrow," John replied.

Informal and brief quotations require no punctuation before the quotation marks:

She said "Why?"

In the United States (but not in most other English-speaking nations), commas and periods at the end of quotations are placed within the quotation marks:

The project engineer reported, "A new factor has been added."
"A new factor has been added," the project engineer reported.

Question marks, dashes, and exclamation points, on the other hand, are placed inside the quotation marks when they apply only to the quotation and outside the quotation marks when they apply to the whole sentence:

He asked, "Did the shipment come in yet?"
Did he say, "This is the limit?"

Note that only one punctuation mark is used at the end of a set of quotation marks:

Incorrect   Did she say, "What time is it?"?
Correct   Did she say, "What time is it?"

Block Quotations

When quotations reach a certain length—generally, more than four lines—writers tend to switch to a block format. In typewritten manuscript, a block quotation is usually

- ☐ indented from the left margin
- ☐ single-spaced
- ☐ typed without quotation marks
- ☐ introduced by a complete sentence followed by a colon

(Different organizations observe their own variations on these basic rules.)

McFarland writes:

> The extent to which organisms adapt to their environment is still being charted. Many animals, we have recently learned, respond to a dry winter with an automatic birth-control chemical that limits the number of young to be born that spring. This prevents mass starvation among the species in that locale.

Hollins concurs. She writes, "Biological adaptation will be a major research area during the next decade."

Mechanics

Ellipses

Ellipses (three spaced periods) indicate the omission of some material from a quotation. A fourth period with no space before it precedes ellipses when the sentence in the source has ended and you are omitting material that follows or when the omission follows a portion of the source's sentence that is in itself a grammatically complete sentence:

> Larkin refers to the project as "an attempt . . . to clarify the issue of compulsory arbitration. . . . We do not foresee an end to the legal wrangling . . . but perhaps the report can serve as a definition of the areas of contention."

The second example has omitted words after "attempt" and after "wrangling." The sentence period plus three spaced periods after "arbitration" indicates that the original writer's sentence ends and the sentence following it has been omitted.

Brackets

Brackets are used in the following instances.

- To indicate words added to a quotation:

  > "He [Pearson] spoke out against the proposal."

  A better approach would be to shorten the quotation:

  > The minutes of the meeting note that Pearson "spoke out against the proposal."

- To indicate parentheses within parentheses:

  > (For further information, see Charles Houghton's *Civil Engineering Today* [New York: Arch Press, 1982].)

Italics

Italics (or underlining) are used in the following instances.

- For words used as words:

    In this report, the word *operator* will refer to any individual who is actually in charge of the equipment, regardless of that individual's certification.

- To indicate titles of long works (books, manuals, etc.), periodicals and newspapers, long films, long plays, and long musical works:

    See Houghton's *Civil Engineering Today.*
    We subscribe to the *New York Times.*

- To indicate the names of ships, trains, and airplanes:

    The shipment is expected to arrive next week on the *Penguin.*

- To set off foreign expressions that have not become fully assimilated into English:

    The speaker was guilty of *ad hominem* arguments.

- To emphasize words or phrases:

    Do not press the ERASE key.

If your typewriter or word processor does not have italic type, indicate italics by underlining.

   Darwin's Origin of Species is still read today.

Numbers

The use of numbers varies considerably. Therefore, you should find out what guidelines your organization or research area follows in choosing between words and numerals. Many organizations use the following guidelines.

- Use numerals for technical quantities, especially if a unit of measurement is included:

    3 feet
    12 grams
    43,219 square miles
    36 hectares

- Use numerals for nontechnical quantities of 10 or more:

    300 persons
    12 whales
    35-percent increase

- Use words for nontechnical quantities of fewer than 10:

  three persons
  six whales

- Use both words and numerals

  □ For back-to-back numbers:

    six 3-inch screws
    fourteen 12-foot ladders
    3,012 five-piece starter units

  In general, use the numeral for the technical unit. If the nontechnical quantity would be cumbersome in words, use the numeral.

  □ For round numbers over 999,999:

    14 million light-years
    $64 billion

  □ For numbers in legal contracts or in documents intended for international readers:

    thirty-seven thousand dollars ($37,000)
    five (5) relays

  □ For addresses:

    3801 Fifteenth Street

Special Cases

- If a number begins a sentence, use words, not numerals:

  Thirty-seven acres was the agreed-upon size of the lot.

  Many writers would revise the sentence to avoid this problem:

  The agreed-upon size of the lot was 37 acres.

- Don't use both numerals and words in the same sentence to refer to the same unit:

  On Tuesday the attendance was 13; on Wednesday, 8.

- Write out fractions, except if they are linked to technical units:

  two-thirds of the members
  3 ½ hp.

- Write out approximations:

  approximately ten thousand people
  about two million trees

- Use numerals for titles of figures and tables and for page numbers:

Figure 1
Table 13
page 261

▪ Use numerals for decimals:

3.14
1,013.065

Add a zero before decimals of less than one:

0.146
0.006

▪ Avoid expressing months as numbers, as in "3/7/92": in the United States, this means March 7, 1992; in many other countries, it means July 3, 1992. Use one of the following forms:

March 7, 1992
7 March 1992

▪ Use numerals for times if A.M. or P.M. are used:

6:10 A.M.
six o'clock

Abbreviations

Abbreviations provide a useful way to save time and space, but you must use them carefully; you can never be sure that your readers will understand them. Many companies and professional organizations have lists of approved abbreviations.

Analyze your audience in determining whether and how to abbreviate. If your readers include nontechnical people unfamiliar with your field, either write out the technical terms or attach a list of abbreviations. If you are new in an organization or are writing for publication in a certain field for the first time, find out what abbreviations are commonly used. If for any reason you are unsure about whether or how to abbreviate, write out the word.

The following are general guidelines about abbreviations.

▪ You may make up your own abbreviations. For the first reference to the term, write it out and include, parenthetically, the abbreviation. In subsequent references, use the abbreviation. For long works, you might want to write out the term at the start of major units, such as chapters.

The heart of the new system is the self-loading cartridge (SLC).

This technique is also useful, of course, in referring to existing abbreviations that your readers might not know:

The cathode-ray tube (CRT) is your control center.

- Most abbreviations do not take plurals:

  1 lb
  3 lb

- Most abbreviations in scientific writing are not followed by periods:

  lb
  cos
  dc

  If the abbreviation can be confused with another word, however, use a period:

  in.
  Fig.

- Spell out the unit if the number preceding it is spelled out or if no number precedes it:

  How many square meters is the site?

Capitalization

For the most part, the conventions of capitalization in general writing apply in technical writing.

- Capitalize proper nouns, titles, trade names, places, languages, religions, and organizations:

  William Rusham
  Director of Personnel
  Quick Fix Erasers
  Bethesda, Maryland
  Methodism
  Italian
  Society for Technical Communication

  In some organizations, job titles are not capitalized unless they refer to specific persons:

  Alfred Loggins, Director of Personnel, is interested in being considered for vice-president of marketing.

- Capitalize headings and labels:

  A Proposal to Implement the Wilkins Conversion System
  Section One
  The Problem
  Figure 6
  Mitosis
  Table 3
  Rate of Inflation, 1980–1990

APPENDIX B

Documentation of Sources

Author/Date Citations
Numbered Citations

Documentation explicitly identifies the sources of the ideas and quotations used in your document.

Complete and accurate documentation is primarily a professional obligation—a matter of ethics. Effective documentation also helps to place the report within the general context of continuing research and to define it as a responsible contribution to knowledge in the field. Failure to document a source—whether intentionally or unintentionally—is plagiarism. At most universities and colleges, plagiarism means automatic failure of the course and, in some instances, suspension or expulsion. In many companies, it is grounds for immediate dismissal.

For your readers, complete and accurate documentation is an invaluable tool. It enables them to find the source you have relied on should they want to read more about a particular subject.

What kind of material should you document? Any quotation from a printed source or an interview, even if it is only a few words, should be documented. In addition, a paraphrased idea, concept, or opinion gathered from your reading should be documented. There is one exception to this rule: if the idea or concept is so well known that it has become, in effect, general knowledge. Examples of such knowledge are Einstein's theory of relativity and the Laffer curve. If you are unsure whether an item is within the public domain, document it anyway, just to be safe.

Many organizations have their own documentation style; others use published style guides, such as the *United States Government Printing Office Style Manual*, the American Chemical Society's *Handbook for Authors*, or *The Chicago Manual of Style*. Learn and abide by your organization's style.

The system of documentation you are probably most familiar with—the notes and bibliography—is used rarely today, especially outside the liberal arts. Even the Modern Language Association, a major organization in the humanities, no longer recommends the use of the notes-and-bibliography style because of the duplication involved in creating two different lists.

Two basic systems of documentation have become standard:

1. author/date citations and bibliography
2. numbered citations and numbered bibliography.

Variations on these systems abound.

Author/Date Citations

In the author/date style, a parenthetical notation that includes the name of the source's author and the date of its publication immediately follows the quoted or paraphrased material:

> This phenomenon was identified as early as 50 years ago (Wilkinson 1943).

Sometimes, particularly if the reference is to a specific fact or idea, the page (or pages) from the source is also listed:

> (Wilkinson 1943: 36–37)

A citation can also be integrated into the sentence:

> Wilkinson (1943: 36–37) identified this phenomenon as early as 50 years ago.

If two or more sources by the same author in the same year are listed in the bibliography, the notation may include an abbreviated title, to prevent confusion:

> (Wilkinson, ''Cornea Research,'' 1943: 36–37)

Or the citation for the first source written that year can be identified with a lowercase letter:

> (Wilkinson 1943a: 36–37)

The second source would be identified similarly:

> (Wilkinson 1943b: 19–21)

For sources written by two authors, cite both names:

(Smith and Jones 1987)

For sources written by three or more authors, use the first name followed by *et al.*

(Smith *et al.* 1987)

The simplicity and flexibility of the author/date system make it highly attractive. Of course, because the citations are minimal and because their form is dictated more by common sense than by a style sheet, a conventional bibliography that contains complete publication information must follow the text. Your obligation when using this system is to remove all doubt about which source you are citing in any particular instance. *The Chicago Manual of Style*, 13th edition (Chicago: University of Chicago Press, 1982), advocates using the author/date system throughout the natural and social sciences and recommends the following order of basic bibliographic information.

For Books
1. author (last name followed by initials)
2. date of publication
3. title
4. place of publication
5. publisher

For Articles
1. author
2. date of publication
3. article title
4. journal or anthology title
5. volume (of a journal) or place of publication and publisher (of an anthology)
6. inclusive pages

The individual entries are arranged alphabetically by author (and then by date if two or more works by the same author are listed). Anonymous works are integrated into the alphabetical listing by title. Where several works by the same author are included, they are arranged by date, beginning with the most recent publication, under the author's name. In such cases, a long dash (10 hyphens on a typewriter) takes the place of the author's name in all entries after the first. The first line of a bibliographic entry is flushed left with the margin; each succeeding line is indented.

Chapman, D. L. 1991. The closed frontier: Why Detroit can't make cars that people will buy. Motorist's Metronome 12 (June): 17–26.
_____. 1989. The driver's guide to evaluating compact automobiles. Athens, Ga.: Consumer Press.

Courting credibility: Detroit and its mpg figures. 1990.
Countercultural Car & Driver (July): 19.

Following are the standard bibliographic forms used with various types of sources.

A Book

Cunningham, W. 1980. Crisis at Three Mile Island. New York:
Madison.

The author's surname is followed by first initials only. The date of publication comes next, followed by the title of the book, underlined (italicized in print). In the natural and social sciences, the style is generally to capitalize only the first letter of the title, the subtitle (if there is one), and all proper nouns. The last items are the location and name of the publisher.

For a book by two or more authors, all the authors are named.

Cunningham, W., and A. Breyer. . . .

Only the name of the first author is inverted.

A Book Issued by an Organization

Department of Energy. 1989. The energy situation in the nineties.
Washington, D.C.: U.S. Government Printing Office, Technical
Report 11346–53.

A Book Compiled by an Editor or Issued Under an Editor's Name

Morgan, D. E., ed. 1991. Readings in alternative energies. Boston:
Smith-Howell.

An Edition Other than the First

Schonberg, N. 1991. Solid state physics. 3d ed. London: Paragon.

An Article Included in a Book

May, B., and J. Deacon. 1990. Amplification systems. In Third
Annual Conference of the American Electronics Association, ed.
A. Kooper, 101–14. Miami: Valley Press.

A Journal Article

Hastings, W. 1990. The space shuttle debate. The Modern Inquirer
13: 311–18.

The title of the journal is underlined (italicized in print). The first letters of the first and last words and all nouns, adjectives, verbs (except the infinitive *to*), adverbs, and subordinating conjunctions are capitalized. After the journal title come the volume and page numbers of the article.

An Anonymous Journal Article

> The state of the art in microcomputers. 1989. <u>Newscene</u> 56: 406–21.

Anonymous journal articles are arranged alphabetically by title. If the title begins with a grammatical article such as *the* or *a*, alphabetize it under the first word following the article (in this case, *state*).

A Newspaper Article

> Eberstadt, A. 1989. Why not an Excel, why not an Eclipse? <u>Morristown Mirror and Telegraph</u>. Sunday 31 July: C19.

A Personal Interview

> Riccio, Dr. Louis, Professor of Operations Research, Tulane University. Personal interview. New York City, 13 July 1986.

The name and professional title of the interviewee should be cited first, then the fact that the source was an interview, and finally the place and date of the interview.

A Questionnaire Conducted by Someone Other Than the Author

> Recycled Resources Corp. Data derived from questionnaire administered to 33 foremen in the Bethlehem plant, June 3–4, 1991.

An individual, an outside firm, or a department might also serve as the questionnaire's "author" for bibliographic purposes. If you are citing data derived from your own questionnaire, a bibliographic entry is unnecessary: you include the questionnaire and appropriate background information in an appendix. When citing your own questionnaire data in the text of your report, refer your readers to the appendix: "Foremen at the Bethlehem plant overwhelmingly favored staggered shifts that would keep the plant in operation 16 hours per day (Appendix D)."

Computer Software

> Block, K. 1991. <u>Planner</u>. Computer software. Global Software. IBM-PC.
> <u>Tools for Drafting</u>. 1990. Computer software. Software International. Apple Macintosh.

Begin the citation with the author of the software, if an author is identified. Underline (or italicize) the name of the program, then identify the program as computer software. List the name of the publisher. Finally, add any identifying information, such as the kind of system or the brand and model of hardware the software runs on.

Information Retrieved from Database Service

Crayton, H. 1991. "A new method for insulin implants." Biotechnology 6:31–42. DIALOG file 17, item 230043 867745.

Information retrieved from a database service is treated like a printed source. At the end of the citation, however, you should list the name of the database service and the necessary identifying information to enable a reader to find the item.

Following is a sample bibliography for use with the author/date citation system:

BIBLIOGRAPHY

Daly, P. H. 1988. Selecting and designing a group insurance plan. Personnel Journal 54: 322–23.

Flanders, A. 1988. Measured daywork and collective bargaining. British Journal of Industrial Relations 9: 368–92.

Goodman, R. K., J. H. Wakely, and R. H. Ruh. 1987. What employees think of the Scanlon Plan. Personnel 6: 22–29.

Trencher, P. 1988. Recent trends in labor-management relations. New York: Madison.

Zwicker, D. Professor of Industrial Relations, Hewlett College. Interview with author. Philadelphia, 19 March 1989.

Numbered Citations

You will occasionally encounter a variation on the author/date notation system, the numbered citation and numbered bibliography system. In this system, the items in the bibliography are arranged either alphabetically or in order of the first appearance of each source in the text and then assigned a sequential number. The citation is the number of the source, enclosed in brackets or parentheses or set as superscript:

According to Hodge [3], "There is always the danger that the soil will shift." However, no shifts have been noted.

As in author/date citations, page numbers may be added:

According to Hodge [3:26], "There is always...."

In this example, 3 means that Hodge is the third item in the numbered bibliography; 26 is the page number.

Except that they are numbered, the individual bibliography entries in this system are identical to those used with author/date.

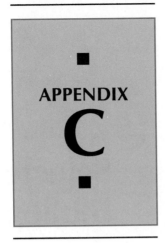

Postal Service Abbreviations for the States

The following abbreviations have been approved by the United States Postal Service. These abbreviations should be used in all correspondence. Note that each abbreviation consists of two capital letters, without periods or spaces. Also note that some of the abbreviations—such as those of Maine, Minnesota, and Maryland—might not be what you would expect.

Alabama	AL	Montana	MT
Alaska	AK	Nebraska	NE
Arizona	AZ	Nevada	NV
Arkansas	AR	New Hampshire	NH
California	CA	New Jersey	NJ
Colorado	CO	New Mexico	NM
Connecticut	CT	New York	NY
Delaware	DE	North Carolina	NC
District of Columbia	DC	North Dakota	ND
Florida	FL	Ohio	OH
Georgia	GA	Oklahoma	OK
Hawaii	HI	Oregon	OR
Idaho	ID	Pennsylvania	PA
Illinois	IL	Puerto Rico	PR
Indiana	IN	Rhode Island	RI
Iowa	IA	South Carolina	SC
Kansas	KS	South Dakota	SD
Kentucky	KY	Tennessee	TN
Louisiana	LA	Texas	TX
Maine	ME	Utah	UT
Maryland	MD	Vermont	VT
Massachusetts	MA	Virginia	VA
Michigan	MI	Washington	WA
Minnesota	MN	West Virginia	WV
Mississippi	MS	Wisconsin	WI
Missouri	MO	Wyoming	WY

APPENDIX
D.

Selected Bibliography

Ethics
Technical Writing
Usage and General Writing
Style Manuals
Technical Manuals
Graphics

Proposals
Oral Presentations
Journal Articles
Word Processing and Computer
   Graphics

Ethics

Beauchamp, T. L., and N. E. Bowie. 1988. *Ethical theory and business.* 3d ed. Englewood Cliffs, N.J.: Prentice-Hall.

Behrman, J. N. 1988. *Essays on ethics in business and the professions.* Englewood Cliffs, N.J.: Prentice-Hall.

Brockmann, R. J., and F. Rook, eds. 1989. *Technical communication and ethics.* Washington, D.C.: Society for Technical Communication.

Chalk, R. 1980. *AAAS professional ethics project: Professional ethics activities in the scientific and engineering societies.* Washington, D.C.: American Association for the Advancement of Science.

Honor in science. 1986. New Haven, Conn.: Sigma Xi. The Scientific Research Society.

Velasquez, M. G. 1988. *Business ethics: Concepts and cases.* 2d ed. Englewood Cliffs, N.J.: Prentice-Hall.

Technical Writing

Currently, more than one hundred technical writing texts and guides are on the market. Following is a representative selection.

Bell, P. 1986. *High tech writing: How to write for the electronics industry.* New York: Wiley-Interscience.

Blicq, R. S. 1986. *Technically—write! Communicating in a technological era.* 3d ed. Englewood Cliffs, N.J.: Prentice-Hall.

Brusaw, C. T., G. J. Alred, and W. E. Oliu. 1987. *Handbook of technical writing.* 3d ed. New York: St. Martin's.

Kolin, P., and J. Kolin. 1985. *Models for technical writing.* New York: St. Martin's.

Mathes, J. C., and D. W. Stevenson. 1991. *Designing technical reports.* 2d ed. New York: Macmillan.

Pickett, N. A., and A. A. Laster, 1988. *Technical English.* 5th ed. New York: Harper & Row.

Samuels, M. S. 1989. *The technical writing process.* New York: Oxford University Press.

Sherman, T. A., and S. S. Johnson. 1990. *Modern technical writing.* 5th ed. Englewood Cliffs, N.J.: Prentice-Hall.

Journals   *IEEE Transactions on Professional Communication*
Journal of Business Communication
Journal of Technical Writing and Communication
Technical Communication
Technical Communication Quarterly

Usage and General Writing

Barzun, J. 1985. *Simple and direct: A rhetoric for writers.* Rev. ed. New York: Harper & Row.

Bernstein, T. M. 1965. *The careful writer: A modern guide to English usage.* New York: Atheneum.

Corbett, E. P. 1990. *Classical rhetoric for the modern student.* 3d ed. New York: Oxford University Press.

Ede, L., and A. Lunsford. 1989. *Singular texts—plural authors: Perspectives on collaborative writing..* Carbondale, Ill.: Southern Illinois University Press.

Flesch, R. 1985. *The art of plain talk.* New York: Macmillan.

Fowler, H. W. 1987. *A dictionary of modern English usage.* 2d ed., rev. by Sir E. Gowers. New York: Oxford University Press.

Hayakawa, S. I., et al. 1978. *Language in thought and action.* 4th ed. New York: Harcourt Brace Jovanovich.

Maggio, R. 1987. *The nonsexist word finder: A dictionary of gender free usage.* Phoenix: Oryx.

Miller, K., and K. Swift. 1988. *The handbook of nonsexist writing for writers, editors, and speakers.* 2d ed. New York: Harper & Row.

Sorrels, B. 1983. *The nonsexist communicator: Solving the problems of gender and awkwardness in modern English.* Englewood Cliffs, N.J.: Prentice-Hall.

Strunk, W., Jr., and E. B. White. 1979. *The elements of style with index.* 3d ed. New York: Macmillan.

Trimble, J. R. 1975. *Writing with style: Conversations on the art of writing.* Englewood Cliffs, N.J.: Prentice-Hall.

Zinsser, W. 1988. *On writing well: An informal guide to writing nonfiction.* 3d ed. New York: Harper & Row.

Style Manuals

American National Standards Institute. 1979. *American national standard for the preparation of scientific papers for written or oral presentation.* ANSI Z39.16—1972. New York: American National Standards Institute.

CBE Style Manual Committee. 1983. *Council of Biology Editors style manual: A guide for authors, editors, and publishers in the biological sciences.* 5th ed. Washington, D.C.: Council of Biology Editors.

The Chicago manual of style. 1982. 13th ed., rev. Chicago: University of Chicago Press.

Dodd, J. S., ed. 1985. *The ACS style guide: A manual for authors and editors.* Washington, D.C.: American Chemical Society.

ORNL style guide. 1974. Oak Ridge, Tenn.: Oak Ridge National Laboratory.

Pollack, G. 1977. *Handbook for ASM editors.* Washington, D.C.: American Society for Microbiology.

Publications manual of the American Psychological Association. 1983. 3d ed. Washington, D.C.: American Psychological Association.

Skillin, M., R. Gay, et al. 1974. *Words into type.* 3d ed. Englewood Cliffs, N.J.: Prentice-Hall.

United States Government Printing Office style manual. 1984. Rev. ed. Washington, D.C.: Government Printing Office.

Also, many private corporations, such as John Deere, DuPont, Ford Motor Company, General Electric, and Westinghouse, have their own style manuals.

Technical Manuals

Brockmann, R. J. 1990. *Writing better computer documentation: From paper to Hypertext.* New York: Wiley-Interscience.

Cohen, G., and D. H. Cunningham. 1984. *Creating technical manuals: A step-by-step approach to writing user-friendly instructions.* New York: McGraw-Hill.

Forbes, M. 1988. *Writing technical articles, speeches, and manuals.* New York: Wiley.

Katzin, E. 1985. *How to write a really good user's manual.* New York: Van Nostrand/Reinhold.

McGehee, B. M. 1984. *The complete guide to writing software user manuals.* Cincinnati, Ohio: Writer's Digest.

Weiss, E. H. 1991. 2d ed. *How to write a usable user manual.* Phoenix: Oryx.

Graphics

Arntson, A. E. 1988. *Graphic design basics.* New York: Holt, Rinehart & Winston.

Beakley, G. C., Jr., and D. D. Autore, 1973. *Graphics for design and visualization.* New York: Macmillan.

Crow, W. C. 1986. *Communication graphics.* Englewood Cliffs, N.J.: Prentice-Hall.

Lefferts, R. 1982. *Elements of graphics: How to prepare charts and graphs for effective reports.* New York: Harper & Row.

MacGregor, A. J. 1979. *Graphics simplified: How to plan and prepare effective charts, graphs, illustrations, and other visual aids.* Toronto: University of Toronto Press.

Morris, G. E. 1975. *Technical illustrating.* Englewood Cliffs, N.J.: Prentice-Hall.

Smith, R. C. 1986. *Basic graphic design.* Englewood Cliffs, N.J.: Prentice-Hall.

Thomas, T. A. 1978. *Technical illustration.* 3d ed. New York: McGraw-Hill.

Tufte, E. R. 1983. *The visual display of quantitative information.* Cheshire, Conn.: Graphics Press.

Turnbull, A. T., and R. N. Baird. 1980. *The graphics of communication: Typography, layout, design, production.* 4th ed. New York: Holt, Rinehart and Winston.

Journals

Graphic Arts Monthly

Graphics: USA

Proposals

Helgeson, D. V. 1985. *Handbook for writing technical proposals that win contracts.* Englewood Cliffs, N.J.: Prentice-Hall.

Lefferts, R. 1983. *The basic handbook of grants management.* New York: Basic Books.

Society for Technical Communication. 1973. *Proposals and their preparation.* Vol. 1. Washington, D.C.: Society for Technical Communication.

Whalen, T. 1987. *Writing and managing winning technical proposals.* Norwood, Mass.: Artech House.

Oral Presentations

Anastasi, T. E., Jr. 1972. *Communicating for results.* Menlo Park, Calif.: Cummings.

Howell, W. S., and E. G. Barmann. 1971. *Presentational speaking for business and the professions.* New York: Harper & Row.

Weiss, H., and J. B. McGrath, Jr. 1963. *Technically speaking: Oral communication for engineers, scientists, and technical personnel.* New York: McGraw-Hill.

Woelfle, R. M., ed. 1975. *A guide for better technical presentations.* New York: Institute of Electrical and Electronics Engineers.

Journal Articles

Carter, S. P. 1987. *Writing for your peers: The primary journal paper.* New York: Praeger.

Day, R. A. 1988. *How to write and publish a scientific paper.* 3d ed. Phoenix: Oryx.

Graham, B. P. 1980. *Magazine article writing: Substance and style.* New York: Holt, Rinehart and Winston.

Katz, M. J. 1986. *Elements of the scientific paper.* New Haven, Conn.: Yale University Press.

Michaelson, H. B. 1990. *How to write and publish engineering papers and reports.* 3d ed. Phoenix: Oryx.

Mitchell, J. H. 1968. *Writing for professional and technical journals.* New York: Wiley.

Journals

The Writer

Writer's Digest

Word Processing and Computer Graphics

Banks, M. A., and A. Dibell. 1989. *Word processing secrets for writers.* Cincinnati, Ohio: Writer's Digest.

Bernstein, S., and L. McGarry. 1986. *Making art on your computer.* New York: Watson-Guptill.

Chicago guide to preparing electronic manuscripts. 1987. Chicago: University of Chicago Press.

Fluegelman, A., and J. J. Hewes. 1983. *Writing in the computer age: Word processing skills and style for every writer.* Garden City, N.Y.: Anchor Press/Doubleday.

Krull, R. 1988. *Word processing for technical writers.* Amityville, N.Y.: Baywood.

Mail, P., and R. W. Sykes. 1985. *Writing and word processing for engineers and scientists.* New York: McGraw-Hill.

Muehlman, S. 1989. *Word processing on microcomputers: Applications and exercises.* Englewood Cliffs, N.J.: Prentice-Hall.

Price, J., and Pinneau Urban, L. 1984. *The definitive word-processing book.* New York: Viking Penguin.

Schwartz, H. J. 1985. *Interactive writing: Composing with a word processor.* New York: Holt, Rinehart and Winston.

Simcoe, A. L. 1988. *Advanced word processing applications.* New York: Wiley.

Stone, M. D. 1984. *Getting on-line: A guide to accessing computer information services.* Englewood Cliffs, N.J.: Prentice-Hall.

Sudol, R. A. 1987. *Textfiles: A rhetoric for word processing.* New York: Harcourt Brace Jovanovich.

Zinsser, W. 1983. *Writing with a word processor.* New York: Harper & Row.

INDEX